Advances in Intelligent Systems and Computing

Volume 828

Series editor

Janusz Kacprzyk, Systems Research Institute, Polish Academy of Sciences,
Warsaw, Poland
e-mail: kacprzyk@ibspan.waw.pl

D0503347

The series "Advances in Intelligent Systems and Computing" contains publications on theory, applications, and design methods of Intelligent Systems and Intelligent Computing. Virtually all disciplines such as engineering, natural sciences, computer and information science, ICT, economics, business, e-commerce, environment, healthcare, life science are covered. The list of topics spans all the areas of modern intelligent systems and computing such as: computational intelligence, soft computing including neural networks, fuzzy systems, evolutionary computing and the fusion of these paradigms, social intelligence, ambient intelligence, computational neuroscience, artificial life, virtual worlds and society, cognitive science and systems, Perception and Vision, DNA and immune based systems, self-organizing and adaptive systems, e-Learning and teaching, human-centered and human-centric computing, recommender systems, intelligent control, robotics and mechatronics including human-machine teaming, knowledge-based paradigms, learning paradigms, machine ethics, intelligent data analysis, knowledge management, intelligent agents, intelligent decision making and support, intelligent network security, trust management, interactive entertainment, Web intelligence and multimedia.

The publications within "Advances in Intelligent Systems and Computing" are primarily proceedings of important conferences, symposia and congresses. They cover significant recent developments in the field, both of a foundational and applicable character. An important characteristic feature of the series is the short publication time and world-wide distribution. This permits a rapid and broad dissemination of research results.

More information about this series at http://www.springer.com/series/11156

Anand J. Kulkarni · Suresh Chandra Satapathy
Tai Kang · Ali Husseinzadeh Kashan
Editors

Proceedings of the 2nd International Conference on Data Engineering and Communication Technology

ICDECT 2017

 Springer

Editors
Anand J. Kulkarni
Symbiosis Institute of Technology
Symbiosis International University
Pune, Maharashtra, India

Suresh Chandra Satapathy
Department of CSE
PVP Siddhartha Institute of Technology
Vijayawada, Andhra Pradesh, India

Tai Kang
School of Mechanical
    and Aerospace Engineering
Nanyang Technological University
Singapore, Singapore

Ali Husseinzadeh Kashan
Faculty of Industrial
    and Systems Engineering
Tarbiat Modares University
Tehran, Iran

ISSN 2194-5357   ISSN 2194-5365   (electronic)
Advances in Intelligent Systems and Computing
ISBN 978-981-13-1609-8   ISBN 978-981-13-1610-4   (eBook)
https://doi.org/10.1007/978-981-13-1610-4

Library of Congress Control Number: 2018948581

This Springer imprint is published by the registered company Springer Nature Singapore Pte Ltd.
The registered company address is: 152 Beach Road, #21-01/04 Gateway East, Singapore 189721, Singapore

Preface

The 2nd International Conference on Data Engineering and Communication Technology (ICDECT 2017) was successfully organized by Symbiosis Institute of Technology, Symbiosis International (Deemed University), Pune, India, on December 15–16, 2017. The objective of this international conference was to provide a platform for academicians, researchers, scientists, professionals, and students to share their knowledge and expertise in the fields of soft computing, evolutionary algorithms, swarm intelligence, Internet of things, and machine learning and address various issues to increase awareness of technological innovations and to identify challenges and opportunities to promote the development of multidisciplinary problem-solving techniques and applications. Research submissions in various advanced technology areas were received, and after a rigorous peer review process, with the help of technical program committee members, high-quality papers were accepted. The conference featured nine special sessions on various cutting-edge technologies which were chaired by eminent academicians. Many distinguished researchers like Dr. Ali Husseinzadeh Kashan, Tarbiat Modares University, Iran; Dr. Jeevan Kanesan, University of Malaya, Malaysia; Dr. E. S. Gopi, National Institute of Technology, Trichy, India, attended the conference.

We express our sincere thanks to all special session chairs: Prof. Atul Magikar, Prof. Nitin Solke, Dr. Anirban Sur, Dr. Arun Kumar Bongale, Dr. Sachin Pawar, Dr. Komal Pradhan, Dr. Husseinzadeh Kashan, Dr. Sanjay Darvekar, and Dr. Pritesh Shah, and distinguished reviewers for their timely technical support. We would like to extend our special thanks to very competent team members for successfully organizing the event.

Pune, India   Dr. Anand J. Kulkarni
Vijayawada, India   Dr. Suresh Chandra Satapathy
Singapore, Singapore   Dr. Tai Kang
Tehran, Iran   Dr. Ali Husseinzadeh Kashan

Organizing Committee

Organizing Chair
Dr. Anand J. Kulkarni, Symbiosis International (Deemed University), Pune, India

Organizing Co-chair
Dr. Tai Kang, Nanyang Technological University, Singapore
Dr. Ali Husseinzadeh Kashan, Tarbiat Modares University, Tehran, Iran

TPC Chair
Dr. Anand J. Kulkarni, Symbiosis International (Deemed University), Pune, India

Publication Chair
Dr. Suresh Chandra Satapathy, Vijayawada, India

Organizing Committee Chairs
Dr. Anand J. Kulkarni, Symbiosis Institute of Technology, Pune, India
Dr. Tai Kang, Nanyang Technological University, Singapore
Dr. Ali Husseinzadeh Kashan, Tarbiat Modares University, Tehran, Iran

Special Session Chair
Dr. Saeid Kazemzadeh Azad, Atilim University, Turkey

Program Committee Chairs
Dr. Suresh Chandra Satapathy, PVP Siddhartha Institute of Technology, Vijayawada, India

Technical Program Committee
Dr. Luigi Benedicenti, University of Regina, Canada
Dr. Anand J. Kulkarni, Symbiosis Institute of Technology, India
Dr. Saeid Kazemzadeh Azad, Atilim University, Turkey
Dr. Siba K. Udgata, University of Hyderabad, India
Dr. Sanjay Darvekar, University of Pune, India
Dr. Ganesh Kakandikar, University of Pune, India
Ganesh Krishnasamy, University of Malaya, Malaysia
Dr. Suwin Sleesongsom, King Mongkut's Institute of Technology, Thailand

Dr. Norrima Binti Mokhtar, University of Malaya, Malaysia
Mandar Sapre, Symbiosis Institute of Technology, India
Apoorva S. Shastri, Symbiosis Institute of Technology, India
Ishaan Kale, Symbiosis Institute of Technology, India
Aniket Nargundkar, Symbiosis Institute of Technology, India
Dr. Pritesh Shah, Symbiosis Institute of Technology, India
Dipti Kapoor Sarmah, Symbiosis Institute of Technology, India
Hinna Shabir, California State University, USA
Dr. Abdesslem Layeb, University Mentouri of Constantine, Algeria
Hai Dao Thanh, Hanoi University of Industry, Vietnam
Dr. Debajyoti Mukhopadhyay, ADAMAS University, Kolkata, India
Dr. Pramod Kumar Singh, ABV-Indian Institute of Information Technology and
    Management, India
Shreesh Dhavle, Symbiosis Institute of Technology, India
Shruti Maheshwari, Symbiosis Institute of Technology, India
Rushikesh Kulkarni, Symbiosis Institute of Technology, India
Dr. Sankha Deb, Indian Institute of Technology Kharagpur, India
Dr. Rene Mayorga, University of Regina, Canada
Dr. S. Gopi, National Institute of Technology, Trichy, India
Mukundraj Patil, University of Pune, India
Dr. Rahee Walambe, Symbiosis Institute of Technology, India
Dr. Ziad Kobti, University of Windsor, Canada
Dr. Zhang Xiaoge, Vanderbilt University, USA
Dr. Cheikhrouhou Omar, Taif University, Saudi Arabia
Dr. Jeevan Kanesan, University of Malaya, Malaysia
Dr. Sanjay S. Pande, Indian Institute of Technology Bombay, India
Wen Liang Chang, University of Malaya, Malaysia
Dr. S. P. Leo Kumar, PSG College of Technology, Coimbatore
Dr. Anirban Sur, Symbiosis Institute of Technology, India
Dr. Sachin Pawar, Symbiosis Institute of Technology, India
Nitin Solke, Symbiosis Institute of Technology, India
Dr. Komal Pradhan, Symbiosis Institute of Management Studies, India
Dr. Arun Kumar Bongale, Symbiosis Institute of Technology, India
Atul Magikar, Symbiosis Institute of Technology, India
Dr. Jayant Jagtap, Symbiosis Institute of Technology, India
Parag Narkhede, Symbiosis Institute of Technology, India
Satish Kumar, Symbiosis Institute of Technology, India

Contents

About the Editors

Anand J. Kulkarni holds Ph.D. in distributed optimization from Nanyang Technological University, Singapore; MS in artificial intelligence from the University of Regina, Canada; bachelor of engineering from Shivaji University, India; and a diploma from the Board of Technical Education, Mumbai. He worked as a research fellow on a cross-border, supply-chain disruption project at Odette School of Business, University of Windsor, Canada. Currently, he is working as Head and Associate Professor at the Symbiosis Institute of Technology, Symbiosis International University, Pune, India. His research interests include optimization algorithms, multi-objective optimization, continuous, discrete, and combinatorial optimization, multi-agent systems, complex systems, cohort intelligence, probability collectives, swarm optimization, game theory, self-organizing systems, and fault-tolerant systems. He is the founder and chairman of the Optimization and Agent Technology (OAT) Research Lab. He has published over 30 research papers in peer-reviewed journals and conferences.

Suresh Chandra Satapathy holds Ph.D. in computer science and is currently working as Professor and Head of the Department of CSE, PVPSIT, Vijayawada, Andhra Pradesh, India. In 2015–17, he was the national chairman of the Computer Society of India's Div-V (Educational and Research), which is the largest professional society in India. He was also Secretary and Treasurer of the Computational Intelligence Society's IEEE—Hyderabad Chapter. He is a senior member of IEEE. He has been instrumental in organizing more than 18 international conferences in India and has been a corresponding editor for over 30 books. His research activities include swarm intelligence, machine learning, and data mining. He has developed a new optimization algorithm known as social group optimization (SGO) published in Springer Journal. He has delivered a number of keynote addresses and tutorials at various events in India. He has published over 100 papers in leading journals and conference proceedings. Currently, he is on the editorial board of IGI Global, Inderscience, and Growing Science journals and is also a guest editor for the *Arabian Journal of Science and Engineering*.

Tai Kang holds bachelor of engineering from the National University of Singapore and Ph.D. from Imperial College, London. He is currently an associate professor at the School of Mechanical and Aerospace Engineering, Nanyang Technological University, Singapore. He teaches various undergraduate and graduate courses in design, design optimization, and finite element analysis. His research interests include optimization, evolutionary computation, collective intelligence, structural topology/shape optimization, compliant/flexural mechanisms, folding/unfolding of 3D folded structures, mathematical modeling of industrial/manufacturing processes, and modeling and vulnerability analysis of critical infrastructure interdependencies.

Ali Husseinzadeh Kashan holds degrees in industrial engineering from Amirkabir University of Technology (Polytechnic of Tehran), Iran. He worked as a post-doctoral research fellow in the Department of Industrial Engineering and Management Systems with the financial support of Iran National Elite foundations. He is currently an assistant professor in the Department of Industrial and Systems Engineering, Tarbiat Modares University, and has been active in the applied optimization research field since 2004. His research focuses on modeling and solving combinatorial optimization problems in areas such as logistics and supply networks, revenue management and pricing, resource scheduling, grouping problems, and financial engineering. As solution methodologies for real-world engineering design problems, he has introduced several intelligent optimization procedures, such as the League Championship Algorithm (LCA), Optics Inspired Optimization (OIO), Find-Fix-Finish-Exploit-Analyze (F3EA) metaheuristic algorithm, and Grouping Evolution Strategies (GESs). He has published over 70 peer-reviewed journals and conference papers and has served as a referee for several outstanding journals such as *IEEE Transactions on Evolutionary Computations, Omega, Computers and Operations Research, Journal of the Operational Research Society, Computers and Industrial Engineering, International Journal of Production Research, Information Sciences, Applied Soft Computing, Ecological Informatics, Engineering Optimization, and Optimal Control and Applications*. He has received several awards from the Iran National Elite Foundation, and in 2016, he was honored by the Academy of Sciences of Iran as the "outstanding young scientist of Industrial Engineering."

Optimization of Constrained Engineering Design Problems Using Cohort Intelligence Method

Apoorva S. Shastri, Esha V. Thorat, Anand J. Kulkarni and Priya S. Jadhav

Abstract This paper proposes Cohort Intelligence (CI) method as an effective approach for the optimization of constrained engineering design problems. It employs a probability-based constraint handling approach in lieu of the commonly used repair methods, which exhibits the inherent robustness of the CI technique. The approach is validated by solving three design problems. The solutions to these problems are compared to those evaluated from Simple Constrained Particle Swarm Optimizer (SiC-PSO) and Co-evolutionary Particle Swarm Optimization based on Simulated Annealing (CPSOSA) (Cagnina et al., Informatica 32(3):319–326, [1]). The performance of Cohort Intelligence method is discussed with respect to best solution, standard deviation, computational time, and cost.

Keywords Cohort intelligence · Constrained optimization · Engineering design problems

A. S. Shastri (✉) · E. V. Thorat · A. J. Kulkarni · P. S. Jadhav
Symbiosis Institute of Technology, Symbiosis International University,
Pune, MH 412115, India
e-mail: apoorva.shastri@sitpune.edu.in

E. V. Thorat
e-mail: eshavt@gmail.com

A. J. Kulkarni
e-mail: anand.kulkarni@sitpune.edu.in; kulk0003@uwindsor.ca

P. S. Jadhav
e-mail: priya.jadhav@sitpune.edu.in

A. J. Kulkarni
Odette School of Business, University of Windsor,
401 Sunset Avenue, Windsor, ON N9B3P4, Canada

A. J. Kulkarni et al. (eds.), *Proceedings of the 2nd International Conference on Data Engineering and Communication Technology*, Advances in Intelligent Systems and Computing 828, https://doi.org/10.1007/978-981-13-1610-4_1

1 Introduction

Cohort Intelligence method proposed by Kulkarni et al. [2] is a bio-inspired optimization technique, in which a cohort refers to a group of candidates/individuals with the shared characteristic of achieving their individual goals [2]. It is based on the natural tendency of an individual to evolve its behavior by observing the behavior of other candidates of the cohort and emulating it. Every candidate follows a certain behavior and imbibes the associated qualities which may improve its own behavior. Roulette wheel approach based on the probability of the behavior being followed is used by the individual candidates to decide the candidate whose behavior is to be followed. It results in every candidate learning from the other and leads to the evolution of the overall behavior of the cohort. In order to arrive at the best solution, every candidate shrinks the sampling interval associated with every variable by using a reduction factor r, which along with the number of candidates is determined based on preliminary trials. The cohort behavior is considered to be saturated if significant improvement in the behavior is not seen or it becomes difficult to distinguish between the behaviors of all the candidates in the cohort. In this case, convergence is said to have occurred and the solution thus obtained is considered to be the optimum solution. The CI methodology was validated by solving four test problems such as Rosenbrock function, Sphere function, Ackley function, and Griewank function.

In [2], the CI methodology was validated as a robust, viable, and a competitive alternative to its contemporaries. However, the computational performance was governed by the reduction factor r and could be improved further to make it solve real-world problems which are generally constrained in nature. Hence for solving constrained optimization problems, probability collectives and a penalty function approach were introduced in [3]. It involves decomposing a complex system into subsystems in order to optimize them in a distributed and decentralized way. The approach produced competitive results, if not better compared to the GA techniques. It was concluded that the approach could be made more generalized and the diversification of sampling, rate of convergence, quality of results, etc., could be improved. Additionally, a local search technique can be incorporated for assisting in a neighborhood search to reach the global optimum [3].

Working on the following lines, CI method with probability-based constraint handing approach was proposed in [4]. Instead of the commonly used repair methods like penalty function method, in this approach a probability distribution is devised for every individual constraint. The lower and upper bounds of the distribution are chosen by finding the minimum and maximum values among all the constraints by substituting the lower and the upper boundaries of the variables in all the constraints. Based on the range in which the value of each of the constraint lies, the probability and the probability score are calculated. Roulette wheel approach based on the probability score is used to select the behavior to be followed, and similar to CI method range reduction is used to arrive at the best solution.

It was observed that a probability-based constraint handling approach was not only robust but also reduced the computational time as compared to Differential

Evolution (DE), Genetic Algorithm (GA), Evolutionary Strategy (ES), and Particle Swarm Optimization (PSO) which employ various repair methods [4]. In this paper, CI method with probability-based constraint handling approach was used to solve three engineering design problems, namely spring design, welded-beam design, and pressure vessel design.

2 Literature Review

Over the years, various optimization approaches have been applied to the selected engineering design problems. Among the commonly used techniques for these problems are GA-based co-evolution model and a feasibility-based tournament selection scheme, an effective Co-evolutionary PSO (CPSO) for constrained engineering design problems and a hybrid PSO (hPSO) with a feasibility-based rule for constrained optimization [5–7].

Cagnina et al. [1] developed a constrained version of PSO for solving engineering optimization problems. In this method, two particles are compared, and if both are feasible the one with better fitness function value is selected. If both are infeasible, the one with lower infeasibility is selected. The approach contains a constraint handling technique as well as a mechanism to update the velocity and position of the particles. The approach was tested by solving four problems, three of the problems have been adopted here (A01, A02, and A03). The best solution they found was compared to that obtained from different methods like Mezura and CPSO. The values obtained by this method are similar to those obtained in [8].

The hybrid CPSO exploits simulated annealing and penalty based method for handling constraints.

The approach was tested by solving three of the problems adopted here (A01, A02, and A03). The best solution they found was compared to that obtained from different methods like CPSO, structural optimization, self-adaptive constraint handling [5, 6, 9–11]. The solution attained in [8] was significantly better for all the problems (A01, A02, and A03) as compared to the other methods.

Cohort Intelligence is a relatively recent technique which has been used to solve engineering design optimization problems in this paper. The performance of CI method is compared to the techniques in hCPSO as they have attained better results as compared to the earlier methods [1, 8].

3 Problem Definition

A01: Tension/compression spring design optimization problem.

The spring design problem is taken from [1] in which a tension/compression spring is designed for minimum weight, subject to constraints of minimum deflection, shear stress, surge frequency, and limits on outside diameter and on design variables. There are three design variables: the wire diameter d (x_1), the mean coil diameter D (x_2), and the active coils P (x_3). The mathematical formulation of this problem is:

Minimize: $f(\vec{x}) = (x_3 + 2)x_2x_1^2$

Subject to:

$$g_1(\vec{x}) = 1 - \frac{x_2^3 x_3}{7178x_1^4} \leq 0$$

$$g_2(\vec{x}) = \frac{4x_2^2 - x_1 x_2}{12{,}566(x_2 x_1^3) - x_1^4} + \frac{1}{5108x_1^2} - 1 \leq 0$$

$$g_3(\vec{x}) = 1 - \frac{140.45x_1}{x_2^2 x_3} \leq 0$$

$$g_4(\vec{x}) = \frac{x_2 + x_1}{1.5} - 1 \leq 0$$

where $0.05 \leq x_1 \leq 2$, $0.25 \leq x_2 \leq 1.3$, and $2 \leq x_3 \leq 15$ (Table 1).

A02: Welded-beam design optimization problem.

The welded-beam design problem is referred to from [12] in which the welded beam is designed for minimum fabrication cost, subject to constraints of shear stress τ, bending stress in the beam σ, buckling load on the bar Pc, and end deflection on the beam δ. Four design variables x_1, x_2, x_3, and x_4 are considered for minimizing the fabrication cost. The mathematical formulation of this problem is:

Minimize: $f(\vec{x}) = 1.10471x_2x_1^2 + 0.04811x_3x_4(14 - x_2)$

Subject to:

$$g_1(\vec{x}) = \tau(\vec{x}) - 13{,}000 \leq 0$$

$$g_2(\vec{x}) = \sigma(\vec{x}) - 30{,}000 \leq 0$$

$$g_3(\vec{x}) = x_1 - x_4 \leq 0$$

$$g_4(\vec{x}) = 1.10471x_1^2 + 0.04811x_3x_4(14 + x_2) - 5 \leq 0$$

$$g_5(\vec{x}) = 0.125 - x_1 \leq 0$$

$$g_6(\vec{x}) = \delta(\vec{x}) - 0.25 \leq 0$$

$$g_7(\vec{x}) = 6000 - Pc(\vec{x}) \leq 0$$

where

$$\tau(\vec{x}) = \sqrt{(\tau')^2 + (2\tau/\tau'')\frac{x_2}{2R} + (\tau'')^2}$$

$$\tau' = \frac{6000}{\sqrt{2}x_1x_2}, \tau'' = \frac{MR}{J}$$

Table 1 Solution vector A01 (spring design)

Parameter	x_1	x_2	x_3	$g_1(\vec{x})$	$g_2(\vec{x})$	$g_3(\vec{x})$	$g_4(\vec{x})$	$f(\vec{x})$
Best solution	0.05157	0.35418	11.43864	−9.00443	−0.13467	−4.04831	−0.72948	0.01266

$$M = 6000\left(14 + \frac{x_2}{2}\right), \quad R = \sqrt{\left(\frac{x_2^2}{4}\right) + \left(\frac{x_1 + x_3}{2}\right)^2},$$

$$J = 2\left\{x_1 x_2 \sqrt{2}\left[\left(\frac{x_2^2}{12}\right) + \left(\frac{x_1 + x_3}{2}\right)^2\right]\right\}$$

$$\sigma(\vec{x}) = \frac{504{,}000}{x_4 x_3^3}, \quad \delta(\vec{x}) = \frac{65{,}856{,}000}{(30 \times 10^6)x_4 x_3^3}$$

$$Pc(\vec{x}) = \frac{4.013(30 \times 10^6)\sqrt{\frac{x_3 x_4^6}{36}}}{196}\left(1 - \frac{x_3\sqrt{\frac{30 \times 10^6}{4(12 \times 10^6)}}}{28}\right)$$

where $0.1 \leq x_1 \leq 2$, $0.1 \leq x_2 \leq 10$, $0.1 \leq x_3 \leq 10$, $0.1 \leq x_4 \leq 2$ (Table 2).
A03: Pressure vessel design optimization problem.

The welded-beam design problem is referred to from [6, 12]. A cylindrical vessel is capped at both ends by hemispherical heads as shown in figure. Using rolled steel plate, the shell is made in two halves that are joined by two longitudinal welds to form a cylinder. The objective is to minimize the total cost, including the cost of the materials forming the welding. The design variables are: thickness x_1, thickness of the head x_2, the inner radius x_3, and the length of the cylindrical section of the vessel x_4. Consider a compressed air storage tank with a working pressure of 3000 psi and a minimum volume of 750 ft^3. The mathematical formulation of this problem is:

Minimize:

$$f(\vec{x}) = 0.6624x_1 x_3 x_4 + 1.7781x_2 x_3^2 + 3.1661x_1^2 x_4 + 19.84x_1^2 x_3$$

Subject to:

$$g_1(\vec{x}) = -x_1 + 0.0193x_3 \leq 0$$
$$g_2(\vec{x}) = -x_2 + 0.00954x_3 \leq 0$$
$$g_3(\vec{x}) = -\pi x_3^2 x_4 - \tfrac{4}{3}\pi x_3^2 + 1{,}296{,}000 \leq 0$$
$$g_4(\vec{x}) = x_4 - 240 \leq 0$$

where $1 \times 0.0625 \leq x_1 \leq 99 \times 0.0625$, $1 \times 0.0625 \leq x_2 \leq 99 \times 0.0625$, $10 \leq x_3 \leq 200$, and $10 \leq x_4 \leq 200$.

As stated in [1], the variables x_1 and x_2 are discrete values which are integer multiples of 0.0625 in. Hence, the upper and lower bounds of the ranges of x_1 and x_2 are multiplied by 0.0625 as shown in Table 3.

Table 2 Solution vector for A02 (welded beam)

Parameter	x_1	x_2	x_3	x_4	$g_1(\vec{x})$	$g_2(\vec{x})$	$g_3(\vec{x})$	$g_4(\vec{x})$	$g_5(\vec{x})$	$g_6(\vec{x})$	$g_7(\vec{x})$	$f(\vec{x})$
Best solution	0.2057	3.4704	9.0366	0.2057	0.0449	0.0924	$-1.8924\mathrm{E}{-12}$	-3.4329	-0.0807	-0.2355	0.0559	1.7248

Table 3 Solution vector for A03 (pressure vessel)

Parameter	x_1	x_2	x_3	x_4	$g_1(\vec{x})$	$g_2(\vec{x})$	$g_3(\vec{x})$	$g_4(\vec{x})$	$f(\vec{x})$
Best solution	0.81249	0.43750	42.09844	1.76636E+02	6.96000E−09	−0.03588	−0.03588	−63.36340	6059.71438

Table 4 Comparison of CI with existing algorithms

Problem		CPSOSA	SiC-PSO	CI
A01	Best	0.01266	0.01266	0.01266
	Mean	0.01267	0.0131	0.01266
	SD	3.68569E−6	4.1E−4	3.4552E−07
A02	Best	1.72485	1.72485	1.72484
	Mean	1.72485	2.0574	1.72484
	SD	1.70514E−05	0.2154	3.61161E−11
A03	Best	6059.7143	6059.714335	6059.71438
	Mean	6059.7143	6092.0498	6059.71438
	SD	2.26415E−06	12.1725	5.02269E−08

4 Results and Discussion

The CI algorithm for the stated engineering design optimization problems was coded in MATLAB 7.8.0 (R2009a), and simulations were run on a Windows platform using Intel Core i5 processor. The suitability of CI algorithm to solve constrained optimization problems was reaffirmed by applying it to engineering design problems. These well-known design problems have been solved by various optimization techniques and solutions from different methods that are available for comparison. Table 1 represents the comparisons between the solutions obtained from CI algorithm and that obtained from SiC-PSO and CPSOSA in terms of the best solution, mean, and standard deviation (SD).

It can be observed from Table 1 that the solution obtained from CI method in terms of the best solution is precisely comparable to those obtained by CPSOSA. Whereas the standard deviation (SD) obtained from CI algorithm is substantially better than those obtained from CPSOSA and SiC-PSO. The values of the mean obtained for CI algorithm are consistent with the best solution. The performance of Cohort Intelligence method was also measured in terms of the computational time, number of function evaluations (FE), and convergence. Parameters such as the number of candidates and reduction factor are also listed. The function evaluations (FE) required for CI algorithm are less as compared to 24,000 objective function evaluations performed by SiC-PSO, as stated in [1]. The time required for computation is also reasonable.

The solution convergence plots for the spring design function, welded-beam design function, and pressure vessel design function are presented in Figs. 1, 2, and 3, respectively, which exhibit the overall evolution of the cohort by improving individual behaviors as shown in Table 4. Initially, the behavior of the individual candidates can be distinguished. As the sampling interval reduces and the cohort evolves, we do not see significant improvement in the behavior of the candidates. At a particular point, it is difficult to distinguish between the behaviors of the candidates. The cohort is considered to be saturated, and convergence is said to have occurred.

Fig. 1 Spring design
function

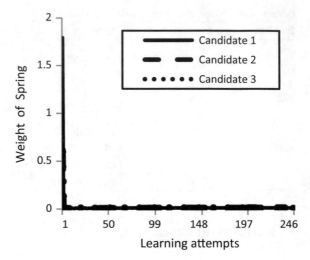

Fig. 2 Welded-beam design
function

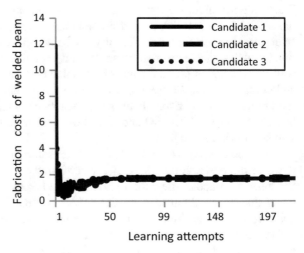

Fig. 3 Pressure vessel
design function

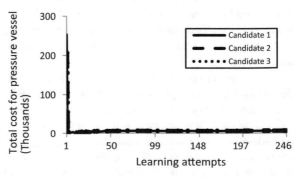

Table 5 Performance of CI algorithm

Problem	Convergence	FE	Time (s)	Parameters (C, r)
A01	190	570	0.12	3, 0.95
A02	118	354	0.13	3, 0.95
A03	133	399	0.24	3, 0.95

5 Conclusion and Future Work

We have presented Cohort Intelligence method with probability-based constraint handling for constrained optimization problems. Cohort Intelligence approach has shown very good performance when applied to the given engineering design optimization problems as shown in Table 5. CI generates results which are comparable precisely to two other algorithms with respect to best solution and the individual variable values. Thus, the approach could be considered as a credible alternative to solving constrained optimization problems due to its robustness and high computational speed.

In the future, CI method could be applied for solving resource utilization-related problems by formulating them as a goal programming problems.

References

1. Cagnina LC, Esquivel SC, Coello CAC (2008) Solving engineering optimization problems with the simple constrained particle swarm optimizer. Informatica 32(3):319–326
2. Kulkarni AJ, Durugkar IP, Kumar M (2013) Cohort intelligence: a self supervised learning behavior. In: 2013 IEEE international conference on systems, man, and cybernetics (SMC). IEEE, pp 1396–1400
3. Kulkarni AJ, Tai K (2011) Solving constrained optimization problems using probability collectives and a penalty function approach. Int J Comput Intell Appl 10(04):445–470
4. Shastri AS, Jadhav PS, Kulkarni AJ, Abraham A (2016) Solution to constrained test problems using cohort intelligence algorithm. In: Innovations in bio-inspired computing and applications. Springer, pp 427–435
5. Coello CAC (2000) Use of a self-adaptive penalty approach for engineering optimization problems. Comput Ind 41(2):113–127
6. Coello CAC, Montes EM (2002) Constraint-handling in genetic algorithms through the use of dominance-based tournament selection. Adv Eng Inform 16(3):193–203
7. He Q, Wang L (2007) A hybrid particle swarm optimization with a feasibility-based rule for constrained optimization. Appl Math Comput 186(2):1407–1422
8. Zhou Y, Pei S (2010) A hybrid co-evolutionary particle swarm optimization algorithm for solving constrained engineering design problems. JCP 5(6):965–972
9. Hernandez Aguirre A, Muñoz Zavala A, Villa Diharce E, Botello Rionda S (2007) COPSO: Constrained optimization via PSO algorithm. Technical report no. I-07–04/22-02-2007. Center for Research in Mathematics (CIMAT)
10. Belegundu A (1982) A study of mathematical programming methods for structural optimization. Ph.D. thesis, Department of Civil Environmental Engineering, University of Iowa, Iowa
11. Arora JS (2004) Introduction to optimum design. Elsevier, Boston
12. Rao SS (1996) Engineering optimization. Wiley, New York

Application of Blowfish Algorithm for Secure Transactions in Decentralized Disruption-Tolerant Networks

Smita Mahajan and Dipti Kapoor Sarmah

Abstract Disruption-tolerant network (DTN) is the network which tends to lose the connection frequently due to many reasons like environmental issues, software/hardware faults. If the DTN system is not applied with any powerful cryptography, then data can be lost and safety of such device can be wondered. The protection of data is very much essential in the fields like military areas and also in rural areas where poor network connectivity is a consistent issue. The existing system for data transferring and retrieving securely, such as ciphertext-policy attribute-based encryption (CP-ABE), i.e., generating and distributing keys on the basis of attributes, has been successfully implemented. However, there can be chance of knowing the attributes such as name and its basic details about file. In the proposed paper, an effort has been made to provide security by using Blowfish algorithm, two-phase key, and two-phase commit protocol (2pc) for secure data transaction.

Keywords Two-phase key · Key tree · Blowfish · Two-phase commit protocol
Disruption-tolerant network · Tile bitmap · Random bit generator (RBG) key
Data side information key

1 Introduction and Related Work

In many military network scenarios, connections of wireless devices [1] carried by soldiers may be temporarily disconnected by jamming, environmental conditions, and mobility, especially when they operate in hostile environments. Disruption-tolerant network (DTN) technologies [6] are becoming successful solutions that allow nodes to communicate with each other in these extreme networking environments [2, 3]. DTN involves primarily three nodes, i.e., storage, sender, and receiver

S. Mahajan (✉) · D. K. Sarmah (✉)
Symbiosis Institute of Technology, Symbiosis International University, Pune, India
e-mail: smita.mahajan@sitpune.edu.in

D. K. Sarmah
e-mail: dipti.sarmah@sitpune.edu.in

© Springer Nature Singapore Pte Ltd. 2019
A. J. Kulkarni et al. (eds.), *Proceedings of the 2nd International Conference on Data Engineering and Communication Technology*, Advances in Intelligent Systems and Computing 828, https://doi.org/10.1007/978-981-13-1610-4_2

nodes. Achieving protection related to these nodes is an important concern. Many researchers have worked in this direction. Ciphertext-policy attribute-based encryption (CP-ABE) cryptographic solution has been applied to resolve access control issue in DTN in a secured way [3]. An effort has been made to overcome the key distribution challenges. We have explored and considered Blowfish algorithm due to its simple structure and its effectiveness. Blowfish creates a very large key, which makes this algorithm secure and makes it more difficult for the cryptanalyst to hack the key values. Blowfish algorithm is resistant to brute force attack; however, it is vulnerable to birthday attacks. Nie et al. [4] proposed a comparative study between Blowfish algorithm and Data Encryption Standard (DES) in his paper. Scope of this paper includes key creation, 2pc protocol, two-phase key creation, tiled bitmap technique, encryption/decryption, and secure transactions. The paper is organized as follows: Sect. 2 describes the design of the proposed system, Sect. 3 describes the implementation of the proposed system, Sects. 4 and 5 discuss the results and conclusions, respectively. References are enlisted at the end.

2 Design of the Proposed System and System Features

This section is further divided into two sections from Sects. 2.1 to 2.2 wherein Sect. 2.1 describes resources requirement and system module, and Sect. 2.2 describes the working of Blowfish algorithm. The proposed system is used to retrieve the data in secure DTN wherein the key handling is done in tree-based manner. We have considered three keys in total as mentioned in Sect. 1. Two-phase key is based on the property information of the file which generates data side information key. The Blowfish algorithm is used to provide higher data security, and 2pc protocol is used for the secure data transaction. The storage node acts as a master in our application and also has a control over the process of generation of keys with the help of other two keys, viz. data information key and RGB key. In the proposed approach, the following steps are observed:

i.   Generation of RBG key—non-deterministic RBG is secure as it is having higher entropy rate. 2pc protocol is used for secure transmission of data. 2pc protocol is implemented in the distribution system with dedicated transaction manager (TM). 2pc is maintaining the sending of data continuously till data reaches to the receiver though the disruption occurs in between. Sender generates the data side information key. This key is generated based on extra information about the file such as type of file, size, location, modified, created, accessed time. Once the RBG and data side information key are formed, a new two-phase key is generated by merging these keys. Purpose of successful file encryption is done by applying Blowfish algorithm. For the security of storage node, where all the keys are stored and maintained, the time-based cryptography is used.

ii. Hash function key is generated as soon as the data comes into the storage node. This key is generated by using time-based key generation technique and is called as current hash key. With the next arrival of the data, correspondingly next hash key would be generated. Each time these keys will get compared before sending to check whether any intruder or third party has changed or modified the key. Advantage of hash function is impossibility to generate, modify a message from its hash, or there is no same hash for two different data. At the end, data is decrypted using Blowfish algorithm. Thus, secure transaction is achieved.

2.1 Resources Requirement and System Modules

The system is developed using open-source software under the GNU general public license. It uses the framework to handle secure data retrieval in DTN using 2pc protocols and high-level encryption mechanism. The proposed work is implemented using Java technology (JDK 1.6 and above), IDE (Netbeans 7.3, 6.9.1), and Database MYSQL version 5.0. Three systems are required to build the setup with the minimum configuration as Processor: Dual Core of 2.2 GHz, Hard Disk: 100 GB, RAM: 2 GB on D-Link 5 port Ethernet network switch using cables CAT 5/6. The main purpose of this proposed system is to secure the transactions in the DTN using multilayer key handling mechanism with strong cryptographic techniques. DTN network users, which include sender, receiver, and storage nodes, are keen to handle the transaction securely. The key authorities as a part of storage node are responsible to generate, distribute, and manage the keys. The further task of the system under various modules is as stated: to produce high security keys, to implement 2pc Routing protocol, to apply the encryption and decryption techniques, and to handle the keys properly in a tree-based manner.

2.1.1 UML Diagrams

In this section, the working of the system and its modules are represented through various diagrams.

Use case diagram: In Use case diagram Fig. 2 describes the relation between user and the system. The system contains three actors, namely storage, sender, and receiver nodes. Their association with the system is mentioned as below:

(a) Storage node: This is a central node which governs and manages the communication between sender and receiver. The overview of the system using these three nodes is shown in Fig. 1.
(b) Sender node: This node is used for creating, encrypting, and sending the message. The responsibilities of this node are create profile, login into the account, update the credential if required, create the message, auto-generation of keys using data side information, two-phase key creation, data encryption, and sending data.

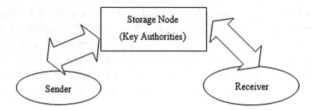

Fig. 1 Overview of node

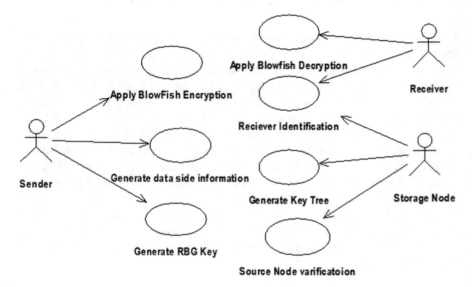

Fig. 2 Use case diagram

(c) Receiver Node: This node receives, validates, and then decrypts the encrypted message.

State Transition diagram, Sequence and Collaboration Diagram: State transition diagram is used to represent the overall behavior of the system as shown in Fig. 3. The purpose of sequence and collaboration diagrams is to find the interactions between different objects in the system as shown in Figs. 4 and 5, respectively.

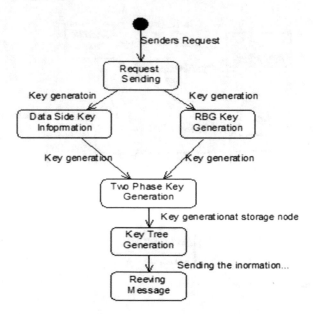

Fig. 3 State transition diagram

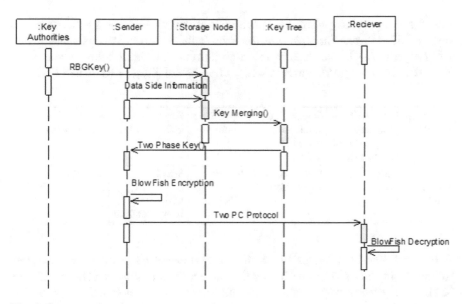

Fig. 4 Sequence diagram

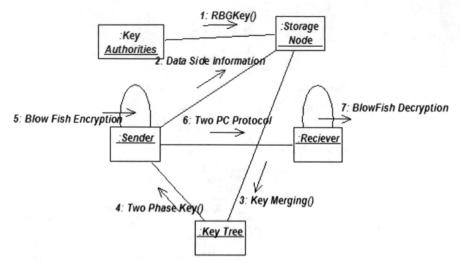

Fig. 5 Collaboration diagram

2.2 Working of Blowfish Algorithm

Blowfish is a popularly utilized symmetric algorithm which takes a variable length key, i.e., 32–448 bits for ciphering the data. This algorithm is considered [5] as a fast algorithm [4]. The steps of this algorithm are shown as below where $P = $ Plain Text, $X = 8$ Byte Stream Block, $x_L = $ Left half of X, $x_R = $ Right half of X.

Step 1: get the Plain text P	Step 2: divide data into 8-byte stream block as X	Step 3: divide X into two halves: x_L and x_R	Step 4: for i = 1–16	Step 5: $x_L = x_L$ XOR P_i	Step 6: $x_R = F(x_L)$ XOR x_R
Step 7: swap x_L and x_R	Step 8: end for	Step 9: $x_R = x_R$ XOR key	Step 10: $x_L = x_L$ XOR key	Step 11: $E = x_R + x_L$ where E : Encrypted Data	Step 12: stop

Blowfish algorithm divides the 64-bit plain text message into two halves, i.e., left half and right half of 32 bits each. The first element of key array is XORed with left 32 bits, and new array is generated P′ on which a transformation function F is applied that generates F′, which is further XORed with the right 32 bits of the message to generate a new message bits. These new message bits will be replaced by the left half of the message. The process will be repeated for fifteen times. The resulting P′ and F′ are then XORed with the last two entries in the P-array (entries 17 and 18) and recombined to produce the 64-bit ciphertext. For more details refer to Ref. [5].

3 Design and Implementation Constraints

System setup of the proposed system is shown in Fig. 6. As described in Sect. 2, there are three nodes sender node, receiver node, and storage node in the system which is connected by a router. The overall process of communication is as below:

1. Once the connection is established between sender and receiver, the receiver node signals the storage node. This signal is time bound, i.e., it will remain with the storage node for specified amount of configurable time. Thus, the message is sent to the receiver. In case, receiver is not available and the 2pc algorithm will be activated and try to send the data until it reaches to the destination.
2. Storage node creates and allocates the RBG key to the sender. Sender generates data side information key, which helps to generate the two-phase key. In order to secure the storage node, where all the keys are stored and maintained, and the time-based cryptography is used. Storage node creates the queue for the next communication which includes list of senders and receivers.
3. By selecting the name of sender and receiver in queue, the intended communication is scheduled by the storage node. Sender node sends data by applying blowfish encryption and receiver receives the encrypted data. 2pc protocol is used for routing the messages and data retrieval.

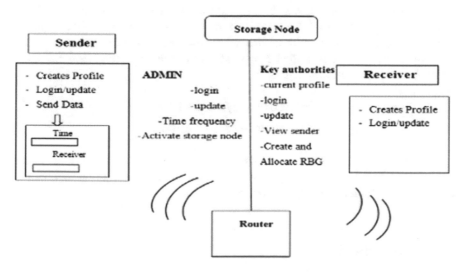

Fig. 6 System setup

Fig. 7 Performance
measurement for different
number of users

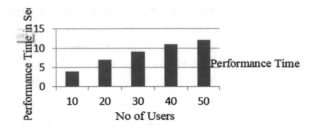

4 Result and Discussion

To show the effectiveness of the proposed system, an experiment is conducted on
Java-based windows machine using Net beans as IDE. Authenticity of the experiment is verified by putting the system under rigorous tests. The relation between
performance time and number of users (Fig. 7) which indicates that the system generates the expected results in military network paradigm in distributed system. To
measure the performance of the system, we set the benchmark by considering the
system with more number of operating nodes for the designed storage node (i.e.,
users). To evaluate the effectiveness of the proposed approach, precision rates and
recall rates are been calculated. Precision can be defined as the ratio of the number
of identified attacks on relevant files to the total number of identified attacks on
irrelevant and relevant files. Recall is the ratio of the number of identified relevant
files to the total number of relevant files on the directory server. Let us assume $A =$
number of relevant files identified, $B =$ number of relevant files not identified, and C
$=$ number of irrelevant files identified.

$$Precision = (A/A + C) * 100 \qquad (1)$$
$$Recall = (A/A + B) * 100 \qquad (2)$$

In Figs. 8 and 9, the tendency of average precision for the relevant files is been
identified is 95.5% and tendency of average recall for the relevant files identified is
about 95.2% which is actually a better recall result.

5 Conclusion

This paper introduces idea of overcoming the flaws more accurately by securing
DTN system using Blowfish algorithm. This also suggests the formation of new key
by using two different keys which gives the better security to the communication
and is considered a very important aspect of our work. Also in this paper, an innovative methodology for data transactions in DTN is discussed and tested for results.
However, more secure DTN system can be developed by using strong cryptographic

Fig. 8 Retrieval average precision of the proposed approach

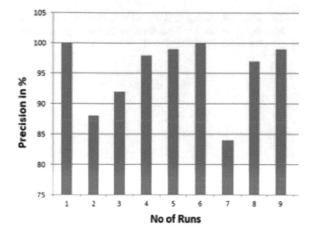

Fig. 9 Retrieval average recall of the proposed approach

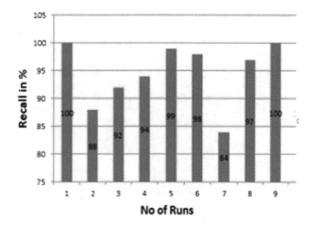

techniques. The comparison analysis of CP-ABE with the proposed work can be considered for future scope.

References

1. Bai Q, Zheng Y (2011) Study on the access control model. In: Cross strait quad-regional radio science and wireless technology conference (CSQRWC), vol 1
2. Chuah M, Yang P (2006) Node density-based adaptive routing scheme for disruption tolerant Networks. In: Proceedings of the IEEE MILCOM, pp 1–6
3. Hur J, Kang K (2014) Secure data retrieval for decentralized disruption-tolerant military networks. IEEE Trans Netw 22(1), ACM
4. Nie T, Song C, Zhi X (2010) Performance evaluation of DES and Blowfish algorithms. In: International conference in biomedical engineering and computer science (ICBECS)
5. Schneier B (1994) The blowfish encryption algorithm. Dr. Dobb's J 19(4):38–40

6. Tian F, Liu B, Xiong J, Gui L (2015) Efficient caching scheme for data access in disruption tolerant networks. In: Broadband multimedia systems and broadcasting (BMSB), 2015 IEEE International Symposium

A Rig-Based Formulation and a League Championship Algorithm for Helicopter Routing in Offshore Transportation

Ali Husseinzadeh Kashan, Amin Abbasi-Pooya and Sommayeh Karimiyan

Abstract Crew must be transported to offshore platforms to regulate the production of oil and gas. Helicopter is the preferred means of transportation, which incurs a high cost on oil and gas companies. So, flight schedule should be planned in a way that minimizes the cost (or equivalently the makespan) while considering all conditions related to helicopter's maximum flight time, passenger and weight capacity, and arrival time of crew members to airport. Therefore, a mathematical formulation is proposed in this paper that has the objective of minimizing the finish time of the final tour of the helicopter (i.e., makespan) while taking the above-mentioned conditions into account. The model is solved for problems that have at most 11 rigs, which results in optimal solution in a reasonable amount of time. For large-size problems for which the computational effort for finding exact solution is not possible in an efficient way, a metaheuristic, namely League Championship Algorithm (LCA), is proposed. Computational experiments demonstrate that LCA can find good solutions efficiently, so it may be employed for large-size helicopter routing problems in an efficient manner.

Keywords Helicopter routing · Mathematical formulation
League championship algorithm

A. Husseinzadeh Kashan (✉) · A. Abbasi-Pooya
Faculty of Industrial and Systems Engineering, Tarbiat Modares University,
Tehran, Iran
e-mail: a.kashan@modares.ac.ir

A. Abbasi-Pooya
e-mail: a.abbasipooya@modares.ac.ir

S. Karimiyan
Young Researchers and Elite Club, Islamic Azad University,
Islamshahr Branch, Islamshahr, Iran
e-mail: S_karimiyan@iiau.ac.ir

© Springer Nature Singapore Pte Ltd. 2019
A. J. Kulkarni et al. (eds.), *Proceedings of the 2nd International Conference on Data Engineering and Communication Technology*, Advances in Intelligent Systems and Computing 828, https://doi.org/10.1007/978-981-13-1610-4_3

1 Introduction

Crew transportation between offshore rigs and onshore airport is one of the major issues to address in operations management. To maintain oil and gas (O & G) production process, crew must be transported daily to the offshore rigs with helicopter. Outsourcing the helicopter service is very costly, and the cost is calculated based on flight hours. This has made the helicopter routing problem extremely important to researchers and practitioners. Therefore, proposing a model and solution approaches that can give satisfactory solutions is very interesting.

There are some researches on the Helicopter Routing Problem (HRP) for offshore O & G rigs. Galvão and Guimarães [1] designed an interactive routing procedure for HRP in a Brazilian oil company. Fiala Timlin and Pulleyblank [2] proposed heuristics for the incapacitated and capacitated problem in a Nigerian company. Sierksma and Tijssen [3] formulated the problem in a Dutch company as a Split Delivery Vehicle Routing Problem (SDVRP) and proposed column-generation and improvement heuristics. The HRP with one helicopter was modeled by Hernád-völgyi [4] as a Sequential Ordering Problem (SOP). Moreno et al. [5] proposed a column-generation-based algorithm for the mixed integer formulation of HRP, and a post-optimization procedure to remove extra passengers in generated schedules. Romero et al. [6] studied the case of a Mexican company through modeling the problem as the Pickup and Delivery Problem (PDP) and solving by Genetic Algorithm. Velasco et al. [7] presented a Memetic Algorithm (MA) with construction and improvement heuristic for the PDP without time windows. Menezes et al. [8] tested their column-generation-based algorithm on 365-day data from a Brazilian company and compared the result with manual flight plans, demonstrating the cost and time savings of the flight plans generated by the algorithm. De Alvarenga Rosa et al. [9] modeled the problem based on Dial-A-Ride Problem (DARP) and designed a clustering metaheuristic. Abbasi-Pooya and Husseinzadeh Kashan [10] proposed mathematical formulation and a Grouping Evolution Strategy (GES) algorithm for the HRP in South Pars gas field. In some other studies, safety was the objective of the problem. Qian et al. [11] formulated the problem where the minimizing the expected number of fatalities is the objective. A tabu search and comparison of different routing methods were given in Qian et al. [12]. Qian et al. [13] considered two non-split and split scenarios and drew the relation between HRP and parallel machines scheduling problem and bin packing problem. Gribkovskaia et al. [14] recommended a quadratic programming model and dynamic programming approaches and analyzed computational complexity of the problem.

This paper offers two contributions. First, it proposes a mathematical formulation for the problem of helicopter routing that takes account of operational rules while minimizing the finish time of the final tour. Second, a metaheuristic solution approach is presented for large-size problems that may not be solved efficiently with the mathematical model. Specifically, a solution approach is proposed that based on the League Championship Algorithm (LCA), which is a population-based algorithm proposed for constrained and unconstrained problems [15–18]. The problem of a case study is solved with both the mathematical model and the proposed algorithm to compare the results.

The remainder of this paper is organized as follows: The problem is described and formulated in Sect. 2. A League Championship Algorithm is proposed for the problem in Sect. 3. Section 4 presents computational experiments and results of solving the problem with the proposed formulation and LCA. The paper is concluded in Sect. 5.

2   Problem Definition and Formulation

The problem that is confronted in offshore platforms is the transportation of crew to and from platforms. The flight schedule of the helicopter must be planned in a way that each crew member is picked up from their respective origin delivered to their respective destination. The main features in a flight schedule are the routes and the passengers in each route. Additionally, the following considerations must be taken into account:

1. One helicopter is available for the daily transportation.
2. The helicopter has maximum weight capacity, passenger capacity, and maximum flight time limitations.
3. The start and finish points of each tour is the onshore airport.
4. The time to land, to pick up, to deliver and to take off of the helicopter should be considered.
5. The helicopter is not working between flights due to refueling, inspection, pilot rest, etc.
6. The crew members arrive at the airport at different times in the day.

Some special characteristics of the problem are as follows: (1) Having one helicopter requires multiple use of the helicopter, which is similar to the VRP with multiple use of vehicles or multi-trip VRP, which was first addressed by Fleischmann [19] for the problems where the fleet or length of the route is not large [20]. (2) Delivery passengers are available at different times at the start of the day. (3) Each passenger cannot be delivered to every delivery node, i.e., the destination of each linehaul passenger is specific.

A small instance of the problem is shown in Fig. 1. As shown in Fig. 1, there are some passengers (d1–d4) at the airport waiting to be delivered to their respective rigs (R1–R4). There are also some passengers that are waiting to be picked up (p1, p2, and p3) from their respective rigs (R1, R2, and R3, respectively) to be transported to the airport (AP). Figure 2 shows the problem as a graph with rigs representing nodes.

Fig. 1 An instance of the
helicopter routing problem

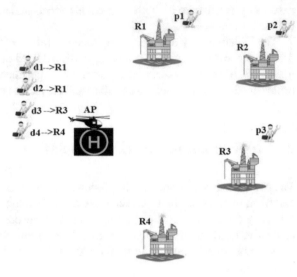

Fig. 2 Graph representing
the problem

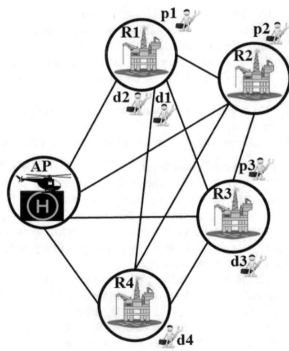

2.1 Notations

Items used in the model are as shown in Table 1.

Table 1 Definition of the items used in RBF

Item	Definition
Indices	
t	Tour counter
i, j	Rig counter ($i = 0$ or $j = 0$ represents the airport)
m	Backhaul passenger
m'	Linehaul passenger
Sets	
$T = \{1, \ldots, NT\}$	The set of the tours in a day
$R = \{1, \ldots, NR\} \cup \{0\}$	The set of rigs, 0 represents the airport
$B = \{p_1, \ldots, p_m\}$	The set of backhaul passengers
$L = \{d_1, \ldots, d_{m'}\}$	The set of linehaul passengers
$C = B \cup L \cup \{0\}$	The set of all passengers, 0 represents the airport
Parameters	
Q	The maximum number of passengers that the helicopter can take
FL	The maximum flight time of the helicopter in each tour
WL	The maximum weight that the helicopter can take
ST	The time required for landing, passenger dropoff, and takeoff
RT	The time between tours that is required for refueling, pilot rest, etc.
NT	Number of tours required in a day
NR	Number of rigs
$M1, M2, M3$	Adequately large numbers
t_{ij}	The flight time between the nodes i and j ($i, j \in R$)
$LB_{m'}$	The arrival time of the passenger m' to the airport ($m' \in L$)
W_m	The weight of backhaul passenger m ($m \in B$)
$Wp_{m'}$	The weight of linehaul passenger m' ($m' \in L$)
pd_{im}	=1 if m should be picked up from rig i, and = 0 otherwise ($i \in R$ and $m \in B$)
$dd_{im'}$	=1 if m' should be delivered to rig i, and = 0 otherwise ($i \in R$ and $m' \in L$)

Using the above notations, the mathematical model is built and presented in the following sections.

2.2 The Rig-Based Formulation (RBF)

Considering all of the aforementioned limitations, in order to minimize the finish time of the final tour, i.e., the makespan, a mixed integer linear formulation is presented. The variables of the model are as follows:

X_{ijt} =1 if helicopter travels from rig i to j in tour t, and =0 otherwise
Y_{it} =1 if helicopter visits rig i in tour t, and =0 otherwise
P_{mt} =1 if passenger m is picked up in tour t, and =0 otherwise
$D_{m't}$ =1 if passenger m' is delivered in tour t, and =0 otherwise
S_t Start time of the tour t
F_t Finish time of the tour t
U_{it} Number of passengers in the helicopter after taking off from rig i in tour t,
HW_{it} Helicopter's passenger weight after taking off from rig i in tour t
A_{it} Variable for elimination of subtour
FF Tour makespan.

The proposed formulation of the problem as a mixed integer linear programming is as follows:

$$\text{RBF:Min } z = FF \tag{1}$$

s.t.

$$FF \geq F(t) \quad \forall t = NT \tag{2}$$

$$\sum_{j \in R - \{0\}} X_{0jt} = 1 \quad \forall t \in T \tag{3}$$

$$\sum_{i \in R - \{0\}} X_{i0t} = 1 \quad \forall t \in T \tag{4}$$

$$\sum_{j \in R} X_{ijt} = Y_{it} \quad \forall i \in R, t \in T \tag{5}$$

$$\sum_{j \in R} X_{ijt} = \sum_{j \in R} X_{jit} \quad \begin{array}{l} \forall i \in R - \{0\}, \\ t \in T \end{array} \tag{6}$$

$$X_{ijt} + X_{jit} \leq 1 \quad \begin{array}{l} \forall i, j \in R - \{0\}, \\ i \neq j, t \in T \end{array} \tag{7}$$

$$\sum_{t \in T} X_{jjt} = 0 \quad \forall j \in R \tag{8}$$

$$A_{it} + 1 \leq A_{jt} + (1 - X_{ijt})M1 \quad \begin{array}{l} \forall i \in R, j \in R - \{0\}, \\ i \neq j, t \in T \end{array} \tag{9}$$

$$P_{mt} \leq \sum_{i \in R} pd_{im} Y_{it} \quad \forall m \in B, t \in T \tag{10}$$

$$\sum_{t \in T} P_{mt} = 1 \quad \forall m \in B \tag{11}$$

$$D_{m\prime t} \leq \sum_{i \in R} dd_{im\prime} Y_{it} \quad \forall m\prime \in L, t \in T \tag{12}$$

$$\sum_{t \in T} D_{m\prime t} = 1 \quad \forall m\prime \in L \tag{13}$$

$$U_{0t} = \sum_{m\prime \in L} D_{m\prime t} \quad \forall t \in T \tag{14}$$

$$U_{it} + \left(\sum_{m \in B} pd_{jm} P_{mt} - \sum_{m\prime \in L} dd_{jm\prime} D_{m\prime t} \right) \leq U_{jt} + (1 - X_{ijt})M2$$
$$\forall i \in R, j \in R - \{0\}, i \neq j, t \in T \tag{15}$$

$$U_{it} \leq Q \quad \forall i \in R, t \in T \tag{16}$$

$$HW_{0t} = \sum_{m\prime \in L} D_{m\prime t} Wp_{m\prime} \quad \forall t \in T \tag{17}$$

$$HW_{it} + \left(\sum_{m \in B} pd_{jm} P_{mt} W_m - \sum_{m\prime \in L} dd_{jm\prime} D_{m\prime t} Wp_{m\prime} \right) \leq HW_{jt} + (1 - X_{ijt})M3$$
$$\forall i \in R, j \in R - \{0\}, i \neq j, t \in T \tag{18}$$

$$HW_{it} \leq WL \quad \forall i \in R, t \in T \tag{19}$$

$$S_t \geq LB_{m\prime} D_{m\prime t} \quad \forall t \in T, m\prime \in L \tag{20}$$

$$S_t + \sum_{i \in R} \sum_{j \in R} (ST + t_{ij}) X_{ijt} = F_t \quad \forall t \in T \tag{21}$$

$$S_t \geq F_{t-1} + RT \quad \forall t \in T, t \neq 1 \tag{22}$$

$$F_t - S_t \leq FL \quad \forall t \in T \tag{23}$$

$$FF, S_t, F_t, U_{it}, HW_{it}, A_{it} \geq 0 X_{ijt}, Y_{it}, P_{mt}, D_{m\prime t} \in \{0, 1\} \tag{24}$$

The objective function (1) is the minimization of the finish time of the last tour. Constraint (2) calculates the value of the objective function based on the last tour's finish time. Constraints (3) and (4) ensure that the helicopter, respectively, begins and ends each tour at the airport. Constraint (5) checks whether each rig is visited in a tour or not. Constraint (6) are flow balance constraints. Backward flights in each tour and loops in rigs are avoided by constraints (7) and (8), respectively. Subtours are eliminated by constraints (9). Constraints (10) state if a rig is visited by a helicopter, the passenger of that rig can be picked up, while Constraints (11) guarantee that all passengers picked up. Similarly, these are ensured for deliveries with constraints (12) and (13). Constraints (14) and (15) calculate the occupied capacity of the helicopter after leaving the airport and rigs, respectively, in each tour. Constraint (16) ensures that capacity requirement is met. Constraints (17)–(19) impose similar conditions to (14)–(16) for helicopter's weight. Constraint (20) calculates the start time of the flight of helicopter in each tour based on the availability time of passengers, while constraint (21) calculates the finish time of each tour. Constraint (22) considers the time between consecutive tours for refueling, etc. Constraint (23) ensures that the maximum flight time of each tour is met. Finally, all variables are declared in (24).

3 League Championship Algorithm for HRP

League Championship Algorithm (LCA) is a population-based metaheuristic proposed for constrained and unconstrained optimization. LCA was first proposed by Husseinzadeh Kashan [15] using the metaphor of sports championship.

The flowchart of LCA is depicted in Fig. 3. The algorithm is initialized by generating a constant-size random population (league) of individuals (teams) and determining the fitness function values (playing strengths). The algorithm then generates a league schedule using single round-robin schedule (see Fig. 4) to plan matches. The winner and the loser of each match are identified stochastically considering the fact that the probability of one team defeating the other is inversely proportional to the difference between that team's strength and the ideal strength. Following winner/loser determination, in each iteration (week), the algorithm moves to the next set of potential solutions (team formations) by using a SWOT matrix (see Fig. 5) that is derived from the artificial match analysis. This analysis is analogous to what a coach typically carries out after a match to determine the team's formation for the next match. Based on the strategy adopted using the SWOT matrix, the equation to move to a new solution (formation) is determined. The procedure is performed iteratively until a termination condition, such as the number of seasons, has reached.

3.1 Idealized Rules of LCA

In the implementation of LCA, there are some idealized rules implied. These rules, which actually idealize some features of normal championships, are:

Rule 1. A team that plays better is more likely to win the game.
Rule 2. The result of each game is unpredictable.
Rule 3. The probability of team i winning team j is the same from both teams' viewpoints.
Rule 4. The result of a match is win or loss (no draw).
Rule 5. If team i beats team j, any strength of team i is a weakness in team j.

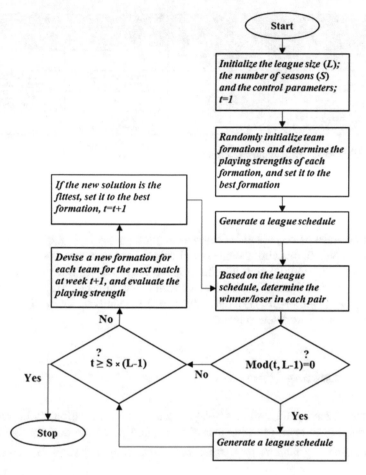

Fig. 3 Flowchart of LCA

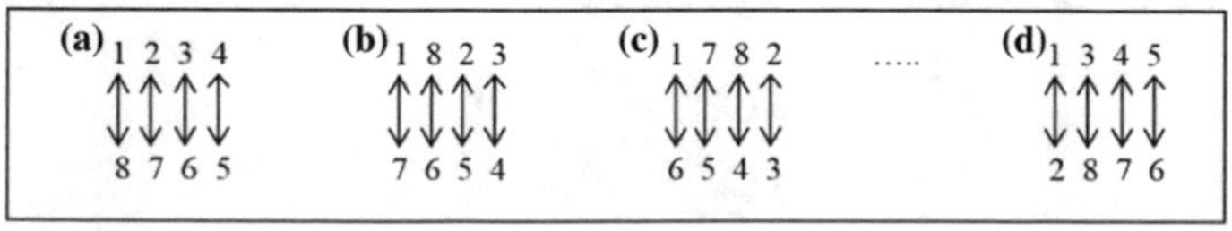

Fig. 4 An example of single round-robin algorithm

Rule 6. Teams only concentrate on their next match without taking account of any of future matches.

	Adopt S/T strategy *i won, l won.* *Focus on …*	Adopt S/O strategy *i won, l lost.* *Focus on …*	Adopt W/T strategy *i lost, l won.* *Focus on …*	Adopt W/O strategy *i lost, l lost.* *Focus on …*
S	own strengths (or weak- nesses of *j*)	own strengths (or weak- nesses of *j*)	–	–
W	–	–	own weaknesses (or strengths of *j*)	own weaknesses (or strengths of *j*)
O	–	weaknesses of *l* (or strengths of *k*)	–	weaknesses of *l* (or strengths of *k*)
T	strengths of *l* (or weaknesses of *k*)	–	strengths of *l* (or weaknesses of *k*)	–

Fig. 5 SWOT matrix for building new formation

3.2 Generating a League Schedule

A schedule must be generated in each season allowing teams (solutions) to compete with each other. To this aim, s single round-robin schedule may be used. An example of this league scheduling algorithm for a sports league of eight teams is depicted in Fig. 4. For the first week (Fig. 4a), 1 plays with 8, 2 with 7, etc. For the second week (Fig. 4b), one team (team 1) is fixed and others are rotated clockwise. This continues until generating a complete schedule.

3.3 Determining Winner/Loser

Given team i having formation X_i^t playing team j with formation X_j^t in week t, let the probability of team i beating team j be p_i^t, and p_j^t be the probability of team j beating team i. If $\hat{f}$ denotes the optimal value, according to the Rule 3, we may write

$$p_i^t + p_j^t = 1 \tag{25}$$

According to the Rule 1, we may also write

$$\frac{f(X_i^t) - \hat{f}}{f(X_j^t) - \hat{f}} = \frac{p_j^t}{p_i^t} \tag{26}$$

From (25) and (26), p_i^t can be obtained. In order to decide about the result of the match, a random number is generated; if it is less than or equal to p_i^t, team i wins and j loses; otherwise j wins and i loses.

3.4 Building a New Team Formation

In order to move to a new set of solutions (population), we have to change the configuration of the solutions (team formations). First, let us define the following indices:

l the team that will play with team i ($i = 1,..., L$) at week $t + 1$ according to the league schedule

j the team that has played with team i ($i = 1,..., L$) at week t according to the league schedule

k the team that has played with team l at week t according to the league schedule

Using the SWOT matrix in Fig. 5, for determining team i's formation to play with l, if i won the previous game and l won, too, then the S/T strategy for team i is to focus on its own strength (or j's weaknesses) and strengths of l (or k's weaknesses). Other cases can be defined in a similar fashion. Based on the strategy, an equation is used to build the new team's formation for the next week. The interested reader is referred to [15] for detailed information.

3.5 Solution Representation

The solution is represented by a permutation of (NP+NT − 1) numbers in which the numbers less than or equal to NP show the passenger number and the numbers greater than NP work as delimiters for tours. A sample solution representation for NP=7 and NT=2 is depicted in Fig. 6, where passengers 3, 5, 7, and 2 are picked up or delivered in tour 1 and passengers 4, 1, and 6 are in tour 2 (8 is the delimiter).

3.6 Objective Function

In order to handle the constraints, penalty function is utilized in the objective function as defined in (27).

$$F(x) = \begin{cases} f(x) + h(l)H(x) & \text{in case of infeasible x} \\ f(x) & \text{otherwise} \end{cases} \tag{27}$$

| 3 | 5 | 7 | 2 | 8 | 4 | 1 | 6 |

Fig. 6 Solution representation for the problem of helicopter routing

where $f(x)$ is the value of the makespan relevant to solution x, $h(l)$ is the penalty value that is dynamically adjusted in the search, l is the iteration counter, and $H(x)$ is a penalty function that is defined in (28).

$$H(x) = \sum_v \theta(q_v(x)) \, q_v(x)^{\gamma(q_v(x))} \tag{28}$$

where $q_v(x)$ is the amount of constraint v's violation, $\theta(q_v(x))$ and $\gamma(q_v(x))$ are functions of the constraint violation. Three cases of constraint violation (or a mixture of them) can happen: (1) violation of maximum flight time, (2) violation of helicopter's capacity limitation, and (3) violation of helicopter's passenger weight limitation, where the amount of violation is equal to $\sum_t \max(0, F_t - S_t - FL)$, $\sum_i \sum_t \max(0, U_{it} - Q)$, and $\sum_i \sum_t \max(0, HW_{it} - WL)$, respectively, (notations are as defined previously).

3.7 Heuristics

Three intra-route heuristics are also incorporated in the body of the algorithm to improve the solution. They are applied to a percentage of generated solutions. The description of these heuristics is as follows.

The swap: Two passengers of a tour are randomly selected and swapped, as demonstrated in Fig. 7.

The inversion: The visiting order of all passengers between two randomly selected passengers is reversed. It is depicted in Fig. 8.

The insertion: This operator moves one portion of a tour to somewhere else in the tour, as shown in Fig. 9.

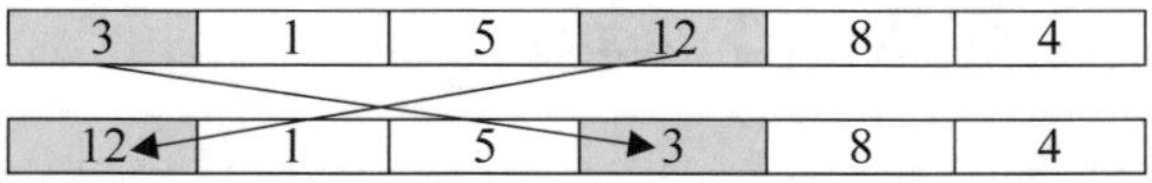

Fig. 7 Swap

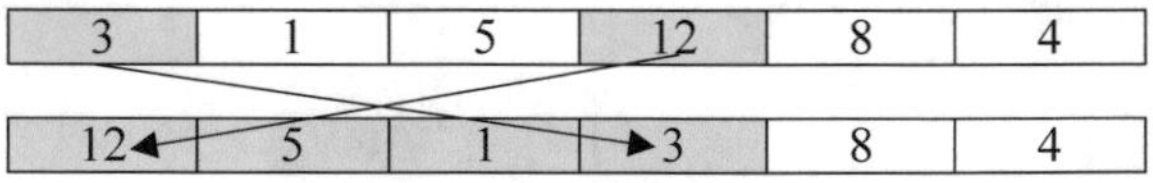

| 3 | 1 | 5 | 12 | 8 | 4 |

| 12 | 1 | 5 | 3 | 8 | 4 |

Fig. 8 Inversion

| 3 | 1 | 5 | 12 | 8 | 4 |

| 12 | 5 | 1 | 3 | 8 | 4 |

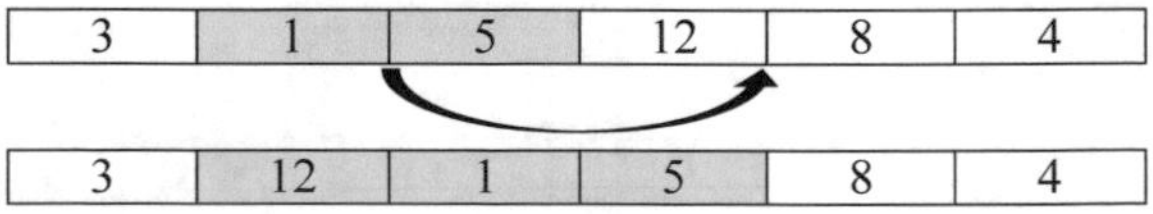

Fig. 9 Insertion

| 3 | 1 | 5 | 12 | 8 | 4 |

| 3 | 12 | 1 | 5 | 8 | 4 |

4 Computational Experiments and Results

To test the performance of the proposed formulation and algorithm, the problems of case study were solved with RBF and LCA and the results were compared as summarized in Table 2. The first section of Table 2 reports the case parameters, namely number of rigs (NR), number of pickups and deliveries, and the number of tours. The second and third sections report the results for solving RBF and LCA, respectively. A comparison of objective value and time between RBF solutions with the ones obtained by LCA is presented in the fourth section. Two metrics are computed to evaluate the mean performance of LCA: Gap_{obj} and Gap_{time}. The relative difference between the average objective value obtained by LCA and the optimum solution is computed based on Eq. (29) and reported under the column Gap_{obj} as a percentage. In Eq. (29), Obj_{LCA} is the objective value obtained by LCA and Obj_{opt} is the objective value obtained from RBF solution. Lower values of Gap_{obj} show better performance of LCA.

$$Gap_{obj} = (Obj_{LCA} - Obj_{Opt})/Obj_{Opt} \qquad (29)$$

The second metric, Gap_{time}, shows the relative difference between the CPU time of RBF and LCA and is computed by Eq. (30). T_{Opt} and T_{LCA} are, respectively, the run times of RBF and LCA.

$$Gap_{time} = (T_{Opt} - T_{LCA})/T_{LCA} \qquad (30)$$

As Table 2 demonstrates, the case study instances were solved to optimality when the number of rigs is between 4 and 11, and the number of passengers is between 43 and 46. The CPU time for solving the problems with RBF is approximately between 7 and 15 min, while it is about between 4 and 6 min for LCA, which proves the efficiency of both approaches. The column Gap_{obj} shows that LCA finds solutions with objective function gaps between 3.539 and 5.757%, which is an acceptable value. The Gap_{time} has positive values, which shows the efficiency of the algorithm compared to the RBF. This demonstrates the efficiency of LCA in providing suboptimal solutions. Therefore, it can be used for problems with large size, where optimal solution cannot be found within an acceptable amount of time.

5 Conclusions and Future Research

This study presented a mathematical model of the problem of routing helicopter to transport crew to and from offshore platforms. Constraints such as passenger and weight capacity of the helicopter, maximum flight time, and delivery passenger's arrival time to the airport were taken into account with the objective of minimizing

Table 2 LCA compared with RBF

Problem parameters				RBF		LCA		Comparison	
NR	No. of pickups	No. of deliveries	NT	Objective value	CPU time (s)	Objective value	CPU time (s)	Gap$_{obj}$(%)	Gap$_{time}$(%)
4	18	27	4	509.187442	0:07:02.781	535.58424	0:03:53.028	5.184	81.12
5	16	30	4	529.515663	0:08:57.437	549.29674	0:03:50.492	3.736	133.48
6	17	26	3	491.601450	0:08:45.907	509.94723	0:03:21.740	3.732	161.19
7	20	25	3	450.870342	0:09:37.052	476.82592	0:04:12.947	5.757	128.97
11	19	27	4	537.152514	0:14:58.505	556.16476	0:05:52.815	3.539	155.11

makespan. The proposed mathematical model is tested using case study instances where it shows efficiency and effectiveness in solving the problems.

Furthermore, an algorithm based on the League Championship Algorithm (LCA) was proposed to solve the problem. The LCA proves its efficiency in producing near-optimal solutions with an acceptable amount of computational effort. Therefore, LCA is fit for solving large problem instances where finding optimal solution is not affordable.

In this research, the objective was to minimize the makespan. Other objectives such as passenger's waiting time may also be considered which will make a multi-objective optimization problem. Furthermore, VRP heuristics may be incorporated with LCA to improve its performance in terms of solution quality. The performance of other metaheuristic algorithms, such as Optics Inspired Optimization (OIO) algorithm [21, 22] is also worth investigating in solving the problem of helicopter routing.

References

1. Galvão RD, Guimarães J (1990) The control of helicopter operations in the Brazilian oil industry: issues n the design and implementation of a computerized system. Eur J Oper Res 49:266–270
2. Fiala Timlin MT, Pulleyblank WR (1992) Precedence constrained routing and helicopter scheduling: heuristic design. Interfaces 22(3):100–111
3. Sierksma G, Tijssen GA (1998) Routing helicopters for crew exchanges on off-shore locations. Ann Oper Res 76:261–286
4. Hernádvölgyi IT (2004) Automatically generated lower bounds for search. Doctoral dissertation, University of Ottawa
5. Moreno L, de Aragao MP, Uchoa E (2006) Column generation based heuristic for a helicopter routing problem. In: Experimental algorithms. Springer, Berlin, pp 219–230
6. Romero M, Sheremetov L, Soriano A (2007) A genetic algorithm for the pickup and delivery problem: an application to the helicopter offshore transportation. In: Theoretical advances and applications of fuzzy logic and soft computing. Springer, Berlin, pp 435–444
7. Velasco N, Castagliola P, Dejax P, Guéret C, Prins C (2009) A memetic algorithm for a pick-up and delivery problem by helicopter. In: Bio-inspired algorithms for the vehicle routing problem. Springer, Berlin, pp 173–190
8. Menezes F, Porto O, Reis, ML, Moreno L, de Aragão MP, Uchoa E, Abeledo H, Nascimento NCD (2010) Optimizing helicopter transport of oil rig crews at Petrobras. Interfaces 408–416
9. de Alvarenga R, Machado AM, Ribeiro GM, Mauri GR (2016) A mathematical model and a clustering search metaheuristic for planning the helicopter transportation of employees to the production platforms of oil and gas. Comput Ind Eng 101:303–312
10. Abbasi-Pooya A, Husseinzadeh Kashan A (2017) New mathematical models and a hybrid grouping evolution strategy algorithm for optimal helicopter routing and crew pickup and delivery. Comput Ind Eng 112:35–56
11. Qian F, Gribkovskaia I, Halskau Ø Sr (2011) Helicopter routing in the Norwegian oil industry: including safety concerns for passenger transport. Int J Phys Distrib Logistics Manage 41(4):401–415
12. Qian F, Gribkovskaia I, Laporte G, Halskau Sr Ø (2012) Passenger and pilot risk minimization in offshore helicopter transportation. Omega 40(5):584–593
13. Qian F, Strusevich V, Gribkovskaia I, Halskau Ø (2015) Minimization of passenger takeoff and landing risk in offshore helicopter transportation: models, approaches and analysis. Omega 51:93–106

14. Gribkovskaia I, Halskau O, Kovalyov MY (2015) Minimizing takeoff and landing risk in helicopter pickup and delivery operations. Omega 55:73–80
15. Husseinzadeh Kashan A (2009) League championship algorithm: a new algorithm for numerical function optimization. In: International conference of soft computing and pattern recognition
16. Husseinzadeh Kashan A (2014) League championship algorithm (LCA): an algorithm for global optimization inspired by sport championships. Appl Soft Comput 16:171–200
17. Husseinzadeh Kashan A, Karimi B (2010) A new algorithm for constrained optimization inspired by the sport league championships. In: Evolutionary computation (CEC), 2010 IEEE congress, pp 1–8
18. Husseinzadeh Kashan A (2011) An efficient algorithm for constrained global optimization and application to mechanical engineering design: league championship algorithm (LCA). Comput.-Aided Des 43:1769–1792
19. Fleischmann B (1990) The vehicle routing problem with multiple use of vehicles. In: Working paper, Fachbereich Wirtschaftswissenschaften, Universitat Hamburg
20. Olivera A, Viera O (2007) Adaptive memory programming for the vehicle routing problem with multiple trips. Comput Oper Res 34(1):28–47
21. Husseinzadeh Kashan A (2015) A new metaheuristic for optimization: optics inspired optimization (OIO). Comput Oper Res 55:99–125
22. Husseinzadeh Kashan A (2015) An effective algorithm for constrained optimization based on optics inspired optimization (OIO). Comput.-Aided Des 63:52–71

Inclusion of Vertical Bar in the OMR Sheet for Image-Based Robust and Fast OMR Evaluation Technique Using Mobile Phone Camera

Kshitij Rachchh and E. S. Gopi

Abstract Optical mark recognition (OMR) is a prevalent data gathering technique which is widely used in educational institutes for examinations consisting of multiple-choice questions (MCQ). The students have to fill the appropriate circle for the respective questions. Current techniques for evaluating the OMR sheets need dedicated scanner, OMR software, high-quality paper for OMR sheet and high precision layout of OMR sheet. As these techniques are costly but very accurate, these techniques are being used to conduct many competitive entrance examinations in most of the countries. But, small institutes, individual teachers and tutors cannot use these techniques because of high expense. So, they resort to manually grading the answer sheets because of the absence of any accurate, robust, fast and low-cost OMR software. In this paper, we propose the robust technique that uses the low-quality images captured using mobile phone camera for OMR detection that gives 100% accuracy with less computation time. We exploit the property that the principal component analysis (PCA) basis identifies the direction of maximum variance of the data, to design the template (introducing the vertical bar in the OMR sheet) without compromising the look of OMR answer sheet. Experiments are performed with 140 images to demonstrate the proposed robust technique.

Keywords PCA · OMR · Skew correction

K. Rachchh · E. S. Gopi (✉)
Department of Electronics and Communication Engineering,
National Institute of Technology Trichy,
Tiruchirapalli 620015, Tamil Nadu, India
e-mail: esgopi@nitt.edu

K. Rachchh
e-mail: rachchhk@gmail.com

© Springer Nature Singapore Pte Ltd. 2019
A. J. Kulkarni et al. (eds.), *Proceedings of the 2nd International Conference on Data Engineering and Communication Technology*, Advances in Intelligent Systems and Computing 828, https://doi.org/10.1007/978-981-13-1610-4_4

1   Introduction

OMR is a widely accepted data gathering technique, which detects the presence of deliberated marked responses by recognizing their darkness on the sheet. A deliberated responses can be filled with either pencil or ballpoint pen. Because of this easiness of this technique, OMR has been worldwide accepted as a direct data collecting technique in many fields, in which the value of a response falls among a limited number of choices. In the field of education, OMR technique is most popularly being used to process MCQs in the competitive examinations [1, 2].

However, current techniques for evaluating the OMR sheets need dedicated scanner, OMR software, high-quality paper for OMR sheet having 90–110 gsm (grams per square metre) and high precision layout of OMR sheet [3]. As these techniques are costly, small institutes, individual teachers and tutors cannot use them because of high expense. As a result of this, they resort to manually grading answer sheets because of the absence of any accurate, robust, fast and low-cost OMR software [4].

The motivation of this paper comes from an idea to simplify OMR evaluation process by providing mobile-based free of cost, fast and robust OMR evaluation technique. Nowadays, smartphones are ubiquitous, so there will be no extra cost associated with this mobile-based solution. PCA is used for skew correction process [5], which makes this solution fastest among all the currently available OMR softwares [6]. Also, there is no need of high-quality paper sheet and high precision layout is not required either, which makes it possible to overcome from all the drawbacks persisted in existing OMR scanner technique.

The rest of this paper is organized as follows. Detailed literature survey is presented in Sect. 2. In Sect. 3, contribution of this paper is explained along with the implemented algorithm. Experiments and Results are discussed in Sect. 4. Finally, the paper is concluded in the Sect. 5.

2   Literature Survey

The invention of OMR scanners has been started in 1950s, which reached in matured state in 1970s when we had desktop sized models of OMR scanner. The novel technology was popularly known as 'mark sensing', and a series of sensing brushes were used to detect graphite particles on a paper, which was passed through the machine [7].

Study of image-based OMR has been evolved in 1999. In [8], the first image-based OMR system was presented, which was based on image scanner and personal computer-type micro-controller. For the validation of a marked answer, the number of black pixels in each answer block is calculated. After this, the difference of the calculated counts between the input and its corresponding reference model is used as a decision criterion. Finally, the database containing a list of students, subjects and corresponding scores can be created.

In [9], the design and implementation of an OMR prototype system for automatically evaluating multiple-choice tests are described in 2003. Different methods for position verification and position recognition have been proposed and implemented in an intelligent line scan camera. The position recognition process is implemented into field-programmable gate array (FPGA). The developed prototype system could read OMR sheets at a speed of 720 forms/h, and the overall system cost was much lower than commercially available OMR products. However, the resolution and overall system design were not fully satisfying and required some further investigation.

The development of a high-speed and low-cost OMR system prototype for marking multiple-choice questions is described in [10]. The uniqueness is the implementation of the broad system into a solitary low-cost FPGA to accomplish the high processing speed. The performance of the algorithm is verified for different marker colours and marking methods.

In 2014, first attempt was made to develop mobile camera-based auto-grading of answer sheets in [11]. They designed their algorithm assuming that the position of circles corresponding to each question is known. The algorithm mainly consists of two parts. The first part of the algorithm is to apply image registration methods to align a captured image with the reference image and compute the transformation matrix. The second part is to apply the transformation matrix to centre pixels of all circles and count the number of black pixels around that region to figure out whether the circle is filled or not.

After applying this algorithm to different images captured by mobile phone, they observed the following conclusion: For well-aligned images with uniform brightness, their algorithm worked well. Their iterative experiments showed almost 100% accuracy in this case. For small tilted images with no rotation, their algorithm still worked fine, but less accuracy than the first case. For images with large tilt and rotation, the accuracy is hampered with up to 26% result being inaccurate. For images with blurred logos at the top or bottom due to bad illumination condition, their algorithm did not work as per expectation, because the logos were key features used for key point matching. In this paper, we propose the algorithm for OMR evaluation based on images captured by mobile phone camera. The developed algorithm is robust, fast and almost zero-cost and is ready to use as a mobile phone application.

3 Contribution of the Paper

This paper suggests to introduce the vertical bar in the OMR sheet, that enables to perform PCA-based skew correction at a faster rate, inspite of large tilt and rotation while capturing the image using the mobile camera. Steps implemented to demonstrate this proposed technique are explained here from user point of view. User needs to capture the image of marked OMR sheet using a mobile phone camera, and the proposed solution displays the total marks scored by a student in average time of 1.15 seconds. Skew correction technique using the entropy method is compared with the PCA-based technique (with the proposed vertical bar) and are reported in the following section.

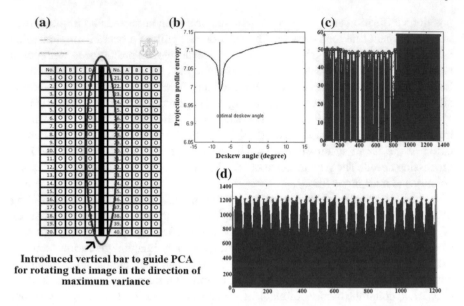

Fig. 1 a Proposed template of OMR sheet. **b** Entropy-based skew correction technique: entropy versus angle. **c** Row-wise addition of an image. **d** Column-wise addition of an image

3.1 Skew Correction Techniques

Entropy-based skew correction technique [12, 13] and PCA-based skew correction technique [14, 15] are chosen from various available techniques for skew correction [16] and applied to 100 images. Black vertical bar is introduced in the OMR sheet design (refer Fig. 1), which helps to remove the problem of large tilt and rotation easily. The width of this vertical bar is further reduced in PCA-based skew correction technique, which guides to decide the maximum variance direction accurately. Steps to implement both skew correction techniques are explained in the following subsections.

1. Extract the coordinates of all zero-valued pixels,which is the actual data from binary converted image of OMR sheet.
2. For each angle ranging from minimum predefined angle to maximum predefined angle(In this technique, minimum and maximum predefined angles are taken as $-15°$ and $+15°$, respectively, and predefined angle interval level is taken as $0.5°$)
   (a) Find the overall entropy from extracted coordinates at that particular angle,
   (b) Increase the angle by predefined angle interval level.
3. Plot the value of entropy at all the predefined angle values (Fig. 1b).
4. Select the angle at which the entropy is minimum.
5. Rotate the image by the angle, which is selected in the previous step.

3.1.1 Steps to Compute PCA-Based Skew Corrections

1. Extract the coordinates of all zero-valued pixels,which is the actual data from binary converted image of OMR sheet.
2. Compute the covariance matrix from those extracted coordinates.
3. Compute the eigenvalues and the eigenvectors of the covariance matrix.
4. Multiply the original data, i.e. extracted coordinates with the eigenvectors.
5. Represent the modified data as a new image.

3.2 Comparison Between PCA and Entropy-Based Skew Correction Technique

Both the skew correction methods are applied to few images, and comparison is shown in Table 1. From the table, we can observe that the PCA technique gives almost 15 times faster results with the introduction of vertical bar in the OMR design. Also, there is no such requirement of defining the minimum and maximum predefined angles and predefined interval level in PCA-based technique as PCA works well for all the resolutions and for all the angles without any limitations.

Since PCA is faster as well as gives 100% accuracy, PCA is implemented in the proposed technique to boost the speed of the evaluation of OMR sheet.

3.3 Implemented Algorithm

1. Design OMR sheet with the proposed vertical bar for robust capturing using the mobile camera to conduct the examination.
2. Capture the image of OMR sheet marked by student using mobile phone camera and transfer the captured image to the computer.
3. Convert the RGB image into greyscale image.

Table 1 Comparison between entropy-based skew correction and PCA-based skew correction

	Entropy method		PCA method	
	Average computation time (s)	Success rate (%)	Average computation time (s)	Success rate (%)
Without vertical bar	13.64	86	0.87	26
With vertical bar	14.01	100	0.91	100

4. Convert the greyscale image to the binary image using Otsu's method [17]. As Otsu thresholding is simplest and fastest adaptive algorithm for thresholding, it is preferred here for selecting appropriate threshold level.
5. Perform PCA-based skew correction technique on the binary image (refer Fig. 2).
6. Perform row-wise addition and detect the dip that detects the black horizontal lines (Fig. 1d).
7. Crop the rows for each question from the detected dip.
8. Perform column-wise addition for each cropped row and detect the dip that detects the black vertical lines (refer Fig. 1c).
9. Crop the options (i.e. option A, B, C, D) for each question from the detected dip and remove the blank circles from unattempted options using connectivity concept [18] (refer Fig. 3).
10. For each question,

   - Set flag = 0. Perform pixel by pixel addition for each cropped option.
   - Detect the option with minimum_sum_ breakvalue. Compare that minimum_sum_value with other three sum_value. If minimum_sum_value < (0.95 * sum_value) increment flag by 1.
   - Check the final value of flag (refer Fig. 3b).
     – If flag == 3, then marked option is detected, having minimum_sum_value.
     – If flag == 0, the question is either unattempted or all the options are marked. (refer Fig. 3c).
     – If flag == 1 or flag == 2, then multiple answers are marked. If flag value is 1, then 3 options are marked, and if flag value is 2, then 2 options are marked (refer Fig. 3d).

11. Calculate the total marks from total number of correct, incorrect, unattempted, and multiple-attempted question (Fig. 3). Here, correction of the answer can be found by either predefined and stored correct answers or by comparing the image of a reference sheet which contains all the correct answers.
12. Display the total marks on graphical user interface (GUI) (refer Fig. 2e).

(a)   (b)   (c)   (d)   (e)

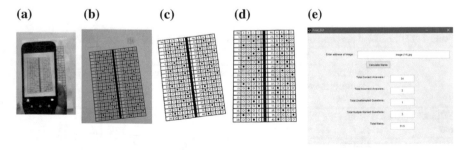

Fig. 2 Demonstration of steps 1–5 of the algorithm. **a** Capturing the image using mobile phone. **b** Captured image of OMR sheet. **c** Converted binary scale image. **d** Skew-corrected image after applying PCA. **e** Displaying output in GUI

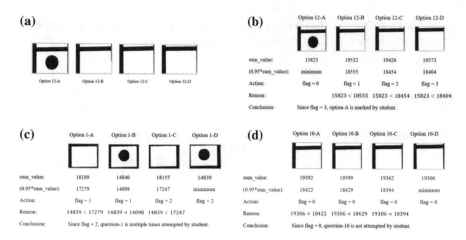

Fig. 3 **a** Cropped option. **b** Procedure to detect attempted question. **c** Procedure to detect unattempted question. **d** Procedure to detect multiple-attempted question

4 Experiments and Results

The proposed algorithm is implemented on 100 images captured using different mobile phone cameras. The resolution of these mobile phone cameras is ranging from 1 to 13 MP. 100% accuracy is achieved by the proposed algorithm. Average time for evaluating an OMR paper is 1.15 s, which is much lower than all the existing techniques. Also, one class test was conducted for checking the real-time performance of this proposed technique and papers were evaluated by capturing the images of OMR sheets using mobile phones and then this algorithm is applied. All the answer sheets are successfully corrected with 100% accuracy.

Database of all the 100 demo images as well as 40 real-time images is uploaded on Website [19]. As this technique is focused for educational institutes only, it is assumed that adequate brightness is available for capturing images. For the case of blurred images also, if the images are clear enough to detect all the horizontal and vertical lines, then this technique works well on blurred images also. It is also possible to merge this solution as a part of one mobile application, so that the mobile application can directly display the total marks scored by the student, immediately after clicking the image of the OMR sheet.

5 Conclusion

In this paper, we proposed a robust and fast image-based OMR evaluation project. We have concluded that the proposed OMR evaluation technique which requires only 1.15 s average time for all the processing is fastest among all the currently available techniques, and it is also the most accurate technique which achieves 100% accuracy. Thus, this technique is ready to use in real-time environment.

References

1. Sattayakawee N (2013) Test scoring for non-optical grid answer sheet based on projection profile method. Int J Inf Educ Technol 3(2):273–277
2. Krishna G, Rana HM, Madan I et al (2013) Implementation of OMR technology with the help of ordinary scanner. Int J Adv Res Comput Sci Softw Eng 3(4):714–719
3. Nalan K (2015) OMR sheet evaluation by web camera using template matching approach. Int J Res Emerg Sci Technol 2(8):40–44
4. Gaikwad SB (2015) Image processing based OMR sheet scanning. Int J Adv Res Electron Commun Eng 4(3):519–522
5. Steinherz T, Intrator N, Rivlin E (1999) Skew detection via principal components analysis. Fifth international conference on document analysis and recognition, Sept 1999:153–156
6. Nirali P, Ghanshyam P (2015) Various techniques for assessment of OMR sheets through ordinary 2D scanner: a survey. Int J Eng Res Technol 4(9):803–807
7. Smith AM (1981) Optical mark reading—Making it easy for users. In: Proceedings of the 9th annual ACM SIGUCCS conference on user services, United States, pp. 257–263
8. Krisana C, Yuttapong R (1999) An image-processing oriented mark reader. In: Applications of digital image processing XXII. Denver CO, pp 702–708
9. Hussmann S, Chan L, Fung C et al (2003) Low cost and high speed Optical mark reader based on Intelligent line camera. In: Proceedings of the SPIE aero sense 2003, optical pattern recognition XIV, vol 5106. Orlando, Florida, USA, pp 200–208
10. Hussmann S, Deng PW (2005) A high speed optical mark reader hardware implementation at low cost using programmable logic. Sci Dir Real-Time imaging 11(1):19–30
11. Chidrewar V, Yang J, Moon D (2014) Mobile based auto grading of answer sheets. Stanford University. https://stacks.stanford.edu/file/druid:yt916dh6570/Moon_Chidrewar_Yang_Mobile_OMR_System.pdf
12. Arvind KR, Kumar J, Ramakrishnan AG (2007) Entropy based skew correction of document images. In: Pattern recognition and machine intelligence. Lecture notes in computer science, vol 4815. Springer, pp 495–502
13. Lathi BP, Ding Z (1998) Modern digital and analog communication systems, 4th edn. Oxford University Press, Oxford
14. Mahdi F, Al-Salbi M (2012) Rotation and scaling image using PCA. Comput Inf Sci 5(1):97–106
15. Basavanna M, Gornale SS (2015) Skew detection and skew correction in scanned document image using principal component analysis. Int. J. Sci. Eng. Res 6(1):1414–1417
16. Sunita M, Ekta W, Maitreyee D (2015) Time and accuracy analysis of skew detection methods for document images. Int J Inf Technol Comput Sci 11:43–54
17. Nobuyuki O (1979) A threshold selection method from grey-level histograms. IEEE Trans Syst Man Cybern 9:62–66
18. Gonzalez RC, Woods RE (1977) Digital image processing, 3rd edn. Pearson Prentice Hall, Upper Saddle River
19. Images Dataset. http://silver.nitt.edu/~esgopi/OMR_Database

Music Composition Inspired by Sea Wave Patterns Observed from Beaches

C. Florintina and E. S. Gopi

Abstract Nature has always been a constant source of inspiration for mankind. This work proposes to harness what nature has to offer through the sea waves to compose music. In this work, the features of the sea waves as observed from beaches are utilized to generate musical semitones, without the influence of any other human-composed music. The musical strains so generated are hence completely unique and could serve as a significant plug-in for a composer.

Keywords Music composition · Sea wave pattern · LDS · Kernel technique

1 Introduction

Composing a unique strain of music has become a highly commendable feat, and composing music which is not entirely predictable being every composers' wish certainly does not alleviate the difficulties faced. It is quite beyond the reach of small-scale companies to compose a unique theme music and small-scale gaming companies to come up with a unique strain of music for each level of their games, due to the cost involved. The above-mentioned causes justify the need for adopting machine intelligence to aid the field of music composition.

It is quite impossible for automatically composed music to possess the finesse of human composition. Muoz et al. [1] tries to alleviate this problem with the aid of unfigured bass technique. Santos et al. [2] gives a brief account of the different evolutionary techniques used for composing music. Engels et al. [3] introduces a tool which could be of use to small-level game designers that automatically generates

C. Florintina · E. S. Gopi (✉)
Department of Electronics and Communication Engineering,
National Institute of Technology Trichy,
Tiruchirapalli 620015, Tamil Nadu, India
e-mail: esgopi@nitt.edu

C. Florintina
e-mail: florintinachaarlas@gmail.com

© Springer Nature Singapore Pte Ltd. 2019
A. J. Kulkarni et al. (eds.), *Proceedings of the 2nd International Conference on Data Engineering and Communication Technology*, Advances in Intelligent Systems and Computing 828, https://doi.org/10.1007/978-981-13-1610-4_5

unique music for each level of game. Roig et al. [4] extracts musical parameters from an existing piece of music and designs a probabilistic model to generate unique music. Merwe et al. [5, 6] use Markov models and [7] uses genetic algorithms to automate music composition. Pearce and Wiggins [8] proposes a framework to evaluate machine-composed music. Florintina and Gopi [9] proposes to design a simple linear discrete system (LDS) as a highly storage and computation-efficient classifier in kernel space.

In the following section, we disclose the methodology used to generate music and the details of the classifier used. The extractable features are presented in Sect. 3. Section 4 talks of the experiments done, and Sect. 5 includes the results and conclusions.

2   Methodology

A video of the sea waves as observed from the beach is chosen. The frames are collected with a gap of one second duration (refer Fig. 1). In order to analyze the wave patterns, initially the waves need to be demarcated. To do so, the linear discrete system (LDS) [9] is chosen to function as a classifier in kernel space. To train the classifier, the first frame is taken and random blocks of the desired region (waves) are manually collected into an image to function as training data for desired class. Similarly, training data for the undesired class is also accumulated as in Fig. 2. These training data are used to train the LDS classifier. By feeding the rest of the frames as input into the trained LDS, wave-demarcated output frames are obtained by assigning value '0' to blocks classified as waves and '1' to the other blocks as depicted in Fig. 2. Upon careful observation, the notable and unique features are acquired from the wave patterns. These features are then mapped into the basic musical semitones following a suitable scale, which could be played using any musical instrument, the notes and rests being left to the musician's choice. This technique is not implemented with the illusion that completely musical audio with appropriate periodic structure would be the outcome. As a matter of fact, it just offers to serve as one of the plug-ins in music composition, upon which the composer could build his composition.

2.1   LDS as a Supervised Classifier in Kernel Space

As the name suggests, the classifier is designed as a LDS, wherein the output of the classifier is obtained by simply convolving the input vector with the impulse response coefficients (**h**) of the LDS. The training of this classifier lies in obtaining the appropriate **h**, with which it becomes feasible to precisely classify the input data. Particle swarm optimization (PSO) technique is used to acquire **h**.

In order to employ the LDS as a classifier, the following algorithm (as depicted in Fig. 3) is to be followed.

1. If the training data is of the form of a image, it is divided into blocks of appropriate size and is arranged as vectors, x_{ij}, where $i = 1, 2, \ldots n$, $j = 1, 2, \ldots m$ and n and m signify the number of training vectors in each class and number of classes.
2. While collecting the training data, it becomes vital that sufficient amount of data belonging to every class is assimilated, so as to efficiently train the classifier.
3. In order to increase the classification accuracy, Gaussian kernel is used to map the input data to higher dimensional space (HDS). The sigma value to be used in the Gaussian kernel is optimized with the aid of PSO, such that the value of excess kurtosis is minimized, homoscedasticity is maximized, and Euclidean distance between the classes is maximized [10]. This is done so that the possibility of the classes overlapping is reduced as they are made into Gaussian distributed classes which possess identical covariance matrices and are as much separated from each other as possible.
4. What vector (target) the LDS should generate when a particular class data is fed is chosen beforehand. Generally, the target vector is segmented into 'm'

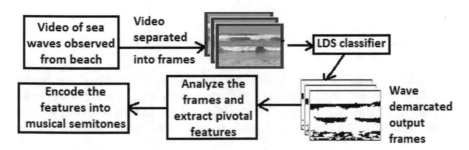

Fig. 1 Methodology adopted to compose music from sea wave patterns

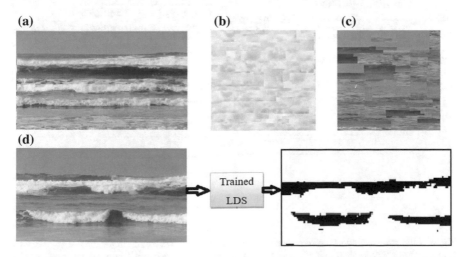

Fig. 2 **a** First frame. **b** and **c** are the training data for the desired and undesired classes, respectively, collected from (**a**). **d** Demarcation of waves

segments. If $\mathbf{t_j}$ is the target for jth class, then it is chosen such that the sum of it's jth segment entries is higher than it's other segments' sums as in Fig. 4.

5. The data mapped into the HDS is then given as input to the LDS, and this vector convolved with the LDS's impulse response coefficients ($\mathbf{h}$) gives the output, $\mathbf{y_{ij}}$. Mean square error (MSE) between $\mathbf{y_{ij}}$ and $\mathbf{t_j}$ is calculated; the values of $\mathbf{h}$ are regulated so as to minimize MSE using PSO.

6. The previous step is repeated till MSE becomes as low as possible.

7. The number of coefficients of $\mathbf{h}$ is varied, the training is repeated every time, and finally, the number of impulse response coefficients is dictated by the scenario which gives the maximum percentage of success.

8. The kernel-mapped testing data is then given as input into the trained LDS, and the output is collected.

9. The output is segmented into 'm' segments and that datum is classified as class j, if the sum of the absolute of the jth segment's entries is higher than the sum of the absolute of any other segments' entries.

3 Extractable Features of Significance

On meticulous analysis of the wave-demarcated frames, it is observed that several spatial and temporal features provide significant information and could be extracted. The spatial features are collected by scrutinizing each and every frame; i.e., the feature extracted from one frame depends neither on the frames preceding it nor

Fig. 3 LDS as a classifier in kernel space

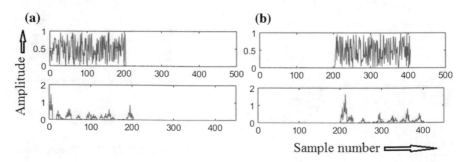

Fig. 4 First row of **a** and **b** are the intended ones or targets predetermined for class 1 (wave) and 2, respectively. The second row of **a** and **b** is output of the LDS for class 1 and 2 data, respectively, trying to imitate their targets

on the ones following it. Number of waves, thickness of the waves, and distance between the waves observed in each frame are some of the spatial features that could be extracted from the wave patterns. If a feature is collected by analyzing a series of frames, belonging to different times, then it is termed as temporal feature. Time duration for which each wave exists, distance travelled by each wave before it ceases to exist and the velocity with which it travels are some of the extractable temporal features.

4 Experimentation

For validating the technique proposed, two videos of sea are used. The first is a video of a turbulent sea shot from near the seaside town of Cannon Beach in the USA, acquired from [11]. The second is a video of a calm sea captured from Marine Street Beach, California, obtained from the source [12].

4.1 Music Composed from Turbulent Sea

The turbulent sea video is broken into frames and carefully scrutinizing them reveals that the waves are quite ferocious and that tracking each and every wave is impossible. Thus, collecting their temporal features would prove to be a futile task but they possess significant spatial information. Wave-demarcated frames are obtained from the trained LDS classifier. Each and every frame is individually analyzed, and the spatial features are collected along a particular reference line, chosen here as the vertical line joining the midpoints of the top and bottom of the frame. The first feature collected is the distance between the first two waves encountered along the reference line measured in pixels (d). The second feature is obtained from the thickness of the first wave encountered along the reference line (t), also measured in terms of pixels. The entire range of d and t is separated into three segments, and these two features are combined together and mapped into the seven basic musical semitones following the conditions in Fig. 5. In some cases, when the feature values of successive frames lie within the same condition zone, same semitones are generated from those consecutive frames and they are grouped together into a single instance of that particular semitone. These are played using any musical instrument. Figure 5 shows the features being extracted and encoded into semitones and the music strain composed from them.

4.2 Music Composed from Calm Sea

When the calm sea video is broken into frames and examined, it is seen that only one or two waves exist at the most at any time. This information signifies that not much

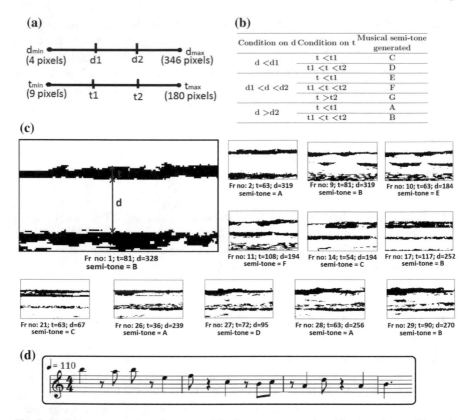

Fig. 5 Turbulent sea: **a** gives the range of the features extracted and **b** gives the conditions for mapping of features extracted from wave patterns into musical semitones. **c** Features (t and d) being extracted from frames 1 to 29 of turbulent sea and mapped into semitones. Frames that are mapped into semitones same as their successive ones are not depicted and are not used. 'Fr no' signifies frame number. **d** is the piece of music composed from the frames

spatial information could be collected, but at the same time, tracking a single wave across the frames and collecting temporal features is not so arduous a task. Wave-demarcated frames obtained as output from the LDS classifier are inspected. The first feature collected is the time of existence of each wave (tw). To do so, the frame number (equivalent to time) at which the existence of a wave is first detected along the mid-reference line is noted and is subtracted from the frame number at which the wave collapses (or exists no more) along the mid-reference line. The second feature collected is the distance travelled by each wave (dw) which is done by subtracting the starting point of the wave and the point where it collapses. The entire range of dw is separated into four segments, and the range of tw being less is segmented into two segments, and the two features are jointly mapped into the seven basic musical semitones (making sure consecutive like-valued semitones are written only once)

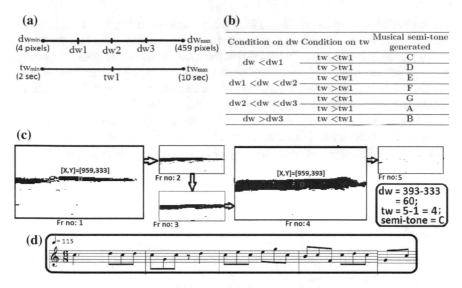

Fig. 6 Calm sea: **a** gives the range of the features extracted and **b** gives the conditions for mapping of features into musical semitones. **c** Features (tw and dw) of the first wave being extracted from frames 1 to 5 and mapped into semitone, and 'Fr no' signifies frame number. **d** is a sample piece of music composed

as in Fig. 6. The procedure adopted to extract the features and encode them into semitones along with a sample strain of music composed is depicted in Fig. 6.

5   Results and Conclusion

The LDS classifier employed to demarcate waves is found to give a percentage of success of 100% in the case of training, for both the turbulent and calm sea datasets. In the case of testing, in both cases, 25 randomly chosen frames are manually compared with the wave-demarcated output frames obtained from the trained LDS and are found to be appropriately classified, with every wave clearly highlighted. These high percentages of success along with the fact that the LDS classifier requires the storage of only four coefficients (by using linear phase property while designing the LDS with seven coefficients) and has no need for dimensionality reduction techniques stand testimony to the fact that the LDS is highly storage and computation efficient without compromising the performance.

In order to evaluate the work done, three evaluation experiments are done by enlisting the aid of 50 people by means of Google form [13]. In the first two tests, they are asked to identify the machine-composed music piece from an already existing human-composed music clipped to the same duration, all of them being played by the same musician, using the same electronic keyboard so as to create a platform for fair evaluation. The fact that 58 and 60% chose wrongly in the two tests indicates that the machine compositions are as musical as human ones and that they are unable to distinguish between them. In the third experiment, random numbers are encoded into musical semitones following the same procedure as the one in this proposal and played in the same instrument, by the same person. In order to analyze the significance of this proposal's contribution to the field of music composition, the subjects are asked to compare this audio with a machine-generated piece and testify to how vastly both vary and that random numbers are incapable of generating music. Their rating of 91.6% reveals that the project is worthy of contributing much to the music composition domain.

References

1. Muoz E, Cadenas JM, Ong YS, Acampora G (2016) Memetic music composition. IEEE Trans Evol Comput 20:2202–2213
2. Santos A, Arcay B, Dorado J, Romero J, Rodriguez J (2000) Evolutionary computation systems for musical composition. Proc Acoust Music: Theory Appl 1:97–102
3. Engels S, Chan F, Tong T (2015) Automatic real-time music generation for games. In: Proceedings of the eleventh AAAI conference on artificial intelligence and interactive digital entertainment
4. Roig C, Tardn Lorenzo J, Barbancho I, Barbancho AM (2014) Automatic melody composition based on a probabilistic model of music style and harmonic rules. Knowl-Based Syst 71:419–434
5. Merwe A, Der V, Schulze W (2011) Music generation with markov models. IEEE Multimedia 18:78–85
6. Conklin D (2003) Music generation from statistical models. In: Proceedings of the symposium on artificial intelligence and creativity in the arts and sciences, pp 30–35
7. Miranda E, Biles J (2007) Evolutionary computer music. Springer, Berlin
8. Pearce M, Wiggins G (2001) Towards a framework for the evaluation of machine compositions. In: Proceedings of the AISB01 symposium on AI and creativity in arts and science, pp 22–32
9. Florintina C, Gopi ES (2017) Constructing a linear discrete system in kernel space as a supervised classifier. IEEE WiSPNET 2017 conference (accepted)
10. Gopi ES, Palanisamy P (2014) Maximizing Gaussianity using kurtosis measurement in the kernel space for kernel linear discriminant analysis. Neurocomputing 144:329–337
11. Breaking waves—1 hour of beautiful Pacific Ocean waves in HD. Retrieved from https://www.youtube.com/watch?v=e_2t83JsbGI. Accessed on 01 May 2017
12. Relaxing 3 hour video of California Ocean waves. Retrieved from https://www.youtube.com/watch?v=zmPzbZVUp3g&t=2s. Accessed on 01 May 2017
13. Evaluation of music composed through computational intelligence inspired by sea waves. Retrieved from https://docs.google.com/forms/d/e/1FAIpQLSd7LOai_W6mSZcDN82rGSBIT8zXodtuACaVIaB2tOBlOQCs8A/viewform?c=0&w=1

Multi-graph-Based Intent Hierarchy Generation to Determine Action Sequence

Hrishikesh Kulkarni

Abstract All actions are result of scenarios. While we analyze different information pieces and their relationships, we notice that actions drive intents and intents drive actions. These relationships among intents and actions can decode the reasoning behind action selection. In the similar way, in stories, news and even email exchanges these relationships are evident. These relationships can allow us to formulate a sequence of mails irrespective of change in subject and matter. The continuity among concepts, drift of concepts, and then reestablishing of original concepts in conversation, email exchanges, or even series of events is necessary to establish relationships among various artifacts. This paper proposes a multi-edge fuzzy graph-based adversarial concept mapping to resolve this issue. Instead of peaks, this technique tries to map different valleys in concept flow. The transition index between prominent valleys helps to decide your order of that particular leg. The paper further proposes a technique to minimize this using fuzzy graph. The fuzziness in graph represents the fuzzy relationships among concepts. The association among multiple such graphs helps to represent the overall concept flow. The dynamic concept flow represents the overall concept flow across the documents. This algorithm and representation can be very useful to represent lengthy documents with multiple references. This approach can be used to solve many real-life problems like news compilation, legal case relevance detection, and associating assorted news from the period under purview.

Keywords Multi-edge graph · Fuzzy graph · Adversarial concept mapping
Data mining · Machine learning · Cognitive sciences

H. Kulkarni (✉)
PVG's College of Engineering and Technology, (SPPU), Pune 411009, India
e-mail: hrishikeshparag@gmail.com

© Springer Nature Singapore Pte Ltd. 2019
A. J. Kulkarni et al. (eds.), *Proceedings of the 2nd International Conference on Data Engineering and Communication Technology*, Advances in Intelligent Systems and Computing 828, https://doi.org/10.1007/978-981-13-1610-4_6

1   Introduction

Every day, we deal with a lot of information. It generally comes in unstructured form. Natural language processing tackles mining unstructured data very elegantly. It can range from parsing, tokenization to disambiguation and what not? Throughout the text, the concepts are flowing. These concepts are sometimes represented partially, while in other cases they go through many transitions. The concepts come and go—sometimes, there is abrupt change in concept. In some other cases, there is smooth transition from one concept to another. Concepts come in context. Context has more life in documents as compared to concept. Multiple concepts converge in a particular context and form a new concept. A particular concept in a given context may mean completely different than in case of some other context.

With reference to this discussion, it becomes very necessary to understand these two terms: concept and context. The context is the situation associated with social environment, whereby knowledge is acquired and processed [1]. We may need to collate multiple concepts in a given context to derive meaning, also to provide meaningful impact, and to help decision-making. Let us define the context:

Context refers to the situation under purview—here, situation is associated with user, actions, and environment. Hence, the perspective of user with reference to environment, action sequence, and flow of events represents a context. Thus, context is typically represented by the properties of situation, relationships among events, and properties of events.

On the other hand, the concept is idea depicted in text, paragraph, or conversation. A text under observation may depict multiple ideas and hence can have multiple concepts. Generally, it is expected to represent associated concepts. Some researchers believe that a concept can have multiple meanings [2]. But we will not go into that. The concept mining focuses on finding out keywords and relationships among them. In this paper, we have redefined the concept as a flow of the idea across the series of sentences trying to convey a particular meaning. There can be multiple ideas in this flow. The relationships among relevant words are used along with information flow and intent mapping to mine the concepts.

The prominent concepts become nodes of the graph. Since concept relationship cannot be defined in crisp way, we have proposed to use fuzzy graph notion for this purpose. The relationships between the two concepts can carry more than one route and also depend on context. To represent this, we have proposed use of multi-edge graph. It represents this context action association. The flow of information across the multiple nodes contributes to overall concept in the documentation where edges to follow are selected based on context.

Concept and flow matching algorithm (CFMA) will study all the nodes and determine flow across the nodes to derive the overall concept transitions. This will also help to determine concept drifts, concept transitions, etc. The association across multiple text artifacts will present the collated concepts with reference to context. This can further be improved to map to user's sensitivity index set.

In fuzzy graph, fuzziness is associated with the edges of the graph. In this case, a new algorithm of Adversarial Concept Mapping (AdCMa) is proposed. Adversarial is typically meant as involving or presenting or characterized by opposition or conflict or contrary view. The adversarial concept mapping does not focus on flow of concept or smoothness but typically marked by peaks and valleys with reference to distortions. The peaks and valleys under observation build the transition nodes for concept mapping and even in case of our fuzzy multi-edge graph. The concept regions are identified using the valleys in the concept flow. The concept is determined based on intent and action association. The change in flow and relationships between dips and peaks is used for concept deriving.

2 Literature Review

Concept and context have been the keywords of unstructured data research for many years. Researchers defined concepts and contexts in different ways and contributed to information decoding. When we need to use information, we need to organize it and derive meaning from it with reference to tasks at hand. Term frequency and inverse document frequency (TFIDF) was a first breakthrough in assigning meaning to unstructured scattered text. The extension of this method to bigram-based methods and trigram-based methods was tested, and the evident conclusion restricted majority of efforts to bigram-based approaches. This helped to achieve clustering of documents in relevant groups and classes [3]. These methods evolved to obvious weighted TFIDF-based methods. Here, calculating weights was the most challenging task and it is simplified by assigning weights based on frequency of occurrence. Later, some additional parameters are also included in weight calculation, like position of the word in document. This is used for concept determination and qualification [4]. In these cases, researchers put efforts in defining concept. Mathematically, it is defined as a corpus of words. It used the association and relationship among the words as deciding factor [5]. The association is further used to derive meaning and knowledge. The association and mapping are used to derive symbolic knowledge from set of documents and information artifacts. Many researchers started working on defining the context. Context-based relationships are used to derive these associations [6].

Researchers used dictionary-based keyword search and bigrams for document mining. Textual association and disambiguation for deriving meaning were also used by many researchers [7]. Then, there was a trend of personalized delivery, personalized processing. The context has got importance, and many researchers turned to context-based processing and classification. Some researcher worked on associating multiple contexts to derive overall meaning. Traditionally, researchers were focused on place, location, and time to derive the context. Slowly, the definition of the context became broader. Some researcher deployed other algorithms like association rule mining and Bayes classification for context determination [8]. The world evolution and learning are always incremental in nature. Building on what you have is the organic property of world. This additional information coming from different sources makes it necessary to correct your learning vectors. Hence, there is pressing need for incremental learning. Researchers used closeness factor-based approach for the same [9]. Semi-supervised learning allows learning from labeled as well as unlabeled data. Use of absolute value inequalities for semi-supervised learning showed a lot of promise [10]. Researchers used Gaussian mixture models for multimodal optimization problems [11]. Population-based meta-heuristic is also used for best path selection [12]. Information never comes as one unit. It becomes available in parts and over the time. Deriving the meaning out of these parts of information is necessary to take right decision. Systemic and multi-perspective learning is required to address this issue. In this work, focus was on building a systemic view [13]. Words and series of words express emotions. Emotions can depend on context, or it can be like an impulse based on intensity and positioning of the word. The simple sentiment for selection of product is derived from feedback from users. The text flow also determines emotions. The sentiment analysis is strongly mapped to decisive and expressive keywords. Researchers and professionals used it for movie, product, and book rankings. For this purpose, crowdsourcing and crowd intelligence are also used [14, 15]. Text is always associated with creativity. In this quest, a few researchers worked for creativity and learnability mapping [16]. Also, researchers used AI and machine learning for different creative activities like expressing in poetic words, writing fictions, and creative assimilations [17]. The learning can happen in compartment, and then combining those learning is a difficult task. Researchers used ε-distance-based density clustering to cluster distributed data [18]. The literature shows an impetus to create a real and intelligent learning through association and inference. The compelling intent of researchers helped to solve some very difficult and challenging problems in the domains of creativity. The emotional tone and intent action can play a major role while solving this problem of completing stories with effective mapping of hidden and missing text [19]. Deriving a concept with reference to context remains a challenge. Researcher even made use of positional significance [20]. This paper proposes a multi-edge graph-based algorithm to derive this mapping.

3 Context-Based Learning

World is about relationships so is mathematics. Different data structures have emerged to represent these relationships. Graphs, trees, and other data structures represent these relationships. When it comes to unstructured data, these structures need to be modified. For the same purpose, mathematicians introduced directed acyclic graphs, fuzzy graphs, multi-edge graphs, and semi-graphs. In this paper to represent this unstructured relationship, we are going to use multi-edge fuzzy graph.

A fuzzy graph is defined as:

A fuzzy set is denoted by $A = (X, m)$. A fuzzy graph $\xi = (V, \sigma, \mu)$ is an algebraic structure of non-empty set V together with a pair of functions $\sigma: V \rightarrow [0,1]$ and $\mu: V \times V \rightarrow [0, 1]$ such that for all $x, y \in V$, $\mu(x, y) \leq \sigma(x) \wedge \sigma(y)$ and μ is a symmetric fuzzy relation on σ.

Majority of the problems in the environment are partially observable. In this environment, we need to use information and intelligence effectively to arrive at the best possible solution. The same analogy can be applied to problems represented in text form. Any paragraph can have multiple concepts, and we are keen to derive meaning of that paragraph in a particular context.

Concepts $=f$ (keywords, associated concept, positions). Hence, document can be represented as a concept map. The document is consumed or used in a particular context:

Context $=f$ (user, situation depicted in document, situation in which document is used).

The continuous text may include different concepts. There can be transition from one concept to another. This transition may be sudden or progressive or smooth. The dominant concept at a particular location may have contextual impact due to description before the occurrence of that sentence. Thus, context may progress with concept and representing this unique association with typical graphs becomes difficult, as it does not follow standard graph rules. What can we do about it? Where is the context here? There are many text artifacts floating related to same concept. Which one is related to prime or dominant concept?

There are totally five core steps to determine concept map:

1. Text organization: Organizing text with flow and indicators.
2. Concept retrieval—first level: The multiple concepts are retrieved from text.
3. Associating concepts: Two or more concepts are associated—this is done using context.
4. Weaving concept to determine concept map: These association and relationships determine concept maps.
5. Determine external parameters and concept maps.

We will consider a paragraph. It is parsed through standard parser, and properties are extracted. The words in paragraph are divided into three types: intent words, action words, and concept words. The core words are identified. These are the words associated with one or more action words. Identify all words, which are action reachable from concept words. The relationships among these words build a concept corpus. There can be multiple concepts in a single paragraph.

Core intent word is a word which irrespective of its frequency of occurrence impacts one or more actions in the region under observation

{Represent it mathematically}

$$\forall w \in ws | w \rightarrow [\text{IA}] \text{ where } [\text{IA}] \neq \Phi$$

$[IA]$ refers to set of impacted prominent actions.

Border word: A word that shows relationship with more than one prominent actions but not directly impact any action.

$$\forall w \in ws | w \leftrightarrow [\text{IA}] \text{ where } [\text{IA}] \neq \Phi \text{ and } w \text{ does not drive } IA$$

Action connected words: Is a set of words connected by one or more actions? Any word in given paragraph is classified into one of these types.

What is an intent action relationship? Intent results in action. Even actions are mapped to intent. Then, concepts are associated to build the central concept and associated concepts. Every additional node/concept added to graph results in information gain. This information gain depends on action reachability of the particular node. We use standard entropy formulation to define information gain. Hence, information gain is defined as joint entropy $H(X, Y)$ of discrete random variable X and Y with probability distribution $p(X, Y)$

$$H(X, Y) = -\sum_{x \in X} \sum_{y \in Y} p(x, y) \log p(x, y) \tag{1}$$

The conditional entropy will be calculated iteratively to represent overall information gain. A concept path in given context is derived, and for that the modified Bayesian is used [19].

$$P(T_{2.1}/(T_1 + T_3)) = \frac{P(T_{2.1}/T_1) * P(T_3/T_{2.1})}{\sum_{i=1}^{n} P(T_{2.i}/T_1) * P(T_3/T_{2.i})} \tag{2}$$

The mutual information due to occurrence of multiple concepts is defined as: [21]

$$I(X; Y) = \sum_{x \in X} \sum_{y \in Y} \frac{p(x, y) \log p(x, y)}{p(x)p(y)} \tag{3}$$

$$= D(p(x, y) \| p(x)p(y)) \tag{4}$$

$$= E_{p(x,y)} \log \frac{p(X, Y)}{p(X)p(Y)} \tag{5}$$

Context helps to decide the context nodes to be traversed. Interestingly, primary concept, secondary concept, action reachable points along with preceding information contribute to this concept traversal. In our case, we have used simple Bayesian to decide the best path for given context.

Let us take simple example where concepts are—demonetization, cash crunch, corruption, digitization, and black money. There can be many other tokens like ATM, Reserve Bank; finance minister can contribute to concept but cannot be concept in itself. Now, context can be derived from core concept, supporting word, prelude, and situational parameters. In this case, context could be:

{Demonetization, corruption, cash crunch, ATM, opposition, procession (date, time, location)}.

The same core concept with different contexts below traverses a different multigraph path:

{Demonetization, corruption, finance minister, ATM, digitization, arrested, action taken (date, time, location)}.

In multi-graph, multiple concepts are connected by zero or more edges. The association is represented through connection. How to decide degree of node is dependent on action reachability of different concepts.

4   Context-Based Learning

4.1   Simple Example

Let us consider an example!

Paragraph 1: RBI Governor told the Parliamentary Standing Committee on Finance that there is a reasonable stock of new 500 and 2000 currency notes. You are already aware of the announcement of demonetization that was made on November 8, 2016. It is about discontinuing old notes from immediate effect.

Paragraph 2: The International Monetary Fund trimmed India's growth forecast for 2017 by 0.4%, citing demonetization effects. Opposition mentioned that there are long queues at ATM and most of the ATMs are without cash. Dry ATMs, no cash, and unavailability of digital resources are posing serious challenge. Is the digitization move working? is a serious question.

Paragraph 3: The Income Tax department carried out over 1100 searches and surveys immediately after demonetisation (claimed as one of the step towards digital economy) and detected undisclosed income of over Rs 5400 crore, Finance Minister told the Rajya Sabha on Thursday. He added that more follow-up action was taken, and 18 lakh people were identified whose tax profiles were not in line with the cash deposits made by them in the demonetization period and on lines responses were sought.

4.2   Multi-edge Graph Context Representation Model

Figure 1 represents core concept node demonetization. There is a context from opposition represented with red edge. This context associates multiple concepts in a particular order. There are multiple edges connecting different concepts. The demonetization is a core concept since there are five action reachable and three direct reachable concepts from this node. It is shown in Table 1. The concept drift or adversarial point is identified based on degree transformation of nodes while traversing. These points are used for concept mapping.

Node is action reachable if there is action edge between two nodes. There can be number of action edges, and it can be action reachable through other nodes. Same is true in direct reachable.

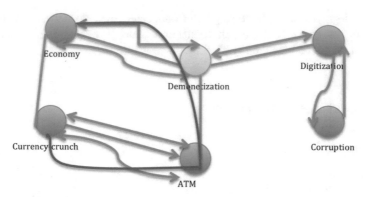

Fig. 1 Opposition context—demonetization

Table 1 Concept nodes

Node	Core	Direct reachable	Action reachable
Demonetization	Y	3	5
Digitization	Y	2	4
ATM	N	1	1
Currency crunch	Y	2	2

$$\text{Concept Relationship Index} = \frac{\sum p - \sum n}{\sum c} \qquad (6)$$

where $\sum c$ is number of concepts in text.

5 Experimental Evaluation

Testing creativity, measuring learnability are always a difficult task. To evaluate the performance of our model, we have created a test set of 110 paragraphs. The frequency is used in the beginning to deciding relative importance of keywords. The concepts are derived based on reachability. Core, primary, and secondary concepts are identified. Based on intent action relationship, these concepts are connected with one or more directed edges. Fuzzy edges with weights are used in case of uncertain relationships. Adversarial nodes are used for mapping concept drifts.

The concept associated with the context and story built based on concept relationships are verified using 110 paragraph data. The observations clearly suggest that in 80% of the cases the concepts mapped to the context were correct. The context even determines the association among concepts, and in 90% of the cases core concept mapping is correct.

Fig. 2 Success for
concept–context mapping

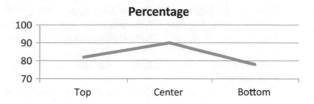

The success is checked for three prominent concepts. When primary concept is at the center of the context, the success is close to 90%. The success with reference to these three categories is depicted in Fig. 2.

6 Conclusion

Identifying the most relevant text, most relevant route, most relevant data, most interesting product is what world is striving for over the years. It is never absolute and depends on the context. There can be context in document, and there can be context associated with user. The concept on one side describes the idea explained—context tries to depict a perspective with which the document is looked upon. This paper proposed a method based on multi-graph and fuzzy graph to represent concept–context mapping. This can help in solving problems in different domains. It can help to identify key concepts. It can be very helpful in literary and creative activities ranging from recreation to building a story. It can help in identifying gaps in the market to decide initiative. The experimentation is carried out on the 110 text paragraphs. The said algorithm showed close to 90% accuracy. Here, the human association of concept to context is assumed as the correct one. The promising results definitely show the possibility of extending this technique to many applications. In future work, concept maps can be applied to larger documents.

Measuring learnability and creativity is always a difficult task. To evaluate the performance of our model, we have created a test set of 110 paragraphs. The frequency is used in the beginning to deciding relative importance of keywords. The concepts are derived based on reachability. Core, primary, and secondary concepts are identified. Based on intent action relationship, these concepts are connected with one or more directed edges. Fuzzy edges with weights are used in case of uncertain relationships. Adversarial nodes are used for mapping concept drifts.

References

1. Rose DE (2012) Context based learning. Springer, Berlin
2. Liu B, Cin C et al Mining topic specific connect and definitions on the web. Retrieved from https://www.cs.uic.edu/~liub/publications/WWW-2003.pdf. 27 May 2017

3. Cutting DR, Pedersen JO, Karger DR, Turkey JW (1992) Scatter/gather: a cluster-based approach to browsing large document collections. In: ACM SIGIR
4. Sundaranar M et al (2013) Quantification of portrayal concepts using TF-IDF weighting. Int J Inf Sci Tech (IJIST) 3(5)
5. Ramos J (2003) Using TF-IDF to determine word relevance in document queries. In: Proceedings, 2003
6. Craven M, DiPasquo D et al (1998) Learning to extract symbolic knowledge from world wide web. In: AAAI-98
7. Fiburger N, Maurel D (2002) Textual similarity based on proper names. In: Workshop on mathematical methods in information retrieval, ACM SIGIR
8. Kulkarni AR, Tokekar V, Kulkarni P (2012) Identifying context of text document using naïve Bayes classification and Apriori association rule mining. In: CONSEG
9. Kulkarni P, Mulay P (2013) Evolve your system using incremental clustering approach. Springer J Evolving Syst 4(2):71–85
10. Mangasarian OL (2015) Unsupervised classification via convex absolute inequalities. Optimization 64(1):81–86
11. Dong W, Zhou M (2014) Gaussian classifier-based evolutionary strategy (GCES) to solve multimodal optimization. IEEE Trans Neural Netw Learn Syst 25(6)
12. Li X, Epitropakis M, Deb K, Engelbrecht A (2016) Seeking multiple solutions, an updated survey on niching solution. IEEE Trans Evol Comput 99:1
13. Kulkarni P (2012) Reinforcement and systemic machine learning for decision making. Book, Wiley/IEEE, Hoboken
14. Agrawal A et al (2011) Sentiment analysis on twitter data. In: Proceedings of the workshop on language in social media. pp 30–38
15. Fang X, Zhan J (2015) Sentiment analysis using product review data. J Big Data 2:5
16. Kulkarni P (2017) Reverse hypothesis machine learning, intelligent systems reference library. Springer, New York
17. Bringsjord S, Ferrucci D (2000) Artificial intelligence and literary creativity. Lawerence Erlbaum Associates Publishers, London
18. Jahirabadkar S, Kulkarni P (2014) Algorithm to determine ε-distance parameter in density based clustering. Expert Syst Appl 41:2939–2946
19. Kulkarni H (2017) Intelligent context based prediction using probabilistic intent-action ontology and tone matching algorithm. International Conference on Advances in Computing, Communications and Informatics (ICACCI)
20. Kulkarni AR, Tokekar V, Kulkarni P, Identifying context of text documents using Naïve Bayes classification and Apriori association rule mining. In: 2012 CSI sixth international conference on software engineering (CONSEG). pp 1–4
21. Cover T, Thomas J (2006) Elements of information theory. Wiley, New York

Predicting the Molecular Subtypes in Gliomas Using T2-Weighted MRI

Jayant Jagtap, Jitender Saini, Vani Santosh and Madhura Ingalhalikar

Abstract Classifying gliomas noninvasively into their molecular subsets is a crucial neuro-scientific problem. Prognosis of the isocitrate dehydrogenase (IDH) mutation in gliomas is important for planning targeted therauptic intervention and tailored treatment for individual patients. This work proposes a novel technique based on texture analysis of T2-weighted magnetic resonance imaging (MRI) scans of grade 2 and grade 3 gliomas to differentiate between IDH1 mutant 1p/19q positive and IDH1 mutant 1p/19q negative categories. The textural features used in the proposed method are local binary patterns histogram (LBPH), Shannon entropy, histogram, skewness, kurtosis, and intensity grading. We discriminate the tumors into their molecular subtypes using standard artificial neural networks (ANNs). LBPH attributes demonstrated maximum discrimination between the two groups followed by Shannon entropy. In summary, the technique proposed facilitates an early biomarker to detect the IDH subtype noninvasively and can be employed as an automated tool in clinics to aid diagnosis.

Keywords Isocitrate dehydrogenase (IDH) · Gliomas · Magnetic resonance imaging (MRI) · Local binary patterns histogram (LBPH) · Artificial neural networks (ANN)

J. Jagtap (✉) · M. Ingalhalikar
Symbiosis Institute of Technology, Symbiosis International University,
Pune 412115, India
e-mail: jayant.jagtap@sitpune.edu.in

M. Ingalhalikar
e-mail: madhura.ingalhalikar@sitpune.edu.in

J. Saini · V. Santosh
National Institute for Mental Health and Neurosciences, Bengaluru 560029, India
e-mail: jitender.s@nimhans.kar.nic.in

V. Santosh
e-mail: vani@nimhans.kar.nic.in

© Springer Nature Singapore Pte Ltd. 2019
A. J. Kulkarni et al. (eds.), *Proceedings of the 2nd International Conference on Data Engineering and Communication Technology*, Advances in Intelligent Systems and Computing 828, https://doi.org/10.1007/978-981-13-1610-4_7

1 Introduction

Gliomas are the most common type of tumor of the central nervous system that are usually life threatening and require early diagnosis for better mortality [1, 2]. Traditionally tumors are classified into grade I to IV subtypes that are based on the microscopic features provided via biopsy procedure. With the advent of computer-aided diagnosis (CAD), the tumors can be marked using magnetic resonance imaging (MRI). The location and size can be accurately mapped; however, the inhomogeneity in intensity and high heterogeneity in each subtype makes it difficult to discriminate the tumors. However, recent genetic molecular advances have contributed to a better understanding of tumor pathophysiology as well as disease stratification. Recent studies have identified novel mutations in the isocitrate dehydrogenase (IDH)1 gene from a genome-wide mutational analysis of glioblastomas [3]. Further, it has been found that IDH1 mutations are present in 50–90% of cases of grade II and III gliomas and are associated with longer survival in glioma patients [4]. However, to identify the IDH1 status, a genome-wide study is required.

The aim of this research is to present an automated noninvasive CAD approach that can categorize brain tumors MRI scans into their molecular subtypes namely oligodendroglioma IDH1 mutant 1p/19q positive and anaplastic astrocytoma IDH1 mutant 1p/19q negative. The proposed approach can be employed as an early molecular biomarker and can therefore aid in the diagnosis and treatment planning of the patient.

Recent work on characterizing IDH1 subtypes includes textural analysis of T2-weighted images by using Shannon entropy and edge maps using Prewitt filtering [5]. The study demonstrated IDH1 wild-type gliomas to have statistically lower Shannon entropy than IDH1-mutated gliomas; however, no significant difference was observed on the edge strength maps. Another recent study employed multimodal approach by using T2-weighted and diffusion MRI images [6]. Texture features that include entropy, skewness, and kurtosis were computed. However, other than entropy none of the features could segregate the molecular subtypes. This work is highly innovative, as we propose to employ multiple texture features that include the local binary patterns histogram (LBPH), Shannon entropy, histogram, skewness, kurtosis, and intensity grading, which provide superior discrimination between the 1p/19q positive and 1p/19q negative molecular subgroups. The classification is achieved by using a simple two-layer feedforward backpropagation artificial neural network (ANN). Despite the inhomogeneity in intensity levels and variability in size and structure, we obtain a classifier with high accuracy, establishing the applicability for early diagnosis of the gliomas.

2 Proposed Method for Glioma Classification

2.1 Overview

Our method of creating texture-based glioma classifiers includes the following steps: (1) manual segmentation of gliomas on T2-weighted images, (2) feature extraction, (3) classifier training and cross-validation. Figure 1 displays a schematic block diagram of the method. In the next few sections, we shall first describe the features involved and the classification technique, followed by the experiment on the T2-weighted images.

2.2 Feature Extraction

The textural features used in the proposed method are local binary patterns histogram (LBPH), Shannon entropy, histogram, skewness, kurtosis, and intensity grading.

1. **Local Binary Patterns**

   Local binary patterns (LBPs) proposed by Ojala et al. [7] are further grouped into 10 codes from 0 to 9 [8]. The histogram of these 10 LBP codes, called as local binary pattern histogram (LBPH), is used as textural features in the proposed method.

2. **Shannon Entropy**

   Shannon entropy is the amount of information carried by an image. The Shannon entropy E is calculated by using Eq. 1 as follows.

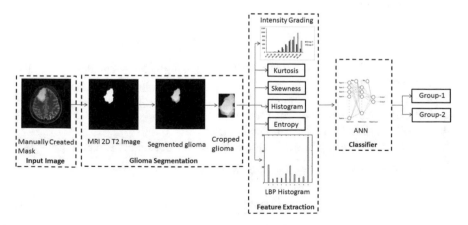

Fig. 1 Work flow diagram of proposed method for glioma classification

$$E = -\sum_{g=0}^{255} f_g log_2(f_g) \tag{1}$$

where f_g denotes the frequency of occurrence of gth the gray level.

3. **Histogram, Skewness, Kurtosis**

   Histogram gives the frequency of occurrence of a particular gray level in an image. Skewness and kurtosis are the third and fourth moment of data distribution, respectively.

4. **Intensity Grading**

   Intensity grading gives the count of number of pixels lies within the specified range of intensities in an image. Intensity grading is also calculated from histogram of an image.

2.3 Classifier

The artificial neural network (ANN) classifier is used to classify the gliomas into one of the two groups. ANN is designed by using multilayer feedforward backpropagation neural networks [9]. It consists of an input layer, a hidden layer, and an output layer. Input layer consists of number of neurons equal to the length of feature vector of an input image. Output layer consists of two neurons excluding bias neuron equal to number of output groups. For the hidden layer, we select number of neurons equal to $\frac{1}{3}$rd of the neurons used in the input layer.

2.4 Algorithm

The steps involved in training and testing of the proposed method are as follows.

1. Let $I(x, y)$ be an input grayscale image, where x and y are the pixel intensities in the range [0, 255].
2. Resize $I(x, y)$ to 256×256.
3. Manually mark glioma in $I(x, y)$.
4. Let $I_C(x, y)$ be the cropped glioma image.
5. Perform texture analysis for feature extraction using LBPH and entropy.

   a. Calculate LBPH of I_C and store in a vector F_{LBPH}, as given in Eq. 2.

   $$F_{LBPH} = [LBPH(I_C)]_{10 \times 1} \tag{2}$$

   b. Calculate the entropy of I_C and store in a vector F_{ENT}, as given in Eq. 3.

   $$F_{ENT} = [Entropy(I_C)]_{1 \times 1} \tag{3}$$

c. Calculate the histogram of I_C and store in a vector F_{HIST}, as given in Eq. 4.

$$F_{HIST} = [Histogram(I_C)]_{255 \times 1} \tag{4}$$

d. Calculate skewness of I_C using F_{HIST} and store in a vector F_{SKEW}, as given in Eq. 5.

$$F_{SKEW} = [Skewness(I_C)]_{1 \times 1} \tag{5}$$

e. Calculate kurtosis of I_C using F_{HIST} and store in a vector F_{KURT}, as given in Eq. 6.

$$F_{KURT} = [Kurtosis(I_C)]_{1 \times 1} \tag{6}$$

f. Calculate intensity grading of I_C using F_{HIST} and store in a vector F_{IG}, as given in Eq. 7

$$F_{IG} = [IntensityGrading(I_C)]_{10 \times 1} \tag{7}$$

g. Concatenate all the textural feature vectors calculated in steps a to f in a vector F_T, called textural feature vector, as given in Eq. 8.

$$F_T = [F_{LBPH}, F_{ENT}, F_{HIST}, F_{SKEW}, F_{KURT}, F_{IG}]_{278 \times 1} \tag{8}$$

h. The normalized feature vector F_{T_n} is calculated by normalizing vector F_T using min-max normalization technique as given in Eq. 9.

$$F_n = \frac{F_{a_n} - min(F_a)}{max(F_a) - min(F_a)} \tag{9}$$

where $F_a = (F_{a_1}, ..., F_{a_n})$, n is the length of feature vector, and F_n is n^{th} normalized data.

6. Apply the normalized feature vector F_{T_n} to ANN classifier, with corresponding group labels in training phase, and without corresponding group labels in testing phase, to classify the input images into one of the two groups, namely Group-1(1p/19q positive) and Group-2(1p/19q negative).

3 Experimental Results

3.1 Dataset

The glioma dataset was acquired in part of the effort at National Institute of Mental Health And Neuro-Sciences (NIMHANS), Bangalore. All the patients were clinically assessed by an experienced oncologist. The subjects, 15 with IDH1 1p/19q positive and 10 with IDH1 1p/19q negative, were recruited. These subjects were in

(a)   (b)

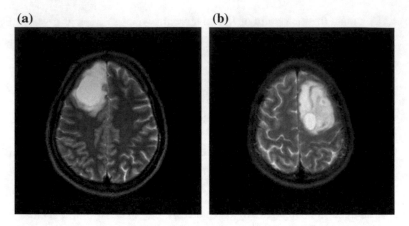

Fig. 2 Sample images from dataset **a** Group-1, **b** Group-2

the age range of 17 years to 58 years with an average age of 38 years. All imaging was performed on the same site, using the Siemens 1.5 T with a 12 channel head coil. The T2-weighted images were acquired with the following parameters: TR/TE = 4500/90 ms, slice thickness = 5 mm, and voxel-size = 1 × 1 × 5 mm. Since these images acquired at a low resolution in the sagittal and coronal planes, we used only the 2D axial slices in this work.

We employed 18 axial slices from 15 subjects in the 1p/19q positive group and 15 axial slices from the 10 subjects in 1p/19q negative group. These slices were chosen at the maximum cross-sectional area of the tumor, with each patient contributing not more than 3 slices. Figure 2 demonstrates the sample slices for the 1p/19q positive and 1p/19q negative groups. The figure is shown to illustrate the complexity of the classification problem.

3.2 Feature Extraction

By using the steps mentioned in Sect. 2, first, the LBPH of glioma-segmented image is calculated and stored in a vector F_{LBPH} of size 10×1. Then, entropy of the glioma segmented image is calculated and stored in a vector F_{ENT} of size 1×1. Next, histogram of the glioma-segmented image is calculated and stored in a vector F_{HIST} of size 255×1. Later on, skewness and kurtosis of the glioma-segmented image are calculated and stored in vector F_{SKEW} of size 1×1 and F_{KURT} of size 1×1, respectively. But last, intensity grading of glioma-segmented image is calculated and stored in a vector F_{IG} of size 10×1. Finally, all the features are concatenated and stored in a vector F_T of size 278×1.

3.3 Classification

The feature vector F_T is normalized by using min-max normalization technique, as discussed in section 2, to get the values of the features in the range [0 1]. The normalized feature vector F_{T_n} is applied to ANN classifier with corresponding group labels in training phase and without corresponding group labels in testing phase to classify the input images into one of the two groups. The ANN is designed by using two-layer feedforward backpropagation algorithm. The ANN consists of an input layer, a hidden layer, and an output layer. The input layer consists of 278 neurons, hidden layer consists of 93 neurons, and the output layer consists of 2 neurons. The values for rest of the parameters of ANN such as number of epochs, training function, error function are fixed experimentally to get maximum classification accuracy.

3.4 Statistical Analysis

Statistical analysis is carried out to measure the performance of the proposed method. The statistical analysis using t-test is carried out for comparing two groups. A Receiver Operating Characteristic (ROC) curve is plotted in turn which shows sensitivity and specificity for estimated groups with respect to predefined groups, as shown in Fig. 3. The Area Under the Curve (AUC) of 0.8993 is obtained from ROC. Sensitivity and specificity of 83.3% and 93.3%, respectively, are achieved by using proposed method to classify the gliomas into two groups. A p-value of <0.05 is considered statistically significant. A p-value of 0.0107 is observed between the feature vectors of two groups. LBPH, entropy, and intensity grading show a significantly lower p-value (<0.05), whereas histogram, skewness, and kurtosis demonstrate insignificant results. Experimentally, it has been observed that LBPH is the most efficient textu-

Fig. 3 ROC for glioma classification into two groups using proposed method

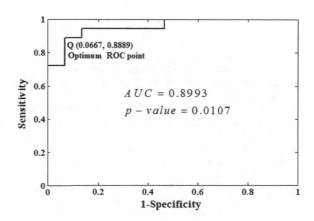

ral feature to discriminate between Group-1 and Group-2 followed by entropy and intensity grading.

The classification accuracy of proposed method is calculated by using tenfold cross-validation. Overall classification accuracy of 87.9% is achieved by using proposed method to classify the glioma images into one of the two groups.

4 Conclusion and Future Scope

We have presented a classification methodology based on meaningful texture features for identification of molecular classes in gliomas. Superior experimental results, with high classification accuracy indicate that our technique, can be successfully employed in computed aided diagnostic systems. Future work includes extending and testing the classifier on a bigger dataset and not restricting the analysis to T2-weighted images, but using multimodal features from other MRI contrasts such as T1-weighted images, T1-contrast enhanced images, fluid-attenuated inversion recovery (FLAIR), and diffusion MRI.

In summary, the paper presents a proof of concept for noninvasively discriminating the gliomas into molecular subtypes and has the potential to be used as a diagnostic biomarker for IDH1.

Ethical Approval

All the subjects had provided written informed consent for their participation in this study and the Institute Ethics Committee of NIMHANS had approved this study.

References

1. Munshi A (2016) Central nervous system tumors: spotlight on India. South Asian J Cancer 5(3):146–147
2. Thakkar J, Dolecek T, Horbinski C et al (2014) Epidemiologic and molecular prognostic review of glioblastoma. Cancer Epidemiol Biomarkers Prev 23(10):1985–1996
3. Cohen A, Holmen S, Colman H (2013) IDH1 and IDH2 mutations in gliomas. Curr Neurol Neurosc Rep 13(5):345
4. Qi S, Yu L, Li H, et al (2014) Isocitrate dehydrogenase mutation is associated with tumor location and magnetic resonance imaging characteristics in astrocytic neoplasms. Oncol Lett 7(6):1895–1902
5. Kinoshita M, Sakai M, Arita H et al (2016) Introduction of high throughput magnetic resonance T2-weighted image texture analysis for WHO grade 2 and 3 gliomas. PLoS ONE 11(10):e0164268
6. Ryu Y, Choi S, Park S et al (2014) Glioma: application of whole-tumor texture analysis of diffusion-weighted imaging for the evaluation of tumor heterogeneity. PLoS ONE 9(9):e108335
7. Ojala T, Pietikainen M, Maenpaa T (2002) Multiresolution gray-scale and rotation Invariant texture classification with local binary patterns. IEEE Trans Pattern Anal Mach Intell 24(7):971–987

8. Maenpaa T, Pietikainen M (2004) Texture analysis with local binary patterns. Handbook of pattern recognition and computer vision. World Scientific, Singapore
9. Zurada JM (1992) Introduction to artificial neural systems. West Publishing Company, St. Paul, US

Modeling Runoff Using Feed Forward-Back Propagation and Layer Recurrent Neural Networks

Dillip Kumar Ghose

Abstract This work describes the function of (i) feed forward-back propagation network (FFBPN) model and (ii) layer recurrent network (LRN), to predict runoff as a function of rainfall, temperature, and evapotranspiration loss. For model architecture, the criteria for evaluation are mean square error training, testing, root mean square error training, testing, and coefficient of determination. Overall results found that LRN performs best as compared to FFBPN for predicting runoff in the watershed. This result will help for planning, design, and management of hydraulic structures in the vicinity of the watershed.

Keywords Runoff · Watershed · Feed forward-back propagation network
Layer recurrent network

1 Introduction

In planning of water resource project, the estimation of the availability of water plays an important role. The first step in the water availability estimation is the computation of runoff resulting from the precipitation and its abstracts along with weather parameters on watershed. The length of the runoff measured in a watershed may be of short period or long period. In such cases, the second step may be the development of precipitation–evapotranspiration–runoff co-relations. These co-relations can be used to predict the runoff from the observed precipitation, temperature, and evapotranspiration. Precipitation is the major input of water to a catchment area and needs careful assessment in hydrological study. Runoff from land surface is the flow of water that comes from excess water from rain, over the Earth's surface. It is a major component in regional and global hydrological cycle. It is crucial to understand complex relationships between rainfall and runoff processes and then to estimate surface runoff for efficient design, planning, and management of watershed. This can be

D. K. Ghose (✉)
Department of Civil Engineering, National Institute of Technology, Silchar, Assam, India
e-mail: dillipghose2002@gmail.com

© Springer Nature Singapore Pte Ltd. 2019
A. J. Kulkarni et al. (eds.), *Proceedings of the 2nd International Conference on Data Engineering and Communication Technology*, Advances in Intelligent Systems and Computing 828, https://doi.org/10.1007/978-981-13-1610-4_8

achieved using hydrological modeling. The common estimating methods include physical, statistical, combined approaches selected according to need and available data. Kite [3] has used the National Oceanic and Atmospheric input–output models for all types and sizes of watersheds. Smith and Eli [6], Minns and Hall [5], Srini-vasulu and Jain [7] have found that the runoff response of a watershed is subjected to rainfall input including storm characteristics, catchment characteristics, percent of the catchment contributing runoff at outlet and climatic characteristics similar to temperature, humidity, and wind characteristics. Chen and Adams [1] have found conceptual rainfall–runoff models employed in hydrological modeling compared with black box techniques and physical models. Zhao [9] has developed the Xinan-jiang (XAJ) model used in catchment rainfall–runoff modeling across China. Xiong et al. [8] have observed rainfall–runoff modeling results perform similar to that of the multilayer perceptron neural network (MLPNN). The development of hybrid neural networks with conceptual model has received considerable attention [2, 4].

Present work motivates to study on correlation of rainfall, runoff, and predica-tion of runoff using precipitation maximum temperature, minimum temperature, and evapotranspiration.

2 Study Area

Dhankauda watershed of Sambalpur, Odisha, India, is taken into consideration for the proposed study area. The study is made for predicating runoff in three watersheds to assess the drainage capacity of watershed during monsoon period ranging from 1993 to 2012. The watersheds are located in the upper part of Hirakund reservoir. The latitude $21° 88' 36''$N and longitude $84° 90' 75''$E are the geocoordinate of the watersheds shown in Fig. 1.

The average monthly precipitation, maximum monthly mean temperature, and minimum monthly mean temperature and evapotranspiration data for monsoon month (May to October) from the period 1997 to 2016 spanning over 20 years are collected from IMD Bhubaneswar. The monthly runoff data are collected from soil conservation office, Sambalpur.

3 Methodology

3.1 Artificial Neural Network

A typical feed forward neural network comprises input layer, hidden layer, and output layer. The number of nodes in the input layer represents the number of input features responsible for the output, and the number of output nodes represents the output features to be predicted. The network receives the data from input layer and transfers

Fig. 1 Study area: Dhankauda watershed (*Source* Google map)

this information to the network with associated weights for dispensation. Hidden layer shares the information and inaudibly performs the information processing and send it to output layer to compile the results.

3.2 Feed Forward-Back Propagation Network (FFBPN)

Back propagation is an organized technique to exercise multilayer artificial neural networks. **FFBPN is trained** using gradient descent-based delta-learning algorithm. The gradient is used for adjusting weights to minimize the error. Back propagation allows quick convergence in networks. The architecture of FFBPN consists of input layer (L), hidden layer (M), and output layer (N), as represented in Fig. 2.

3.3 Layer Recurrent Network (LRN)

The network comprises an input layer with four nodes, a hidden layer with three nodes, a context layer with three units, and an output layer with one node as shown in Fig. 3. The connections allow the context units to store the outputs of the hidden

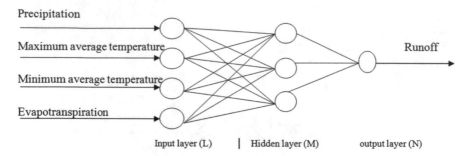

Fig. 2 Architecture of FFBPN model

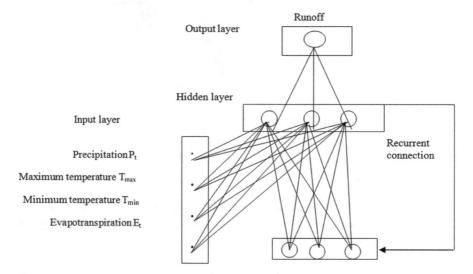

Fig. 3 Architecture of recurrent layer network

nodes at each time step; then the fully distributed upward links feed them back as additional inputs. Therefore, the recurrent associations permit the hidden units to reprocess the information over manifold time steps, to discover temporal information enclosed in the sequential input and applicable to the target.

3.4 Processing and Preparation of Data

The monthly rainfall, maximum monthly temperature, minimum monthly temperature are collected from India meteorological department and daily runoff data are collected from soil conservation department Sambalpur, India, for the period of the monsoon months (May to October), from 1997 to 2016. The data from 1997 to 2010 are used for training and data from 2011 to 2016 are used for testing the network.

Daily data are converted into monthly data, which is used for developing the model. The input and output data are scaled in such a way that each data plunge within a specified range before training. The process involved is called normalization so that the normalized values are bounded within the range of 0–1. The normalization equation used for scaling the data is

$$X_t = 0.1 + \frac{X - X_{min}}{X_{max} - X_{min}} \tag{1}$$

where X_t = transformed data series, X = original input data series, X_{min} = minimum of original input data series, X_{max} = maximum of original input data series.

3.5 Evaluating Criteria

The evaluating criteria to ascertain the best model are coefficient of determination, mean square error, and root mean square error. To choose the ideal model for this study area, the condition is MSE, RMSE should be least and coefficient of determination should be highest.

$$\text{Co-efficient of determination} \left(R^2\right) = \left[\frac{n \sum xy - (\sum x)(\sum y)}{\sqrt{[n \sum x^2 - (\sum x^2)] - [n \sum y^2 - (\sum y)^2]}} \right]^2 \tag{2}$$

The value of coefficient of determination indicates the percent of variation in one variable explained by the other variable.

$$\text{Mean squared error (MSE)} = \frac{1}{N} \sum_{i=1}^{N} (\widehat{y_i} - y_i)^2 \tag{3}$$

where

$\widehat{y_i}$ Predicted value of runoff
y_1 Actual value of runoff

$$\text{Root mean squared error (RMSE)} = \left[\frac{1}{N} \sum_{i=1}^{N} (\widehat{x_i} - x_i)^2 \right]^{1/2} \tag{4}$$

RMSE is the root of mean squared error. RMSE has also non-negative value and closer to zero is the best performance.

4 Results and Discussions

4.1 Results at Dhankauda

Here to evaluate the model efficiency of various architectures, different transfer functions like tangential sigmoidal and logarithmic sigmoidal are used to establish the model performance. For every architecture, the criteria for evaluation are mean square error training, testing, root mean square error training, testing, and coefficient of determination. In Table 2 for Tansig function in FFBPN, 4-2-1, 4-3-1, 4-5-1, and 4-9-1 architectures are taken into consideration for computation of performance. For Tansig function, the best model architecture is found to be 4-2-1 which possesses MSE training architecture, MSE training value 0.000808, MSE testing value 0.004119, RMSE training value 0.028392, RMSE testing value 0.064165, and coefficient of determination for training 0.9914 and testing 0.9255. For Logsig 4-2-1, 4-3-1, 4-5-1, and 4-9-1, architectures are taken into consideration for computation of performance. For Logsig function, the best model architecture is found to be 4-5-1 which possess MSE training architecture MSE training value 0.000591 MSE testing value 0.004029, RMSE training value 0.024271 RMSE testing value 0.063460 and coefficient of determination value training 0.9915, testing value 0.9417. The comprehensive results are accessible in Table 1.

Similarly for the layer recurrent network, the results are discussed below for Dhankauda station. For Tansig 4-2-1, 4-3-1, 4-5-1, and 4-9-1 architectures are taken into consideration for computation of performance. For Tansig function, the best model architecture is found to be 4-5-1 which possesses MSE training architecture MSE training value 0.000483, MSE testing value 0.001025, RMSE training value 0.02316 RMSE testing value 0.03085 and coefficient of determination value training 0.9925, testing value 0.9611. For Logsig 4-2-1, 4-3-1, 4-5-1, and 4-9-1 architectures are taken into consideration for computation of performance. For Logsig function, the best model architecture is found to be 4-2-1 which possesses MSE training architecture MSE training value 0.001236, MSE testing value 0.001236, RMSE training value 0.03514, RMSE testing value 0.05299, and coefficient of determination value training 0.9861, testing value 0.9591 presented in Table 2.

5 Simulation

The graphs with best values for runoff from precipitation, maximum temperature, minimum temperature, evapotranspiration using feed forward-back propagation (FFBPN) (refer to Fig. 4) and layer recurrent network (LRN) with Tansig and Logsig transfer function at Dhankauda (refer to Fig. 5). The best value for each evaluating criteria are represented. The graphs below show how these best values result in the variation between the observed runoff and predicted runoff.

Table 1 Results of FFBPN at Dhankauda

Model input	Sigmoid function	Architecture (L-M-N)	MSE		RMSE		R²	
			Training	Testing	Training	Testing	Training	Testing
Precipitation Maximum temperature Minimum temperature Evapotranspiration	Tansig	**4-2-1**	**0.000808**	**0.004119**	**0.028392**	**0.064165**	**0.9914**	**0.9255**
		4-3-1	0.000797	0.005759	0.028197	0.075876	0.9857	0.9075
		4-5-1	0.12449	0.26765	0.35282	0.51734	0.8691	0.800
		4-9-1	0.000681	0.00309	0.02607	0.05542	0.9889	0.9103
	Logsig	4-2-1	0.000465	0.001167	0.02153	0.034134	0.984	0.9390
		4-3-1	0.001195	0.004088	0.03455	0.06394	0.9793	0.8114
		4-5-1	**0.000591**	**0.004029**	**0.024271**	**0.063460**	**0.9915**	**0.9417**
		4-9-1	0.00120	0.003739	0.03437	0.061133	0.9783	0.9167

Table 2 Results of LRN at Dhankauda

Model input	Sigmoid function	Architecture (L-M-N)	MSE		RMSE		R²	
			Training	Testing	Training	Testing	Training	Testing
Precipitation Maximum temperature Minimum temperature Evapotranspiration	Tansig	4-2-1	0.019289	0.001298	0.13890	0.03601	0.8188	0.9270
		4-3-1	0.000751	0.003339	0.02737	0.05799	0.9875	0.9724
		4-5-1	**0.000483**	**0.001025**	**0.02316**	**0.03085**	**0.9925**	**0.9611**
		4-9-1	0.000394	0.000958	0.01980	0.03093	0.9761	0.9596
	Logsig	**4-2-1**	**0.001236**	**0.002808**	**0.03514**	**0.05299**	**0.9861**	**0.9591**
		4-3-1	0.000940	0.004980	0.03064	0.07057	0.9678	0.9112
		4-5-1	0.015516	0.00255	0.1248	0.05021	0.7127	0.9536
		4-9-1	0.001437	0.00656	0.03789	0.08083	0.979	0.9019

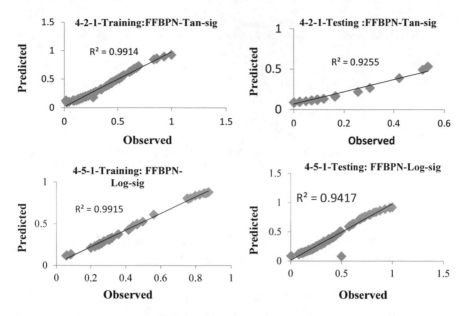

Fig. 4 Observed versus predicted runoff using FFBPN

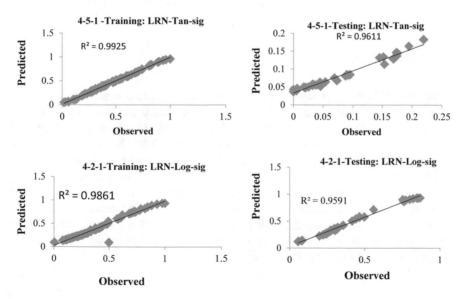

Fig. 5 Observed versus predicted runoff using LRN

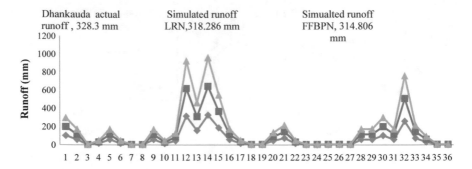

Fig. 6 Actual versus simulated runoff using FFBPN, and LRN at Dhankauda in testing phase

5.1 Assessment of Actual Runoff Versus Simulated Runoff at Dhankauda During Testing Phase

The variation of actual runoff versus simulated or predicted runoff is exposed in Fig. 6. Results show that the estimated peak runoffs are 318.286 and 314.806 mm for LRN and FFBPN against the actual peak 328.3 mm for the watershed Dhankauda.

6 Conclusions

Study with different parameters like precipitation, evapotranspiration, and temperature has been considered for predicting runoff. At Dhankauda watershed, among two neural networks with the evaluation criteria MSE, RMSE, and coefficient of determination, LRN performs best with architecture 4-5-1 following Tansig transfer function by considering both training and testing performance criteria. Similarly FFBPN performs best at architecture 4-5-1 with all criteria considered for model performance. This work will help for estimating the parameters related to watershed, for planning, designing, and management of the watershed. The results suggest that both BPNN and LRN are suitable methods for estimating runoff in the watershed of arid region. However, the combination technique needs to be investigated for improving integrated model techniques for future research.

References

1. Chen JY, Adams BJ (2006) Integration of artificial neural networks with conceptual models in rainfall-runoff modeling. J Hydrol 318(1–4):232–249. https://doi.org/10.1061/(ASCE)HE.194 3-5584.0000970

2. Jain A, Srinivasulu S (2006) Integrated approach to model decomposed flow hydrograph using artificial neural network and conceptual techniques. J Hydrol 317(3–4):291–306. https://doi.or g/10.1016/j.jhydrol.2005.05.022292
3. Kite GW (1991) A watershed model using satellite data applied to a mountain basin in Canada. J Hydrol 128:157–169. https://doi.org/10.1016/0022-1694(91)90136-6
4. Lee DS, Jeon CO, Park JM, Chang KS (2002) Hybrid neural network modeling of a full-scale industrial wastewater treatment process. Biotechnol Bioeng 78(6):670–682. https://doi.org/10.1002/bit.10247
5. Minns, AW, Hall MJ (1996) Artificial neural networks as rainfall-runoff models. Hydrol Sci 41(3):399–417, Oxford, England. https://doi.org/10.1080/09715010.2009.10514969
6. Smith J. Eli RB (1995) Neural-network models of rainfall runoff process. J Water Resour Plann Manage 121(6):499–508. ASCE
7. Srinivasulu S, Jain A (2006) A comparative analysis of training methods for artificial neural network rainfall-runoff models. Appl Soft Comput 6(3):295–306. https://doi.org/10.1016/j.aso c.2005.02.002296
8. Xiong L, Shamseldin AY, O'Connor KM (2001) A non-linear combination of the forecasts of rainfall-runoff models by the first-order Takagi-Sugeno fuzzy system. J Hydrol 245(1–4):196–217. https://doi.org/10.1016/j.jhydrol.2003.08.011
9. Zhao RJ (1992) The Xinanjiang model applied in China. J Hydrol 135(1–4):371–381

Classification of EEG Signals in Seizure Detection System Using Ellipse Area Features and Support Vector Machine

Dattaprasad A. Torse, Veena Desai and Rajashri Khanai

Abstract Epilepsy is a brain disorder, characterized by transitory and impulsive electrical signal of the brain. The electroencephalogram (EEG)-based seizure detection system used for automated diagnosis of epilepsy requires optimum classification of signals. This paper presents an optimized method for classification as normal and epileptic signals using Empirical-Mode Decomposition (EMD) technique. The Intrinsic-Mode Functions (IMFs) are few symmetric and band-limited signals obtained by applying EMD to the signals. However, optimum selection of IMF features is crucial step in deciding feature set for classification. The 95% confidence ellipse area is calculated from the Second-Order Difference Plot (SODP) of selected IMFs to form features for classification. The feature space is used by the Cosine Similarity Measure Support Vector Machine (CSM-SVM) classifier with optimum feature selection. It is observed that the features formed using ellipse area of dissimilar combination of IMFs have given superior classification performance on EEG data available from the Bonn University, Germany.

Keywords Empirical-mode decomposition · Intrinsic-mode function · Ellipse area · Cosine similarity measure support vector machine

D. A. Torse (✉) · V. Desai
Department of Electronics and Communication Engineering, KLS Gogte Institute of Technology, Belagavi, Karnataka 590008, India
e-mail: datorse@git.edu

V. Desai
e-mail: veenadesai@git.edu

R. Khanai
Department of Electronics and Communication Engineering, KLE's Dr. M.S.S.C.E.T, Belagavi, Karnataka 590008, India

© Springer Nature Singapore Pte Ltd. 2019   87
A. J. Kulkarni et al. (eds.), *Proceedings of the 2nd International Conference on Data Engineering and Communication Technology*, Advances in Intelligent Systems and Computing 828, https://doi.org/10.1007/978-981-13-1610-4_9

1 Introduction

Epilepsy is an acute and recurring neurological confusion generally evident by frequent seizures which has an effect on about 1% of the world's population [1]. It results in worsening of consciousness and may show random and recurrent body convulsions. The developing countries contribute about 85% of the 60 million people affected by epilepsy worldwide. EEG recording characterizes electrical activities of the brain by measuring electrical potential of neurons. The EEG is recorded a prominent tool in the detection of epileptic seizures and includes recognition of spikes in the epilepsy identification [2]. Further, the accurate diagnosis of epilepsy helps in deciding the course of the antiepileptic medication. The automated seizure detection system generally involves three steps, namely preprocessing, signal transformation feature extraction and classification [3]. In joint time-frequency techniques, one finds two-dimensional (2D) time-frequency representation of filtered EEG [4], using Hilbert–Huang Transform (HHT) and its variant Empirical-Mode Decomposition (EMD) [5].

The proposed method employs two signal processing techniques for segregation of normal and epileptic EEG signals. In the first step, the EMD method is applied on the EEG signals for the extraction of the IMFs followed by SODP which measures the rate of variability of individual IMFs [6]. In the second step, the 95% confidence ellipse areas of SODPs of two IMFs is computed and used as a feature set for categorization of the EEG signal information using Cosine Similarity Measure Support Vector Machine (CSM-SVM) classifier [7].

The remaining paper is ordered as follows: Sect. 2 explains the data set, EMD, SODP and ellipse area formulation, selection of features based on CSM, CSM-SVM-based classification, and estimation of statistical parameters. The result and discussion are presented in Sect. 3. Conclusion has been drawn in Sect. 4.

2 Methodology

2.1 Data Set

The EEG data set information that is accessible online publicly and explained in [8] is used to validate the results of this work. The data set with normal and seizure EEG signals of five sub-signals is represented as A, B, C, D and E, each having 100 numbers of single-channel EEG signals. The duration of each sample is 23.6 s with sampling rate of 173.61 Hz. The subsets A and B are recorded with international 10–20 standard electrode placement scheme with surface EEG recordings. The set A is for healthy subjects with eyes open and B for eyes closed, the subsets C and D are recorded in normal intervals from five patients in the epileptogenic zone (subset C) and from the hippocampal formation of the opposite hemisphere of the brain (subset D). The typical EEG signal samples of normal and seizure are shown in Fig. 1.

(a)   **(b)**

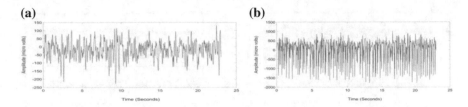

Fig. 1 EEG signals **a** normal and **b** seizure

2.2 Empirical-Mode Decomposition

EMD is a data-driven technique that splits a signal into a finite set of amplitude and frequency modulated (AM-FM) oscillating components. These components are called as IMFs. An IMF is a function which satisfies the conditions as follows: (1) number of maxima and minima are either equal or vary almost by one. (2) The mean value of envelopes, characterized by local maxima and minima, is zero. This signal-dependent decomposition is adaptively done for predefined levels. Moreover, the decomposition of long duration signals is done without a stationary and linearity condition of the signal. The EMD algorithm is used to obtain number of band-limited functions from a signal $x(t)$. The EMD algorithm for the signal $x(t)$ is depicted below [5]:

1. Original signal is set as $x(t) = P_1(t)$.
2. Both the maximum and minimum values for $P_1(t)$ are estimated.
3. The cubic spline interpolation is used to establish the upper and lower envelopes represented as $env_{upper}(t)$ and $env_{lower}(t)$, respectively. To achieve this, the maximum and minimum points are joined.
4. The local mean is computed as $m(t) = (env_{lowet}(t) + env_{lower}(t))/2$.
5. The signal $m(t)$ is subtracted from the given signal as $P_1(t) = P_1(t) - m(t)$ means IMF should have zero local mean.
6. To $P_1(t)$ is tested for above defined conditions to check if it is an IMF or not.
7. The steps (2)–(6) are repeated and when an IMFs are computed, the process is stopped.

The sifting is continued until the last IMF is generated. The computation functions and the residues are given by:

$$r_1(t) - P_2(t) = r_2(t), \ldots, r_{IMF-1}(t) - P_{IMF}(t) = r_{IMF}(t) \tag{1}$$

where $r_{IMF}(t)$ is called the final signal remained. The final stage of decomposition results is the signal $x(t)$ given by:

$$x(t) = \sum_{p=1}^{M} P_p(t) + r_{IMF}(t) \tag{2}$$

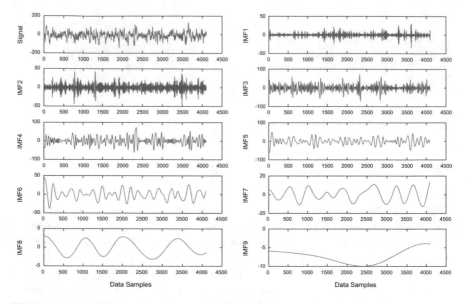

Fig. 2 Intrinsic-mode functions of normal EEG signal

where the number of IMFs is represented by M and $r_{\mathrm{IMF}}(t)$ is the final signal remained.

The implementation of the EMD algorithm is done using MATLAB. The implementation of EMD on the 23.6 s EEG signals for 4097 samples is shown in Fig. 2.

2.3 Second-Order Difference Plot and Calculation of Ellipse Area

A nonlinear signal can be viewed with a different perspective, and a continuous chaotic modelling is an effective tool in understanding the nonlinearity in long data series. To classify normal and seizure EEG signals, the SODP of IMFs of EEG signals present important parameters.

For the signal $x(n)$, the SODP is achieved by plotting $X(n)$ versus $Y(n)$ which are defined as [6]:

$$X(n) = x(n + 1) - x(n) \tag{3}$$
$$Y(n) = x(n + 2) - x(n + 1) \tag{4}$$

The chaotic equations are employed in generating graphs, which are known as Poincare plots. The equation below represents a Poincare equation, which is an example of a chaotic system.

$$a_n = Aa_{n-1}(1 - a_{n-1})2 \leq A \leq 4 \tag{5}$$

where A is a constant. The performance of the function is depending on this value.

Recently, the variability measured from the SODP has been used for analysis of EEG signals [6]. The 95% confidence ellipse area measured from the SODP of IMFs of EEG signal data represents a set of parameters for segregation of normal and seizure EEG signals. Figure 2 shows the IMFs of normal signal patterns. It motivates to work out the ellipse area of "SODP of IMFs" for separation of EEG signals. The method to calculate "95% confidence ellipse area" from the SODP approach is presented as follows [6]: The mean value of $X(n)$ and $Y(n)$ is computed as:

$$S_X = \sqrt{\frac{1}{N}\sum_{n=0}^{N-1} X(n)^2} \quad S_Y = \sqrt{\frac{1}{N}\sum_{n=0}^{N-1} Y(n)^2} \quad S_Y = \frac{1}{N}\sqrt{\sum X(n)Y(n)} \tag{6}$$

The D value is computed as:

$$D = \sqrt{\left(S_X^2 + S_Y^2\right) - 4\left(S_X^2\, S_Y^2 - S_{XY}^2\right)} \tag{7}$$

The ellipse area is formulated as:

$$A_{\text{ellipse}} = P_i\,(ab); \quad a = \sqrt{3 \times \left(S_X^2 + S_Y^2 + D\right)}; \quad b = \sqrt{3 \times \left(S_X^2 + S_Y^2 - D\right)} \tag{8}$$

2.4 Cosine Similarity Measure Support Vector Machine (CSM-SVM) Classifier

The SVM is a powerful machine learning algorithm for regression and classification which produces very precise classification results. SVMs are considered as an important example of "kernel methods", one of the key areas in machine learning, and the Radial Basis Function (RBF) kernel has been employed in this work. The RBF is given by:

$$K\left(x_i, x_j\right) = \exp\left(-\left(\left\| x_i - x_j \right\|^2 / 2\,\rho^2\right)\right) \tag{9}$$

where $\rho > 1$ is the parameter controlling the width of the kernel. The Cosine Similarity Index (CSI) between two data vectors is a standard criterion for finding the distance between two data samples. CSI determines the cosine of the angle between the data samples. To construct the cosine resemblance equation, the equation of the dot product for the $\cos\theta$ is to be solved as:

$$\vec{a}.\vec{b} = \left\| \vec{a} \right\| \left\| \vec{b} \right\| \cos\theta; \quad \text{CSI} = \cos\theta = \frac{\vec{a}.\vec{b}}{\left\| \vec{a} \right\| \left\| \vec{b} \right\|} \tag{10}$$

where *CSI* represents the whitened cosine similarity value between vector *a* and *b* [9].

2.5 Performance Evaluation

The assessment of the SVM-based classifier for classification of seizure and non-seizure EEG signals is carried out by calculating the sensitivity, specificity and accuracy. The sensitivity (SE), specificity (SP) and accuracy (AC) can be defined as:

$$SE = \frac{TP}{TP + FN} \times 100\%; \quad SP = \frac{TN}{TN + FP} \times 100\%; \tag{11}$$

$$AC = \frac{TP}{TP + FN + FP + FN} \times 100\% \tag{12}$$

where TP and TN signify the overall number of appropriately detected true positive patterns and true negative patterns, respectively. The FP signifies overall number of erroneously positive patterns, and FN signifies erroneously negative patterns. The positive and negative patterns signify detected seizure and detected normal EEG signals, respectively.

3 Results and Discussion

In the first step, the results are obtained by decomposing EEG signals from data sets B and E by EMD method to obtain nine IMFs as shown in Fig. 2. The first four IMFs are selected to compute SODP, and 95% ellipse area as the maximum frequency variation and nonlinearity of the signal is obtained from first few IMFs. The classification of normal and seizure EEG signals is performed using the ellipse area parameters of SODP for first four IMF's pairs. Figure 3 shows plot of SODP for all the IMFs achieved by EMD process. The compute 95% ellipse area of SODP of combination of initial four IMFs, the six pairs has been selected as IMF12, IMF13, IMF14, IMF23, IMF24, and IMF34.

The plot of 95% confidence ellipse area of these six pairs is shown in Fig. 4. Generally, first pair of IMFs represents frequency variation with more values. The 95% confidence area as a statistical feature is computed from the pair of IMFs as it clearly represents the nature of the underlying EEG signal. The maximum and minimum scaled values of 95% confidence are shown in Table 1 for normal and seizure data samples. The values for IMF12 pair are 19,642.13 and 123.92 respectively. The first IMF and its associated SODP show that the feature set can distinguish between normal and seizure signals. The elliptical shape of SODP of IMFs encourages computation of 95% confidence ellipse area of SODP of IMFs of EEG signals. As with the seizure-free case, similar observation can be drawn for the

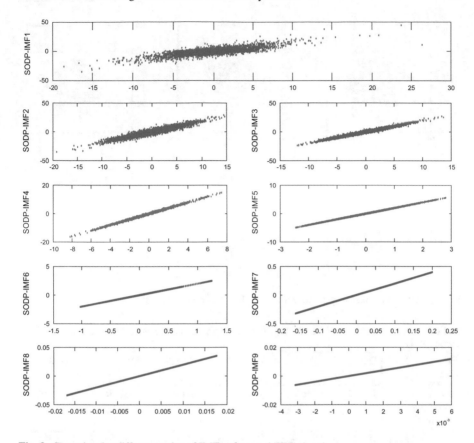

Fig. 3 Second-order difference plot of IMFs of normal EEG signal

case of ictal EEG signal also. The 95% of confidence ellipse area parameters has been computed for ictal and non-ictal classes using SODP of IMFs for a range of data values of the signal (500, 1000, 1500 and 4097 samples). The class discriminating performance of ellipse area feature is verified using box plots.

The IMF's ellipse area parameters of SODP of IMFs are considerably diversified for the two classes of EEG signals (normal and seizure). The MATLAB function "*tcdf*" is used to compute the p-value of the data classified with defined degree of freedom. It is found that $p \leq 0.01$ indicates the 1%, i.e. less than 1 in 100 chance of being wrong (refer to Fig. 4).

The comparison of classifier performance reveals that the division of EEG time series into four sub-bands results in varied performance parameters. The performance for 1500 and 4097 data sample is depicted in Table 2. The ellipse area features obtained from shape of the SODP of IMFs result in more suitable features for a data segment of 1500. Figure 5 shows the box plot of ellipse area of SODP of IMFs for

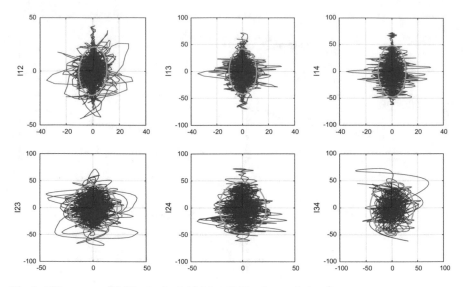

Fig. 4 Ellipse area of IMF pairs for initial four IMFs of normal signal

Table 1 95% confidence area for six IMF pairs for normal and seizure EEG data (80 samples of normal EEG)

EEG data	IMF pairs	IMF12	IMF13	IMF14	IMF23	IMF24	IMF34
Normal data (80 samples)	Min.	123.92	246.12	241.89	450.45	580.36	1280.46
	Max.	19,642.13	26,486.63	29,694.69	46,342.25	51,957.78	69,503.2
Seizure data (80 samples)	Min.	7933.89	8723.08	11,605.05	10,254.90	16,020.05	16,305.7
	Max.	3,884,243.3	3,231,613.7	1,783,224.9	3,346,937.3	1,547,209.4	1,795,701.2

normal and seizure EEG signals. The experimental results are found noteworthy for inequity among normal and ictal signals.

The CSM-SVM is an optimized form of SVM which uses wrapper method to select features suitable for classification and implemented in MATLAB. The toolbox is constructed around a fast LS-SVM training and simulation algorithm [10]. The LS-SVM functions have been used for classification as well as for plotting the performance. A fast LS-SVM training and simulation algorithm is present in the MATLAB toolbox. The LS-SVM functions have been used for categorization in addition to that for plotting the performance.

Table 2 The performance of LS-SVM classifier for 1500 and 4097 data samples

EEG data	Performance parameters	IMF12	IMF13	IMF14	IMF23	IMF24	IMF34
1500 data samples	SE	95.2	94.2	96.3	95.4	92.1	88.3
	SP	91.5	93.2	84.2	87.3	79.5	64.3
	AC	92.1	98.5	91.2	90.1	50.2	65.5
	AVG	96.2	95.5	75.5	88.5	91.5	90.3
4097 data samples	SE	96.3	95.1	90.2	89.3	85.5	79.3
	SP	95.1	90.5	95.3	94.1	94.4	91.7
	AC	91.2	90.3	40.5	82.5	92.5	50.1
	AVG	90.3	88.2	39.2	81.3	89.5	54.3

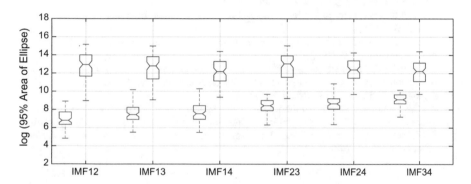

Fig. 5 Box plots of ellipse area of SODP of IMFs for normal and seizure EEG signals

4 Conclusion

Epilepsy is a chronic disorder, and EEG-based automated seizure detection system is evolving as an alternative diagnosis option where the main requirement is optimum signal processing algorithms. This paper presents the performance of classifier using ellipse area of SODP of IMFs as a feature set. For categorization of ictal and non-ictal EEG signals, the ellipse area is obtained using EMD algorithm to generate optimum feature space. The ellipse area parameters of first four IMFs are used in generating six pairs. The ellipse areas of first and second IMFs have provided better classification accuracy. The seizure EEG signals have significant ellipse area as compared to the normal EEG signals. The increased variation in seizure signals results into the desired variation in ellipse area of SODP of IMF pairs. The CSM-SVM provided highest classification accuracy of 96.4% for first pair of IMFs. The application of this method in automated seizure detection system will be tested on out-of-sample data sets. The system may also assist caretakers of epileptic patients by detecting occurrence of seizure signal.

References

1. Pati S, Alexopoulos AV (2010) Pharmacoresistant epilepsy: from pathogenesis to current and emerging therapies. Cleve Clin J Med 77(7):457–467
2. Sharma M, Pachori RB, Rajendra Acharya U (2017) A new approach to characterize epileptic seizures using analytic time-frequency flexible wavelet transform and fractal dimension. Pattern Recogn Lett
3. Acharya UR, Sree SV, Swapna G, Martis RJ, Suri JS (2013) Automated EEG analysis of epilepsy: a review. Knowl Based Syst 45:147–165
4. Torse DA, Desai VV (2016) Design of adaptive EEG preprocessing algorithm for neuro-feedback system. In: 2016 international conference on communication and signal processing (ICCSP). IEEE, pp 0392–0395
5. Rilling G, Flandrin P, Goncalves P (2003) On empirical mode decomposition and its algorithms. In: IEEE-EURASIP workshop on nonlinear signal and image processing, vol 3. IEEER, Grado, Italy
6. Pachori RB, Patidar S (2014) Epileptic seizure classification in EEG signals using second-order difference plot of intrinsic mode functions. Comput Methods Programs Biomed 113(2):494–502
7. Chen Gang, Chen Jin (2015) A novel wrapper method for feature selection and its applications. Neurocomputing 159:219–226
8. Andrzejak RG, Lehnertz K et al (2001) Indications of nonlinear deterministic and finite-dimensional structures in time series of brain electrical activity: dependence on recording region and brain state. Phys Rev E 64(6):061907
9. Machine Learning: Cosine Similarity for Vector Space Models (Part III): http://blog.christian perone.com/2013/09/machine-learning-cosine-similarity-for-vector-space-models-part-iii/
10. Suykens JAK, Vandewalle J (1999) Least squares support vector machine classifiers. Neural Process Lett 9:293–300

Explore Web Search Experience for Online Query Grouping and Recommendation by Applying Collaborative Re-ranking Algorithm

Jyoti Anpat and Rupali Gangarde

Abstract Millions of users are Web dependent to search relevant information of complex tasks. Query clustering is a collection of relevant queries which are previously searched and related to currently issued query. A complex task is divided into a number of smaller tasks, and multiple queries are searched for each task repeatedly. Searching task-related information online is still textually based. But it suffers from the problem of polysemy and synonymy queries. K-Means algorithm solves problem of textual similarity, but there is no flexibility to increase the number of clusters. Graph-based query clustering method is used to detect similarity between current query and existing query group by exploring collaborative search history. Online clustering algorithm provides facility to create query group dynamically. Collaborative re-ranking algorithm improves search performance by recommending highly relevant searched results by ranking queries in query group. Several experimental results indicate the proposed system has higher precision and recall values.

Keywords Web search experience · Search history · Query recommendation
Online query grouping · Query fusion graph · Collaborative re-ranking

1 Introduction

Query clustering in data mining allows a query to become part of an existing query group based on similarity relevance value. Otherwise, a new query group is created dynamically by online clustering algorithm. Query group recommends the user by providing previously searched relevant search results for the current query. Collabo-

J. Anpat (✉)
Computer Engineering, D. Y. Patil College of Engineering, Savitribai Phule Pune University, Akurdi, Pune, India
e-mail: anpat.jyoti@gmail.com

R. Gangarde
Symbiosis Institute of Technology, Symbiosis International University, Pune, India
e-mail: rupali.gangarde@sitpune.edu.in

© Springer Nature Singapore Pte Ltd. 2019   97
A. J. Kulkarni et al. (eds.), *Proceedings of the 2nd International Conference on Data Engineering and Communication Technology*, Advances in Intelligent Systems and Computing 828, https://doi.org/10.1007/978-981-13-1610-4_10

rative search history helps to understand the previous search experience of all users for the current query and also stores clicked URLs and consecutive count of current query helpful in the calculation of similarity relevance value. The user searches for informational, navigational, and transactional queries online. Searches are divided into specific type according to the intention of query. In informational search queries, the user is collecting certain information for complex tasks. Search for complex task is divided into multiple steps, and each step issues number of queries to gather required information. For example, the query "swine flu" needs to search causes, symptoms, precautions, treatment, etc. Navigational queries (e.g., "Pune university website" and "Passport login") navigate to a relevant Web site for the inserted query. Transactional search queries indicate user wish to execute a transaction. For example, "purchase Dell Laptop" or "Buy Samsung S8".

Disease detection, tourism planner, and event organizer are a few examples of complex tasks. Complex task is a collection of divided informational queries as broader information required to collect relevant information from Web for such task. String similarity functions suffer from the problem of polysemy and synonymy. In K-Means algorithm, there is no flexibility to increase the number of clusters, if new data objects arise. Session-based agglomerative algorithm solves problem of polysemic queries, but synonymic queries are unaddressed. Monte Carlo Tree Search (MCTS) method finds out a global optimal solution using random sampling. Query Fusion Graph (QFG) [1] resolves problem ambiguous queries.

Online question answering, query suggestion, query alteration, collaborative search and sessionization are few search engine components where query grouping is helpful. In the proposed system, collaborative search history is collected as searched session of all users. Proposed system provides facility to create query groups dynamically where collected queries are syntactically and semantically related. Query image and context vector use fusion relevance vector to calculate similarity relevance value. Query group is a collection of recommended queries for the current query while user searching online. Re-ranking algorithm improves search performance by ranking and recommending queries from query group. Related work is discussed in Sect. 2. Architecture of proposed system is explained in Sect. 3. Graphical presentation to form query grouping and re-ranking is explained in Figs. 1 and 2. Improvement of the proposed system is shown in Sect. 4. Improvement of re-ranking is graphically presented in Figs. 3 and 4. Section 5 represents conclusion and future scope of the proposed system.

2   Related Work

2.1   String Similarity Functions

Text, time, Jaccard, Levenshtein, and CoR (Co-Retrieval) are string similarity functions [1] used to calculate the similarity between different queries. Similarity func-

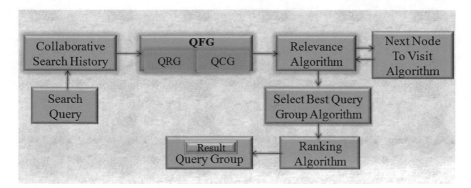

Fig. 1 System architecture

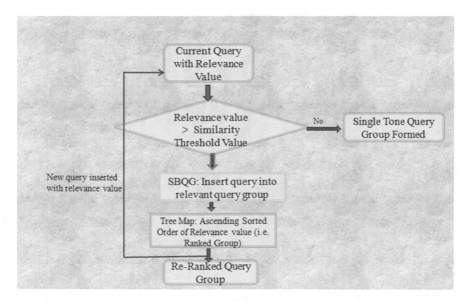

Fig. 2 Flow chart for query group re-ranking

tions face problem of polysemy and synonymy queries. Jaccard Similarity between existing query q_i and current query q_c is fraction of common words to the total set of words. Similarity between q_c and each of q_i by Levenshtein distance is calculated by Eq. (1).

$$d_{leven} = 1 - dist_{edit}(q_i, q_j) \tag{1}$$

where edit distance is the number of character substitution, deletion, insertion required to transform one sequence of character to another. CoR similarity is the fraction of commonly retrieved pages by q_c and q_i to the totally retrieved pages.

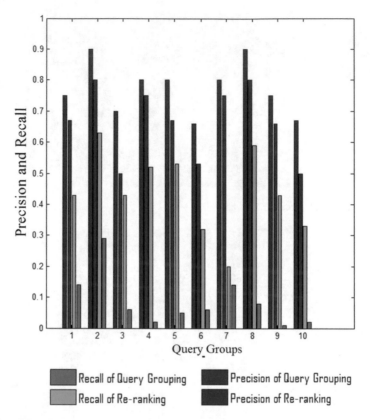

Fig. 3 Precision and recall

2.2 K-Means Algorithm and Modified K-Means Algorithm

In K-Means algorithm [2], a number of clusters and data set containing a number of data object points are taken as input from the user. There is no flexibility to increase the number of clusters if more data objects arise. K-Means reflects only locally optimal not globally optimal. In modified K-Means algorithm [2], there is flexibility to re-assign each object to the cluster to which the object is similar, as per mean value.

2.3 Bipartite Graph Construction Method

Click through data is a set of tuples containing user ID, query, URL selected by user, rank of URL document, and time. Bipartite graph is a set of vertices that consist of queries and corresponding clicked URLs. Agglomerative Iterative Clustering Algo-

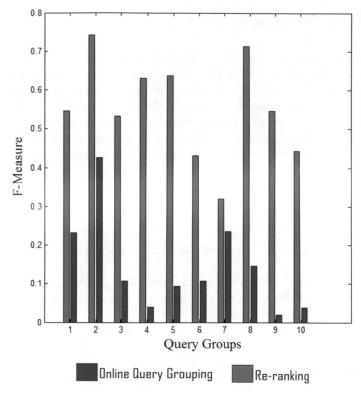

Fig. 4 F-measure comparison

rithm [1] clusters queries together from a bipartite graph if two different queries click on one of the URL maximum number of times.

2.4 MCTS

MCTS [3] method resolves problem of local optimality by using random sampling. Default policy means non-terminal state production by value estimation. MCTS is implemented by selection, expansion, simulation, and back propagation. Selection and expansion work is based on tree policy. Simulation refers default policy to select the node which will insert new node into the tree.

2.5 Online Clustering Algorithm

Online clustering algorithm [4] is useful for updating a cluster by making the relationship between all data objects and all cluster means. To overcome local optimality problem and to maintain global optimality, query grouping uses an online query clustering algorithm [1]. In Select Best Query Group (SBQG) algorithm [1], the existing query group is searched for each newly inserted query by exploring search history. It maintains dynamic clustering [5], and also a global optimal solution is provided. Data replication problem is solved by fragmentation (i.e., clustering based) [6].

3 Implementation Details

System Architecture: Architecture of the query group re-ranking is presented in Fig. 1.

Search engine collects the Web search experience of all users and which is stored in the form of search history. To search existing query group or to create a new query group, system is divided into a number of steps.

3.1 Collaborative Search History

The first step of the system consists of collection of queries, URLs, snippets and list of commonly clicked URLs searched online by different users and stored in MySQL database server 5 (i.e., collaborative search history). For each newly inserted query, URLs are fetched from the Google's first Web page. The JSoup HTML parser is used to fetch the URLs and to parse an HTML page. Page fetcher is used to fetch the title (entered query), URLs along with snippets from the Google's first Web page.

3.2 QFG

Query fusion graph consists of a set of queries and edges between them. The relative contribution of weight of common click and weight of query reformulation in the query fusion graph is controlled by alpha (α). QFG is the second stage of system design shown in Fig. 1. Edge between two queries (i.e., vertices) exists either because of reformulation or common click. Edge weight (w) between the currently issued query to each of existing queries sequentially is calculated by Eq. (2).

$$w_f(q_i, q_j) = \alpha \times w_r(q_i, q_j) + (1 - \alpha) \times w_c(q_i, q_j) \tag{2}$$

3.2.1 Query Reformulation Graph (QRG)

Consecutive count represents queries which are searched next to each other more than two times. Queries are in the reformulation with each other only if consecutive count between these queries is at least two. Edge weight between current query to each of its consecutive queries is calculated by Eq. (3).

$$w_r(q_i, q_k) = \frac{\text{count}_r(q_i, q_j)}{\sum_{(q_i, q_k) \in \varepsilon_{QR}} \text{count}_r(q_i, q_k)} \tag{3}$$

3.2.2 Query Click Graph (QCG)

Edge exists between two queries only if while searching online user clicked on the same URL for these two textually different queries. In QCG, edge represents the weight of common click between two queries; the u_k represents the URL clicked by user while searching online which is calculated by Eq. (4).

$$w_c(q_i, q_j) = \frac{\sum_{u_k} \min(\text{count}_c(q_i, u_k), \sum_{u_k} \text{count}_c(q_j, u_k))}{\sum_{u_k} \text{count}_c(q_i, u_k)} \tag{4}$$

3.3 Query Relevance Algorithm

Next Node to Visit algorithm [1] uses the approach of MCTS to select next query by applying random walk on QFG. The relevance between a set of $((q_c, q_{i1})(q_c, q_{i2}), \ldots$ $(q_c, q_{in}))$ is calculated by relevance algorithm. In query relevance algorithm, default values are chosen by repeating experiments with different values for parameters of default values and selected the ones that allow algorithm to achieve the best performance. Next Node to Visit algorithm decides next query from QFG to calculate relevance. Inputs to relevance algorithm are the current query, total number of random walks, i.e., random walk size (50), the size of the neighborhood (10), damping factor (0.6) determines the probability of random walk restart at each node, and alpha (0.3) decides relative contribution of QCG and QRG.

3.4 Select Next Node to Visit Algorithm

At each simulation of the random walk, if damping factor is greater than the random value, then first neighborhood from the list of QFG having highest weight value is taken, at next iteration second highest and so on. If damping factor is less than random value, then a query from QFG is selected containing jump vector value one. Jump vector value is one only if two queries are textually same as searching query.

3.5 Context Vector

The relevance score of current query to the query group is calculated by aggregating the fusion relevance values of the queries and clicks in group "s" and this relevance score is a context vector value. Default value consideration $w_{recency} = (0.9)$. For example, context vector of recently added query is calculated by Eq. (5).

$$cxt_s = w_{recency} \sum_{j=1}^{k} (1 - w_{recency})^{k-j} rel_{\left(q_{s_j}, clk_{s_j}\right)} \qquad (5)$$

3.6 Query Image

Query image is used to choose highly relevant queries for the current query (i.e., queries having a relevance value greater than 0.80). The relevance between the user's current singleton query group s_c (q_c, clk_c) in each of existing query group is calculated by Eq. (6).

$$sim_{rel}(s_c, s_i) = \sum_{q \in I(s_c) \cap I(s_i)} rel_{(q_c, clk_c)} * \sum_{q \in I(s_c) \cap I(s_i)} cxt_{s_i}(q) \qquad (6)$$

Value of sim_{rel} is used in a SBQG algorithm to decide whether to create a new query group or to add query into the existing query group.

3.7 SBQG Algorithm

The current query and its click, set of existing query groups and relevance (i.e., calculated by query image) are the input parameters for SBQG algorithm. Smaller the similarity threshold value (i.e., 0.01), there are more chances to add query into the existing query group. If query image is greater than the threshold, then the current query is getting added to its relevant existing group otherwise new group will be created. It is the second last stage used to form a query grouping as shown in Fig. 1.

3.8 Collaborative Re-ranking

Tree map, class of java, uses a value comparator method and stores data in sorted order (i.e., ascending). Queries with their relevance values are stored in hash map according to sorted order. Whenever a new query enters, it will get added into a group according to its relevance value (i.e., sorted rank value). Insertion of new query into query group and process of query group re-ranking are explained in Fig. 2.

To rank queries in the query group, program query group re-ranking is used and it is explained as follows:

```
Program Re-Ranking Query Group (Output)
 {Re-Ranking query according to similarity relevance
 value (i.e. calculated by query image and context
 vector)};
 const similarity threshold =0.01;
 var Query : query₁,..queryₙ;
var Relevance : relevane₁,..relevanceₙ;
begin
  Value comparator = null,
 Tree Map = null;
 repeat
 for each new query;
 if (relevance value > similarity threshold)
  Insert query into relevant group;
  value comparator(relevance);
 TreeMap((query_r₁,relevance_r₁),..(query_rn, relevance_rn));
 Ranked((query_r₁,relevance_r₁),..(query_rn, relevance_rn));
  Re-Ranked(Hash Map: Ascending sorted order of
  relevance);
 else
  Single tone query group formed;
 until query group = ranked(queryₙ, relevanceₙ);
 end.
```

4 Results

4.1 Database

Existing query groups created are 150, number of queries searched by different users are 2826. 3606 URLs and snippets are fetched from Google's first page. Commonly clicked URL count is 1202. Uniquely searched queries are 800. Precision, recall, and F-Measure are calculated for the evaluation of system performance. Precision is calculated by taking a fraction of retrieved queries those are relevant to the current query, and recall is calculated by the fraction of relevant queries those are retrieved. Precision, recall, and F-Measure formulas are applied on query grouping and query re-ranking to compare the existing system with the proposed system. For example, Query group for "Hill Stations in Maharashtra" is shown in Table 1.

Table 1 Query group containing recommended queries and its ranking order

Query group	Re-ranked query group	QFG of ranked group
Anandvan resort	Hill stations in maharashtra	0.02
3 Star hotel at lonavala	Cars on rent in matheran	0.7
Matheran	Anandvan resort	0.17
Cars on rent in matheran	Mahabaleshwar	0.11
Tourism in maharashtra	Hotels in mahabaleshwar	0.02
Hotels in mahabaleshwar	Tourism in maharashtra	0.06
Mahabaleshwar	Matheran	0.08
Hill stations in maharashtra	3 Star hotel at lonavala	0.03

Context vector and relevance value are 0.62 and 0.18. Comparison of precision, recall of 10 groups for query grouping and query re-ranking is shown in Fig. 3.

Figure 4 shows F-measure comparison between re-ranking and online query grouping.

5 Conclusion

For query grouping, different data mining techniques have been proposed, but it faces the problem of semantic ambiguity. Query grouping uses graph-based query clustering for similarity checking. Query grouping and query re-ranking algorithm resolves problem of semantic ambiguity. Query re-ranking provides improved and more relevant previously searched results for the current query.

Acknowledgements I would like to thank the researchers and for making their resources available. I would like to thank all friends and guides for giving their valuable guidance.

References

1. Hwang H et al (2012) Organizing user search histories: IEEE Trans Knowl Data Eng 24: 912–925
2. Qin L, Yuhong G, Jie W, Guojun W (2012) Dynamic grouping strategy in cloud computing. In: Second international conference on cloud and green computing 2012. IEEE, pp 59–66
3. Cameron BB, Edward P, Daniel W, Simon ML, Peter IC, Philipp R, Stephen T, Diego P, Simon C, Spyridon S (2012) A survey of monte carlo tree search methods. IEEE Trans Comput Intell AI Games 4(1):1–43
4. Minh NH, Nguyen, Chuan P, Jaehyeok S, Choong SH (2016) Online learning-based clustering approach for news recommendation systems. In: The 18th Asia-Pacific network operations and management symposium (APNOMS). IEEE Conference, Kanazawa, Japan
5. Shangfeng M, Hong C, Yinglong L (2011) Clustering-based routing for top-k querying in wireless sensor networks. EURASIP J Wireless Commun Network. Springer
6. Lena W (2014) Clustering-based fragmentation and data replication for flexible query answering in distributed databases. J Cloud Comput Adv Syst Appl. Springer

S-LSTM-GAN: Shared Recurrent Neural Networks with Adversarial Training

Amit Adate and B. K. Tripathy

Abstract In this paper, we propose a new architecture *Shared-LSTM Generative Adversarial Network* (S-LSTM-GAN) that works on recurrent neural networks (RNNs) via an adversarial process and we apply it by training it on the handwritten digit database. We have successfully trained the network for the generator task of handwritten digit generation and the discriminator task of its classification. We demonstrate the potential of this architecture through conditional and quantifiable evaluation of its generated samples.

Keywords Generative adversarial networks · Recurrent neural networks · Adversarial training · Handwritten digit generation · Deep learning

1 Introduction

Generative adversarial networks (GANs) are a relatively new category of neural network architectures which were conceptualized with the aim of generating realistic data [1]. Their method involves training two neural networks, architectures with contrasting objectives, a generator, and a discriminator. The generator tries to produce samples that look authentic, and the discriminator tries to differentiate between the generated samples and real data. This methodology makes it possible to train deep models without expensive normalizing constants, and this framework has proven to produce highly realistic samples of data [2].

The GAN framework is the most popular architecture with successes in the line of research on adversarial training in deep learning [3] where the two-player minimax

A. Adate (✉) · B. K. Tripathy
VIT University, Vellore, India
e-mail: email2amitadate@gmail.com; adateamit.sanjay2014@vit.ac.in

B. K. Tripathy
e-mail: tripathybk@vit.ac.in

© Springer Nature Singapore Pte Ltd. 2019
A. J. Kulkarni et al. (eds.), *Proceedings of the 2nd International Conference on Data Engineering and Communication Technology*, Advances in Intelligent Systems and Computing 828, https://doi.org/10.1007/978-981-13-1610-4_11

game is crafted carefully so that the convergence of the networks attains the optimal criteria. The initial work on GANs focused on generating images [4], however, GANs have a range of application domains like feature learning [5] and improved image generation [2].

In this work, we propose a hybrid model called as S-LSTM-GAN (Shared-LSTM-GAN) and we investigate the viability of using adversarial training on it for the task for handwritten digit generation. We would demonstrate the workings of two closely related neural networks, both variants of the recurrent neural unit, long short-term memory (LSTM) cells.

2 Background: LSTM

LSTM is a variant of the RNN and was introduced in [6]. The fundamental idea behind LSTM cell was the use of memory cell for retaining data for longer time and overcoming the limitations of recurrent neural networks. RNNs have problem with recognizing long-term dependencies like recognizing sequences of words which are quite apart from each other, and this problem is also referred to as the vanishing gradient problem. More technically speaking, the values in the matrix and multiple matrix multiplication are diminishing or becoming closer to zero and after a few time steps they vanish completely [7].

At far away time steps gradient is zero, and these gradients are not contributing to learning. In fact, vanishing gradient problem is not only limited to RNN. They are also observed in case of deep feed-forward neural networks.

3 S-LSTM-GAN: Shared Recurrent Networks with Adversarial Training

The proposed model consists of two different deep recurrent neural models with adversarial training. The adversaries are made up of two different architectures, but both sharing their weights. The generator (G) is trained to generate pixel values that are similar to the real data, while the discriminator (D) is trained to identify the generated data. The training can be modeled as a two-player minimax game for which the equilibrium is reached when the generator can consistently generate digits which the discriminator cannot identify from the real data. We define the following loss functions for the discriminator and generator, respectively, L_D and L_G

$$L_G = \frac{1}{m} \sum_{i=1}^{m} \log(1 - D(G(z^{(i)}))) \tag{1}$$

$$L_D = \frac{1}{m} \sum_{i=1}^{m} \left[-\log(D(x^{(i)})) - (\log(D(G(z^{(i)})))) \right] \tag{2}$$

where $(z^{(i)})$ is the sequence of input vectors and $x^{(i)}$ is a sequence from the training data. k is the dimensionality of the input data.

The overall model is depicted in Fig. 1. In the figure, the pink portion of the model is the generator and the brown portion is the discriminator. For reasons of clarity, the image is split into quadrants here, but in our experiments the aim was to split the image into pixels in an effort to create a generator that could create digits by each pixel using the long-range memory of the LSTM cells. The architectures of the generator and the discriminator are elaborated in Figs. 2 and 3, respectively.

The input data for each cell in G comes from a random vector, merged with the output of the previous cell, similar to application in [8]. The dimensionality of the input data set is defined as the number of sections we have sampled the dataset into, we are essentially splitting the image into k sections and then feeding them to the LSTM layers sequentially.

The discriminator consists of a series of LSTM cells all sharing the weights with their adversaries, and they feed forward into a linear neural network which is twice the size of the individual LSTM cells. Further, a softmax layer with a linear score function was applied to perform the task of classification.

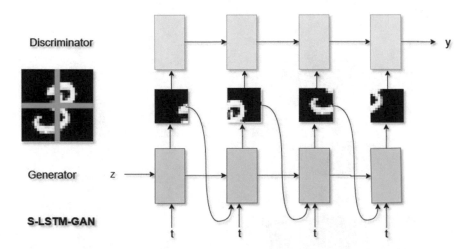

Fig. 1 S-LSTM-GAN: two recurrent neural sharing weights and training in parallel

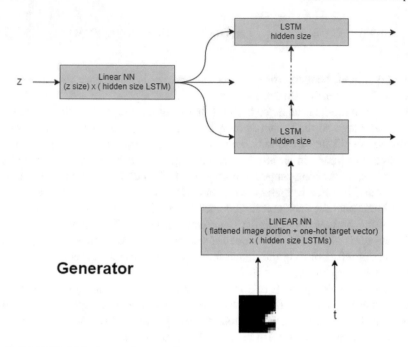

Fig. 2 S-LSTM-GAN: the generator architecture

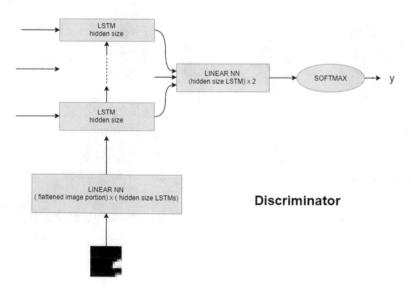

Fig. 3 S-LSTM-GAN: the discriminator architecture

4 Experimental Setup

We evaluated the above-mentioned architecture on the MNIST dataset [4] for two separate instances. Firstly with dividing it into four segments, refer Fig. 1 and secondly by dividing into 16 segments.

Model Layout Details: The number of layers in the LSTM network in the generator is six, while in the discriminator is two. They have 600 and 400 internal (hidden) units, respectively. Both architectures have unidirectional layout. In D, the cells feed forward to a fully connected network, with weights that are shared across each time step. The activation function used for each cell is sigmoid.

Dataset: We have imported the MNIST dataset, and its images contain gray levels due to anti-aliasing technique used by the normalization algorithm. The images were centered in an image by computing the center of mass of the pixels and translating the image so as to position this point at the center of the 28×28 field. We segment that image as per our instance, and this sets the dimensionality k for our model. We operate directly on the pixel values considering each image as a 28×28 matrix.

Training: Backpropagation and mini-batch stochastic gradient descent were used. For the initial instance of 4 segments, learning rate of 0.0001 for the second instance of 16 segments, learning rate of 0.0002 was used. The GPU used was Quadro P6000, and we trained the model for 10^7 epochs. For the generator, we recorded the loss as it changed alongside the increment in epochs. And for the discriminator, the metric we recorded was classification accuracy, best being one, against the rise in epochs. Refer Fig. 4.

Fig. 4 S-LSTM-GAN: classification versus epochs during training with instance of four segments

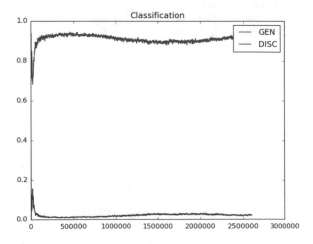

At four segments, we had to train for the initial for 5×10^5 epochs to see the generator able to generate image that were giving a steady loss, and then the spike in loss started for the generator while the discriminator converges at loss $= 0.25$. We are generating a sample at every 10^5 epoch. Hence, our sample at the third checkpoint was having some distinguishable features compared to the rest.

At 16 segments, we got better results than the previous instance, and here the classification was comparatively cleaner, with just a few spikes. The generator showed signs of heavy spikes in losses in early epochs, but later converged upon a value. It is noted that the loss is higher than the previous instance, but it still yields a cleaner classification. Due to the high variance, no singular checkpoint was distinguishable but compared to its predecessor, and they were easily recognizable to the human eye.

5 Results

The results of our experimental study are presented in Figs. 8 and 9 for 4 time steps and 16 time steps, respectively. We have chosen a very small learning rate to help the model learn the pixel values with more variability and larger intensity span. Allowing multiple layers for the generator and allowing it to train for these many epochs help to generate handwritten digits with a high degree of classification.

Figure 5 reveals that to attain convergence we can reduce the number of iterations by 2×, for the task of building the classifier net only. Figure 7 displays the variance of the same model over the initial iterations of turbulent variance in the accuracy

Fig. 5 S-LSTM-GAN: classification versus epochs during training with instance of 16 segments

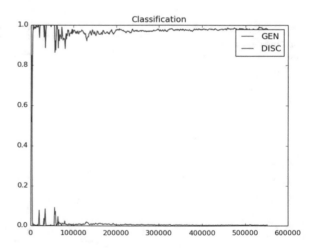

Fig. 6 S-LSTM-GAN: loss versus epochs during training with instance of four segments

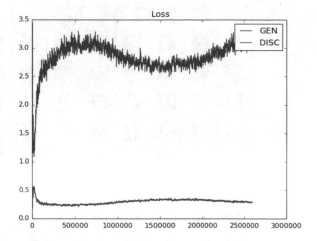

Fig. 7 S-LSTM-GAN: loss versus epochs during training with instance of 16 segments

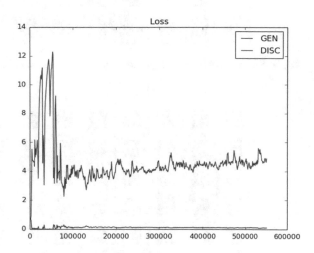

of its digit generation. We maintain generator accuracy by varying dropout and the network depth of six layers.

Unlike the graph presented in Fig. 6, we have found that beyond the 16-segment instance, the generator would collapse at early iterations. We believe this is because the generator has to reduce the variance of the minimax objective over the iterations, and it will not be able to continue to do the following as implied in Fig. 7. Hence, if we were to segment the image ahead of 16 segments, the proposed model would fail.

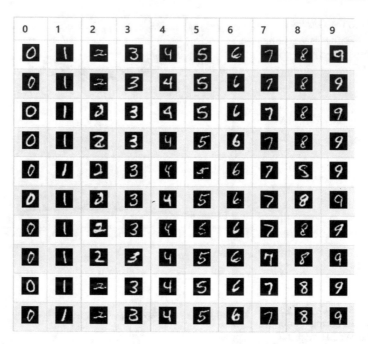

Fig. 8 S-LSTM-GAN: images generated during training with instance of four segments

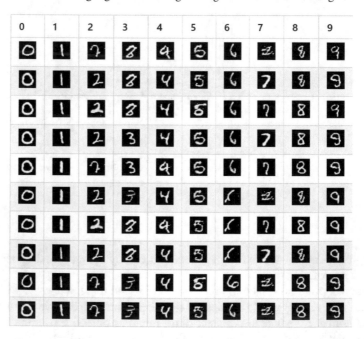

Fig. 9 S-LSTM-GAN: images generated during training with instance of 16 segments

We also found that beyond the 16-segment instance, the generator would collapse at early iterations. We believe this is because the generator has to reduce the variance of the minimax objective over the iterations, and it will not be able to continue to do the following as implied in Fig. 7. Hence, if we were to segment the image ahead of 16 segments, the proposed model would fail.

To view our code and the images generated by S-LSTM-GAN: https://github.com/amitadate/S-LSTM-GAN-MNIST.

6 Conclusion

In this paper, we have proposed a neural network model for the task of learning hand-written digits trained using an application based on generative adversarial networks. We believe that training could be accelerated greatly by developing better techniques for synchronizing the generator and the discriminator or by forming better partitioning of the data sample before or during training.

In future work, we will look at more sophisticated mechanisms for applying the similar approach toward the task of handwriting synthesis. For that amount of features, introducing multiple generators would be ideally the next step, and the discriminator would also be needed to be broadened as per the feature–label pairs.

References

1. Goodfellow I, Bengio Y, Courville A (2016) Deep learning. MIT Press. Book in preparation for MIT Press
2. Ledig C, Theis L, Huszar F, Caballero J, Aitken AP, Tejani A, Totz J, Wang Z, Shi W (2016) Photo-realistic single image super-resolution using a generative adversarial network. CoRR, arXiv:1609.04802
3. Ganin Y, Ustinova E, Ajakan H, Germain P, Larochelle H, Laviolette F, Marchand M, Lempitsky V (2016) Domain-adversarial training of neural networks. J Mach Learn Res 17(1):2030–2096
4. LeCun Y, Cortes C (2010) MNIST handwritten digit database
5. Donahue J, Jia Y, Vinyals O, Hoffman J, Zhang N, Tzeng E, Darrell T (2013) Decaf: A deep convolutional activation feature for generic visual recognition. CoRR, arXiv:1310.1531
6. Graves A, Schmidhuber J (2005) Framewise phoneme classification with bidirectional LSTM and other neural network architectures. Neural Netw. 18:602–610
7. Greff K, Srivastava RK, Koutník J, Steunebrink BR, Schmidhuber J (2015) LSTM: a search space odyssey. CoRR, arXiv:1503.04069
8. Hochreiter S, Schmidhuber J (1997) Long short-term memory. Neural Comput 9(8):1735–1780

Hybrid Approach for Recommendation System

Rohan Passi, Surbhi Jain and Pramod Kumar Singh

Abstract The primary objective of recommendation systems (RSs) is to analyze user's fondness and taste and recommend similar items to him/her. There exist various methods, e.g., user/item collaborative filtering (CF), content-based filtering (CBF), association rule mining (ARM), hybrid recommender system (HRS), for recommendations. Though these methods possess excellent characteristics, they are inefficient in providing good recommendations in particular situations. For example, item CF produces recommendations for cold-start objects; however, it typically has low accuracy compared to user CF. Conversely, user CF often provides more accurate recommendations; however, it fails to provide recommendations for cold-start objects. The hybrid methods aim to combine different approaches coherently to yield better recommendations. This paper presents an HRS based on user CF, item CF, and adaptive ARM. The proposed HRS employs ARM as a fundamental method; however, it considers only a set of users who are nearest to the target user to generate association rules (ARs) among items. Also, the support levels to mine associations among items are adaptive to the number of rules generated. Results of the study indicate that the proposed HRS provides more personalized and practical suggestions compared to the traditional methods.

Keywords Hybrid recommendation system · User collaborative filtering · Item collaborative filtering · Adaptive association rule mining

R. Passi (✉) · S. Jain · P. K. Singh
Computational Intelligence and Data Mining Research Laboratory,
ABV-Indian Institute of Information Technology and Management,
Gwalior 474015, India
e-mail: rohanpassi94@gmail.com

S. Jain
e-mail: surbhijain.iiitm@gmail.com

P. K. Singh
e-mail: pksingh@iiitm.ac.in

© Springer Nature Singapore Pte Ltd. 2019   117
A. J. Kulkarni et al. (eds.), *Proceedings of the 2nd International Conference on Data Engineering and Communication Technology*, Advances in Intelligent Systems and Computing 828, https://doi.org/10.1007/978-981-13-1610-4_12

1 Introduction

RSs are information agents that mine patterns to provide recommendations for items that could be of interest to a user. Various approaches, e.g., user CF, item CF, CBF, ARM, HRS, have been proposed for recommendations. However, each one has its own benefits and shortcomings. Nonetheless, here, we discuss user CF, item CF, and ARM in brief which are of interest for this work.

1.1 *User Collaborative Filtering*

User CF attempts to discover users who are a close match to the target user (user for which the recommendations are being made) and uses their ratings to compute recommendations for the target user. Similarity among users is calculated using similarity scores, e.g., Euclidean distance, Pearson's correlation coefficient [1]. Various issues with user CF are as follows.

Sparsity in database: Most RSs have a significant number of users and items. However, on an average, a user rates only a small fraction of items. It leads to s sparsity in the user–item matrix. Therefore, user-based CF approach would be unable to provide any recommendations for some users [2].

False nearest neighbor: False nearest neighbor is one of the nearest neighbors who tend to be the closest match to user's preferences. However, there exists some other user who indeed is an exact match to user's behavior, and any metric used for computing nearest neighbor does not evaluate this user as the nearest neighbor. As a result, recommendations provided by the system will not be of high quality [3].

Gray sheep: Gray sheep refers to the users who have unusual taste compared to the rest of the users. They have very low correlations with other users, and their opinions do not persistently agree or disagree with any other group of users. Due to this, the gray sheep users may not receive good recommendations and may even have an adverse impact on the recommendations of other users [4].

1.2 *Item Collaborative Filtering*

In the item CF, the similarity between each item is computed by applying one of the similarity metrics, and then, these similarity values are used to make recommendations for a user [5]. Various issues with the item CF are as follows.

Similar recommendations: Only products similar to the input product are recommended. However, it is highly unlikely that a user shall buy a similar product again. Suppose a user buys X, then it is highly unlikely that he/she will again buy X. However, item CF would only recommend similar products; that is, it will only recommend laptops of another company [6].

Non-personalized recommendations: No preferences from the user are taken into account in the recommendation process. It will recommend a product irrespective of whether the user liked that product or not. As a result, the recommendations provided may not be of the user's interest [5].

1.3 Association Rule Mining

The ARM is used to obtain Frequent Patterns (FPs), associations, and correlations among the set of objects contained in a transaction database (DB) [7]. The ARs are generated using criteria such as support, confidence. Various issues with the ARM are as follows.

Computationally inefficient: In the case of Apriori algorithm, the cost of computation is high as it requires DB scan and candidate set generation at every iteration for finding FPs. Though FP-growth algorithm is far better than Apriori, computational time increases as the size of transaction DB increases. Therefore, FP-growth algorithm is also inefficient for large DB [7].

Weak recommendations: Sometimes very few or no recommendations are provided corresponding to a particular product. This is because very few rules are generated with high confidence, and it becomes highly infeasible to recommend a product from the rules generated [7].

Non-personalized recommendations: The ARM provides recommendations based on the previous transactions of all the users in the DB. It does not take into account the ratings given by users to different items. It evaluates the recommendations to a user by taking the transactions of those users who are conflicting to that user's behavior [7].

2 Related Work

The CF system is the earliest and the most successful recommendation method [5]. This RS is extensively used in many e-commerce industries such as Amazon and Netflix [8]. This system recommends objects to a user based on the objects that are not rated by that user but have been rated by some other similar users. In [5], the author described CF system as a person-to-person correlation system because they recommend items to a user based on the degree of interconnection between the given user and other users who have bought that item in the past. CF systems are implemented in various domains such as newsgroup articles domain—Grouplens; music domain—Ringo; movies domain—Bellcore's Video Recommender; books domain—Amazon.com; and other product domains. One of the shortcomings of the CF approach is that it must be initialized with huge data containing user preferences to provide meaningful recommendations [5]. As a consequence, this approach faces the cold-start problem when new items or new users enter the system [9].

Item-based RS is an item-to-item correlation system as it recommends similar items based on the content similarity to the items that the user liked before [10]. Examples of such systems are newsgroup filtering system—NewsWeeder; Web page recommendation system—Fab and Syskill & Webert; book recommendation system—Libra; funding recommendation system—ELFI. One of the drawbacks of item-based recommendation approach is a new user problem [11, 12]. Another drawback is over-specialization; that is, item-based approach favors objects that are similar to the objects rated by that user in the past [10]. Singh and Gupta [13] asked the user for the required features of an item to solve cold-start problem and recommended items close to their choice.

Chellatamilan and Suresh proposed a plan of recommendations for e-learning system using ARM, offering students the exceptional selection of e-learning resources [14]. Bendakir and Aïmeur presented a course RS based on ARM. The system integrates a data mining procedure with user ratings in recommendations. This system allows students to assess the previous recommendations for system enhancement and rules up gradation. On the other side, this system does not take into account student's academic background [15].

The hybrid RS integrates multiple approaches to improve recommendation performance and avoid the weaknesses of a single recommendation approach [16]. Content/collaborative hybrids are widely deployed because the rating data is already available or can be inferred from data [17]. However, this combination does not mitigate the cold-start problem as both the methods rely on rating DB given by the user. Ansari et al. [18] proposed a system based on Bayesian network approach that incorporates CF with item-based filtering approach. A Bayesian approach is also used in the system presented in [19] that recommends movies by taking into account information such as the actors starred in each movie. The advantage of this RS is that it can supply personalized recommendations even if it is deprived of a large DB of previous transactions [9]. The ARM has been widely applied in CF system to enhance the recommendation results and to solve the existing system's problems. For example, Garcia et al. [20] combined the ARM and CF recommendation methods to improve e-learning courses.

3   Architecture of the Proposed Recommendation System

The proposed approach works in four phases: computing nearest neighbors to target user, computing the similarity matrix for all the items, generating ARs, and finding recommendations based on generated ARs.

3.1   Computing Nearest Neighbors Corresponding to Target User

The first phase of the architecture is based on the user CF. It involves computation of nearest neighbors of the target user using K-Nearest Neighbor (KNN) algorithm.

The similarity between each user is measured by applying Pearson's correlation coefficient on the item ratings given by the user. It tends to obtain better outcomes in situations even when the data suffers from grade inflation and is not well-normalized [1]. Consider this dataset $x_1, \ldots, x_n$ and another dataset $y_1, \ldots, y_n$, both containing n values, then the formula for Pearson's correlation coefficient, represented by r, is given by Eq. (1).

$$r = \frac{\sum_{i-1}^{n} x_i y_i - n\bar{x}\bar{y}}{\sqrt{\sum_{i=1}^{n} x_i^2 - n\bar{x}^2} \sqrt{\sum_{i=1}^{n} y_i^2 - n\bar{y}^2}} \tag{1}$$

All the items rated by a user are considered as transactions corresponding to that user. Next, the transactions of all the nearest neighbors are combined with the transaction of the target user to form the transaction DD for the target user. Thus, the transaction DB formed contains the transactions of only the relevant users.

3.2 Computing Similarity Between Items in the Database

Next step is to compute the similarity of each item with every other item in the transaction DB formed in the above phase using Jaccard index and store these values in a form of the similarity matrix. The Jaccard index is a statistic measure used for comparing the similarity and variability of sample sets. Let U and V be two finite sets, then Jaccard index is defined as shown in Eq. (2).

$$J(U, V) = \frac{|U \cap V|}{|U \cup V|} = \frac{|U \cap V|}{|U| + |V| - |U \cap V|} \tag{2}$$

3.3 Association Rule Mining to Generate Association Rules

Next phase of the proposed approach is to obtain the required ARs via FP-growth approach. The input to the FP-growth algorithm includes the transaction DB of the target user, minimum support count, and minimum confidence. Since it is tough to choose a proper minimum support count before the mining process, adaptivity in the support levels has been introduced. The algorithm adjusts the support threshold such that an adequate number of rules are generated.

3.4 Providing Recommendations to Target User

Once the strong ARs are generated, they are sorted in the descending order of their confidence values. To provide recommendations, consequents of those ARs are added to recommendation list whose antecedent completely match with the input item list. Finally, if the number of recommendations obtained from ARs is less than the required recommendations, the remaining places are filled in with the recommendation list based on the similarity values computed in phase 2. To provide recommendations, similarity values of items in the input list are added corresponding to every item in the DB. Consequently, items with more similarity values are appended to the recommendation list generated through ARs.

4 Experimental Analysis

For evaluation purpose, MovieLens dataset containing *100,000* ratings given by *943* users to *1682* movies has been used. This section illustrates the proposed algorithm for providing recommendations with an example. Suppose there are *7 users*, namely $U_1, U_2 \ldots U_7$ and *7 movies*, $I_1, I_2 \ldots I_7$. The matrix representation of user–item DB is presented in Table 1, where each row represents a distinct user and each column represents a distinct movie. Each cell contains the rating of the movie on the scale of [*1–5*], where *1* and *5* are lowest and highest rating, respectively. Since all the users do not rate all the items, some cells are vacant. To start with, we first find the nearest neighbors to all the users using Pearson's correlation coefficient. Using Eq. (1), the matrix shown in Table 2 is computed, where each cell represents the similarity value between users. If K is *2* in KNN, nearest neighbors to all the users shall be as below.

$$U_1 = \{ U_4, U_2 \} \quad U_3 = \{ U_2, U_5 \} \quad U_5 = \{ U_3, U_4 \} \quad U_7 = \{ U_2, U_4 \}$$
$$U_2 = \{ U_7, U_4 \} \quad U_4 = \{ U_7, U_2 \} \quad U_6 = \{ U_7, U_1 \}$$

Transaction DB for target user U_1 will be the transactions of his *2* nearest neighbors along with his transaction as shown in Table 5. Let the movies be classified into *4* genres, namely *A*, *B*, *C*, and *D*, Table 3 shows the classification of each movie, where *1* and *0* indicate whether the movie belongs to that genre or not, respectively.

Table 1 Ratings matrix

	I_1	I_2	I_3	I_4	I_5	I_6	I_7
U_1	5	2	–	4	–	1	3
U_2	3	3	4	–	–	3	–
U_3	2	4	3	–	5	–	–
U_4	4	–	5	3	–	2	4
U_5	–	5	1	2	4	–	5
U_6	–	–	–	5	1	5	2
U_7	3	–	4	3	2	–	–

Table 2 Similarity matrix of users

	U_1	U_2	U_3	U_4	U_5	U_6	U_7
U_1	1	0.19	−0.98	0.62	−0.79	−0.28	0
U_2	0.19	1	−0.09	0.75	−0.96	−1	0.94
U_3	−0.98	−0.09	1	−0.56	0.59	−1	−0.64
U_4	0.62	0.75	−0.56	1	−0.37	−0.89	0.89
U_5	−0.79	−0.96	0.59	−0.37	1	−0.84	−0.75
U_6	−0.28	−1	−1	−0.89	−0.84	1	0.79
U_7	0	0.94	−0.64	0.89	−0.75	0.79	1

Table 3 Movies genre matrix

	A	B	C	D
I_1	0	1	0	0
I_2	1	0	1	0
I_3	0	1	1	1
I_4	0	1	0	1
I_5	0	0	1	0
I_6	1	1	0	0
I_7	1	0	1	0

Table 4 Similarity matrix of movies

	I_1	I_2	I_3	I_4	I_5	I_6	I_7
I_1	1	0	0.33	0.5	0	0.5	0
I_2	0	1	0.25	0	0.5	0.33	1
I_3	0.33	0.25	1	0.67	0.33	0.25	0.25
I_4	0.5	0	0.67	1	0	0.33	0
I_5	0	0.5	0.33	0	1	0	0.5
I_6	0.5	0.33	0.25	0.33	0	1	0.33
I_7	0	1	0.25	0	0.5	0.33	1

Table 5 Transaction DB of U_1

User	Transaction
U_1	I_1, I_2, I_4, I_6, I_7
U_4	I_1, I_3, I_4, I_6, I_7
U_2	I_1, I_2, I_3, I_6

Next step is to calculate the similarity between all the movies using Jaccard index. Similarity matrix shown in Table 4 is computed using Eq. (2). Let input item set of user U_1 is { I_1, I_2 } then ARs generated for user U_1 using *minimum support count = 3* and *minimum confidence = 0.7* are: $I_1 \rightarrow I_6$ and $I_6 \rightarrow I_1$. Suppose the number of rules should fall in the range [10, 20]. Since number of rules generated for *support count threshold = 3* do not fall in the specified range, decrease the value of minimum support count to 2 and minimum confidence is still *0.7*. Then, following rules are generated:

$I_1 \rightarrow I_6$	$I_1, I_2 \rightarrow I_6$	$I_4, I_6 \rightarrow I_7$	$I_7 \rightarrow I_4, I_6$
$I_3 \rightarrow I_6$	$I_1, I_3 \rightarrow I_6$	$I_4, I_7 \rightarrow I_6$	$I_3, I_6 \rightarrow I_1$
$I_6 \rightarrow I_1$	$I_1, I_4 \rightarrow I_6$	$I_6, I_7 \rightarrow I_4$	$I_2, I_6 \rightarrow I_1$
$I_2 \rightarrow I_1, I_6$	$I_1, I_7 \rightarrow I_6$	$I_4 \rightarrow I_6, I_7$	
$I_1, I_6, I_7 \rightarrow I_4$	$I_1, I_4, I_6 \rightarrow I_7$	$I_1, I_4, I_7 \rightarrow I_6$	

To provide recommendations to target user U_1, first, consequents of those rules are added whose antecedents completely matches with the items in the input list. Therefore, I_6 is added to recommendation list owing to the presence of rule $I_1, I_2 \rightarrow I_6$. Then, if any of the subsets of input item set is present in the antecedent of the rules, its consequents are added to the recommendation list. Since rules $I_1, I_4, I_6 \rightarrow I_7$ and $I_1, I_6, I_7 \rightarrow I_4$ contain I_1 in their antecedents, now, the recommendation list shall be { I_6, I_7, I_4 } . Since 4 movies are to be recommended to the user U_1, for every item in input item set, their respective rows of Jaccard index obtained in Table 4 corresponding to every movie are added. The list thus obtained is sorted in the descending order of the total value of Jaccard index of each movie. The remaining places in recommendation list are filled up by the items having highest total value. Adding values of Jaccard index for item I_1 and I_2 will result in values as shown in Table 6. Sorting the movies according to their total value of Jaccard index for items in the input list will result in the following order of movies.

$$\{ \{ I_1 : 1 \}, \{ I_2 : 1 \}, \{ I_7 : 1 \}, \{ I_6 : 0.83 \}, \{ I_3 : 0.58 \}, \{ I_4 : 0.5 \}, \{ I_5 : 0.5 \} \}$$

Though movies I_1 and I_2 have the maximum similarity value, they are not added to recommendation list as they belong to the input set. As movies I_7 and I_6 are already present in the recommendation list, movie I_3 is added to the list. Therefore, final recommendations provided to the user U_1 are { I_6, I_4, I_7, I_3 }.

Table 6 Jaccard coefficient for I_1 and I_2

	I_1	I_2	I_3	I_4	I_5	I_6	I_7
I_1	1	0	0.33	0.5	0	0.5	0
I_2	0	1	0.25	0	0.5	0.33	1
Total	1	1	0.58	0.5	0.5	0.83	1

5   Results

To compare the accuracy, the ranking of various items have been considered rather than rating prediction of the items [21]. In the test data, the ratings given by a user to a particular item are already known. A list of items $list_1$ based on the ratings given by the user to the items is generated in descending order. Another ranked list of items $list_2$ is also generated according to the recommendations provided by the proposed system. Then, two sets S_1 and S_2 are made from the items in the upper half of the ranked list $list_1$ and $list_2$, respectively. A new set S_3 is generated from the intersection of the two sets S_1 and S_2 as shown in Eq. (3).

$$S_3 = S_1 \cap S_2 \tag{3}$$

Consider the cardinality of the set S_3 be n. Now, the accuracy of a RS may be defined as shown in Eq. (4).

$$\text{Accuracy} = \frac{n}{N} \times 100 \tag{4}$$

Figure 2 compares the accuracy of the proposed RS with other existing RSs.

5.1   Comparison with User Collaborative Filtering

The proposed architecture deals with the problem of a false nearest neighbor not only by merely depending upon the nearest neighbors' ratings to a product. But also, the items reviewed by the nearest neighbors have been combined to form a transaction DB of the target user. Then, the ARM has been applied on the relevant set of transactions to get FPs and associations for a particular item. Thus, the results obtained are efficient and more personalized. The proposed architecture also handles the gray sheep problem effectively. The partial behavior of all those users who somewhat resembles the gray sheep user has been considered properly. On an average, the user CF took 0.44 s, while the proposed system took 0.57 s to provide recommendations.

5.2   Comparison with Item Collaborative Filtering

The transactions of nearest neighbors of the target user have been taken into account to solve the problem of non-personalized recommendations. The user's transaction set includes only those users' transactions who are highly coupled to the user's behavior. The issue of vague recommendations has also been solved as every user has its unique transaction set. The recommendations are generated after carefully studying the transaction DB of the user. As the last phase of the proposed system involves ARM, the problem of recommending only similar products is also solved. Recommendations provided are based on what a customer would need next if she has already bought a product, i.e., user's history played a key role in generating recommendations. The results reveal that the proposed architecture outperforms the item CF system regarding accuracy and personalization and, however, has a slight overhead in computation time. On an average, item CF took *0.31* s for generating recommendations, while the proposed RS took *0.57* s.

5.3   Comparison with Association Rule Mining

In the proposed architecture, the DB size has been significantly reduced by using the transaction DB of nearest neighbors and the target user only. This improvement in reduction of DB significantly improves the performance of the proposed RS; the average computation time required by traditional ARM algorithms was *3.23* s while the proposed RS took only *0.57* s. Also, the proposed system takes into account only relevant data; therefore, it generates more efficient and personalized recommendations which are relevant to the context of the user's preferences. Unlike existing ARM that requires minimum support threshold of the rules to be specified previously before the mining process, the proposed RS adjusts the value of threshold at the time of mining so that the number of ARs generated falls within the specified range. This ensures that enough rules are generated so that recommendations are always provided to the user.

Table 7 compares the time required by traditional RS architectures and the proposed RS based on the number of items present in the input item list. Figure 1 compares the average time required by the existing RSs and the proposed RS. The proposed RS takes a little more time than user CF and item CF; however, the recommendations provided are more personalized and accurate. Compared to RSs that uses traditional ARM to provide recommendations, the proposed RS is computationally more efficient and requires less time to provide recommendations to the users (Fig. 2).

Table 7 Time required by recommendation systems

	User CF system (s)	Item CF system (s)	ARM system (s)	Proposed system (s)
1 Items	0.33	0.20	3.12	0.49
2 Items	0.46	0.32	3.25	0.56
3 Items	0.53	0.41	3.32	0.67
Avg. time	0.44	0.31	3.23	0.57

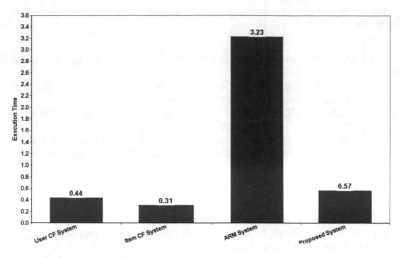

Fig. 1   Average time of recommendation systems

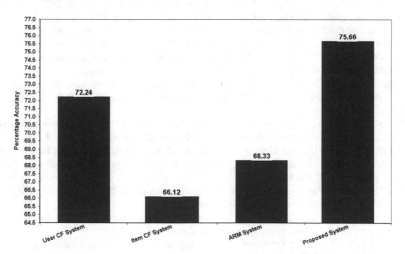

Fig. 2   Accuracy of different recommendation systems

6   Conclusion

All traditional RSs are incapable to the user's taste and preferences. Therefore, an architecture has been proposed that is adequate, as it deals with the problems incurred by all the traditional RSs. The proposed RS combines the positive parts of traditional RSs to overcome the gray sheep problem, the problem of false nearest neighbor, the effect of irrelevant users and recommendations of similar items only. The experiment reveals that the proposed RS successfully overcomes the problems associated with the traditional RSs.

References

1. Ricci F, Rokach L, Shapira B (2011) Introduction to recommender systems handbook. Springer, Berlin
2. Guo G, Zhang J, Thalmann D (2014) Merging trust in collaborative filtering to alleviate data sparsity and cold start. Knowl-Based Syst 57:57–68
3. Liu H, Hu Z, Mian A, Tian H, Zhu X (2014) A new user similarity model to improve the accuracy of collaborative filtering. Knowl-Based Syst 56:156–166
4. Srivastava A (2016) Gray sheep, influential users, user modeling and recommender system adoption by startups. In: Proceedings of the 10th ACM conference on recommender systems, pp 443–446
5. Shi Y, Larson M, Hanjalic A (2014) Collaborative filtering beyond the user-item matrix: a survey of the state of the art and future challenges. ACM Comput Surv (CSUR) 47(1):3
6. Poriya A, Bhagat T, Patel N, Sharma R (2014) Non-personalized recommender systems and user-based collaborative recommender systems. Int J Appl Inf Syst 6(9):22–27
7. Witten IH, Frank E, Hall MA, Pal CJ (2016) Data mining: practical machine learning tools and techniques. Morgan Kaufmann, Cambridge
8. Herlocker JL, Konstan JA, Borchers A, Riedl J (1999) An algorithmic framework for performing collaborative filtering. In: Proceedings of the 22nd annual international ACM SIGIR conference on Research and development in information retrieval. ACM, pp 230–237
9. Lika B, Kolomvatsos K, Hadjiefthymiades S (2014) Facing the cold start problem in recommender systems. Expert Syst Appl 41(4):2065–2073
10. Ekstrand MD, Riedl JT, Konstan JA et al (2011) Collaborative filtering recommender systems. Found Trends® Hum Comput Interact 4(2):81–173
11. Felfernig A, Burke R (2008) Constraint-based recommender systems: technologies and research issues. In: Proceedings of the 10th international conference on electronic commerce. ACM
12. Wei K, Huang J, Fu S (2007) A survey of e-commerce recommender systems. In: 2007 International Conference on service systems and service management. IEEE, pp 1–5
13. Singh PK, Gupta N (2016) Recommendation model for infrequent items. In: Proceedings of the first New Zealand text mining workshop. TMNZ (Online). Available: http://tmg.aut.ac.nz/tmnz2016/papers/Pramod2016.pdf
14. Chellatamilan T, Suresh R (2011) An e-learning recommendation system using association rule mining technique. Eur J Sci Res 64(2):330–339
15. Bendakir N, Aïmeur E (2006) Using association rules for course recommendation. In: Proceedings of the AAAI workshop on educational data mining, vol 3
16. Sharma L, Gera A (2013) A survey of recommendation system: research challenges. Int J Eng Trends Technol (IJETT) 4(5):1989–1992
17. Nilashi M, bin Ibrahim O, Ithnin N (2014) Hybrid recommendation approaches for multi-criteria collaborative filtering. Expert Syst Appl 41(8):3879–3900
18. Ansari A, Essegaier S, Kohli R (2000) Internet recommendation systems. J Mark Res 37(3):363–375
19. Beutel A, Murray K, Faloutsos C, Smola AJ (2014) Cobafi: collaborative Bayesian filtering. In: Proceedings of the 23rd international conference on World wide web. ACM, pp 97–108
20. García E, Romero C, Ventura S, De Castro C (2009) An architecture for making recommendations to courseware authors using association rule mining and collaborative filtering. User Model User-Adap Inter 19(1–2):99–132
21. Steck H (2013) Evaluation of recommendations: rating-prediction and ranking. In: Proceedings of the 7th ACM conference on recommender systems. ACM, pp 213–220

Discussion on Problems and Solutions in Hardware Implementation of Algorithms for a Car-type Autonomous Vehicle

**Rahee Walambe, Shreyas Nikte, Vrunda Joshi, Abhishek Ambike,
Nimish Pitke and Mukund Ghole**

Abstract The self-driving cars will bring fundamental change in the transportation industry. Many hazardous situations like land-mine detection, war zones, nuclear decommissioning highlights the need of autonavigation in open terrain. In the last few decades, due to the increasing interest in mobile robots, a number of autonavigation algorithms have been developed. The autonomous vehicles are used extensively in different domains from passenger car to the hazardous applications. These autonomous cars extensively use mathematical calculations and machine intelligence. In order for the car-type vehicle to manoeuver smoothly in a given workspace, accurate planning (motion and path) algorithms are essential. As part of the research, our group has developed the nonholonomic motion planning algorithms for the car-type vehicle based on differential flatness approach. In the previous work, the hardware realization of these algorithms is presented. This paper discusses the hardware implementation issues that we have faced during this work. Hardware implementation comes with various inherent challenges, such as manufacturing error in the car hardware components, physical limitation of the component, limited processing power of low-power onboard computer, accuracy of data from sensors in diverse conditions.

Keywords Nonholonomic motion planning · Car type robot · Hardware implementation · Open loop

1 Introduction

Latombe [1] has discussed the motion planning and path planning methods in detail. Although these methods are suitable for specific tasks with holonomic motion plan-

R. Walambe (✉)
Symbiosis Institute of Technology, Pune, India
e-mail: rahee.walambe@sitpune.edu.in

S. Nikte · V. Joshi · A. Ambike · N. Pitke · M. Ghole
PVG's College of Engineering and Technology, Pune, India

© Springer Nature Singapore Pte Ltd. 2019
A. J. Kulkarni et al. (eds.), *Proceedings of the 2nd International Conference on Data Engineering and Communication Technology*, Advances in Intelligent Systems and Computing 828, https://doi.org/10.1007/978-981-13-1610-4_13

ning problems, in order to manoeuver nonholonomic entities like car-type vehicle, motion planning algorithms developed by Fliess [2] and Agarwal [3] can be effectively used. In [4], differential flatness-based nonholonomic motion planner for a car-type mobile robot is presented. Walambe et al. [5] shows how the spline-based optimization can be incorporated in this motion planner to avoid the singularities and generate the shortest path which also satisfy the nonholonomic and curvature constraints. In [6], implementation of sensor data acquisition system on hardware for feedback and real-time localisation of a car-type vehicle is shown. The hardware implementation of motion planning algorithm and sensor data acquisition was an important step in evaluating our autonavigation algorithm. Hardware implementation, though important, is hindered by various issues such as manufacturing error in the car hardware components, physical limitation of the component, limited processing power of low-power onboard computer, accuracy of data from sensors in diverse conditions. These hardware selection criteria and implementation issues are discussed in the following article. In this paper, the work done towards the development and implementation of the open-loop and closed-loop motion planning of a 1:10 scale down vehicle platform is presented. This includes selection of appropriate car model and interfacing of sensors and other computer systems with this hardware which are essential for implementation of autonavigation algorithm. As main purpose of 1:10 scale down model of car is to reduce cost of the hardware implementation for research, cost-effectiveness and adequate performance of the hardware model are the key requirements while designing the hardware. This paper also touches various hardware implementation issues that our team had to address during this work. The contribution of this paper is mainly in two areas: (1) to list and discuss the various issues that were faced in the hardware implementation and (2) to identify and implement the solution to these issues. Authors hope that this paper will benefit the people working in the hardware realization domain. The kinematic model of hardware model is discussed in Sect. 2. In Sect. 3, description of hardware model is provided. Also selection criteria for the particular car model is discussed. Section 4 covers the selection of onboard computer and issues faced with initial selection of computer. In Sect. 5, issue related to PWM generation technique used in Raspberry Pi boards is discussed. The solution for the issue is stated and results are presented. Section 6 deals with issues related with the default Electronic Speed Controller (ESC) and implemented solution for the problem. Section 7 covers issues related with default NiMH battery and problem related with the use of IC LM7805, a common IC used for 5 V supply. Section 8 discusses the sensor data acquisition process and corresponding challenges.

2 System Modelling

Let A be a car-type mobile robot capable of both forward and reverse motion. It is modelled as a rigid rectangular body moving in a planar (two-dimensional) workspace (refer Fig. 1). The rear wheels are fixed and aligned with the car while

Fig. 1 Kinematic model of a car

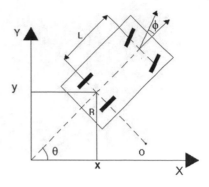

the front wheels are steerable; i.e., they are allowed to spin about the vertical axis with constraints on the maximum angle [7]. The wheelbase of A (distance between the front and rear axles) is denoted by l. The car is assumed to be rear wheel driven, and hence R, the reference point of the car is the midpoint of the rear axle.

A configuration of A is defined by the posture (x, y, ϕ, θ) where (x, y) are the coordinates of R in the Cartesian plane. The orientation of the vehicle is θ, which is the angle between the x-axis and the main axis of the vehicle $(-\pi \leq \theta \leq \pi)$. The steering of angle of the vehicle, which is the angle made by the steerable front wheels with respect to the main axis of the car, is denoted by ϕ.

The kinematic model of the car where (x, y, ϕ, θ) are the state variables is shown below

$$\begin{bmatrix} \dot{x} \\ \dot{y} \\ \dot{\theta} \\ \dot{\phi} \end{bmatrix} = \begin{bmatrix} \cos\theta \\ \sin\theta \\ \frac{\tan\phi}{l} \\ 0 \end{bmatrix} u_1 + \begin{bmatrix} 0 \\ 0 \\ 0 \\ 1 \end{bmatrix} u_2 \tag{1}$$

where u_1 is driving velocity and u_2 is steering velocity also known as control inputs. Development of motion planning algorithm is discussed in [1].The motion planning algorithm was based on this nonholonomic model of a car and employs the differential flatness-based approach for planning the motion and is shown in Algorithm 1.

3　Hardware Model of the Car

After development of the algorithms in simulations, the next step was to port these algorithms on the car prototype. The major criteria for selection of the car model was to get scaled down model of a real car which is functionally as similar as possible. Hence, with the above requirement RTR Sprint 2 Drift model manufactured by HPI racing Ltd is selected. The RTR Sprint 2 Drift chassis is an exactly 1:10 scaled down model of the actual sports car Chevrolet Camaro and is a four wheel drive (4WD). The hardware model shown in Fig. 2 is interfaced with Raspberry Pi 3B which is main

Fig. 2 Hardware car model

Algorithm 1 Autonavigation in static obstacle environment

1: Input $\gamma(0)$, $\gamma(1)$ and obstacles
2: Determine obstacle free path $\gamma(p)$ from geometric planner
3: **if** $\gamma(p)$ exists **then**
4:      continue
5: **else**
6:      exit
7: **end if**
8: $p_0 \leftarrow 0$
9: $p_f \leftarrow 1$
10: **while** $p_0 \neq 1$ **do**
11:      **Generate** path between $\gamma(p_0)$ to $\gamma(p_f)$ considering nonholonomic constraints
12:      **Optimize** path
13:      **if** Path is optimized **then**
14:          Check path for collisions with obstacles
15:          **if** Collision detected **then**
16:              **jump** to line:**22**
17:          **else**
18:              $p_0 \leftarrow p_f$
19:              $\gamma(p_0) \leftarrow \gamma(p_f)$
20:          **end if**
21:      **else**
22:          $p_f^{new} \leftarrow (p_0 + p_f)/2$
23:          $\gamma(p_f) \leftarrow \gamma(p_f^{new})$
24:      **end if**
25: **end while**

---

computer of the system. Arduino Uno is employed for interfacing the motors with the system. Hercules NR-MDR-003 Motor driver is used to control the power given to the driving motor. Optical encoder is used for measuring the distance travelled by wheels, potentiometer is used to measure steering angle, and magnetometer is used for measurement of orientation angle. The system is powered by 5000 mAh 7.2 V Lithium ion battery. A 5 V, 3 A buck converter is used to power devices which require 5 V DC as input.

4 Selection of Onboard Computer

For autonavigation, it is necessary to have a powerful computer platform which can be mounted on the car for implementation of complicated algorithms. At the start of each test case, mathematical calculations are carried out for optimum motion planning using algorithm. Hence, adequate computation power is required. Due to powerful processor and large community support, Raspberry Pi 3B and Beaglebone Black are major SBC platforms. The Raspberry Pi and Beaglebone black are roughly about the size of a credit card and run variety of Linux distributions. One of the biggest downsides to the Beaglebone is that it only has 1 USB port compared to the RPis 4. Hence for Beaglebone Black, creating multiple SPI interface with the USB ports requires additional USB hub. Also, Pi has large number of add-on boards due to the larger community support. Beaglebone Black has more powerful processor and more GPIO pins, although the price of the Beaglebone Black is roughly twice the price of Raspberry Pi 3B, In addition, it is equipped with a Wi-fi port, 40 GPIO pins, I2C, SPI and HDMI interfaces. MCP3008, an external 8-bit ADC chip is required for reading variable voltage of potentiometer. It has a HDMI port, a display interface (DSI) and a camera interface (CSI) for future expansion of the project. Hence, Raspberry Pi 3B was the best choice for the application.

5 PWM Generation on Raspberry Pi 3B

Raspberry Pi 3B has one hardware PWM pin and two software PWM pins. The software PWM frequency for raspberry Pi is derived from the pulse time. Essentially, the frequency is a function of the range and this pulse time. The total period will be (range * pulse time in S), so a pulse time of 100 and a range of 100 gives a period of 100 * 100 = 10,000 S which is a frequency of 100 Hz. It is possible to get a higher frequency by lowering the pulse time; however, CPU usage will skyrocket as WiringPi uses a hard loop to time periods under 100 S—this is because the Linux timer calls are not accurate and have an overhead. Another way to increase the frequency is to reduce the range, however that reduces the overall resolution of PWM signal to an unacceptable level (refer Fig. 3).

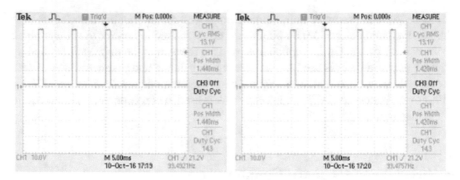

Fig. 3 PWM output for 14.1% dutycycle (left) and 14.7% dutycycle (right) on Raspberry Pi 3B

This characteristics of software PWM function of Raspberry Pi creates problem for the driving motor PWM as well as servo PWM signals. As the vehicle is tuned for lower speeds (0–12% duty cycle), resolution of PWM signal should be high enough to get full control of the above range of duty cycle. This will result in lowering of the PWM frequency as discussed above. Also, modern servo position is not defined by PWM, but only by the width of the pulse. Hence for the frequencies above 100 Hz, very low resolution software PWM is generated on Raspberry Pi. In other words, smaller set of on times are available for the servo control. Conversely, if the frequency of the PWM signal is lowered, we can get higher resolution of PWM and hence more number of pulses in the operating range of the servo motor. But it is advised not to run servo motor on such low frequencies. Hence, additional PWM generating device for one of the motors was required.

Arduino Uno development board was chosen for controlling servo motor due to above-mentioned reasons. Raspberry Pi and Arduino Uno are interfaced by SPI using USB ports. The pair acts as master–slave configuration as Raspberry Pi sends the servo angle command to the Arduino. Arduino then maps the angle into the corresponding pulse width and sends it to the servo motor. Hence, PWM issue was resolved by adapting the above method.

6 Replacing the Radio Control

The vehicle platform is designed to be a racing car. For the development, implementation and obtaining the proof of concept of the algorithms, the radio control was replaced with the controller board. The onboard Electronic Speed Control (ESC) mechanism was provided for control of the steering motor and the driving motor. In order to avoid interference of receiver and transmitter signals with that of other vehicles, frequency switching technique is employed in electronic speed control. Due to this, the system response of ESC for driving control is time variant, i.e. shifting of dead band. This car is originally designed to run at a speed of about 3300 rpm (maximum speed). It was difficult for us to implement and observe the behaviour of the algorithms and car at such high speeds. Therefore, a null signal is passed to the driving channel of ESC (primary switching for steering control) and chopper drive is used to eliminate ESC for control of driving velocity at lower speed range of the prototype.

The peak current of driving motor is 11 A for full loading and voltage supplied by power supply (Li-Po battery) is 7.4–8.4 V and minimum safe operation is 6 V, and therefore, voltage requirement is greater than 6 V.

Logic levels of PWM from Raspberry Pi is 3.3 V. The driver with these specifications is provided by NEX robotics named as Hercules-NR-MDR-003. This driver is rated as 6–36 V, 15 A with a peak current of 30A and can be operated up to 10 KHz PWM. It can be interfaced with 3.3–5 V logic levels. The driver has terminal block as power connector and 7-pin 2510 type connector for logic connections.

Hence, Hercules-NR-MDR-003 is selected as a driver by bypassing the default ESC of the driving motor.

7   Power Supply for the Hardware

These issues are twofold, i.e. issues with the default NiMH battery and issues with using LM7805 IC for 5 V power supply.

The RTR Sprint 2 Drift model came with 7.2 V 2000 mAH NiMH battery. The capacity of battery was not enough for daily testing of 1–2 h. Also, NiMH battery has higher self-discharging rate. The number is around 5% on the first week after the charge and about 50% on the first month. Charging time battery is also long. Hence, it was not suitable for the purpose. On the other hand, Li ion battery is more reliable, smaller and can be recharged faster. Hence, 7.4 V, 5000 mAH Lithium ion battery is chosen to power circuitry.

Although most of the electronic circuitry requires 5 V power supply, this voltage is required to supply all the sensors like magnetometer, encoder, and potentiometer, steering servo motor, Raspberry Pi and Arduino Uno. So, initially in order to provide required 5 V supply, LM 7805 IC was used. But due to higher power demands from the new added circuitry, 7805 IC was not reliable power solution for long term. It was observed that, the voltage across 7805 IC was dropping below acceptable value due to higher current requirements. Hence, above-mentioned equipments got affected at each voltage sag, viz. Raspberry Pi often used to reset, also inaccurate sensor data was acquired, especially from the sensors which are the function of input voltage, such as potentiometer.

Hence after recalculating the power requirement for the electronic components, LM7805 circuit was replaced by 5 V 3 A DC-DC buck converter with over-temperature and short-circuit protection. These power circuit boards are cheap and readily available in the market.

8 Sensor Data Acquisition

According to the open loop results in [5] it is seen that, there are deviations from the open loop simulated trajectory and its hardware implementation. This is primarily due to the fact that the kinematic model does not take into consideration dynamic state variables. Hence, we are currently working on the development of feedback algorithm based on the real-time sensor data. The sensor data acquisition posed a challenging problem in selection, interfacing and placement of the sensors. This has been discussed in details in [6]. Magnetometer, optical encoder and potentiometer are used for sensor data acquisition.

9 Conclusion

This paper presented a viable hardware solution for testing various nonholonomic motion planning algorithms. By addressing the above-mentioned issues and providing the solution for these problems, we were able to create robust and reliable vehicle platform with required sensors and onboard computers needed for autonavigation purpose. In future, advanced sensors like camera and LIDAR can be included on this platform for creating real-time mapping system for AI-based or mathematical algorithm development.

Acknowledgements This project is funded under the WOS-A scheme (SR/ET/WOS-34/2013-14) of Department of Science and Technology, Government of India.

References

1. Latombe JC (1991) Robot motion planning. Kluwer Academic Publishers, Boston
2. Rouchon P, Fliess M, Levine J, Martin P (2011) Flatness and motion planning: the car with n trailers. Automatica
3. Ramirez HS, Agarwal S (2004) Differentially flat systems. CRC Press, Boca Raton
4. Agarwal N, Walambe R, Joshi V, Rao A (2015) Integration of grid based path planning with a differential flatness based motion planner in a non-holonomic car-type mobile robot. Inst Eng Annu Tech J 39 (November)
5. Walambe R, Agarwal N, Kale S, Joshi V (2016) Optimal trajectory generation for car-type mobile robot using spline interpolation. In: Proceedings of 4th IFAC conference on advances in control and optimization of dynamical systems ACODS 2016, Tiruchirappalli, India, 1–5 Feb 2016
6. Walambe R, Dhotre A, Joshi V, Deshpande S (2016) Data acquisition and hardware setup development for implementation of feedback control and obstacle detection for car type robot. In: International conference on advancements on robotics and automations(ICAARS), Coimbatore, India, June 2016
7. Laumond J-P, Jacobs PE, Taix M, Murray RM (1994) A motion planner for nonholonomic mobile robots. IEEE Trans Robot Autom 10:577–592

Software Test Case Allocation

Amrita and Dilip Kumar Yadav

Abstract To allocate optimal number of test cases among functions in order to reduce the expected loss of resources like time, cost etc., and a method has been proposed in this paper. Test cases play an important role in software testing. Allocation of test cases should be done according to the function's significance. There can be many engineering and non-engineering attributes which can provide assumption for function's priority for test case allocation. In this paper, complexity, cost, criticality, and occurrence probability are taken into consideration for optimal test case allocation problem. Previously, test case allocation was performed only using occurrence probability and criticality. We present a new method to allocate number of test cases among various software operations, by fixing total number of test cases. Proportionality factor in terms of failure probability is derived based on the above-identified allocation factors. Fuzzy numbers are used to represent allocation factors. Centroid's method of defuzzification has been used for the valuation of failure probability weight. Test case allocation is performed based on the weightage of operation's failure probability.

Keywords Software operational profile · Test case · Occurrence probability
Criticality

1 Introduction

Due to risk of failure of software and its damage to users, researchers investigated this problem and identified that these risks can be minimized by performing effective testing. Testing process can be accomplished efficiently by optimal usage of resources as a large amount of resources is consumed in this process. According to IEEE, testing is the process of exercising or evaluating a system or system component by manual

Amrita (✉) · D. K. Yadav
Department of Computer Applications, National Institute of Technology,
Jamshedpur 831014, India
e-mail: saiamrita27@gmail.com

© Springer Nature Singapore Pte Ltd. 2019
A. J. Kulkarni et al. (eds.), *Proceedings of the 2nd International Conference on Data Engineering and Communication Technology*, Advances in Intelligent Systems and Computing 828, https://doi.org/10.1007/978-981-13-1610-4_14

or automated means to verify that it satisfies specified requirements [1]. Moreover, it is often seen that all functions of software are not of equal importance. Therefore, the main goal of testing must be to ensure allocating optimal number of test cases to functions having higher weightage comparing to others.

Previously, researchers provide various methods for test case allocation. Cai identified the use of operational profile in partition testing. However, in his work test case is generated based on total probability formula.

They had not taken operations as an individual parameter for test cases. In 1997, Leung [2] also discovered operational profile-based testing and found it more effective for operations that have high occurrence probability as well as operations that are critical.

Musa [3] has identified that allocation of test cases should be done based on its probability of occurrence and criticality of operation. There are different cases discussed by Musa, among which there may be the case in which operation's priority is higher in spite of low frequency. In Musa's paper, equal number of test cases is assigned to all operations. Kumar and Misra [4] have categorized operations in four parts which are based on four possibilities of frequency and criticality.

In 1997, Leung and Wong presented a framework, in which main emphasis is on operational profile, i.e., operation's occurrence probability. They used weighted criticality method in order to obtain total weight. Arora et al. in 2005 [5] have considered low occurrence and high severity for test case allocation.

A framework for test case allocation is given by Takagi [6]. He used decision table to describe possible allocation. In 2017, Dalal [7] has given an approach for test case allocation. He considered usage, criticality, and interaction. There are some more design and engineering attributes that are required to consider for optimized test case allocation. Previously, no research work considers design attributes for test case allocation like cost complexity, size. In our paper, these design attributes along with occurrence probability and criticality, and a proportionality factor for operations is derived which utilizes scale-based measurement and conventional mathematics. We are taking cost and complexity factor along with occurrence probability and criticality. Cost of an operation is associated with its development cost. Cost of software product will be the cost of its function, thereby cost of program and modules. Moreover, we are considering here only cost of operations at design level as we have to predict test case allocation in early stage. Complexity of an operation is about its functionality. As we need to allocate test cases in early phase, therefore expert assessment is employed to derive subjective assessment for the allocation factors cost (CO), complexity (CM), criticality (Cr), and occurrence probability (OP).

In Sect. 2, proposed methodology is described. Results and validation are discussed in Sect. 3. Conclusion and future scope are presented in Sect. 4.

2 Proposed Methodology

The architecture of proposed methodology is shown in Fig. 1. Step-by-step procedure is discussed for optimal test case allocation. We considered complexity, cost, occurrence probability, and criticality of the function as an allocation factor for obtaining possible failure probability (FP). Based on FP, test cases can be allocated in effective manner. A function having higher failure probability should allocate more number of test cases, and function having lower failure probability should allocate less number of test cases.

2.1 *Expert Judgment on Allocation Factors*

In this step, expert judgment is used for providing subjective assessment of allocation factors. Expert judgment is the consultation of one or more experts. It is assumed that expert judgment-based estimation is the involvement of the experience of one or more people who are aware of the development of software applications similar to that currently being sized.

Here, a fuzzy scale is provided for allocating fuzzy numbers according to the linguistic state of allocation factors. Factor's state can be decided based on differ-

Fig. 1 Architecture of proposed methodology

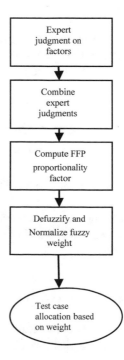

ent ways. History-based project, market survey, similar projects, etc., are the ways through which factor's state can be decided subjectively. In this paper, team of experts is taken for the assessment. Each team has their weight given.

2.2 Combine Expert Judgments

In this step, opinions of all team members are combined. Suppose there are k modules having total n operations, o_1, o_2 ... o_n, that may be evaluated by a team of experts consisting of m members. Factors require a membership function to represent their linguistic state. There are many types of membership functions; however, triangular membership function is best suitable for our purpose. Let CM, CO, Cr, and OP be the allocation factors. We need to represent these factors in the form of triangular membership function. Three linguistic states are defined for each of the allocation factors. These are low, medium, and high. Team consists of m members and satisfies $\sum_{j=1}^{m} hj = 1$ and $hj > 0$.

$$CM = \sum_{j=1}^{m} hjCMij$$

$$= \sum_{j=1}^{m} hjCMijl \sum_{j=1}^{m} hjCMijm \sum_{j=1}^{m} hjCMijh \qquad (1)$$

Equation (1) is used for combining expert opinion on complexity factor. The same equation will be used for remaining factors, and opinions of experts are combined using this equation.

2.3 Fuzzy Failure Probability Proportionality Factor Calculation

2.3.1 Complexity (CM)

Complex operations often get less testing. This increases high chances of failure. It concludes that higher the complexity, it will have high failure probability and vice versa.

2.3.2 Cost (CO)

It is not economical to provide high reliability to the costly operations. Therefore, failure probability of such operations will be high.

2.3.3 Criticality (Cr)

If operation is critical, it must go through efficient testing and therefore less failure probability. Failure probability is inversely proportional to criticality of the function.

2.3.4 Occurrence Probability (OP)

High the occurrence probability, lesser will be failure probability. In a software system, if the operation has high occurrence probability, so it will need high testing; therefore, there are less chances for the failure.

From the above observations, failure probability is given below. This fuzzy failure probability proportionality factor is used to obtain fuzzy weight for operations and given in Eq. 2.

$$\text{FFPP} = \frac{\text{CM} * \text{CO}}{\text{Cr} * \text{OP}} \tag{2}$$

2.4 Defuzzification

For obtaining weight of each operation, we need value in terms of crisp quantity. Conversion of fuzzy into crisp is called defuzzification process [8, 9]. Centroid's method is found to be most effective defuzzification process [10]. Therefore, in this paper for obtaining final output centroid's method is used.

2.5 Test Case Allocation

This is the final step of proposed methodology. Test case allocation is performed with the weight obtained from previous step. In our methodology, it is assumed that total number of test cases is fixed.

$$\text{TCA} \left(O_i \right) = w_i * T \tag{3}$$

$$w_i = \frac{Fi}{\sum_{i=1}^{k} Fi} \tag{4}$$

where

w_i Weightage of ith operation
T Total number of test cases
k Number of operations

This paper aims to allocate test case to that operation that has high probability of failure. It is performed to improve testing as we can predict the priorities before the actual operation.

3 Results and Validation

3.1 Results

This section presents results obtained from the proposed algorithm. A case study has been taken, considering four operations of a software system. These operations are of most importance. Test cases are assigned to operations based on the allocation factors and their expert assessment.

It is clear from Table 1 that failure probability of O3 is highest comparing to other operations; therefore, maximum number of test cases is given to O3.

3.2 Validation

Table 2 shows the comparison among different models. Two models are compared with the proposed model (refer to Fig. 2). There is shifting in precedence of test cases to functions as different models use different parameters for test case allocation; however, total numbers of test cases are fixed. Musa model allocates test cases based on occurrence probability only. In Leung model, test cases are allocated based on occurrence probability and criticality. Proposed model is considering complexity, cost, criticality, and cost of the function. Therefore, by selecting more attributes there will be variation in number of test cases along with possibility of covering maximum aspect.

Table 1 Test case allocation

Operations	Weight (w_i)	No. of test cases
O1	0.29824	149
O2	0.28620	143
O3	0.30464	153
O4	0.110897	55
Σ	1	500

Table 2 Comparison among models

Operations	Occurrence Probability	Criticality	Complexity	Cost	Musa Model	Leung Model	Proposed Model
O1	0.3	0.3	0.2	0.4	15	19	10
O2	0.4	0.2	0.3	0.3	20	17	12
O3	0.1	0.4	0.1	0.2	5	9	6
O4	0.2	0.1	0.4	0.1	10	5	22
Σ	1				50	50	50

Fig. 2 Comparison among different models

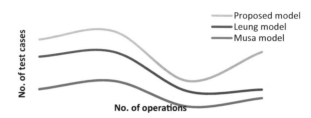

4 Conclusions

A test case allocation methodology has been presented, which allocates test cases based on the weightage of failure probability. Subjective assessment helps to find number of test cases in early phase. This helps to make reliable and quality product. By considering more attributes, we can cover maximum aspects. Quality of test case is also an important aspect. In our future work, we will try to automate our present work and to include other important parameters.

References

1. IEEE Standard Glossary of Software Engineering Terminology, ANSI/IEEE STD-729 (1991)
2. Leung HKN, Wong PWL (1997) A study of user acceptance tests. Software Qual J 6(2):137–149
3. Musa JD (1992) The operational profile in software reliability engineering: an overview. In Proceedings of the 3rd international conference on symposium on software reliability engineering, Research Triangle Park, NC, pp 140–154
4. Kumar KS, Misra RB (2007) Software operational profile based test case allocation using fuzzy logic. Int J Autom Comput 4:388–395
5. Arora S, Misra RB, Kumre VM (2005) Software reliability improvement through operational profile driven testing. In: Proceedings of the 53rd annual reliability and maintainability symposium, pp 621–627
6. Takagi T, Faculty of Engineering, Kagawa University, Takamatsu, Japan, Furukawa Z, Machida Y (2013) Test strategies using operational profiles based on decision tables. In: 2013 IEEE 37th annual computer software and applications conference (COMPSAC), pp 722–723
7. Dalal P, Shah A (2017) Allocation of test cases using severity and data interactivity. Int J Modern Trends Sci Technol 3(1)
8. Ross TJ (2009) Fuzzy logic with engineering applications, 2nd edn. Wiley, New York

9. Amrita, Yadav DK (2016) Software reliability apportionment using fuzzy logic. Indian J Sci Technol 9(45)
10. Bhatnagar R, Ghose MK, Bhattacharjee V (2011) Selection of defuzzification method for predicting the early stage software development effort using Mamdani FIS. In: CIIT, pp 375–381

Seamless Vertical Handover for Efficient Mobility Management in Cooperative Heterogeneous Networks

Mani Shekhar Gupta, Akanksha Srivastava and Krishan Kumar

Abstract The inter-networking cooperation among different wireless networks will help the operators to improve the overall throughput and resource utilization of the wireless networks. During this inter-networking, a mobile node switches among various access networks to satisfy QoS as well as QoE. The next-generation networks require seamless handover among different networks to maintain the connection. In future wireless networks, a mobile node with a multi-interface may have network access from separate service providers using different protocols. Thus, with this heterogeneity environment, spectrum handover decision making must be introduced to take benefit of cooperative support for handover and mobility management in heterogeneous networks. In first stage, an extensive review of the heterogeneous network environment and vertical handover is presented. In second stage, system functionalities and spectrum switching decision making for the vertical handover process are explained. In third stage, architecture and the simulation testbed used for system validation and implementation of the system modules in QualNet simulator are presented.

Keywords Vertical handover · Heterogeneous networks · Mobility management
Inter-networking scenario · Decision making schemes

M. S. Gupta (✉) · K. Kumar
Electronics and Communication Engineering Department,
NIT Hamirpur, Hamirpur, HP, India
e-mail: mi_sr87@yahoo.com

K. Kumar
e-mail: krishan_rathod@nith.ac.in

A. Srivastava
Electronics and Communication Engineering Department, PSITCOE, Kanpur, UP, India
e-mail: akankshasrivastava30@gmail.com

© Springer Nature Singapore Pte Ltd. 2019
A. J. Kulkarni et al. (eds.), *Proceedings of the 2nd International Conference on Data Engineering and Communication Technology*, Advances in Intelligent Systems and Computing 828, https://doi.org/10.1007/978-981-13-1610-4_15

1 Introduction

The users are witnessing the requirement for a network that has capability of handling the traffic in a different perspective. Thus, a pressure is put on the homogeneous network operators to provide seamless handover to the users with both maintained Quality of Service (QoS) and Quality of Experience (QoE). Since wireless network technologies are now becoming a compulsory part of daily work, so this pressure will continue in the coming future. Vertical handover must be performed regardless of the data transmission through network. Service provider may have to combine several networks seamlessly in order to provide uninterrupted services to the users with anywhere, any time, and anyone support. High competitive market competition for users, along with ever-increasing availability of unlicensed networks, like WiFi, MANET, will give mobile users the freedom for network selection to make cost-effective handover decision making. The presence of a heterogeneous wireless network is related with low-cost deployment of low power access points (called relay node) and service provider's approach to provide coverage in small areas at minimum cost [1]. This femtocell-based approach enhances indoor coverage and maintains reliable connectivity without the requirement for the infrastructure-based network which may be cost-inefficient [2, 3]. In highly populated regions, mobile nodes very commonly identify signals from other service providers and access points. In future, these highly dense areas will be served by a combination of networks and form heterogeneous network architecture. Thus, vertical handover and seamless mobility are needed to maintain the same QoS parameters in different–different networks [4].

Different networking protocols and features need specific handover techniques. For example, a mobile node having high mobility, received signal strength is one of the main parameter for triggering handover process. On the other way, for less mobility node, other factors like bit error rate, cost, QoS will be probable factors to initiate handover process.

The rest of this paper is structured as follows. Section 2 explains mobility environment to support vertical handover stages and its technical aspects used for network switching decision making. In Sect. 3, QualNet-simulated sample scenario for mobility management in heterogeneous network and its results are discussed. Finally, the conclusion is drawn from the work given in Sect. 4.

2 Mobility Environment in Heterogeneous Network

The all-IP backbone provides the paths to merge different wireless networks to form heterogeneous network for coming generation. This supports seamless mobile and ubiquitous communication. In a cooperative heterogeneous networks environment, an intelligent mobile node having multiple interfaces can perform handover process seamlessly among different networks to support multimedia real-time services. Thus, future network's mobile node would be capable to move freely across different

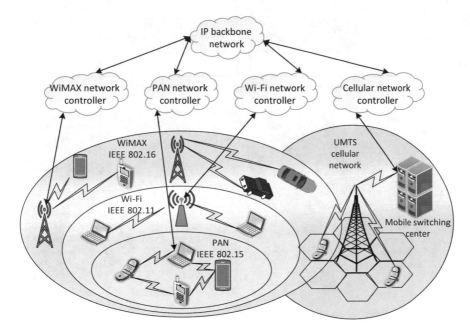

Fig. 1 Cooperative heterogeneous networks environment

wireless access networks like cellular, WiFi, WiMAX, MANET and Bluetooth [5]. Figure 1 represents a heterogeneous roaming scenario network which consists of several different wireless access networks.

2.1 Vertical Handover Phases

Seamless transfer of user's service from existing network to a new network bearing dissimilar radio access technology and protocols is called vertical handover (VHO) [6]. Figure 2 represents three main phases in handover process, namely handover initiation, handover triggering, and handover execution. In first phase, mobile node or an access point starts searching for new available network link; if detected, it

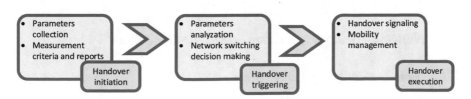

Fig. 2 Handover phases

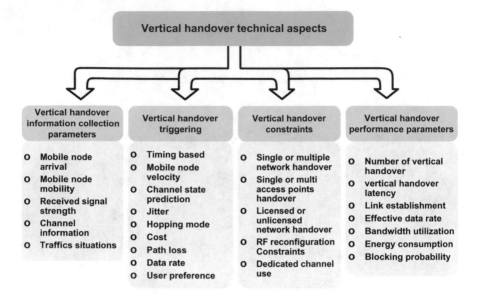

Fig. 3 Various technical aspects of vertical handover

makes the measurements for initiating a handover toward a new network or new access point. In second phase, measurement information is compared with fixed threshold value to take network switching decision whether to perform the handover or not. In third phase, new network or access point is added, all the parameters are adjusted, and active set is updated for mobility management.

2.2 Vertical Handover Technical Aspects

The technical aspects deal with vertical handover-related issues in heterogeneous networks [7]. The vertical handover is the integrating part of mobility management framework. The occurrence of vertical handover initiates with movement of mobile node. The technical aspects related to vertical handover are summarized into four main subgroups as shown in Fig. 3.

2.3 Network Switching Decision-Making Schemes

2.3.1 Simple Additive Weighting (SAW)

SAW is very popular and mostly used in network switching decision making using cost/utility function as an attribute [8]. It can be calculated as

$$K_{\text{SAW}} = \sum_{j=1}^{M} w_j d_{ij} \qquad (1)$$

where w_j represents the weight of jth attribute and d_{ij} denotes the value of jth attribute of the ith network.

2.3.2 Technique for Order Preference by Similarity to an Ideal Solution (TOPSIS)

TOPSIS scheme includes the range from evaluated network to one or several reference networks [9]. The cost or utility function for this scheme is calculated as

$$K_{\text{TOPSIS}} = \frac{T^{\alpha}}{T^{\alpha} + T^{\beta}} \qquad (2)$$

where

$$T^{\alpha} = \sqrt{\sum_{j=1}^{M} w_j^2 \left(d_{ij} - D_j^{\alpha} \right)^2} \quad \text{and} \quad T^{\beta} = \sqrt{\sum_{j=1}^{M} w_j^2 \left(d_{ij} - D_j^{\beta} \right)^2}$$

This equation represents the Euclidean distances from the typical network to the worst and best reference networks, where D_j^{α} and D_j^{β} denote the measurement of jth attribute of the worst and best reference networks, respectively.

2.3.3 Gray Relational Analysis (GRA)

The GRA is an analytical process derived from the Gray system theory to analyze discrete data series [10]. The main difference between TOPSIS and GRA is that, for the calculation of utility function, GRA uses only the best reference network. A utility function is given as

$$K_{\text{GRA}} = \frac{1}{\sum_{j=1}^{M} w_j \left| d_{ij} - D_j^{\beta} \right| + 1} \qquad (3)$$

Data processing, mathematical modeling, decision making, control, prediction, and systems analysis are the main fields of Gray relational analysis.

3 Deployment and Implementation of Scenarios in QualNet

In this work, vertical handover is performed considering two different heterogeneous network scenarios, namely UMTS–WiMAX and WiFi–MANET–WiMAX. Simulation is performed in QualNet software. By calculating the quantity of transferred packets from one mobile node to another mobile node, the performance of transmission control protocol (TCP)-based application is observed. If a large number of packets are transferred, then it shows the high performance of TCP at network layer. For monitoring the performance of user datagram protocol (UDP), the packet loss and end-to-end packets delay variation (called jitter) are analyzed. If these two has minimum value, then it shows the better performance of the UDP.

3.1 Scenario-1 (UMTS–WiMAX Inter-networking)

Figure 4a represents the scenario that consists of 12 nodes. Node 6, 7, 8, and 11 are mobile nodes, and remaining nodes are with dual-radio interfaces acting as a part of UMTS and WiMAX networks. For example, node 3 is represented as radio network controller (RNC), node 12 as a mobile switching center (MSC). All the

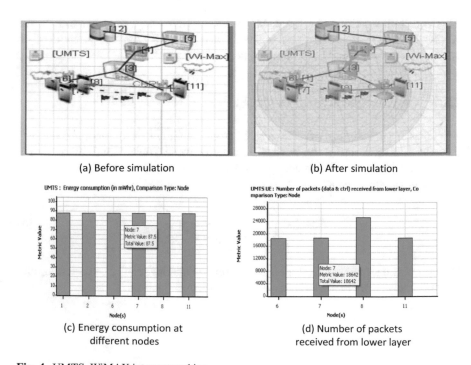

(a) Before simulation   (b) After simulation

(c) Energy consumption at   (d) Number of packets
different nodes   received from lower layer

Fig. 4 UMTS–WiMAX inter-networking

points of attachment are connected through a wired backbone. Node 11 maintains a constant bit rate (CBR) application with node 8. This scenario was simulated as shown in Fig. 4b in order to demonstrate the effectiveness of the implementation. The routing protocol used is Bellman–Ford which is a reactive routing protocol, meaning that it provides a route to a destination node only on demand. Also, the connection setup delay is minimum. Node 11 roams from the area covered by node 8 which is connected with UMTS network. Figure 4c shows that each node consuming almost same energy level which makes it energy-efficient. During this time, vertical handover exists, so each time the packets are received and updated at lower layer as shown in Fig. 4d. In this, RNC node receives these handover requests and triggers a local link-layer command to find out the channel parameters, and when it receives the confirm primitive, it forwards the real-time channel occupancy information to the MSC. For now, the handover is executed by MSC based on the information solely on channel occupancy and received signal strength (RSS).

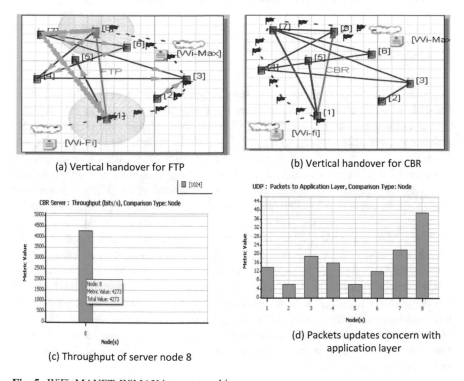

(a) Vertical handover for FTP

(b) Vertical handover for CBR

(c) Throughput of server node 8

(d) Packets updates concern with application layer

Fig. 5 WiFi–MANET–WiMAX inter-networking

3.2 Scenario-2 (WiFi–MANET–WiMAX Inter-networking)

In this scenario, node 8 act as a destination, node 1 as a source, and it uses FTP protocol between them as shown in Fig. 5a. Node 8 is connected to WiMAX network. Node 1 which is connected to WiFi sends data to mobile node 8 via MANET which is formed by node 2-3-4-5-6-7. Once this scenario has established, then mobile node 1 travels toward WiMAX coverage area and vertical handover takes place between WiFi and WiMAX. Now data transmission takes place from node 1 to node 8. In Fig. 5b, same model is simulated for CBR application. Figure 5c represents a throughput measured at node 8. This throughput graph indicates a slight declining trend in case of the traditional vertical handover approach which is desirable for the faster handover. Figure 5d represents the packet updates concerned with application layer. Therefore, all the nodes are participating in packet reception and forwarding to the next node and these nodes are connected with different–different networks. This shows that vertical handover is taking place among the different access networks.

4 Conclusion

Cooperative heterogeneous networks have a great potential to handle inevitable demands of extra spectrum in next-generation wireless networks. In the upgradation to future networks, convergence of IP-based core networks with heterogeneous network is inevitable. Thanks to the rapid growth in communication technologies, the users are becoming more dominant in their traditional choice. In this paper, an overview of heterogeneous network and vertical handover with network switching decision-making schemes is introduced. This work considers a mobility management framework in which two different inter-networking scenarios have been analyzed for the purpose of comparing the performance of vertical handover in UMTS–WiMAX and WiFi–MANET–WiMAX hybrid integrated architectures. Furthermore, empowering best nomadic applications via plug-and-play-type connectivity and always best connected (ABC) network in an autonomous fashion is a new research direction for seamless mobility management in heterogeneous network environment. It is believed that cooperative heterogeneous networks need to learn about network parameters and proper coordination among different networks to work correctly and efficiently with cooperation. Therefore, a common infrastructure is required to provide such heterogeneous networks with information and technology at low cost for real-time multimedia applications.

References

1. Sydir J, Taori R (2009) An evolved cellular system architecture incorporating relay stations. IEEE Commun Mag. 47:115–121 (IEEE)

2. Yeh S, Talwar S, Kim H (2008) WiMAX femtocells: a perspective on network architecture, capacity and coverage. IEEE Commun Mag 46:58–65 (IEEE)
3. Chandrasekhar V, Andrews J, Gatherer A (2008) Femtocell networks: a survey. IEEE Commun Mag 46:56–67 (IEEE)
4. Chuah M (2010) Universal Mobile Telecommunications System (UMTS) Quality of Service (QoS) supporting variable QoS negotiation. Google Patents
5. Buddhikot M, Chandranmenon G, Han S, Lee Y, Miller S, Salgarelli L (2003) Design and implementation of a WLAN/CDMA2000 interworking architecture. IEEE Commun Mag 41:90–100 (IEEE)
6. Lee S, Sriram K, Kim Y, Golmie N (2009) Vertical handoff decision algorithms for providing optimized performance in heterogeneous wireless networks. IEEE Trans Veh Technol 58:865–881 (IEEE)
7. Krishan K, Arun P, Rajeev T (2016) Spectrum handoff in cognitive radio networks: a classification and comprehensive survey. J Netw Comput Appl Elsevier 61:161–181
8. Ramirez C, Ramos V (2013) On the effectiveness of multi-criteria decision mechanisms for vertical handoff. In: IEEE international conference, pp 1157–1164
9. Bari F, Leung V (2007) Multi-attribute network selection by iterative TOPSIS for heterogeneous wireless access. In: Consumer communications and networking conference. IEEE, pp 808–812
10. Song Q, Jamalipour A (2005) Network selection in an integrated wireless LAN and UMTS environment using mathematical modeling and computing techniques. IEEE Wirel Commun 12:42–48 (IEEE)

Sentence Similarity Estimation for Text Summarization Using Deep Learning

Sheikh Abujar, Mahmudul Hasan and Syed Akhter Hossain

Abstract One of the key challenges of natural language processing (NLP) is to identify the meaning of any text. Text summarization is one of the most challenging applications in the field of NLP where appropriate analysis is needed of given input text. Identifying the degree of relationship among input sentences will help to reduce the inclusion of insignificant sentences in summarized text. Result of summarized text always may not identify by optimal functions, rather a better summarized result could be found by measuring sentence similarities. The current sentence similarity measuring methods only find out the similarity between words and sentences. These methods state only syntactic information of every sentence. There are two major problems to identify similarities between sentences. These problems were never addressed by previous strategies provided the ultimate meaning of the sentence and added the word order, approximately. In this paper, the main objective was tried to measure sentence similarities, which will help to summarize text of any language, but we considered English and Bengali here. Our proposed methods were extensively tested by using several English and Bengali texts, collected from several online news portals, blogs, etc. In all cases, the proposed sentence similarity measures mentioned here was proven effective and satisfactory.

Keywords Sentence similarity · Lexical analysis · Semantic analysis
Text summarization · Bengali summarization · Deep learning

S. Abujar (✉) · S. A. Hossain
Department of Computer Science and Engineering, Daffodil International University,
Dhanomondi, Dhaka 1205, Bangladesh
e-mail: sheikh.cse@diu.edu.bd

S. A. Hossain
e-mail: aktarhossain@daffodilvarsity.edu.bd

M. Hasan
Department of Computer Science and Engineering,
Comilla University, Comilla 3506, Bangladesh
e-mail: mhasanraju@gmail.com

© Springer Nature Singapore Pte Ltd. 2019
A. J. Kulkarni et al. (eds.), *Proceedings of the 2nd International Conference on Data Engineering and Communication Technology*, Advances in Intelligent Systems and Computing 828, https://doi.org/10.1007/978-981-13-1610-4_16

1 Introduction

Text summarization is a tool that attempts to provide a gist or summary of any given text automatically. It helps to understand any large document in a very short time, by getting the main idea and/or information of entire text from a summarized text. To produce the proper summarization, there are several steps to follow, i.e., lexical analysis, semantic analysis, and syntactic analysis. Possible methods and research findings regarding sentence similarity are stated in this paper. Bengali language has very different sentence structure and analyzing those Bengali alphabets may found difficult in various programming platforms. The best way of starting for preprocessing both Bengali and English sentences, initially need to convert into Unicode [2]. Sentence could be identified in a standard form, and it will help to identify sentence or words structure as needed. The degree of measuring sentence similarity is being measured by method of identifying sentence similarity as well as large and short text similarity. Sentence similarity measures should state information like: If two or more sentences are either fully matched in lexical form or in semantic form, sentence could be matched partially or we could found any leading sentence. Identifying centroid sentence is one of the major tasks to accomplish [1]. Few sentences can contain some major or important words which may not be identified by words frequency. So, only depending on word frequency may not always provide the expected output, though several times most frequent words may relate to the topic models. Meaningfully same but structurally different sentences have to avoid while preparing a better text summarizer [3]. But related or supporting sentences may add a value to the leading sentences [4]. Finally, most leading sentence and relationship between sentences could be determined.

In this paper, we have discussed several important factors regarding assessing sentence and text similarity. Major findings are mentioned in details, and more importantly potential deep learning methods and models were stated here. Several experimental results were stated and explained with necessary measures.

2 Literature Review

The basic feature of text summarization would be either abstractive or extractive approach. Extractive method applies several manipulation rules over word, sentence, or paragraph. Based on weighted values or other measures, extractive approach chooses appropriate sentence. Abstractive summarization requires several weights like sentence fusion, constriction, and basic reformulation [5].

Oliva et al. [6] introduced a model SyMSS, which measure sentence similarity by assessing, how two different sentences systaltic structure influence each other. Syntactic dependence tree help to identify the rooted sentence as well as the similar sentence. These methods state that every word in a sentence has some syntactic connections and this will create a meaning of every sentence. The combination of LSA

[7] and WordNet [9] to access the sentence similarity in between every word was proposed in Han et al. [8]. They have proposed two different methods to measure sentence similarity. First one makes a group of words—known as the align-and-penalize approach, and the second one is known as SVM approach, where the method applies different similarity measures using n-gram and by using support vector regression (SVR), and they use LIBSVM [10] as another similarity measure.

A threshold-based model always returns the similarity value between 0 and 1. Mihalcea et al. [11] represent all sentences as a list of bag of words vector, and they consider first sentence as a main sentence. To identify word-to-word similarity measure, they have used highest semantic similarity measures in between main sentence and next sentence. The process will continue repeated times until the second main sentence could be found, during this process period. Das and Smith introduced a probabilistic model which states syntax and semantic-based analysis. Heilman and Smith [12, 13] introduce as new method of editing tree, which will contain syntactic relations between input sentences. It will identify paraphrases also. To identify sentence-based dissimilarity, a supervised two-phase framework has been represented using semantic triples [14]. Support vector machine (SVM) can combine distributional, shallow textual, and knowledge-based models using support vector regression model.

3 Proposed Method

This section represents a new proposed sentence similarity measuring model for English and Bengali language. The assessing methods, sentence representation, and degree of sentence similarity have been explained in detail. The necessary steps required especially for Bangla language have been considered while developing the proposed model. This model will work for measuring English and Bengali sentence similarity. The sentence structure and lexical form are very different for Bangla language. The semantic and syntactic measures also can add more values in this regard. The concept of working with all those necessary steps will help to produce better output, in every aspect. In this research, lexical methods have been applied and untimely a perfect expected result has been found.

3.1 Lexical Layer Analysis

The lexical layer has few major functions to perform, such as lexical representation and lexical similarity. Both of these layers have several other states to perform. Fig. 1 is the proposed model for lexical layer.

Figure 1 introduces the sentence similarity measures for lexical analysis. Different sentences will be added into a token. A word-to-word and sentence-to-sentence analyzer will perform together. An order vector will add all those words and/or sentences

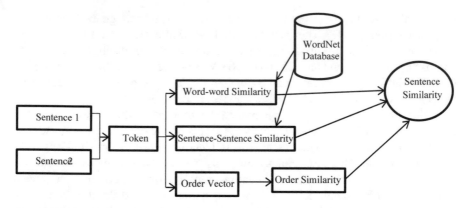

Fig. 1 Lexical layer analysis model introduces different layers of proposed methods

order in a sequence based on similarity measures. With the reference of weighted sum, the order of words and sentence will be privileged. A WordNet database will send lexical resources to word-to-word and sentence-to-sentence processes. Ultimately based on the order preference, the values from three different states (word–word similarity, sentence–sentence similarity, and order similarity) will generate the similar sentence output. The methods were followed by one of the popular deep learning algorithm—text rank.

(a) *Lexical Analysis*: This state splits sentence and words into different tokens for further processing.
(b) *Stop Words Removal*: Several values hold representative information such as article, pronoun. These types of words could be removed while considering text analysis.
(c) *Lemmatization*: This is a step to convert and/or translates each and every token into a basic form, and exactly from where it belongs to the very same verb form in the initial form.
(d) *Stemming*: Stemming is the state of word analysis. Word–word and sentence-to-sentence both methods need all their contents (text/word) in a unique form. Here every word will be treated as a rooted word such as play, player—both words are different as word though in deep meaning those words could be considered as branch words of the word "Play." By using a stemmer, we could have found all those texts in a unique form before further processing. The confusion of getting different words in structure but same in inner meaning will reduce. So, it is a very basic part of text preprocessing modules.

Figure 2 states how lexical steps had been processed with appropriate example. All the necessary processes as lexical analysis, stop words removal, and stemming had been done as per the mentioned process. Those sentences will be used for further experiments in this paper.

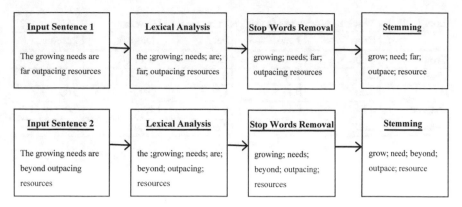

Fig. 2 Lexical layer processing of input sentences. It was clearly shown how multiple process handle single input data

3.2 Sentence Similarity

Path measure helps to sense the relatedness of words from the hierarchies of WordNet. It calculates and replies to the path distance between two words. Path measure will be used to identify similarity scores between two words. Path measure could be calculated through Eq. (1).

$$Path\_measure(\text{token1}, \text{token2}) = 1/Path\_length(\text{token1}, \text{token2}) \qquad (1)$$

Path_measure will send two different tokens as token1 and token2. Both tokens are assigned the value of a single sentence after splitting. *Path_length* will return the distance of two different concepts from WordNet.

Levenshtein Distance (Lev.) algorithm has been used to identify the similarity matrix in between of words. To identify the sentence similarity, measuring words similarly pay more importance. Lev. counts the minimum number of similarity required for the operation of insertion, deletion, and modification of every character which may require transforming from a sentence to another sentence. Here it was used to identify distance and/or similarity measure between words. Longest common subsequences (LCS) have also implemented though expected output was found using Lev. Here LCS does not allow substitutions. The distance of sentences followed by Lev. will be calculated based on Eq. (2).

$$LevSim = 1.0 - (\text{Lev.Distance}(W1, W2)/\text{maxLength}(W1, W2)) \qquad (2)$$

The degree of relationship helps to produce a better text summarizer by analyzing text similarity. The degree of measurement could be word–word, word–sentence, sentence–word, and sentence–sentence. In this research, we had discussed the similarity between two different words. Such as there are a set of words (after splitting

Table 1 Similarity score between words using path measure and LevSim

	Grow	Need	Far	Outpace	Resource
Grow	1.00	0.25	0.00	0.14	0.00
Need	0.25	1.00	0.11	0.17	0.14
Far	0.00	0.11	1.00	0.00	0.09
Outpace	0.14	0.17	0.00	1.00	0.00
Resource	0.00	0.14	0.09	0.00	1.00

every individual sentence): $W = \{W1, W2, W3, W4, ..., Wn\}$. Lev. distance calculates the distance between two words: $W1$ and $W2$, and max length will reply the score of maximum character found in between $W1$ and $W2$. Only similarity will be checked between two different words. The similarity between words could be measured by Algorithm 1.

Algorithm 1 Similarity between Words
```
1:      W1= Sentence1.Split(" ")
2:      W2= Sentence2.Split(" ")
3:              if Path_measure(W1,W2) < 0.1 then
4:                      W_similarity= LevSim(W1,W2)
5:      else
6:                      W_similarity = Path_measure(W1,W2)
7:      end if
```

In Algorithm 1, the value of path will be dependent of distance values and Lev Similarity (LevSim) value could be found from Eq. 1. The words similarity score less than 0.1 will be calculated through the LevSim method, else the score will be accepted form the path measure algorithm. W_similarity will receive similarity score of between two words. The range of maximum and minimum score is in between $\{0.00 - 1.00\}$. Table 1 represents the similarity value of words from sentence 1.

Wu and Palmer measure (WP) use the WordNet taxonomy to identify the global depth measures (relatedness) of two similar or different concepts or words by measuring edge distance as well as will calculate the depth of longest common subsequences (LCS) value of those two inputs. Based on Eq. (3), WP will return a relatedness score if any relation and/or path exist in between on those two words, else if no path exist—it will return a negative number. If the two inputs are similar, then the output from synset will only be 1.

$$\text{WP\_Score} = 2 * \text{Depth(LCS)} / (\text{depth}(t1) + \text{depth}(t2)) \qquad (3)$$

In Eq. 3, $t1$ and $t2$ are token of sentence 1 and sentence 2. Table 2 states the WP similarity values of given input (as mentioned in Fig. 2).

Lin measure (Lin.) will calculate the relativeness of words or concepts based on information content. Only due to lack of information or data, output could become zero. Ideally, the value of Lin would be zero when the synset value is the rooted node.

Table 2 Similarity score between words using Wu and Palmer measure (WP)

	Grow	Need	Far	Outpace	Resource
Grow	1.00	0.40	0.00	0.25	0.00
Need	0.40	1.00	0.43	0.29	0.57
Far	0.00	0.43	1.00	0.00	0.38
Outpace	0.25	0.29	0.00	1.00	0.00
Resource	0.00	0.57	0.38	0.00	1.00

Table 3 Similarity score between words using Lin measure (Lin.)

	Grow	Need	Far	Outpace	Resource
Grow	1.00	0.40	0.00	0.25	0.00
Need	0.40	1.00	0.43	0.29	0.57
Far	0.00	0.43	1.00	0.00	0.38
Outpace	0.25	0.29	0.00	1.00	0.00
Resource	0.00	0.57	0.38	0.00	1.00

But if the frequency of the synset is zero, then the result will also be zero but the reason will be considered as lack of information or data. Equation (4) will be used to measure the Lin. value, and Table 3 will state the output values after implementing the input sentences on Lin. measures.

$$\text{Lin\_Score} = 2 * \text{IC(LCS)} / (\text{IC}(t1) + \text{IC}(t2)) \tag{4}$$

In Eq. 4, IC is the information content.

A new similarity measure algorithm was experimented where all those mentioned algorithm and/or methods will be used. Equation (5) states the new similarity measure process.

$$\text{total\_Sim}(t1, t2) = (\text{Lev\_Sim}\ (t1, t2) + \text{WP\_Score}(t1, t2) + \text{Lin\_Score}(t1, t2)) / 3 \tag{5}$$

In Eq. 5, a new total similarity values will be generated based on all mentioned lexical and semantic analysis. Edge distance, global depth measure, and analysis of information content are very much essential. In that purpose, this method has applied and experimented out is shown in Table 4.

Algorithm 2 A proposed similarity algorithm

1: *matrix=newmatrix(size(X)*size(Y))*
2: *total_sim=0*
3: *i=0*
4: *j=0*
5: *for i∈ A do*

Table 4 New similarity score

	Grow	Need	Far	Outpace	Resource
Grow	1.00	0.21	0.00	0.13	0.00
Need	0.21	1.00	0.18	0.15	0.34
Far	0.00	0.18	1.00	0.00	0.15
Outpace	0.13	0.15	0.00	1.00	0.00
Resource	0.00	0.34	0.15	0.00	1.00

```
 6:  for j ∈ B do
 7:      matrix(i, j)=similarity_token(t1,t2)
 8:  end for
 9:  end for
10:  for has_line(matrix) and has_column(matrix) do
11:      total_Sim=(Lev_Sim(matrix)+WP_Score(matrix)+Lin_Score(matrix))/3
12:  end for
13:  return total_Sim
```

The Algorithm 2 receives the token on two different X, Y as input text. Then it will create a matrix representation of $m * n$ dimensions. Variable total_sim (total similarity) and i, j (which are the values for iteration purpose) will initially become 0. Initially, matrix(i, j) will generate the token matrix, where values will be added. The variable total_sim will record and update calculate the similarity of pair of sentences based on token matrix—matrix(i, j).

4   Experimental Results and Discussion

Several English and Bengali texts were tested though the proposed lexical layer to find out the sentence similarity measure. Texts are being collected from online resource, for example, wwwo.prothom-alo.com, bdnews24.com, etc. Our python Web crawler initially saved all those Web (html content) data into notepad file. We have used Python—Natural Language Toolkit (NLTK: Version–3) and WS4J (a Java API, especially developed for WordNet use). All the experimented results are stated in this section.

Tables 1, 2, and 3 state that the experimented result of similarity measure by using path measure and LevSim, Wu and Palmer measure (WP), and Lin measure (Lin.) consecutively. All those methods are either applied in lexical analysis or semantic analysis. In this research article, the proposed method of identifying sentence similarity using a hybrid model is being stated in Table 4.

This method was also applied in Bengali language using Bengali WordNet. Experimented results are shown in Table 5.

Table 5 New similarity score (applied in Bengali sentence)

	ট্রেন	সিট	ভাড়া	গন্তব্য
ট্রেন	1	0.78	0.88	0.16
সিট	0.78	1	0.31	0.24
ভাড়া	0.88	0.31	1	0.23
গন্তব্য	0.16	0.24	0.23	1

5 Conclusion and Further Work

This paper has presented sentence similarity measure using lexical and semantic similarity. Degree of similarity was mentioned and implemented in the proposed method. There are few resources available for Bengali language. More development on Bengali language is just more than essential. Bengali WordNet is not stable as like other WordNet available for English language. This research found suitable output in the unsupervised approach, though a huge dataset will be required to implement the supervised learning methods. There are other sentence similarity measures and could be done by more semantic analysis and syntactic analysis. Both of these analyses if could be done together including lexical similarities, a better result could be found. More importantly, for a better text summarizer, we need to identify the leading sentences. Centroid sentences could optimize the analysis of post-processing of text summarization. Evaluating system developed summarizer before publishing as final form is more important. Backtracking methods could possibly be a good solution in his regards.

Acknowledgements We would like to thanks Department of Computer Science and Engineering of two universities: Daffodil International University and Comilla University, Bangladesh, for facilitating such joint research. Special thanks to DIU-NLP and Machine Learning Research Laboratory for providing research facilities.

References

1. Ferreira R et al (2013) Assessing sentence scoring techniques for extractive text summarization. Expert Syst Appl 40:5755–5764 (Elsevier Ltd.)
2. Abujar S, Hasan M (2016) A comprehensive text analysis for Bengali TTS using unicode. In: 5th IEEE international conference on informatics, electronics and vision (ICIEV), Dhaka, Bangladesh, 13–14 May 2016
3. Abujar S, Hasan M, Shahin MSI, Hossain SA (2017) A heuristic approach of text summarization for Bengali documentation. In: 8th IEEE ICCCNT 2017, IIT Delhi, Delhi, India, 3–5 July 2017
4. Lee Ming Che (2011) A novel sentence similarity measure for semantic-based expert systems. Expert Syst Appl 38(5):6392–6399
5. Mani I, Maybury MT (eds) (1999) Advances in automatic text summarization, vol 293. MIT Press, Cambridge, MA

6. Oliva, J et al (2011) SyMSS: a syntax-based measure for short-text semantic similarity. Data Knowl Eng 70(4):390–405
7. Deerwester S, Dumais ST, Furnas GW, Landauer TK, Harshman R (1990) Indexing by latent semantic analysis. J Am Soc Inf Sci 41(6):391–407
8. Han L, Kashyap AL, Finin T, Mayfield J, Weese J (2013) UMBC EBIQUITY-CORE: semantic textual similarity systems. Volume 1, Semantic textual similarity, Association for Computational Linguistics, Atlanta, Georgia, USA, June, pp 44–52
9. Miller GA (1995) Wordnet: a lexical database for English. Commun ACM 38:39–41
10. Chang C-C, Lin C-J (2011) LIBSVM: a library for support vector machines. ACM Trans Intell Syst Technol 2(3):27, 1–27
11. Mihalcea R, Corley C, Strapparava C (2006) Corpus-based and knowledge-based measures of text semantic similarity. National conference on artificial intelligence, vol 1. AAAI Press, Boston, MA, pp 775–780
12. Heilman M, Smith NA (2010) Tree edits models for recognizing textual entailments, paraphrases, and answers to questions. In: Annual conference of the North American chapter of the association for computational linguistics. Association for Computational Linguistics, Stroudsburg, PA, USA, pp 1011–1019
13. Heilman M, Smith NA (2010) Tree edit models for recognizing textual entailments, paraphrases, and answers to questions. In: Human Language Technologies, Stroudsburg, PA, USA, pp 1011–1019
14. Qiu L, Kan M-Y, Chua T-S, (2006) Paraphrase recognition via dissimilarity significance classification, EMNLP. Association for Computational Linguistics, Stroudsburg, PA, USA, pp 18–26

Minimization of Clearance Variation of a Radial Selective Assembly Using Cohort Intelligence Algorithm

Vinayak H. Nair, Vedang Acharya, Shreesh V. Dhavle, Apoorva S. Shastri and Jaivik Patel

Abstract Cohort intelligence is a socio-inspired self-organizing system that includes inherent, self-realized, and rational learning with self-control and ability to avoid obstacles (jumps out of ditches/local solutions), inherent ability to handle constraints, uncertainty by modular and scalable system and robust (immune to single point failure). In this method, a candidate self-supervises his/her behavior and adapts to the behavior of another better candidate, thus ultimately improving the behavior of the whole cohort. Selective assembly is a cost-effective approach to attaining necessary clearance variation in the resultant assembled product from the low precision elements. In this paper, the above-mentioned approach is applied to a problem of hole and shaft assemblies where the objective is to minimize the clearance variation and computational time. The algorithm was coded and run in MATLAB R2016b environment, and we were able to achieve convergence in less number of iterations and computational time compared to the other algorithms previously used to solve this problem.

Keywords Selective assembly · Clearance variation · Cohort intelligence Optimization

1 Introduction

Variability in manufacturing process and production is inevitable. Assembling parts are known as mating parts. There are two mating parts in a radial assembly, namely male and female parts. However, in a complex radial assembly, more than one hole or shaft can be present. In this study, a hole (male) and a shaft (female) are considered.

V. H. Nair · V. Acharya · S. V. Dhavle (✉) · A. S. Shastri · J. Patel
Symbiosis Institute of Technology, Symbiosis International University, Pune 412115, Maharashtra, India
e-mail: shreesh.dhavle@sitpune.edu.in

V. H. Nair
e-mail: vinayak.nair@sitpune.edu.in

© Springer Nature Singapore Pte Ltd. 2019   165
A. J. Kulkarni et al. (eds.), *Proceedings of the 2nd International Conference on Data Engineering and Communication Technology*, Advances in Intelligent Systems and Computing 828, https://doi.org/10.1007/978-981-13-1610-4_17

For interchangeable manufacturing, mating parts are assembled in random order, which results in their clearance variation being the sum of tolerances of both parts. Optimization is done for finding the best grouping of selective assembly. In order to minimize the clearance variation conventionally, manufacturing tolerances needed to be reduced by improving the process or the machine. Therefore, selective assembly is considered to be the best possible method for obtaining minimum clearance variation, as it is economical and quite simple.

Mansoor [15] categorized and minimized the problem of mismatching for selective assembly. Pugh [17] recommended a methodology for segregating population from the mating parts. Pugh [18] presented a technique to shortlist components having large variance. He identified and analyzed the effects of three sources of error produced while applying this method. Fang and Zhang [3] proposed an algorithm by introducing two prerequisite principles to match probabilities of mating parts. They found an effective approach to avoid the unforeseen loss of production cost. This method is appropriate when the required clearance is more than the difference in standard deviations of parts. Chan and Linn [1] developed method for grouping the parts having dissimilar distribution to ensure that mating parts' probability is equal to the equivalent groups maintaining dimensional constraints. It also minimizes the production of excess parts and uses another concept of skipping certain portions of shafts or holes so that more mating groups can be made.

Kannan and Jayabalan [4] analyzed a ball bearing assembly with three mating parts (inner race, ball, and outer race) and suggested a method of grouping complex assemblies to minimize excess parts and to satisfy the clearance requirements. Kannan and Jayabalan [5] investigated linear assembly having three parts and suggested method for finding smaller assembly variations having minimum excessive parts. Kannan et al. [7] proposed two methods (uniform grouping and equal probability) to achieve group tolerances which are uniform and to avoid surplus parts in method 1. The group tolerances are designed such that they satisfy the clearance requirements, and the excess parts are minimalized in method 2. Kannan et al. [6] have used GA to select the best combination to minimize the clearance variation and excess parts when the parts are assembled linearly for selective assembly.

This paper is arranged as follows: Section 2 provides the details about mathematical model of selective assembly system that has been used for optimization. Section 3 provides the framework of CI algorithm. Section 4 comprises of the results and discussion that are obtained by optimization of selective assembly. Conclusion and the future scope are given in Sect. 5.

2 Selective Assembly System

Selective assembly is a cost-effective approach to attaining necessary clearance variation in the resultant assembled product from the low precision elements. More often, due to manufacturing limitations and errors, the individual elements have high tolerance limits. Clearance is the difference in the dimensions of the elements. Clearance is one of the zenith priorities for the determination of the quality of the output product. The mating parts characteristically have different standard deviations as a virtue of their manufacturing processes. When from these elements, high-precision assemblies need to be manufactured, and the concept of selective assembly is adopted. The most important benefit of this concept is the low-cost association with the process.

Pugh [17] proposed the idea of partitioning the universal set of the mating part population into a number of groups. Generally, the mating parts include a female part and a male part (hole and shaft respectively in this case). After the creation of groups of these parts, assembly is done by the combination of these individual groups, thereby reducing the overall clearance variation of the assembly. It is a known fact that the actual manufactured product dimensions generally have a normal distribution with the standard deviation of 6σ. However, for very high-precision applications like aerospace. 8σ may also be used [6].

For minimizing clearance variation and reducing computational time, a new system of selective assembly is proposed here to solve the problem of hole and shaft assemblies. Instead of assembling corresponding groups, different combinations of the selective groups are assembled, and the best combination is found using CI.

Each of the groups will have a specific tolerance corresponding to them for female and male elements. The mathematical formulation of the desired clearance is provided below:

$$\text{Max. Clearance} = (N \times (\text{Group clearance width of hole}))$$
$$+ (M \times (\text{Group clearance width of shaft})) \tag{1}$$

$$\text{Min. Clearance} = ((N - 1) \times (\text{Group clearance width of hole}))$$
$$+ ((M - 1) \times (\text{Group clearance width of shaft})) \tag{2}$$

$$\text{Overall Clearance} = \text{Max. Clearance} - \text{Min. Clearance} \tag{3}$$

where N represents the group number of the hole used for combining and M represents the group number of the shaft used for combining, for example, for group 3:

$$\text{Max. Clearance} = (3 \times 2) + (3 \times 3) = 15 \tag{4}$$

$$\text{Min. Clearance} = (2 \times 2) + (2 \times 3) = 10 \tag{5}$$

Similarly, we find the remaining clearance ranges as follows:

Group 1—5 mm; group 2—10 mm; group 3—15 mm; group 4—20 mm; group 5—25 mm; group 6—30 mm.

In selective assembly (refer to Fig. 1), the groups made can be assembled interchangeably [6] (i.e., first group of holes may be assembled with fifth group of shafts). In our case, we have considered six group categorization and group width of 3 and 2 mm for hole and shaft, respectively.

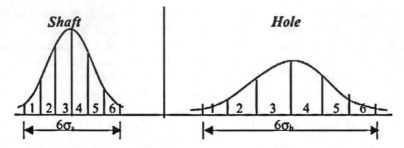

Fig. 1 Traditional method of selective assembly

3 Cohort Intelligence Algorithm

Cohort intelligence is a recent ramification of artificial intelligence. The algorithm is inspired from the behavior of self-supervised candidates such as natural and social tendency in cohort, which was represented in Kulkarni et al. [13]. The algorithm refers to learning of self-supervised behavior of candidates in a cohort and adapts to behavior of other candidates, which it intends to follow. In this, interaction of candidates takes place, which helps the candidates to learn from one another. The candidates update/improve their behavior, ultimately improving behavior of entire candidates in cohort by embracing qualities of other candidates. Roulette wheel approach is used for the selection of candidate to be followed, interact, and compete with. This in turn allows the candidates to follow candidates having either best or worst behavior. The convergence of candidates in cohort takes place when all the candidates reach to the optimal solution using the above technique. The detailed procedure of CI algorithm can be found in Dhavle et al. [2] and Kulkarni et al. [13].

Krishnasamy et al. [8] have used a data clustering, hybrid evolutionary algorithm combining K-mean and modified cohort intelligence. In UCI Machine Learning Repository, various data sets were considered and its performances were noted down. Using hybrid K-MCI, they were able to achieve promising results having quality solutions and good convergence speed. Kulkarni et al. [9, 11] proposed and applied metaheuristic of CI to two constraint handling approaches, namely static (SCI) and dynamic (DCI) penalty functions. Also, mechanical engineering domain-related three real-world problems were also improved to obtain optimal solution. The robustness and applicability of CI were compared with the help of different nature-inspired optimization algorithm. CI has recently been applied to fractional proportional-integral-derivative (PID) controller parameters. Shah et al. [20] designed the fractional PID controller with the help of CI method and compared the validation of the result with the help of existing algorithms such as PSO, improved electromagnetic algorithm, and GA, in which best cost function, function evaluation, and computational time in comparison with other algorithms, which inherently demonstrated the robustness of CI method in control domain also, is yielded. Patankar and Kulkarni [16] proposed seven variations of CI and validated for three uni-modal unconstraint and seven multimodal test functions. This analysis instigated the strategy for working in a cohort. The choice of the right variation may also further opens doors for CI to solve different real-world problems. Shastri et al. [21] used CI approach for solving probability-based constrained handling approach and some inequality-based constrained handling problems. CI was found to be highly efficient by comparing its performance for robustness, computational time, and various static parameters and its further usage in solving more real-world problems. Kulkarni and Shabir [12] have applied CI to solve NP-hard combinatorial problem, in which the problem was successfully solved showing its efficiency and improved rate of convergence. Hence, it has also been used to solve some combinatorial problems. Dhavle et al. [2] applied the CI method for optimal design and inherently reducing the total cost of the shell and tube heat exchanger. Promising results were obtained for all the three cases, which were considered for optimizing the total cost of the complex design of STHE and obtaining least computational time and function evaluation in comparison with other algorithms, showing its applicability and robustness for different complex systems in mechanical engineering domain. Sarmah and Kulkarni [19] applied CI in two steganographic techniques, which use JPEG compression on gray scale image to hide secret text. In this, CI provides good result, in comparison with other algorithms, and reveals the hidden secret text. Kulkarni et al. [10] have applied CI algorithm to various problems such as travelling salesman problem, sea cargo mix problem, and cross-border shippers problem. The flowchart of the CI algorithm applying for selective assembly is given in Fig. 2.

4 Results and Discussions

Two methods were used to obtain the optimized clearance variance. One was adopting CI while using roulette wheel (Table 1), and the other was without using the roulette wheel (Table 2). For comparison, four cases are considered for each method, and these four cases included variation in the number of candidates that are 5, 10, 15, and 20. The number of iterations is kept 500 as constant throughout the study. The algorithm code was written in MATLAB R2016b running on Windows 10 operating system using Intel Core i3, 2.7 GHz processor with 4 GB of RAM.

In the case of non-roulette wheel, the candidates follow the best-performing candidate. In case their performance does not improve, the candidate goes back to its original behavior; i.e., if the updated combination of the candidate yields worse clearance behavior, it goes back to its original combination. However, in the case of using roulette wheel, there are no criteria to follow the best-performing candidate. Therefore, we may get a lower value of clearance variation in one solution, but because of randomization, the next solution may tend to increase the clearance variation value. In spite of this, convergence is achieved fairly faster than before and is highly reliable for higher number of candidates. Please note that the element of a candidate to be changed are randomised in both cases.

Fig. 2 Flowchart for
application of CI to selective
assembly

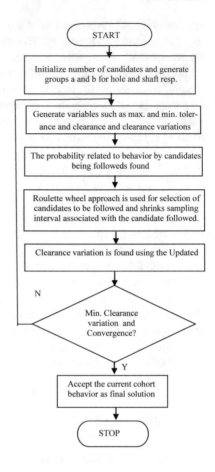

Table 1 Using roulette wheel concept

No. of candidates	Average clearance variation	Average time (s)	Standard deviation of clearance variation	Standard deviation time (s)
5 candidates	16	0.09291	4.31643	0.01197
10 candidates	12	0.23381	1.80642	0.01138
15 candidates	10.7	0.39031	0.86451	0.07526
20 candidates	10.45	0.63186	0.88704	0.09519

In case of using the roulette wheel, we can see that in Fig. 3, the initial value of minimum clearance variation is 19. However, because of no limiting criteria and randomization, this value shoots up and down before converging at the lowest possible value of 10.

In the case of not using the roulette wheel, we can see that in Fig. 4, the initial value of minimum clearance variation is 21. However, because the algorithm follows the

Table 2 Without roulette wheel concept

No. of candidates	Average clearance variation	Average time (s)	Standard deviation of clearance variation	Standard deviation of time (s)
5 candidates	11.7	0.08654	1.55932	0.03138
10 candidates	10.3	0.12579	0.57124	0.10052
15 candidates	10	0.16623	0	0.00829
20 candidates	10	0.23111	0	0.03728

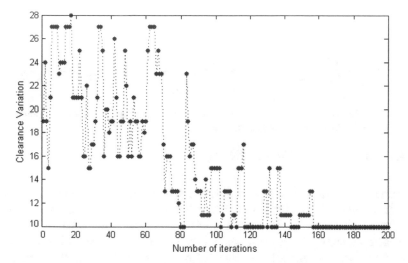

Fig. 3 Graph depicting minimum value of clearance variation for each iteration using roulette wheel for 20 candidates and 200 iterations

best-performing candidate, we see that the solution improves with each successive iteration and converges at the lowest possible value of 10. Moreover, because of this, the convergence is achieved in lesser number of iterations. We can clearly infer from the tables and the figures that as we keep on increasing the number of candidates, we tend to get a near-perfect value of clearance variation at convergence. One of the primary reasons for this occurrence is that the solution depends on having a candidate with low clearance variation in the initial set. Because of randomization, the chances of getting this are lower if you have lower number of candidates. Therefore, as we keep on increasing the number of candidates, the chances of having a lower clearance variation candidate increases, hence improving our overall solution. However, in reality, we will be solving this model with hundreds or thousands of candidates, therefore making it a very reliable tool to use.

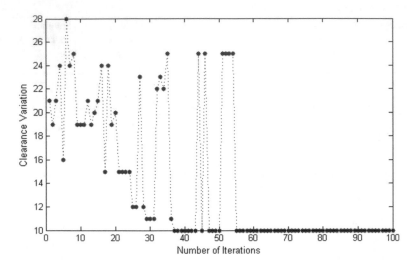

Fig. 4 Graph depicting minimum value of clearance variation for each iteration without using roulette wheel for 20 candidates and 100 iterations

5 Conclusion and Future Scope

Two variations of cohort intelligence are used to obtain the optimized clearance variation. It is noteworthy that this minimum clearance is a function of group width of both, male and female mating elements of the assembly. This is better than interchangeable assembly, as in that case, it is the sum of both tolerances which will always be greater than the solution obtained by this method. One of the most important aspects of this method is the extent up to which it curbs the computational time. It was also observed that on increasing the number of candidates, premature convergence is avoided, thus resulting in better outputs showing the robustness of CI algorithm. In the future, algorithms can be modified to achieve optimal solution using lesser number of candidates. This method can further be extended for linear selective assemblies. Also, further variations of CI can be applied to solve similar problems.

References

1. Chan KC, Linn RJ (1998) A grouping method for selective assembly of parts of dissimilar distributions. Qual Eng 11(2):221–234
2. Dhavle SV, Kulkarni AJ, Shastri A, Kale IR (2016) Design and economic optimization of shell-and-tube heat exchanger using cohort intelligence algorithm. Neural Comput Appl 1–15
3. Fang XD, Zhang Y (1995) A new algorithm for minimising the surplus parts in selective assembly. Comput Ind Eng 28(2):341–350
4. Kannan SM, Jayabalan V (2001) A new grouping method to minimize surplus parts in selective assembly for complex assemblies. Int J Prod Res 39(9):1851–1863

5. Kannan SM, Jayabalan V (2002) A new grouping method for minimizing the surplus parts in selective assembly. Qual Eng 14(1):67–75
6. Kannan SM, Asha A, Jayabalan V (2005) A new method in selective assembly to minimize clearance variation for a radial assembly using genetic algorithm. Qual Eng 17(4):595–607
7. Kannan SM, Jayabalan V, Jeevanantham K (2003) Genetic algorithm for minimizing assembly variation in selective assembly. Int J Prod Res 41(14):3301–3313
8. Krishnasamy G, Kulkarni AJ, Paramesran R (2014) A hybrid approach for data clustering based on modified cohort intelligence and K-means. Expert Syst Appl 41(13):6009–6016
9. Kulkarni AJ, Baki MF, Chaouch BA (2016) Application of the cohort-intelligence optimization method to three selected combinatorial optimization problems. Eur J Oper Res 250(2):427–447
10. Kulkarni AJ, Krishnasamy G, Abraham A (2017) Cohort intelligence: a socio-inspired optimization method. Intelligent Systems Reference Library, vol 114. https://doi.org/10.1007/978-3-319-44254-9
11. Kulkarni O, Kulkarni N, Kulkarni AJ, Kakandikar G (2016) Constrained cohort intelligence using static and dynamic penalty function approach for mechanical components design. Int J Parallel Emergent Distrib Syst 1–19
12. Kulkarni AJ, Shabir H (2016) Solving 0–1 knapsack problem using cohort intelligence algorithm. Int J Mach Learn Cybernet 7(3):427–441
13. Kulkarni AJ, Durugkar IP, Kumar M (2013) Cohort intelligence: a self-supervised learning behavior. In: 2013 IEEE international conference on systems, man, and cybernetics (SMC). IEEE, pp. 1396–1400 (October)
14. Kulkarni AJ, Baki MF, Chaouch BA (2016) Application of the cohort-intelligence optimization method to three selected combinatorial optimization problems. Eur J Oper Res 250(2):427–447
15. Mansoor EM (1961) Selective assembly—its analysis and applications. Int J Prod Res 1(1):13–24
16. Patankar NS, Kulkarni AJ (2017) Variations of cohort intelligence. Soft Comput 1–17
17. Pugh GA (1986) Partitioning for selective assembly. Comput Ind Eng 11(1–4):175–179
18. Pugh GA (1992) Selective assembly with components of dissimilar variance. Comput Ind Eng 23(1–4):487–491
19. Sarmah DK, Kulkarni AJ (2017) Image Steganography Capacity Improvement Using Cohort Intelligence and Modified Multi-random start local search methods. Arab J Sci Eng 1–24
20. Shah P, Agashe S, Kulkarni AJ (2017) Design of fractional PID controller using cohort intelligence method. Front Inf Technol Electron, Eng
21. Shastri AS, Jadhav PS, Kulkarni AJ, Abraham A (2016) Solution to constrained test problems using cohort intelligence algorithm. In Innovations in bio-inspired computing and applications. Springer International Publishing, pp 427–435

M-Wallet Technology Acceptance by Street Vendors in India

Komal Chopra

Abstract The research paper investigates the factors that hinder the acceptance of mobile (M) wallet technology acceptance by street vendors in India. The study is exploratory in nature where the variables under study have been taken from review of the literature and informal interaction with street vendors and mobile wallet technology firms. The responses for the study were collected through a structured questionnaire that was administered to the street vendors. The responses were collected from five cities in India, i.e. Mumbai, Pune, Nasik, Nagpur and Aurangabad. The total sample size was 551. The hypotheses were tested using Friedman test and Kruskal–Wallis test which is a nonparametric statistic since rank order scale was used. The results indicate that privacy of data, secure transaction and trust are the top three factors that hinder the acceptance of M-wallet payment technology. The factors are common across the three types of street vendors considered for study, i.e. milk vendors, vegetable vendors and roadside service repair shops. The contribution to the existing knowledge is in the form of validating the existing technology models in the context of street vendors in India.

Keywords Mobile · Payment · Unorganized retail · Technology

1 Introduction

A study by researcher [16] on technology acceptance in mobile (M) wallet by customers has shown that security, trust, social influence and self-efficacy are most important to influence consumer attitude when using technology. Prominent researchers [8] in the study on mobile consumers have revealed that consumers do not want to share their personal information and hence are apprehensive to use tech-

K. Chopra (✉)
Symbiosis Institute of Management Studies (SIMS), Symbiosis International University (SIU), Range Hills Road, Khadki, Pune 411020, Maharashtra, India
e-mail: chopra.k@sims.edu

© Springer Nature Singapore Pte Ltd. 2019
A. J. Kulkarni et al. (eds.), *Proceedings of the 2nd International Conference on Data Engineering and Communication Technology*, Advances in Intelligent Systems and Computing 828, https://doi.org/10.1007/978-981-13-1610-4_18

nology. Researchers [5] in their study on online shopping have shown that mobile payments have played a very important role in fuelling the growth of retail industry.

The retailers have introduced mobile payment technologies so that the consumers have more options in payments. However, the retailers are not concerned about the risks involved in the use of technology [17]. However, in future, the retailers will need to address the concerns regarding mobile wallet security which may ensure greater acceptance by consumers [3]. Another study by researchers [9] has mentioned that consumer convenience apart from security aspect plays an important role in deciding use of mobile payments.

The study on technology acceptance has been done by several researchers. One such model is technology acceptance model [6] which was developed to understand the intentions of using technology. Another model was theory of planned behaviour (TPB) [2]. A comparison between TAM and TPB was done [12]. The study revealed that TAM had more advantages compared to TPB and had higher acceptability amongst the consumers [12]. The review of the literature has shown that the technology acceptance models have not been tested in the Indian environment with reference to the unorganized segment and specifically street vendors. The research on this topic investigates the factors that concern consumers in acceptance of M-wallet technology in India. The consumers in the current study are street vendors. The Government of India called for demonetization, and this resulted in businesses to move to cashless payments [1].

2 Review of Literature

The technology acceptance model [6] has revealed that behavioural aspects in the use of technology depend on convenience and intentions of the users. Researchers [18] did a further study on this model and added more factors such as employment relevance, result demonstration, experience and image of the firm. Trust factor and risk are another set of factors that determine acceptance of TAM [15]. This study was done with reference to e-commerce. The researcher identified different types of risks which hinder technology acceptance. These are economic risk, personal risk, privacy risk and seller performance risk. Another study with reference to gender [7] reveals that risk perceptions differ with gender. A research [9] in the area of M-wallets has shown that age has a moderating effect in case of technology adoption with respect to trust and ease of use.

The literature of technology adoption models [14] has found out various other models of technology acceptance such as diffusion of innovation theory (DOI), technology–organization–environment (TOE) framework, interorganizational systems (IOs) and institutional theory. The published literature on M-wallets has shown that TAM has been widely accepted to study consumer adoption of technology in making payments with modifications made by various researchers [18] in their research integrated TAM, DIO, perceived risk and cost in one model and for mobile payments and named it as extended TAM. Finding from the research concluded that all the attributes of TAM and DIO had impact on behaviour of user in decision-making for mobile payments. Two researchers [10] checked the technology fit (TTF) model for e-commerce which shows that how a particular technology fits a specific task. However, this study is limited to online shoppers. The benefits and risks associated with mobile payments also become an important factor in adoption of technology [17].

Published research work reveals that research work on technology adoption is in the field of IT employees and shoppers. A study on M-wallet technology adoption [11] by retailers has shown that the objective of enhancing sales and reduction in the cost of payment processing are the major reasons for acceptance, and the factors which hinder the adoption of technology for payments are the complexity of the systems, unfavourable revenue sharing models, lack of critical mass, and lack of standardization, trust and security. A report in Business Standard [1] highlighted that 68% payments by retailers in India are done in cash. Rural retailing in India operates on cash and carry, and the unorganized segment here is in the forefront of cash transactions [4].

2.1 Research Gap

The various research gaps in the review of the literature highlight that models of technology adoption models have not been tested in India specially in the unorganized retail segment. The current study aims to test the technology adoption attributes in Indian context with respect to unorganized retail segment, i.e. street vendors. Based on the research gap, the following research questions were framed.

2.2 Research Questions

1. What are the important factors (top three factors) for acceptance of M-wallet technology by street vendors?
2. Is there a difference in opinion of street vendors (milk vendors, vegetable vendors, service repair shops) with respect to factors for acceptance of M-wallet technology?

Based on the research questions, the following null hypothesis was framed:

H01: There is no significant difference in the response of street vendors with respect to acceptance of payment technology.
H02: There is no significant difference in response of the three sample sets (milk vendors, vehicle repair shops and vegetable vendors) with respect to factors that hinder the acceptance of M-wallet technology.

3 Research Methodology

In the first stage, exploratory research was done by reviewing the existing literature on technology adoption and also interacting with street vendors and technology experts in the area. The exploratory research helped in understanding the research gap and generating the variables for the study. Then, a structured questionnaire was designed which consisted of demographic variables and variables related to technology adoption.

In the second stage, the questionnaire was pilot tested in Mumbai City of India which has the largest population amongst all cities in the country and the largest concentration of street vendors in India about 2,50,000 [13]. The responses were collected after personal interaction with street vendors such as milk vendors, vehicle repair shops and vegetable vendors. The number of responses collected was 82. The vendors were randomly selected from two areas of Mumbai which had the highest density of street vendors. The questions were asked relating to the following attributes: ease of use, secure transaction, social influence, trust, privacy of data, increase in sales, standardization and cost, and hence, content validity was tested. The respondents were asked to rank their responses in the order of preference on a scale of 1–7 where 1 stands for the highest rank and 7 stands for the lowest rank. The pilot testing helped to make improvements in the questionnaire.

In the third stage, the final questionnaire was administered to 1000 respondents in five metropolitan cities of India, i.e. Pune, Mumbai, Nasik, Nagpur and Aurangabad. A metropolitan city has a population of 100,000 or more [1]. The largest concentration of street vendors exists in the metropolitan cities [13], and hence, these cities were selected for study. Five hundred and fifty-one responses were received. The data was analysed using statistical tools. Friedman test was administered to find out top three attributes for acceptance of technology. The next step was to find out whether there was a significant difference in responses of street vendors with respect to acceptance of technology. Kruskal–Wallis test was administered for this purpose. The hypotheses were tested at a significant value of less than 5%. To carry out reliability test, Kendall's W was administered to find out the internal consistency.

4 Data Analysis

Out of the total responses that were collected, 35% were vehicle repair shops, 34% were milk vendors, and 36% were vegetable vendors. The monthly turnover of all the respondents was less than 50,000 INR. Forty-eight per cent of the vendors were in their business for 1–3 years, whereas 52% were in their business from 3 to 5 years.

Table 1 Testing of null hypothesis 1 (top three factors for acceptance of technology using Friedman test, i.e. calculating the mean ranks)

Technology concerns	Pune	Mumbai	Nasik	Nagpur	Aurangabad
Privacy of data	3.07	3.24	3.44	3.94	3.17
Standardization	5.40	4.96	4.86	4.65	4.94
Trust	4.00	3.76	3.68	3.32	3.85
Increase in sales	5.03	5.19	5.10	4.81	5.29
Social influence	5.11	4.80	5.21	4.55	4.60
Secure transaction	3.11	3.34	3.61	3.95	3.26
Ease of use	4.82	5.01	4.61	5.71	4.89
Technology concerns	Pune	Mumbai	Nasik	Nagpur	Aurangabad
	Rank order	Rank order	Rank order	Rank order	Rank order
Privacy of data	1	1	1	2	1
Standardization	7	5	5	4	5
Trust	3	3	3	1	3
Increase in sales	5	7	6	6	7
Social influence	5	4	7	5	4
Secure transaction	2	2	2	3	2
Ease of use	4	6	4	8	6

4.1   Results of Hypothesis Testing

Inference

Table 1: There is a significant difference between the street vendor responses to different variables under study at a significance level of $p < 0.05$. Hence, the null hypothesis is rejected. The rank order of variables indicates that privacy of data, secure transactions and trust in technology rank amongst the top three factors that hinder the acceptance of technology.

Table 2: The results of Table 2 indicate the number of respondents (N) for the five cities and the level of significance for conduct of Friedman test. The results of the five cities have shown a significance value of less than 5%, indicating rejection of null hypotheses for the five cities.

Table 2 Test of significance for five cities

Test statistics

City	Pune	Mumbai	Nasik	Nagpur	Aurangabad
N	158	221	57	74	41
Chi-square	208.713	227.372	51.314	52.9	48.82
Df	7	7	7	7	7
Asymp. Sig.	0.00	0.00	0.00	0.00	0.00

Table 3 Reliability test

Test statistics

N	5
Kendall's W^a	0.853
Chi-square	29.867
Df	7
Asymp. Sig.	0.000

a = W i.e. Kendall's coefficient of concordance

Table 4 Testing of null hypothesis 2 using Kruskal–Wallis test

Test statistics

	Privacy of data	Standardization	Trust	Increase in sale	Social influence	Secure transaction	Ease of use
Chi-square	1.462	0.243	0.369	1.075	0.396	1.615	0.769
Df	2	2	2	2	2	2	2
Asymp. Sig.	0.482	0.885	0.832	0.584	0.820	0.446	0.681

Inference

From Table 3, the data analysis indicates that top three attributes are privacy concerns, security and trust while accepting mobile technology. The Kendall's W is 85.3% which indicates significant agreement between the respondents of different cities.

Inference

The results of Kruskal–Wallis test (Table 4) indicate that there is no significant difference between responses of the different respondents across different cities at a significance value of probability $p < 0.05$. Hence, the null hypothesis is accepted.

4.2 Validity Testing

4.2.1 Content Validity

The validity was tested by borrowing the variables from review of the literature and also validating it from experts in mobile payment technologies.

4.2.2 Construct Validity

The results of Friedman test also indicate that the variables are independent of each other, and hence, discriminant validity is tested.

5 Conclusion and Discussion

From the results of Table 1, it is clear that trust, secure transaction and privacy of data are the top three factors that hinder the acceptance of M-wallet technology in India. The results match with the findings of extension model of TAM [18]. However, their study was done on the acceptance of technology in the manufacturing industry. The results of the study also match with the study on gender [9]. Table 4 highlights that the concerns regarding digital payments are the same irrespective of the type of street vendor. This study was done on the mobile payment industry in Spain. Hence, this study validates the study done in Spain. The paper has strong implications for the mobile wallet industry who want to expand their business taking advantage of demonetization in India. It will be important for them to address the concerns raised by small businesses. The contribution to the existing knowledge is in the form of validating the existing technology models in the context of street vendors in India.

References

1. 68% of transactions in India are cash-based: CLSA. Business Standard India (2016), November 14
2. Ajzen I (1991) The theory of planned behavior. Organ Behav Hum Decis Process 50(2):179–211
3. Caldwell T (2012) Locking down the e-wallet. Comput Fraud Secur 2012(4):5–8
4. Chopra K (2014) Ecopreneurship: is it a viable business model? Archers Elevators Int J Manage Res 2(3) (March)
5. Chopra K, Srivastava A (2016) Impact of online marketing in molding consumer behaviour. Int J Eng Manage Res 6(1):478–486 (2016) (January–February)
6. Davis FD Jr (1986) A technology acceptance model for empirically testing new end-user information systems: theory and results. Doctoral dissertation, Massachusetts Institute of Technology
7. Gefen D, Karahanna E, Straub DW (2013) Trust and TAM in online shopping: an integrated model. MIS Q 27(1):51–90

8. Hoofnagle CJ, Urban JM, Li S (2012) Mobile payments: consumer benefits & new privacy concerns

9. José Liébana-Cabanillas F, Sánchez-Fernández J, Muñoz-Leiva F (2014) Role of gender on acceptance of mobile payment. Ind Manage Data Syst 114(2):220–240

10. Klopping IM, McKinney E (2004) Extending the technology acceptance model and the task-technology fit model to consumer e-commerce. Inf Technol Learn Perform J 22(1):35

11. Mallat N, Tuunainen VK (2008) Exploring merchant adoption of mobile payment systems: An empirical study. E-service J 6(2):24–57

12. Mathieson K (1991) Predicting user intentions: comparing the technology acceptance model with the theory of planned behavior. Inf Syst Res 2(3):173–191

13. National policy for Urban Street vendors. Ministry of Urban Employment and Poverty Alleviation, Government of India. http://muepa.nic.in/policies/index2.htm

14. Oliveira T, Martins MF (2011) Literature review of information technology adoption models at firm level. Electr J Inf Syst Eval 14(1):110–121

15. Pavlou PA (2003) Consumer acceptance of electronic commerce: integrating trust and risk with the technology acceptance model. Int J Electron Commer 7(3):101–134

16. Shin DH (2009) Towards an understanding of the consumer acceptance of mobile wallet. Comput Hum Behav 25(6):1343–1354

17. Taylor E (2016) Mobile payment technologies in retail: a review of potential benefits and risks. Int J Retail Distrib Manage 44(2):159–177

18. Venkatesh V, Davis FD (2000) A theoretical extension of the technology acceptance model: Four longitudinal field studies. Manage Sci 46(2):186–204

Explore-Exploit-Explore in Ant Colony Optimization

Parth A. Kulkarni

Abstract As the real-world systems are getting more complex, we must come up with decentralized, individually smart, and collectively intelligent systems. To achieve this, a meta-heuristic algorithm was successfully tested against Sphere Test Optimization Function. In this algorithm, the ants use each other to reach the destination. Initially, they make three random moves. Afterward, a set percentage of ants explore (make another random move), and the others follow the ant with the most pheromone. Pheromone is given to an ant when it makes a move approaching the destination and taken away from each ant after every move. The results of the experiment will be discussed later in this paper.

Keywords Decentralized · Meta-heuristic algorithm · Ant colony optimization

1 Introduction

In past few years, several nature-inspired optimization algorithms have been proposed. As the real-world systems get more complex, many shortcomings of a centralized optimization can be seen vividly. This makes companies spend more money and waste the valuable computing and natural resources. To solve this, we must come up with decentralized, individually smart, and collectively intelligent systems. Such systems are easy to manage, reduce risk of failure, and provide practically feasible solutions. Things that can be done with these decentralized systems are: optimization of telecom networks by exploiting decentralized SDN (software-defined networks), build medicines that can be released only when and where required, and predict housing market variations, etc.

The notable methods are Genetic Algorithm (GA) [1], Probability Collectives (PC) [2], Cohort Intelligence (CI) [3, 4], Ideology Algorithm (IA) [5], Firefly Algorithm (FA) [6], Cuckoo Search (CS) [7], Grey Wolf Optimizer (GWO), Artificial

P. A. Kulkarni (✉)
Strawberry Crest High School, 4691 Gallagher Road, Dover, FL 33527, USA
e-mail: toparth@gmail.com

© Springer Nature Singapore Pte Ltd. 2019   183
A. J. Kulkarni et al. (eds.), *Proceedings of the 2nd International Conference on Data Engineering and Communication Technology*, Advances in Intelligent Systems and Computing 828, https://doi.org/10.1007/978-981-13-1610-4_19

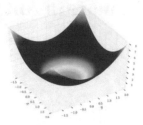

Fig. 1 Sphere optimization function graph

Bee Colony (ABC) [8], Seed Disperser Ant Algorithm (SDAA) [9, 10], etc., which are few notable algorithms. Ant Colony Optimization (ACO) is one of the popular swarm-based methods. The ACO is a probabilistic problem-solving technique. The algorithm was originally proposed by Dorigo [11] and then further modified in 1996. The algorithm is inspired from foraging behavior of ants. It models the indirect communication among ants using pheromone trails, which lead them to find shortest possible path between food source and the nest. The ACO and its variations as well as hybridized methods have been applied for solving a variety of applications such as Traveling Salesman Problem [11–13] and Vehicle Routing Problems [14, 15], Quadratic Assignment Problem [16, 17], Scheduling Problem, Graph Coloring Problem [18], Data Mining [19], Cell Placement in Circuit Design [20], Industrial Problems [21, 22], Multiobjective Problems [23].

2 Literature Review

Decentralization is a main driver behind ACO. Decentralization is when a group of objects (or ants in this case) does not take and execute orders for each individual step from a common leader, but instead work within themselves to achieve the same goal.

The proposed work will be to find the number of ants that should explore after each move using a decentralized ACO method. The algorithm created has not yet been applied to any field, but it is projected to be applied for intelligent medicines, telecom networks, and many other real-world problems.

3 Explore-Exploit-Explore Algorithm Using Ant Colony Optimization

To investigate the ACO, two-dimensional Sphere function with global minimum $x_i = 0$, $\forall i = 1, 2$; $f^*(x) = 0$ was chosen. It is a continuous, convex, and unimodal function. Refer to following equation and graph (Fig. 1):

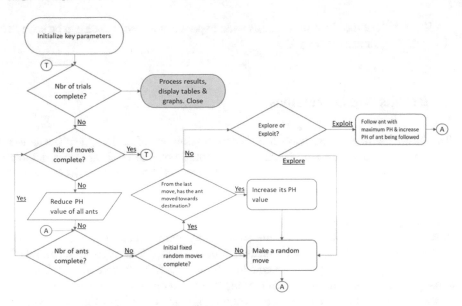

Fig. 2 Explore-Exploit-Explore algorithm flowchart

$$f(x) = \sum_{i=1}^{2} x_i^2 \, , \, x_i \in [-5.12, 5.12] \quad \forall i = 1, 2$$

The objective of testing the algorithm against this function is to see whether the algorithm is truly optimized. The closer the ants reach to the bottom of this graph [(0, 0, 0)], the more optimized the algorithm is. In the experiment conducted, the destination was set to [(0.01, 0.01, 0.01)]. However, when using the algorithm, the destination can be set to whatever you wish.

3.1 Procedure

The ACO algorithm was coded in Adobe ColdFusion 13 on macOS High Sierra (10.13) platform with Intel i7, 2.6 GHz processor speed, and 16 GB RAM. The steps for testing the algorithm to see which percentage of a group of ants should explore after each move to make the algorithm truly optimized are as follows:

Step 1: Initialize key parameters; Ants, Moves, Range, Exploration Ratio, and Trial Count
Step 2: Run the program (refer to the Fig. 2 below to algorithm flow)
Step 3: Record data by saving HTML code onto a computer
Step 4: Repeat steps 2 and 3, but this time change Exploration Ratio.
Step 5: Analyze the results

The details of the algorithm such as what happens after each move can be seen in the following flowchart (Fig. 2).

4 Results and Discussions

The best exploration ratio was 45%. When the number of ants exploring was too low, there was one major problem; the ants could not find the best path available since very few of the ants were exploring. When tested above 75%, it was too much, and the ants were not making any logical moves; just random moves. So, letting 45% of the ants explore allows the ants to reach the destination within 25–43 moves. Previously the destination was set at 0.1 (distance from zero). The ants successfully reached the destination. To see whether it was truly optimized, the destination was set to 0.01. This made the destination much harder to get to, but the algorithm continued to demonstrate a convergence. In conclusion, using ACO we were able to find the most practically optimized way for a group of ants to reach a given destination. We were able to optimize the process by changing the number of ants that will explore and finding the perfect ratio so that ants will use each other like they do in the real world to reach the destination.

4.1 Stagewise Improvements

ACO Improvements in Stage I: In proposed work, initially the ants make three random moves to place themselves somewhere on the plane. In the future, there aren't any changes we can make to this stage.

ACO Improvements in Stage II: In the future, we can place multiple destinations for the ants. This could further test the optimization of this algorithm. Also, it will open more doors for applications to the real world.

ACO Improvements in Stage III: We can measure and optimize number of moves required to reach back to nest. Furthermore, we can also record how much "food" each ant brought back to see whether the ants chose the destination with the most food.

4.2 Graphical and Statistical Analysis

Table 1 as depicted above shows the exploration ratio (how many of the ants were exploring), how long it took for each run, average pheromone count (to show how many times the ant got closer to the destination, the higher the better), and the best/median/worst ending points. This table helps us see exploration ratio between 45 and 75% which is best for running the algorithm.

Table 1 Comparison among results of different exploration ratios

Exploration ratio	Average function evaluation	Average time (s)	Average pheromone added count	Best/median/worst
0	370	2.68	20	0.00/0.01/0.02
0.15	241	2.02	20	0.00/0.03/0.07
0.25	425	2.05	21	0.00/0.01/0.03
0.35	349	2.18	20	0.00/0.02/0.04
0.45	367	3.3	20	0.00/0.005/0.01
0.75	720	2.12	20	0.00/0.005/0.01
1	482	2.05	40	0.00/0.010/0.04

Fig. 3 Ants convergence plot (0% of ants exploring)

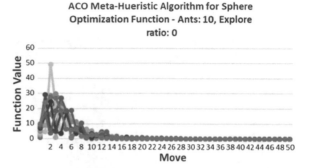

ACO Meta-Hueristic Algorithm for Sphere Optimization Function - Ants: 10, Explore ratio: 0

With 0% of the ants exploring, we can see above (Fig. 3) that the convergence does not start until 41st move. This proves that the less ants that are exploring, the slower time it takes to reach convergence.

With 45% of the ants exploring (refer to Fig. 4), we have convergence at a much earlier point. This proves that if we have the perfect number of ants exploring, we are making the process of reaching the destination much more efficient. We were able to find the best possible routes and have all the ants reach there at a faster pace.

In this scenario, we have 100% of the ants exploring (refer to Fig. 5), and none of the ants are following the ant with the highest pheromone. This is very inefficient since essentially everyone is making random moves till they all reach the destination. Although not seen in this graph in great detail, most of the time ants will not reach the destination in a set number of moves because there is no communication between the ants in order to find the best possible route. This results in the ants not able to reach the destination making it very inefficient.

Fig. 4 Ants convergence
plot (45% of ants exploring)

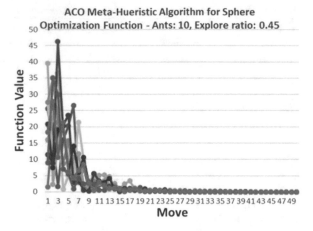

Fig. 5 Ants convergence
plot (100% of ants exploring)

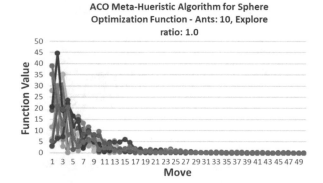

5 Conclusions and Future Directions

In conclusion, using ACO we were able to find practically the most optimized way for a group of ants to reach a given destination by making random moves. We were able to optimize the process by changing the number of ants that will explore and finding the perfect ratio so that ants will use each other like they do in the real world to reach the destination. We can further improve the algorithm and apply to real-world optimization problems.

References

1. Deb K (2000) An efficient constraint handling method for genetic algorithms. Comput Methods Appl Mech Eng 186:311–338
2. Kulkarni AJ et al (2015) Probability collectives: a distributed multi-agent system approach for optimization. In: Intelligent systems reference library, vol 86. Springer, Berlin. https://doi.org/10.1007/978-3-319-16000-9. ISBN: 978-3-319-15999-7

3. Kulkarni AJ, Durugkar IP, Kumar M (2013) Cohort intelligence: a self supervised learning behavior. In: Proceedings of IEEE international conference on systems, man and cybernetics, Manchester, UK, 13–16 Oct 2013, pp 1396–1400
4. Kulkarni AJ, Krishnasamy G, Abraham A (2017) Cohort intelligence: a socio-inspired optimization method. Intelligent systems reference library, vol 114. Springer, Berlin. https://doi.org/10.1007/978-3-319-44254-9. ISBN: 978-3-319-44254-9
5. Teo TH, Kulkarni AJ, Kanesan J, Chuah JH, Abraham A (2016) Ideology algorithm: a socio-inspired optimization methodology (In Press: Neural Comput Appl. https://doi.org/10.1007/s00521-016-2379-4)
6. Yang XS (2010) Firefly algorithm, Levy flights and global optimization. In: Research and development in intelligent systems XXVI, pp 209–218. Springer, London, UK
7. Yang X-S, Deb S (2009) Cuckoo search via Lévy flights. In: World congress on nature & biologically inspired computing (NaBIC 2009). IEEE Publications, pp 210–214 (December)
8. Karaboga D, Basturk B (2007) A powerful and efficient algorithm for numerical function optimization: artificial bee colony (ABC) algorithm. J Global Optim 39(3):459–471
9. Chang WL, Kanesan J, Kulkarni AJ, Ramiah H (2017) Data clustering using seed disperser ant algorithm (In Press: Turk J Electr Eng Comput Sci)
10. Chang WL, Kanesan J, Kulkarni AJ (2015) Seed disperser ant algorithm: ant evolutionary approach for optimization. Lecture notes in computer science, vol 9028. Springer, Berlin, pp 643–654
11. Dorigo M (1992) Optimization, learning and natural algorithms. Ph.D. thesis, Dipartimento di Elettronica, Politecnico di Milano, Italy (in Italian)
12. Dorigo M, Maniezzo V, Colorni A (1991) Positive feedback as a search strategy. Technical Report 91-016, Dipartimento di Elettronica, Politecnico di Milano, Italy
13. Stützle T, Hoos HH (2000) MAX–MIN ant system. Future Gener Comput Syst 16(8):889–914
14. Gambardella LM, Taillard ÉD, Agazzi G (1999) MACS-VRPTW: a multiple ant colony system for vehicle routing problems with time windows. In: Corne D, Dorigo M, Glover F (eds) New ideas in optimization. McGraw-Hill, London, pp 63–76
15. Reimann M, Doerner K, Hartl RF (2004) D-ants: savings based ants divide and conquer the vehicle routing problems. Comput Oper Res 31(4):563–591
16. Maniezzo V (1999) Exact and approximate nondeterministic tree-search procedures for the quadratic assignment problem. INFORMS J Comput 11(4):358–369
17. Maniezzo V, Colorni A (1999) The ant system applied to the quadratic assignment problem. IEEE Trans Data Knowl Eng 11(5):769–778
18. Costa D, Hertz A (1997) Ants can color graphs. J Oper Res Soc 48:295–305
19. Parpinelli RS et al (2002) Data mining with an ant colony optimization algorithm. IEEE Trans Evol Comput 6(4):321–332
20. Alupoaei S, Katkoori S (2004) Ant colony system application to marcocell overlap removal. IEEE Trans Very Large Scale Integr (VLSI) Syst 12(10):1118–1122
21. Bautista J, Pereira J (2002) Ant algorithms for assembly line balancing. In: Dorigo M, Di Caro G, Sampels M (eds) Ant algorithms—Proceedings of ANTS 2002—third international workshop. Lecture notes in computer science, vol 2463. Springer, Berlin, pp 65–75
22. Corry P, Kozan E (2004) Ant colony optimisation for machine layout problems. Comput Optim Appl 28(3):287–310
23. Guntsch M, Middendorf M (2003) Solving multi-objective permutation problems with population based ACO. notes in computer science, vol 2636. Springer, Berlin, pp 464–478

An Attention-Based Approach to Text Summarization

Rudresh Panchal, Avais Pagarkar and Lakshmi Kurup

Abstract Text summarization has become increasingly important in today's world of information overload. Recently, simpler networks using only attention mechanisms have been tried out for neural machine translation. We propose to use a similar model to carry out the task of text summarization. The proposed model not only trains faster than the usually used recurrent neural network-based architectures but also gives encouraging results. We trained our model on a dump of Wikipedia articles and managed to get a ROUGE-1 f-measure score of 0.54 and BLEU score of 15.74.

Keywords Summarization · Attention · Encoder–decoder
Recurrent neural networks

1 Introduction

Text summarization refers to the practice of condensing or shortening text in order to increase its human readability and make it possible to convey useful information with greater ease. This has become increasingly useful in today's world of an information overload, where people cannot sift through large pieces of information but need to prioritize their information intake.

Traditionally an NLP challenge, various machine learning, particularly deep learning-based models are increasingly being applied to carry out this task.

Text summarization can mainly be classified into two types, extractive text summarization and abstractive text summarization. Extractive summarization involves ranking/labelling the sentences in a context based on their importance, with the most

R. Panchal (✉) · A. Pagarkar (✉) · L. Kurup
Dwarkadas J Sanghvi College of Engineering, Mumbai, Maharashtra, India
e-mail: rudreshpanchal09@gmail.com

A. Pagarkar
e-mail: avaispagarkar@gmail.com

L. Kurup
e-mail: lakshmidkurup@gmail.com

© Springer Nature Singapore Pte Ltd. 2019
A. J. Kulkarni et al. (eds.), *Proceedings of the 2nd International Conference on Data Engineering and Communication Technology*, Advances in Intelligent Systems and Computing 828, https://doi.org/10.1007/978-981-13-1610-4_20

important ones being extracted as the summary of the context. Since this approach directly lifts sentences from a corpus, there is little or no language modelling that has to be done. This approach for years has been very popular given its relative simplicity. Since it involves directly extracting sentences/phrases from the context to generate the summary, it often loses out on key ideas and points and faces the challenges like appropriately coreferencing entities for pronouns in the sentences extracted. These challenges make extractive approaches difficult to generalize as these challenges are very often domain specific and have to be dealt with accordingly.

The other approach, abstractive text summarization, involves paraphrasing a large context into a summary. This task involves significant language modelling as the target summary is not merely lifted from the context but needs to be generated, and this makes it a difficult task. But in the recent years with the advent of deep learning-based techniques which can carry out language modelling accurately, various approaches have been employed for abstractive summarization. Because of this, abstractive approaches usually involve some form of neural network-based architecture which trains to convert a context into the target summary. The architecture of these models is usually very similar to that of the ones used for other text conversion problems like neural machine translation and overtimes models have been created for NMT and have been used for text summarization and vice versa.

In this paper, we propose to employ a similar approach. We seek to use a model which works solely based on attention mechanisms. Since this model does not employ any deep neural networks usually associated with such models, their training time reduces manyfold. The proposed model was originally created and tested on various neural machine translation tasks, but we have modified the same to carry out text summarization.

2 Background

Statistical feature-based extractive text summarization has been the most prevalent technique of automatic text summarization until recently. Some of the notable ones include a lexical centrality-based approach(LEXRANK) by Erkan et al. [1] and graph-based ranking model proposed (TextRank) proposed by Mihalcea et al. [2]. These methods employ various statistical metrics like the term frequency-inverse document frequency (TF-IDF) scores to determine their importance and extract the same accordingly. But with the advent of various deep learning-based techniques which managed to carry out complex language modelling accurately, various abstractive approaches to text summarization have become popular.

Sutskever et al. [3] proposed the sequence to sequence to model in 2014. The model uses a LSTM to map the input sequence to vector having a fixed dimension which is followed by another LSTM as a decoder to generate the target from the vector. A problem with this model was that the encoder contained information if the last part of the input sequence and hence is biased towards it. Bahdanau et al. [4] tackled this problem by introducing an attention mechanism. This allowed the decoder to check the input while decoding. Bahdanau et al. [4] then applied the model to machine translation and achieved encouraging results. This paved the way for the recent discoveries in abstractive summarization since machine translation and summarization are similar tasks.

Rush et al. [5] proposed a local attention-based model which generated each word of the summary based on the fixed window of the input sequence. The authors used an attention-based feedforward neural network along with a contextual encoder for the language model. Chopra et al. [6] modified this model to use RNNs in the decoder. Nallapati et al. [7] then tried various methods one of which was using an attentional sequence to sequence-based model. This model employed an attentional encoder–decoder RNN which was originally designed to carry out tasks of neural machine translation. This model achieved encouraging results and outperformed state-of-the-art results on two different datasets. The drawback of these neural network-backed models was that they were resource intensive and had very long training cycles.

Vaswani et al. [8] proposed a model that uses a self-attention mechanism while eliminating the need to use RNNs in the encoder and decoder and thus used a much simpler network. This lack of extensive neural networks in this model reduced training time significantly. The authors trained this model on two machine translation tasks and found that the model outperformed many other state-of-the-art models.

We propose to employ the model proposed by Vaswani et al. to carry out summarization tasks.

3   Dataset Used

We have used Matt Mahoney's large text compression benchmark [9], the "enwik9" dataset, to train our model. The dataset is a dump of first 10^9 bytes of the English Wikipedia XML dumps.

For a given article, we treat the introduction section as the summary. The rest of the article is combined to form the context. Our system takes the context as input and generates a summary. This system-generated summary is compared with the introduction section of the given article during training to minimize the loss and during evaluation to evaluate how well our system performs. The dataset had about 77,000 context–summary pairs which were further filtered according to our requirements.

We chose Wikipedia as our dataset because of the wide variety of articles across diverse domains. This would help us in truly generalizing our model to the open domain.

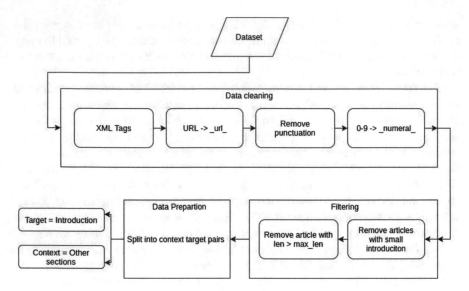

Fig. 1 A data flow diagram of the preprocessing steps

4 Preprocessing

We employed an approach similar to the one adopted by Prasoon Goyal and his colleagues at NYU [10].

Figure 1 shows the flow of data during the preprocessing steps.

The steps are:

A. Data Cleaning:

1. Removed XML tags. The Wikipedia dumps are in the XML format and need to be parsed to extract the articles individually. We also converted the entire corpus to lower case.
2. Replaced all URLs with a placeholder ("_url_"). Wikipedia articles are filled with links to other articles for reference. They were replaced with the placeholder rather than being removed completely in order to preserve the language order.
3. Removed all punctuation, to prevent them from introducing unnecessary dependencies. The punctuation, if not removed, will be treated as separate words.
4. Replaced all numbers with a "_numeral_" tag. This was done to prevent the numbers from generating a large numbers of unknowns while training and evaluation.

B. Filtering:

1. Removed all articles whose introduction section was not comparable to the rest of the article. This was done to weed out articles whose summaries may not be of the appropriate length when compared to the length of the article.
2. Removed articles whose length exceeded the maximum length.

C. Data preparation:

1. We parsed the corpus to split the data into a context source and the target summary.
2. The first section of the article (the introduction of the article) was extracted and labelled as target summary while the rest of the article was labelled as the context.

5 Model

Our work is built on top of the "Transformer" model proposed by Vaswani et al. [8] This model uses stacked self-attention layers for both the encoder and the decoder. Each layer of the encoder has two sub-layers, a multi-head self-attention layer and a simple fully connected feed-forward network. Both these sub-layers are connected, and then, the layer is normalized. The output of each sub-layer is Layer_Norm(x + Sub-layer(x)), where the function of each layer is sub-layer of x. To aid the connection, all the sub-layers and the embedding produce output having a dimension—d-model.

In case of the decoder, it has an extra sub-layer. This sub-layer employs a multi-head attention mechanism over the output embedding produced by the encoder.

In addition to the sub-layers, each layer contains a fully connected feed-forward neural network which is applied to each position. It uses the ReLU activation function along with two transformations and is represented as

$$FFN(x) = \max(0, x\text{W1} + \text{b1})\text{W2} + \text{b2}$$

Each layer uses different parameters.

ReLU is used over other activation functions like sigmoid to reduce the likelihood of the function facing the "Vanishing Gradient" problem.

The attention mechanism used is called "Scaled Dot-Product Attention". Several of such attention layers are put together to form a multi-head attention mechanism. This multi-head attention produces multiple linear projections having different dimensions. The attention function is applied on each of these projections, and then, the results are concatenated. This allows access to information stored in one projection but not in the other.

$$MultiHead(Q, K, V) = Concat(\text{head1}, \dots, \text{head}_b)\text{WO}$$

where head is the single-headed attention.

Like in the original paper, we used eight parallel attention layers [8].

In case of self-attention, all the information comes from the same place. Thus, each position in the encoder and decoder has access to all positions in the previous layer of the encoder and decoder. Additionally, the decoder also has access to the output of the encoder thus allowing the decoder to attend to all positions present in the input sequence.

To make use of the order of the sequence, the model adds some information regarding the relative and absolute position of the tokens present in the sequence. This is done by adding positional encodings to the input embeddings at the bottom of the two stacks. The input embeddings (context) and the target output embeddings (target summary) are fed into the model, and their output probabilities are computed.

The embeddings are generated by concatenating vector representations of the words. A mapping is done between the vector representation generated and the word that they correspond to. A reverse mapping is also generated which is used to decode the vector representation. These vector representations of words are generated from dictionary of words generated during preprocessing. The words which are not in the dictionary are replaced by the <UNK> token, and subsequently its vector representation.

We trained our model using a network with four encoder–decoder blocks and 512 hidden units. We used a drop out rate of 0.1. While the original paper uses varying learning rate dictated by the step number and the number of warm-up steps, we have used a constant learning rate. The Adam optimizer gives us the added advantage of us not having to carry out hyperparameter tuning on the learning rate. We have built our model on top of Kyubyong Park's [11] Tensorflow-based implementation of the transformer model. Figure 2 shows a high-level architecture of the model.

6 Results

After just two epochs of training on the dataset, the model showed some promising results. The model took less than 4 h to train on an NVIDIA Tesla K80 GPU. Figure 3 shows the increase in training accuracy compared to the number of iterations processed. This was with a batch size of 2 and each batch containing about twenty-two thousand articles and a learning rate of 0.1. As seen in Fig. 4, the model averaged at approximately 1.26 global steps per second.

Tables 1 and 2 show an example where our model performs well. Table 3 shows an example of when the model gives unsatisfactory performance.

An example of when the model fails to generalize.

The model seemed to carry out the language modelling of the context pretty well. However, one of the major shortcomings of the results seems to be their failure to retain proper nouns.

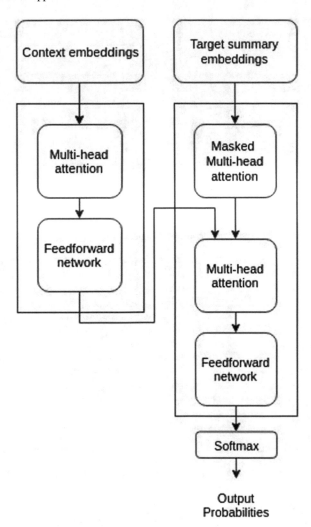

Fig. 2 Model

We used the following metrics to evaluate performance:

1. BLEU: A metric defined by Papinen et al. [12], usually used to evaluate performance of machine translation models.
2. ROUGE-L: A metric based on the longest common subsequence between target summary and generated summary [13].

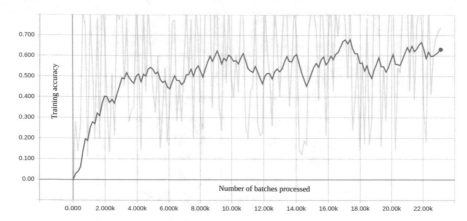

Fig. 3 Training accuracy versus number of batches processed (smoothing factor of 9)

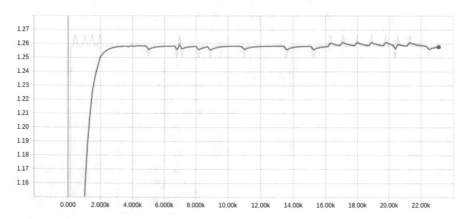

Fig. 4 Global steps/second versus number of batches processed (smoothing factor of 7)

Table 1 Sample summary set one

Target summary	Generated summary
Elburn is a village located in Kane County Illinois as of the numeral census the village had a total population of numeral	Champaign is a village located in Monroe County Illinois as of the numeral census the village had a total population of numeral

3. ROUGE-1: A metric based on the overlapping of one gram (1 g) or single words between target summary and generated summary [13].
4. ROUGE-2: A metric based on the overlapping of bigrams or two words between target summary and generated summary [13].

Since the precision and recall is generated for the scores, we take the f-measure to evaluate overall performance. We used a train dataset of 108 generated-target summary pairs to evaluate performance. The scores generated are given in Table 4.

Table 2 Sample summary set two

Target summary	Generated summary
Hartsburg is a town located in Boone County Missouri as of the numeral census the town had a total population of numeral	Paulding is a town located in Jefferson county Indiana as of the numeral census the town had a total population of numeral

Table 3 Sample summary set three

Target summary	Generated summary
Ko Phi Phi Le or Ko Phi Phi Ley is an island of the Phi Phi Archipelago in the Andaman Sea it belongs to the Krabi province of Thailand	the following is a list of the USA

Table 4 Metrics with the computed score

Metric name	Score
BLEU	15.7452234978
ROUGE-L f-measure	0.48
ROUGE-1 f-measure	0.54
ROUGE-2 f-measure	0.40

7   Conclusion and Future Work

We used the transformer model to generate abstractive summaries for short Wikipedia articles. We achieved an appreciable ROGUE score. The reduced training time is an added advantage. However, while the model produces good summaries, it is yet to be seen how it generalizes to a larger dataset containing a greater number of articles and how accurately and how fast does it train on articles of greater length. We also found the model to be unable to retain proper nouns.

The next steps would be to:

- Decode using beam search. Evaluations carried out using beam search are usually faster and also addresses label bias and exposure bias.
- Larger dataset: To see how well the model generalizes for a wide range of topics. The current experiment generalized only across a small set of articles.
- Longer articles: To see if the model still generates correct summaries even if it has additional information about the input.

References

1. Erkan G, Radev DR. LexRank: graph-based lexical centrality as salience in text summarization. https://www.cs.cmu.edu/afs/cs/project/jair/pub/volume22/erkan04a-html/erkan04a.html
2. Mihalcea R, Tarau P. TextRank: bringing order into texts. https://web.eecs.umich.edu/~mihalcea/papers/mihalcea.emnlp04.pdf

3. Sutskever I, Vinyals O, Le QV (2014) Sequence to sequence learning with neural networks. arXiv:1409.3215 [cs.CL]
4. Bahdanau D, Cho K, Bengio Y (2014) Neural machine translation by jointly learning to align and translate. arXiv:1409.0473v7 [cs.CL]
5. Rush AM, Chopra S, Weston J (2015) A neural attention model for abstractive sentence summarization. arXiv:1509.00685v2 [cs.CL]
6. Chopra S, Auli M, Rush AM (2016) Abstractive sentence summarization with attentive recurrent neural networks. In: HLT-NAACL
7. Nallapati R, Zhou B, Nogueira dos santos C, Gulcehre C, Xiang B. Abstractive text summarization using sequence-to-sequence RNNs and beyond. arXiv:1602.06023v5 [cs.CL]
8. Vaswani A, Shazeer N, Parmar N, Uszkoreit J, Jones L, Gomez AN, Kaiser L, Polosukhin I. Attention is all you need. arXiv:1706.03762v4 [cs.CL]
9. Mahoney M. Wikipedia dump. https://cs.fit.edu/~mmahoney/compression/textdata.html
10. Goyal P, Goel S, Sethia K. Text summarization for wikipedia articles. http://www.cs.nyu.edu/~pg1338/course_projects/text_summarization.pdf
11. Park K. A TensorFlow implementation of the transformer: attention is all you need. https://github.com/Kyubyong/transformer
12. Papineni K, Roukos S, Ward T, Zhu W-J. BLEU: a method for automatic evaluation of machine translation. http://www.aclweb.org/anthology/P02-1040.pdf
13. Lin C-Y. ROUGE: a package for automatic evaluation of summaries. http://anthology.aclweb.org/W/W04/W04-1013.pdf

Enhancement of Security for Cloud Data Using Partition-Based Steganography

D. Suneetha and R. Kiran Kumar

Abstract Data security is a major issue in computer science and information technology. In the cloud computing environment, it is a serious issue because data is located in different places. In the cloud, environment data is maintained by the third party so it is harder to maintain security for user's data. There is a prominent need for security for cloud data, so we proposed an approach which provides better results as compared to previous approaches. In this work, we tried to secure the data by using image steganography partition random edge-based technique. In the next step, partition the original image into a number of parts, apply edge-based algorithms, select random pixels (prime number-based pixels) from each part, and insert the data into them. The performance of this proposed analysis is calculated by using MSE and PSNR values. Results are good as compared to several existing edge-based algorithms of image steganography.

Keywords Steganography · PSNR · MSE

1 Introduction

Cloud computing is an emerging and trending technology nowadays. It provides several services to the different sectors like IT, medical, business. In the cloud computing environment, data security is a serious and prominent issue because data is located in different places from multiusers. In a cloud, environment data is maintained by third the party so it is harder to maintain security for user data. Generally, in a cloud environment security is concerned at three different places. So many techniques are proposed for data security in the cloud environment. In that,

D. Suneetha (✉)
Krishna University, Machilipatnam, Andhra Pradesh, India
e-mail: sunithadavuluri8@gmail.com

R. Kiran Kumar
Department of CSE, Krishna University, Machilipatnam, Andhra Pradesh, India
e-mail: kirankreddi@gmail.com

© Springer Nature Singapore Pte Ltd. 2019
A. J. Kulkarni et al. (eds.), *Proceedings of the 2nd International Conference on Data Engineering and Communication Technology*, Advances in Intelligent Systems and Computing 828, https://doi.org/10.1007/978-981-13-1610-4_21

steganography is also one of the prominent techniques for providing security for our data.

Steganography is a technique whereby data is encoded which is hidden; visibility of the resultant data might not be visible to the typical human eye. In a similar way, cryptography techniques are used where the data is encrypted by generating a secure key. Whereas steganography technique is the challenging one because the resultant file is not noticeable by typical human eye, in cryptography techniques, if u know the key then the unauthorized person easy to decode the data. There are three different types of methods in steganography, i.e., steganography in the form of images, steganography in the form of audio and video files. Due to the growing need of security, steganography in the form of pictures is very popular nowadays. In this technique, data is embedding in the image that resultant image is stored in the cloud environment instead of storing original data in them. Security of any steganography techniques is based on the selection of pixels for inserting secret message. Therefore edge pixels a better idea to place the secret data. Instead of taking the whole image as input, partition the image into multiple parts so we can hide a huge amount of data.

In this paper, we have proposed an image-based steganography technique which can protect and hide the secret data from unauthorized persons by placing it only in the partition-based random edge pixels. This technique has produced excellent security for user data against steganalysis attacks. The performance of this proposed technique is evaluated with MSE and PSNR values and compared with various image steganography algorithms.

The rest of the paper is organized as follows. Section 2 discusses some previous image-based steganography techniques and various observations. Section 3 discusses an efficient partition-based random steganography technique. Section 4 shows experimental results, and Sect. 5 gives the conclusions and future scope.

2   Literature Review

There exist various image based steganographic techniques are present to insert data securely from unauthorized users [1]. The image-based techniques can be classified into two categories of domains: spatial and frequency domains. In most of the spatial domain, steganography techniques are based on the least significant bit (LSB) substitution in which the LSB of the pixels is selected to embed the secret message. In this, we have two different classifications: LSB replacement and LSB matching [2]. In LSB replacement technique, the LSB bit of each pixel is replaced with secret data. In LSB matching, the LSB bit is replaced with the next bit of the LSB; if those are not matched, we cannot insert the data. In both the techniques, we have multiple combinations of LSB bit; finally, we have multiple ways of an embedded secret message [3, 4].

Another embedding technique, known as pixel value difference technique (PVD), has been proposed [5]. In this technique, the cover image is divided into number of blocks which are adjutant to each other and finally select random adjust pixels to embed the original data.

In hiding behind corners (HBC) [6] technique, corner pixels are used to embed the secret message. In this technique, read the input image, and convert the secret image into binary. First, select the corner pixels of an image; after selecting, directly embed secret data into the selected pixels. HBC [7] also leads to poor security because attackers easily identify the corner pixels of an image.

Some of the image steganography techniques are based on LSB matching revisited technique [8]. In this, we calculate the different threshold values of original image and stego image. If the threshold values are matched or up to some measurement inserts secret data into those pixels directly. In this, everything is based on the formulas which are used to calculate the threshold values. If an attacker grabs the formulas, then easily decode the original data. In some of the algorithms, use Markov transition matrix to compute various features of an image [9]. But it may have minor degrading in performance.

3 Observations

This section discusses some observations that are used to select edge pixels for embedding secret data. In Fig. 1, it is seen that embedding in edge pixels leads to changes in edges of the stego image.

In order to keep no changes in edge pixels before and after the insertion of data, the least significant bit of the original image is masked, and after that, apply edge detection techniques for converting original cover image to edge-based image. In Fig. 2 it is seen that embedding in edge pixels after applying masking process.

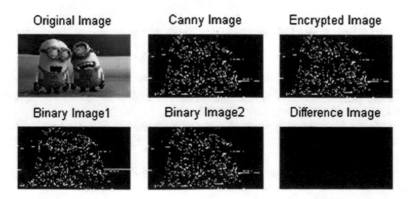

Fig. 1 Changes in embedding in edge pixels [2]

Fig. 2 Changes in embedding in edge pixels

4 Proposed Technique

In this proposed method, we use partition-based random edge pixels of an image. First, the original image is converted into grayscale image. Then, we partition the original cover image into nine partitions because in RGB plane we have three colors so multiple of 3 is 9. Then, we identify edge pixels of a cover image by using Canny edge detection method for each partition. After obtaining edge image of each partition selection prime number based random pixels of an image. Secondly, read the original file which consists of original data and encrypt the data by using encryption algorithm and obtain the key, and then we hide the key in those selected pixels.

At the decrypter side, the stego object is again masked, and then the edge pixel algorithm is applied to identify the edge pixels and then will get same edge pixels at the sender and receiver. Thus, we identify where the key is hidden and then decrypt the data based on the key by using decryption algorithm.

5 Algorithm for Server Side

Input: Input image, input message.
Output: Stego object.
Step 1: Start.
Step 2: Import image using imread() function.
Step 3: Select the image and convert into gray scale (original image and grayscale image are same in size) using formula rgb2gray.
Step 4: Divide the original cover image into nine parts.
Step 5: Read the original text and encrypt original data using RSA algorithm and obtain the public key.
Step 6: Obtain edge-based images for nine parts using Canny edge detection algorithm.
Step 7: Mask the LSB values of each part to avoid difference of pixels in between the original image and edge image.
Step 8: Select prime number-based random pixels from each part of a Canny image.
Step 9: Hide the key secret message or the into the selected pixels.

Fig. 3 Cover image

Step 10: Obtain the stego object.
Step 11: Stop.

6 Algorithm for Receiver Side

Input: Stego object.
Output: Secret message.
Step 1: Start.
Step 2: Import stego image using imread() function.
Step 3: Identify edge pixels from grayscale image using Canny edge detection algorithm.
Step 4: Identify prime number-based random edge pixels from Canny edge image.
Step 5: Read the hiding data from the image and decrypt original data and obtain key.
Step 6: Decrypt original text based on key using decryption algorithm.
Step 7: Stop.

7 Experimental Results and Analysis

The proposed algorithm is used to embed secret data in the position of selected pixels, which meets the requirements of both perception and robustness and produces good results. The proposed algorithms for encoding and decoding processes are used, respectively. The images are taken from the data set http://sipi.usc.edu/database/. We have used different grayscale images for justifying the process. Figure 3 shows the original input image; Fig. 4 shows the original input image after dividing into nine equal parts. Figure 5 shows the cover image after applying edge detection algorithm, and the final stego image after applying the proposed algorithm is shown in Fig. 6.

The proposed algorithm is also applied for another image. The original image is shown in Fig. 7. Figure 8 shows the original image after dividing into nine equal parts. Applying Canny edge detection algorithm for the input image is shown in Fig. 9. Figure 10 shows the final image after applying the proposed algorithm.

Fig. 4 Cover image after
dividing into parts

Fig. 5 Cover image after
applying edge detection
algorithm

Fig. 6 Stego image

Fig. 7 Cover image

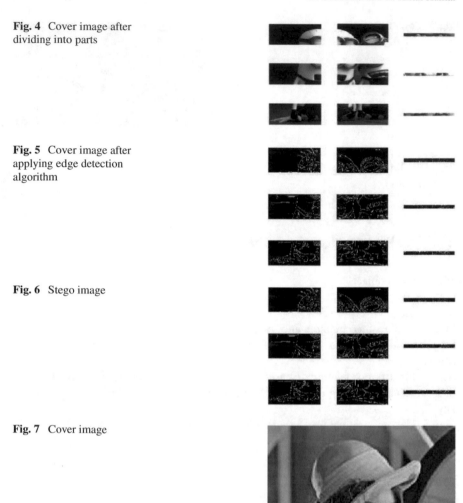

8 Calculation of PSNR (Peak Signal-to-Noise Ratio) and MSE (Mean Square Error)

The peak signal-to-noise ratio and mean square error are two important measurements for calculating picture quality and noise ratio.

The PSNR is used as a quality measurement between the original and a processed image. The higher the PSNR, the better the quality of the reconstructed image. The

Fig. 8 Cover image after dividing into parts

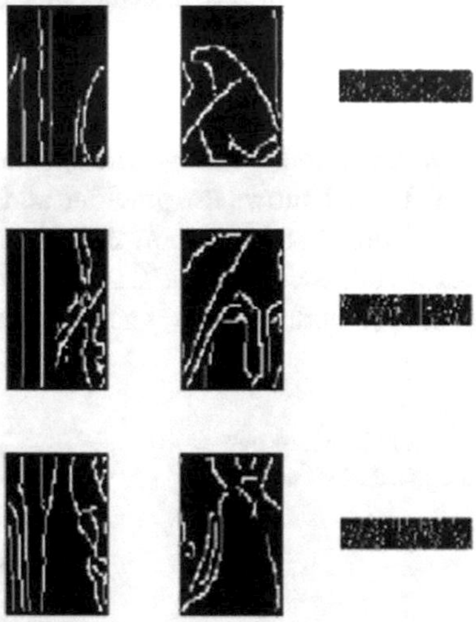

Fig. 9 Cover image after applying edge detection algorithm

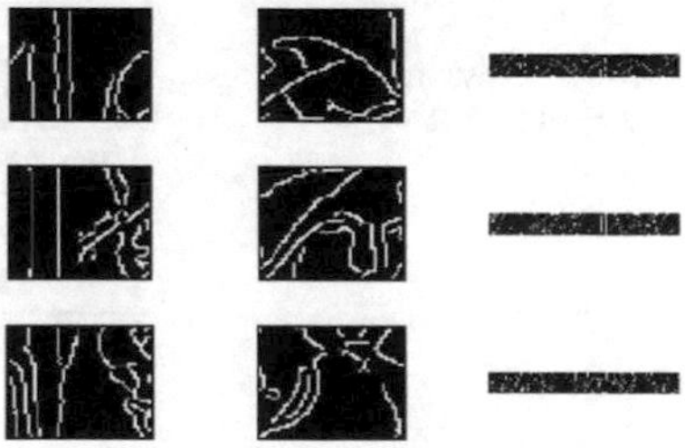

Fig. 10 Stego image

MSE is used as a cumulative squared error between the original and a processed image.

$$\text{PSNR} = 10\log_{10}(\text{MAX}_i^2)/\text{MSE}. \tag{1}$$

$$\text{MSE} = \sum M, N[I_1(m, n) - I_2(m, n)]^2/M * N. \tag{2}$$

9 Comparison Table and Chart

This section contains a comparison between the previous work with various algorithms and proposed algorithm calculated values of both PSNR and MSE. It clearly is seen that the calculated values show some significant decrement which suggests that the proposed algorithm is slightly better than the previous approaches. MATLAB tools are used for evaluating the results of the output image.

Previous MSE means the obtained MSE values using Fibonacci edge-based algorithms, and both previous and proposed MSE values are calculated for same size of images.

Previous PSNR means the obtained PSNR values using Fibonacci edge-based algorithms, and both previous and proposed PSNR values are calculated for same size of images. Table 1 shows the previous and proposed MSE values of various images, and Table 2 shows the previous and proposed PSNR values of various images.

Figures 11 and 12 show the comparative analysis of previous values and calculated values of PSNR and MSE values where it clearly shows the proposed algorithm has produced better results when compared to previous approaches.

Table 1 Previous and proposed MSE values

Cover image	Previous MSE	Proposed MSE
Lena	0.0046	0.0042
Mini	0.0043	0.0041
Cameraman	0.0020	0.0017
Baboon	0.0004	0.0003

Table 2 Previous and proposed PSNR values

Cover image	Previous PSNR	Proposed PSNR
Lena	73.01	75.67
Mini	74.89	76.89
Cameraman	82.98	83.99
Baboon	83.01	84.02

Fig. 11 Comparison chart for MSE value

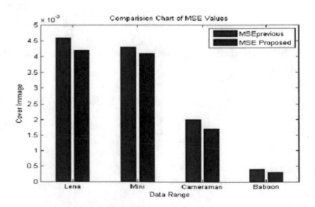

Fig. 12 Comparison chart
for PSNR value

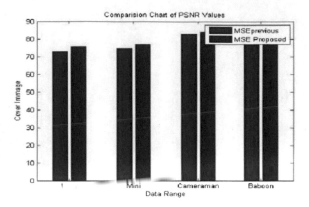

10   Conclusion and Future Scope

In this paper, original user secret data is inserted into the selected pixels based
on the partition-based random prime number pixels. This proposed algorithm has
produced better significant results when compared to the previous approaches. This
proposed algorithm works on various types of data like textual, images, and videos for
embedding and producing better results. In this approach, for obtaining secret public
key use RSA algorithm; hence, the hidden data is also damaged by the third party,
so if we hide the data with new algorithm it improves the performance measurement
of entire process.

References

1. Anderson RJ, Petitcolas FAP (1998) On the limits of steganography. IEEE J Sel Areas Commun
   16(4):474–481
2. Chandramouli R, Memon N (2001) Analysis of LSB based image steganography techniques.
   In: Proceedings 2001 international conference on image processing, vol 3. IEEE, New York
3. Christina L, VS Joe Irudayaraj (2014) Optimized blowfish encryption technique. IJIRCCE
   2:2320–9798
4. Provos N, Honeyman P (2003) Hide and seek: an introduction to steganography. IEEE Secur
   Priv 99(3):32–44
5. Patil P et al (2016) A comprehensive evaluation of cryptographic algorithms: DES, 3DES, AES,
   RSA and Blowfish. Procedia Comput Sci 78:617–624
6. Awwad YB, Shkoukani M (2017) The affect of genetic algorithms on blowfish symmetric
   Algorithm. IJCSNS 17(3):65
7. Sumathi CP, Santanam T, Umamaheswari G (2014) A study of various steganographic techniques
   used for information hiding. arXiv preprint arXiv:1401.5561
8. Van Cleeff A, Pieters W, Wieringa RJ (2009) Security implications of virtualization: a literature
   study. In: International conference on computational science and engineering, CSE'09, vol 3.
   IEEE, New York
9. Islam S, Modi MR, Gupta P (2014) Edge-based image steganography. EURASIP J Inf Sec
   2014(1):8

Large Scale P2P Cloud of Underutilized Computing Resources for Providing MapReduce as a Service

Shashwati Banerjea, Mayank Pandey and M. M. Gore

Abstract In today's world, data is growing at an unprecedented rate. This vast amount of data brings in the challenge of efficient storage and processing. The common solution adopted by organizations is to hire the service of some cloud service provider. However, the cost of cloud computing is not negligible. The machines present in the government offices, homes, and other organizations are well equipped with storage and processing power and are connected through high-speed fiber optical LAN. Generally, the machines are used for Internet surfing and basic utilities. This requires little amount of storage and processing capabilities. Thus, most of the computing and storage capabilities remain underutilized. The motivation of the current work is to harness the underutilized resources present across different educational institutes and develop a framework that can provide Computation-as-a-Service. We have conducted simulation of the algorithms of our proposed model and compared it with existing algorithms. The simulation results prove the effectiveness of the proposed algorithm.

Keywords MapReduce · Chord · DHT · Publish/Subscribe

1 Introduction

Current times are witnessing an unimaginable growth of digital data. The availability of this huge data helps data analytics to provide inputs for correct prediction of trends and more accurate decision making. However, this large amount of data poses the

S. Banerjea (✉) · M. Pandey · M. M. Gore
MNNIT Allahabad, Allahabad, India
e-mail: shashwati@mnnit.ac.in

M. Pandey
e-mail: mayankpandey@mnnit.ac.in

M. M. Gore
e-mail: gore@mnnit.ac.in

© Springer Nature Singapore Pte Ltd. 2019
A. J. Kulkarni et al. (eds.), *Proceedings of the 2nd International Conference on Data Engineering and Communication Technology*, Advances in Intelligent Systems and Computing 828, https://doi.org/10.1007/978-981-13-1610-4_22

challenge of its efficient storage, retrieval, and analysis. A common solution adopted by organizations is to hire the services of some cloud service providers.

Hiring the services from cloud service providers presents a number of advantages. The clients need not worry about the maintenance of the software and hardware. The clients only have to pay for the service that they intend to use. However, in order to meet the ever-increasing demand of customers, the cloud service providers are forced to continuously increase the storage and processing capacity of their data centers. This leads to a situation where service providers have to incur a lot of expenditure toward maintaining huge data centers which ultimately makes the cost of hiring cloud services too high for the customers.

While large data centers are built at a great cost to gather computational resources at a single location, millions of computers present in educational institutes, organizations, and homes are just being utilized for Internet surfing and executing basic software utilities. These activities require minimal amount of computational resources making the machines underutilized most of the time. We have conducted an experiment to estimate the amount of underutilized resources present in our institute. The details of the experiment can be found at [1].

The situation of underutilization of resources is more or less same in all the higher educational institutes. We intend to aggregate the underutilized resources present across different institutes to form an integrated peer-to-peer cloud (P2P-Cloud) that can provide Computation-as-a-Service. MapReduce is a popular paradigm for large-scale data processing. However, deploying MapReduce in an integrated environment poses a number of challenges. The first challenge is communication between the nodes present in two different institutes. The nodes internal to the institute are situated behind NATs/proxies and hence cannot establish direct TCP link with nodes outside the institute. This makes these resources hidden from the public domain. The second challenge is unpredictable nature of nodes. The nodes in a peer-to-peer environment cannot be and should not be forced to stay in the system. If the nodes leave the system in the middle of task execution, that task may need to be re-executed and this may happen repeatedly. The third challenge is heterogeneity. Both the nodes and jobs are heterogeneous in terms of storage and computational requirements. Further, the computational capability of nodes keeps changing with task assignment, task execution, and task migration. The current work presents the design of a network substrate for aggregating underutilized computing resources present across different institutes. Furthermore, we have conducted a simulation to compare the performance of the proposed work with the existing algorithms. The paper is structured as follows. Section 2 sheds light on the related work. Section 3 presents the proposed P2P-Cloud model. Section 4 presents the simulation and comparison of P2P-Cloud with existing algorithms. Section 5 concludes the paper.

2 Related work

The computational resources have a set of attributes. Some research efforts [2–4] have been made to provide support for searching resources with multiple attributes over DHT-based P2P overlay. In [4–6], all the attribute information is stored on a single overlay. In these approaches, a job specification with m attributes requires m messages to search a resource. This requires merging a tremendously high volume of information to derive the outcome of the job specification. Further in [2, 3, 7, 8], multiple overlays have been used for storing each of the attributes. Processing such a query requires processing each sub-query in parallel on different overlays and then concatenating the results like a database join operation. This approach is not scalable as multiple overlays require high maintenance which increases with increase in churn.

Another limitation of these approaches is that there is no provision to store the incoming resource requests which can be handled at a later stage. When a new job request arrives, the node storing the resource information checks for the available set of resources. If the requested resources become available, the job request is served; otherwise, there is no other way but to drop the request. Also, it may happen that just after dropping the request, the resources are available. In addition to this, the resource profile of a node changes with task assignment, task migration, and task completion. The repository storing the resource information has to be updated continuously. This demands a network substrate which can adapt to the dynamic resource profile of nodes, subscribe the incoming job requests, and notify the subscribers upon availability of desired resources.

3 P2P-Cloud Model

The higher educational institutes across India are connected through high-speed (10 Gbps) IP-based backbone called National Knowledge Network (NKN). NKN provides a handful of public IP addresses to every institute. Each institute configures some of the nodes using the public IP address that can be reached globally. The common assumption while designing a large-scale P2P-Cloud is that the nodes forming the overlay can establish direct TCP links between them. However, the nodes present in an institute are situated behind NAT/proxy. Generally, NATs/proxy does not allow two nodes from different private networks to establish end-to-end transport layer links [9, 10].

In order to handle the problem of NATs/proxies, we have used the NKN nodes to act as bridge between the public and private domain. Our proposed P2P-Cloud model consists of two overlay layers (tiers) (see Fig. 1). The upper tier is formed by the NKN nodes which are equipped with two Network Interface Cards (NICs) also known as LAN cards. One LAN card is configured with private IP range of institute, and the other LAN card is configured with one of the public IP assigned to institute. Using the

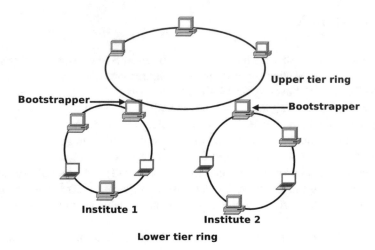

Fig. 1 P2P-Cloud model

public IP, the nodes establish direct TCP connection with other such types of nodes present in different organizations thus forming upper tier of DHT. The private IP is used for communication with the nodes present in the institute. Essentially, these upper tier nodes act as bridge between the private and public domain. The internal nodes of the institute form the lower tier ring of the P2P-Cloud.

3.1 Publication Installation

The volunteers of the P2P-Cloud are usually the internal nodes present in the institute. After joining, a new node publishes its available computational capability. The publication is in the form of attribute–value pairs. The attributes are unused RAM, processor model, number of cores, idle CPU time, storage, and available number of instances. The publication is directed to a rendezvous node which stores related publication and subscription. The rendezvous node present in the lower tier ring of the institute is called *local class node* and the one in the upper tier ring is called *global class node*.

Figure 2 presents an example of installation of computational power publication in P2P-Cloud. There are two chord applications executing on the bootstrapper. Thus, the bootstrapper has two ids which are depicted as NodeId 78 for upper tier and NodeId 89 for lower tier. A new node with NodeId 51 joins the lower tier ring of Institute 1. The node publishes its available computational power which gets stored on *local class node*. The *local class node* (NodeId = 21) periodically sends the number of available instances belonging to this class to the bootstrapper (NodeId = 89). If the computational power of a node changes, the node re-publishes its changed computational capability at the appropriate *local class node*.

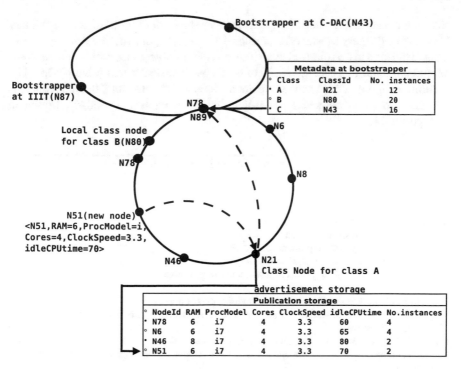

Fig. 2 Publication installation in P2P-Cloud

The bootstrapper keeps a record of the number of instances present in each *local class node*. The bootstrapper further forwards this information to the *global class node*. The number of *global class nodes* is same as the *local class nodes* present in each lower tier ring. *Global class nodes* maintain a metadata of the number of instances present in each of the participating institute for the corresponding class. This scheme ensures that even if the number of nodes present in an institute is less than the desired number of nodes, the job can be executed by routing it to some other institute with the required number of nodes.

3.2 Job Subscription Installation

The job is submitted at one of the bootstrappers. The job subscription specifies the minimum computational capability required for executing each map/reduce task and the total number of tasks. The number of attributes specified in job subscription is same as a node's computational capability publication. Unlike publication of node's computational power, the job subscription can be routed from the host institute (the institute responsible for job management) to other institutes. The bootstrapper extracts the attribute values from job subscription. If sufficient instances are available in the institute, the bootstrapper calculates the hash of jobId to get the Primary Master. The Primary Master is responsible for job management. The Primary Master assigns the map tasks followed by reduce tasks to reliable nodes with sufficient computational capability. The reliability of nodes has been estimated using EWMA [1]. If sufficient number of instances are not available in the institute, the job request is forwarded to *global class node*. The algorithm for new job subscription is presented in Algorithm 1.

Algorithm 1 Job subscription installation in P2P-Cloud

INPUT: Task specification $S = \{(a_1, v_1), (a_2, v_2), \ldots (a_x, v_x), instances\}$
OUTPUT: Job subscription installation at local and global class node
1: Extract the attribute values from subscription S
2: Determine the local class node from the attributes
3: **if** sufficient instances found **then**
4:    Direct the job request to primary master
5:    Primary master: $Hash(className) \rightarrow n_i$
6:    Notification message to primary master consisting of nodeId
7: **else**
8:    Host institute routes the subscription to global class node
9:    Global class node forwards the IP address of bootstrappers of institutes with sufficient instances to host bootstrapper
10:   Host bootstrapper forwards job request to peer bootstrapper
11:   Peer bootstrappers acknowledge the host bootstrapper by sending the available number of instances
12:   Execute job at different institutes depending on the number of available instances
13: **end if**

---

4 Simulation

We have simulated our work on PeerSim [11] simulator. The goal of the simulation is to compare the slave selection algorithm of P2P-Cloud with the slave selection algorithms of P2P-MapReduce [12], P2P-CaaS [1], and FCFS. The simulation has been carried out on a network of size 7000 nodes. The performance of the algorithms is evaluated under different churn conditions. In order to simulate churn, we have defined a joining and leaving rate. We have used four values of churn, viz. 0.2, 0.6, 0.8, and 0.9 to evaluate the system under different churn conditions. These values are expressed as percentage of network size. For example, if the churn rate is 0.2 and network size is 7000, it implies 14 nodes are leaving per minute. We have kept the joining and leaving rate to be same. We have considered short jobs with input file sizes 512, 768, 1024, and 1280 MB. We have compared the number of task re-executions (Fig. 3) and total computing time (Fig. 4) in each of the algorithms when subjected to different churn rates. The scheduling strategy of each of the algorithms is as follows.

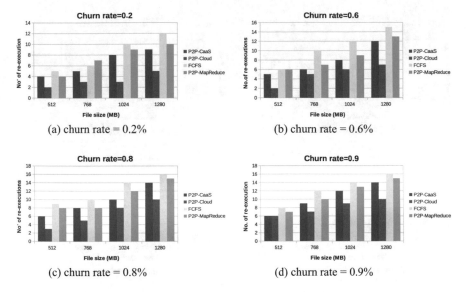

Fig. 3 Comparative study of number of task re-executions

1. P2P-MapReduce [12] assigns tasks to nodes with the lowest workload. The nodes with higher values of idle CPU time denote the lightly loaded nodes. Thus, task assignment policy is to assign tasks to nodes in the decreasing order of their idle CPU time.
2. The First Come First Served (FCFS) scheduling algorithm is based on assigning task to nodes in the order of their joining.
3. P2P-CaaS [1] assigns tasks only on the basis of the estimated success rates.
4. The scheduling strategy of P2P-Cloud assigns tasks to nodes by considering both the computing capability and reliability of nodes.

The observations are as follows.

1. The number of task re-execution in case of P2P-Cloud is less than P2P-MapReduce, P2P-CaaS, and FCFS (see Fig. 3).
2. Decrease in the number of task re-execution leads to decrease in the total computing time (see Fig. 4).

There are two reasons for task re-executions. The first reason is the assignment of tasks to nodes with insufficient computational capability. The task will fail repeatedly until it is assigned to node with sufficient computational power. The second reason for task re-execution is abrupt leaving of node in the middle of task execution. The scheduling algorithm of P2P-MapReduce and FCFS neither considers the heterogeneity nor the abrupt leaving of nodes. This increases the total number of task re-executions which in turn increases the total computing time. P2P-CaaS considers the abrupt leaving of nodes and assigns tasks to nodes on the basis of their estimated success rates. However, the heterogeneity of nodes and jobs is not taken into

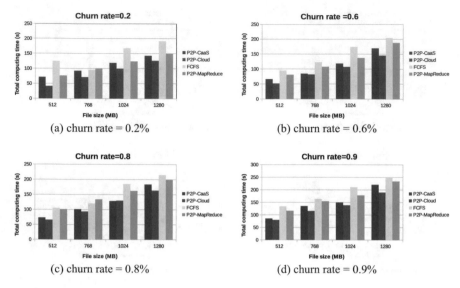

Fig. 4 Comparative study of total computing time

account. P2P-Cloud considers both the available computational nodes and reliability before assigning task. Thus, the number of task re-executions is reduced which in turn minimizes the total computing time.

5 Conclusion

We have designed and implemented a Computation-as-a-Service framework over underutilized resources present across different institutes (P2P-Cloud). We have proposed a publish/subscribe-based model for asynchronous matching between resource publisher and job subscriber. Furthermore, in order to reduce the number of task re-executions occurring due to abrupt leaving of nodes, we have designed a slave selection algorithm that selects nodes on the basis of their estimate success rates before task assignment. The simulation results prove the effectiveness of the proposed model.

References

1. Banerjea S, Pandey M, Gore MM (2016) Towards peer-to-peer solution for utilization of volunteer resources to provide computation-as-a-service. In: 30th IEEE international conference on advanced information networking and applications, AINA 2016, Crans-Montana, Switzerland, 23–25 March, 2016, pp 1146–1153

2. Andrzejak A, Xu Z (2002) Scalable, efficient range queries for grid information services. In: Proceedings second international conference on peer-to-peer computing, pp 33–40

3. Bharambe AR, Agrawal M, Seshan S (2004) Mercury: supporting scalable multi-attribute range queries. SIGCOMM Comput Commun Rev 34(4):353–366

4. Cai M, Frank M, Chen J, Szekely P (2004) MAAN: a multi-attribute addressable network for grid information services. J Grid Comput 2(1):3–14

5. Cai M, Hwang K (2007) Distributed aggregation algorithms with load-balancing for scalable grid resource monitoring. In: 2007 IEEE international parallel and distributed processing symposium, pp 1–10

6. Oppenheimer D, Albrecht J, Patterson D, Vahdat A (2004) Scalable wide-area resource discovery, technical report TR CSD04-1334

7. Ratnasamy S, Hellerstein JM, Hellerstein JM, Shenker S, Shenker S (2003) Range queries over DHTs

8. Talia D, Trunfio P, Zeng J, Hörqvist M (2006) A DHT-based peer-to-peer framework for resource discovery in grids

9. Huang Y, Fu TZ, Chiu DM, Lui JC, Huang C (2008) Challenges, design and analysis of a large-scale P2P-VoD system. SIGCOMM Comput Commun Rev 38(4):375–388

10. Li B, Xie S, Qu Y, Keung GY, Lin C, Liu J, Zhang X (2008) Inside the new coolstreaming: principles, measurements and performance implications. In: IEEE INFOCOM 2008 - The 27th conference on computer communications

11. Peersim. https://peersim.sourceforge.net. Accessed 27 Nov 2016

12. Marozzo F, Talia D, Trunfio P (2012) P2P-MapReduce: parallel data processing in dynamic cloud environments. J Comput Syst Sci 78(5):1382 – 1402 (JCSS Special Issue: Cloud Computing 2011)

Topic Modelling for Aspect-Level Sentiment Analysis

Pratik P. Patil, Shraddha Phansalkar and Victor V. Kryssanov

Abstract The field of sentiment analysis identifies and analyzes the public's favorability in terms of an entity. The advent of social media websites and e-commerce applications has enhanced the significance of analysis of user reviews for the product manufacturers to obtain insights into the acceptability of products. Aspect-level sentiment analysis helps them gain better understanding of the acceptance of different features of the products in specific. This work presents a workflow model of topic extraction (aspect detection) and sentiment detection of user reviews using frequency-based approach and unsupervised machine learning methods over 7.8 million reviews of Amazon users. The work also employs methods to estimate the accuracy of the analytical model using the classical methods like logistic regression, support vector machine, and naïve Bayesian approach.

Keywords Sentiment analysis · Linguistics · Machine learning · Aspect level
Topic modelling · Amazon reviews · Unsupervised machine learning

1 Introduction

In the past few years, there has been tremendous increase in social media websites, e-commerce websites, blogs, and personalized websites, which in turn has popularized the platforms/forums for the expression of public opinions. Nowadays, social media websites like Facebook, Twitter, Pinterest, and LinkedIn, and review websites like Amazon, IMDB, and Yelp have become popular sources to retrieve public

P. P. Patil (✉) · S. Phansalkar
Symbiosis International University, Pune, India
e-mail: pratikpatil1592@gmail.com

S. Phansalkar
e-mail: shraddhap@sitpune.edu.in

V. V. Kryssanov
Ritsumeikan University, Kusatsu, Japan
e-mail: kvvictor@is.ritsumei.ac.jp

© Springer Nature Singapore Pte Ltd. 2019
A. J. Kulkarni et al. (eds.), *Proceedings of the 2nd International Conference on Data Engineering and Communication Technology*, Advances in Intelligent Systems and Computing 828, https://doi.org/10.1007/978-981-13-1610-4_23

opinions [1]. Sentimentally analyzing this huge corpus of opinions can help the manufacturers in realizing the public opinion of the products and the users in buying the right product. The user opinions are always in textual form. Aspect-level sentiment analysis can further help the manufacturers realize the public opinion for finer granularity of different parts (components). In this paper, the proposed system implements and validates a workflow of topic modelling for aspect-level sentiment analysis. The experiments are conducted on Amazon.com reviews dataset and validated at Ritsumeikan University, Japan.

1.1 Sentiment Analysis

Sentiment analysis, also known as opinion mining, is a study in the field of computational linguistics and natural language processing (NLP), to recognize the sentiment, opinion, attitude, and emotion of the user with respect to an entity [2, 3]. Entity here depicts any event, individual, or any topic. Sentiment analysis is useful in determining the sentiment of the text with respect to polarity (positive, negative, or neutral). Sentiment analysis is spread across all the domains such as social media, health care, management [4].

1.1.1 Levels of Sentiment Analysis

Sentiment analysis can be considered as a classification process involving three main classification levels—document level, sentence level, and aspect level [4]. This classification is based on the artifacts that are analyzed and their outcomes. Document-level sentiment analysis finds out the overall sentiment of the complete document. Sentence-level sentiment analysis finds out the sentiment of a sentence. Aspect-level sentiment analysis is when the topics (aspects) are extracted from the text using various algorithms and then sentiment for each topic is evaluated. It can also be known as entity/feature-level sentiment analysis. Sentiments of multiple entities present in a single sentence can be determined. For example, *"The food at the hotel is fantastic, however, the ambiance is not so great"*.

1.1.2 Aspect-Level Sentiment Analysis

Aspect-level sentiment analysis is the study of finding the sentiment involved with respect to the aspects. Aspects can also be considered as features or entities. Aspect-level sentiment analysis can be achieved with three processing steps as mentioned in [5], which are identification, classification, and aggregation. The first step involves in the identification of the sentiment-target pairs contained in the textual data. The second step is the classification step, where these sentiment-target pairs are classified based on the polarity, viz positive, negative, and neutral. In some cases, the classifi-

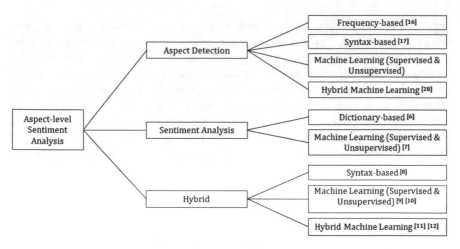

Fig. 1 Taxonomy for aspect-level sentiment analysis approaches

cation has been known to be performed based on the aspects as well. In the final step, the available classified data are aggregated to get a short but descriptive overview. There are some core important components like flexibility, robustness, and speed [6] (Fig. 1).

As described by Schouten et al. in [6], aspect-level sentiment analysis can be performed by three methods, viz aspect detection, sentiment analysis, and hybrid approach which is the combination of the above two methods. This can be made by detecting the most frequently appearing words (frequency-based) [7], checking syntax of the text (syntax-based) [8], using various machine learning (supervised and unsupervised) approaches like latent Dirichlet allocation (LDA), latent semantic indexing (LSI), and K-means for detection and clustering of the aspects, and lastly by using a combination of multiple approaches (hybrid). Hybrid approach can be achieved by using serial hybridization as mentioned in [9]. Sentiment analysis is the second-mentioned approach which involves in detecting the sentiment(s) of the aspects that are identified in aspect detection phase. This involves dictionary-based approach [10], in which a dictionary is a retained-containing set of positive, negative, and neutral words and phrases, which can be compared to the textual data in order to find the right sentiment polarity. WordNet is a very commonly used lexical database, widely known to group the English words into groups synonym called synsets. Secondly, machine learning (supervised and unsupervised)-based algorithms can also be used. In [11], a neural network-based supervised machine learning approach is implemented involving a perceptron method called PRanking. This PRanking algorithm is used on the topics clustered using multi-grained LDA (MG-LDA), to perform sentiment analysis on the topics extracted, while in unsupervised approach, a syntax-based approach is used in which the sentiment of the topic modelled/phrase is determined by checking the nearby vicinity of the phrase. The polarity is assigned using relaxation labelling to the phrase determined. The hybrid approach is the third-mentioned approach, which involves the combination of both aspect detection and sentiment analysis. Syntax-based approach which is explained in [12] had a limitation which could only detect one aspect per sentence. To overcome this limitation, in [13], a conditional random field (CRF) model was proposed to detect the multiple aspects in a single sentence. Aspect-sentiment pairs were found in [14], where an unsupervised probabilistic approach was modelled and performed on restaurant data, two approaches which involve LDA and MaxEnt classifier working together are defined in [15], the LDA model is enriched by MaxEnt classifier. However, in [16], LDA and MaxEnt classifier work together to optimize the words that influence the process of generation of words.

2 Proposed Approach

In this work, the product reviews have been used for topic modelling and for prediction of the review sentiment accuracy, and was developed at Ritsumeikan University. The proposed system uses frequency-based and unsupervised machine learning approaches to identify and cluster the respective set of words with respect to topics. For example, the stemmed words related to mobile phone were clustered in one topic, while those of washing machine were clustered in a separate topic. The proposed system was designed in python using NLTK and Scikit [17] packages. LDA and k-means algorithms have been used to extract topics from the given dataset. Accuracy of the model was then calculated with respect to dataset using logistic regression (LR), support vector machine (SVM), and naïve Bayes (NB) algorithms. The proposed system worked as a recommender system using aspect-level sentiment analysis. Figure 2 shows the workflow of proposed system with respect to phases divided into steps:

Fig. 2 Proposed system workflow

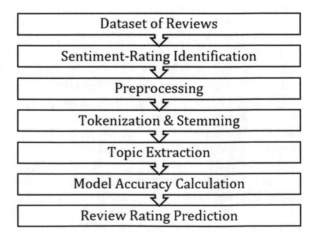

Dataset of Reviews

Sentiment-Rating Identification

Preprocessing

Tokenization & Stemming

Topic Extraction

Model Accuracy Calculation

Review Rating Prediction

Table 1 Rating-polarity assignment for Amazon.com reviews

Ratings	★	★★	★★★	★★★★	★★★★★
Polarity	Negative		Neutral	Positive	

a. Data Preparation and Preprocessing

Review Dataset: Experimentation is performed on the review data collected from Amazon.com by Julian McAuley at University of California, San Diego, introduced and explained in [18, 19]. The complete Amazon.com dataset consists of approximately 143 million reviews involving various categories like books, electronics, cell phone, movies and TV clothing, and home and kitchen. The electronics (7.8 million) reviews dataset has been used as a dataset for the proposed system and was in Json format.

Rating-Sentiment Identification: Amazon.com reviews are rated in an ordinal of 1–5 star(s), where 1 being the lowest and 5 being the highest. We map the Amazon.com ratings with the sentiment polarity as shown in Table 1.

Preprocessing also involves two very important processes, viz stemming and tokenization.

Tokenization and Stemming are very important steps when preprocessing data for performing topic modelling and aspect-level sentiment analysis. Important aspect words, sentiment words, and background words can be identified and distinguished. To perform tokenization and stemming, NLTK (natural language toolkit) [20] library was used. Tokenization is the process of tokenizing each sentence by spaces and letters, and ignoring special symbols and numbers. Tokenizing also removes the stop words, like "a", "an", "this", and "the". Important words like nouns, common nouns, and adjectives are identified. Stemming is the process of reducing derived word(s) to its main stem word (otherwise known as seed words) by eliminating extensions like "ing", "tion", and "ful" from the words.

b. Topic Extraction

After the data are preprocessed, cleansed, tokenized, and stemmed, the topics are extracted from the input data. The topics/aspects are mostly nouns or common nouns present in the text. Topics have been modelled by calculating and clustering the most frequently appearing terms relative to a specific field. LDA and k-means algorithms have been used to extract and cluster topics. When LDA is used to assign K topics for N documents, then each document can have one or more topics involved. While, k-means does the opposite, assigns a topic for one or more documents. Hence, LDA gives much more better and accurate results than k-means. The topics extracted consists of N number of words specified earlier which helps in labelling and describing the topic(s).

c. Model Accuracy Calculation

After topic extraction, the model accuracy is calculated to know how the sentiment polarity works for the textual data. The words present in the sentence are taken into consideration when it comes to finding the sentiment of the sentence. To calculate the model accuracy, three models were developed, one for each polarity. Three approaches were used to calculate model accuracy— logistic regression (LR), support vector machine (SVM) and naïve Bayes (NB). Cross-validation is used in prediction and to estimate the working of a predictive model. In the presence of less data, tenfold cross-validation [21] is used to split data and measure accuracy. The model accuracy results for the datasets are mentioned in Sect. 3.

3   Results

The proposed system is domain independent, but for illustration only the review data of electronics are considered. The number of topics/clusters and stemmed words (aspects) present inside every topic have been defined earlier and can be changed as and when required. Every cluster can be labelled as per which entity it represents (like *wheels*, *engine*, and *oil* in automobiles). The results obtained after clustering using k-means are better than that using LDA and are easy to label the topics/clusters appropriately. Singular value decomposition (SVD) algorithm [22] is used for dimensionality reduction. To visualize these high-dimensional data, t-distribution stochastic neighbor embedding (t-SNE) [23, 24] was used. The data were plotted using a 2D scatter plot using t-SNE. As shown in Fig. 3, the different colors represent different clusters which represent a certain entity, as shown in Table 2.

The model was trained, and dictionaries containing the keywords representing polarities were created. The candidate review text(s) and their predicted polarities identified are shown in Table 3. The model calculates the result and predicts the polarities with respect to a sentence with single aspect. Currently, we are focused on enhancing the model with multi-aspects sentences.

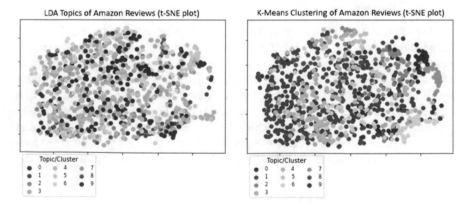

Fig. 3 Visualization of topics extracted using LDA and k-means (electronics)

Table 2 Topics derived electronics dataset

Cluster #0	Computer accessories	Cluster #5	Television and accessories
Cluster #1	Purchase online	Cluster #6	Computer hardware and software
Cluster #2	Apple accessories	Cluster #7	Cable connectors
Cluster #3	Screen protector	Cluster #8	Charger
Cluster #4	Headphones	Cluster #9	Camera

Table 3 Review polarity prediction (electronics)

Review text	Negative (%)	Neutral (%)	Positive (%)
"I love this mobile phone. Camera is awesome"	1.7	6.4	96.7
"I ordered a pair of headsets, but returned as they arrived in damaged condition"	92.7	6.4	2.5
"I hated the design of the laptop. But loved the configuration"	39.5	15.2	41.1

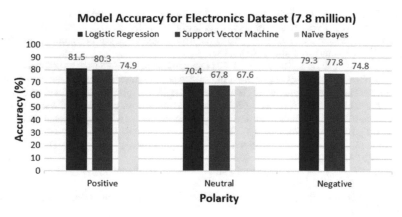

Fig. 4 Model accuracy for electronics dataset

Figure 4 represents the graph showing the model polarity accuracy calculated using the electronics dataset containing 7.8 million reviews in total. Model accuracy shows how accurately the model performs when detecting the polarities using various algorithms. Logistic regression, support vector machine, and naïve Bayes were used to calculate the accuracy.

4   Conclusion

This paper focusses on the topic modelling and how these topics can be used in order to calculate the accuracy of model and to identify sentiment of the review text accurately. Proposed system used Amazon.com review dataset to identify the key aspects and topics and to perform sentiment analysis on it. For topic extraction and clustering, LDA and k-means were used. LDA was used as it is widely used by many researchers for topic modelling, is easy to understand and implement, and gives realistic results. Post-topic extraction, model accuracy was calculated using logistic regression, support vector machine, and naïve Bayes as they are straight forward and most frequently used predictive models. The proposed system, at the moment, is predicting the polarity at document level and sentence level, where a single review is taken as input to model. As a future work, sarcasm detection and spam detection are very interesting and future course of action. Also, system can be tested against plenty of other algorithms like latent semantics analysis [25, 26], multi-grain LDA, Pachinko allocation [27].

References

1. Tang D, Qin B, Liu T (2015) Deep learning for sentiment analysis: successful approaches and future challenges. Wiley Interdisc Rev: Data Min Knowl Discovery 5(6):292–303
2. Medhat W, Hassan A, Korashy H (2014) Sentiment analysis algorithms and applications: a survey. Ain Shams Eng J 5(4):1093–1113
3. Shelke N, Deshpande S, Thakare V (2017) Domain independent approach for aspect oriented sentiment analysis for product reviews. In: Proceedings of the 5th international conference on frontiers in intelligent computing: theory and applications. Springer, Singapore, pp 651–659
4. Shivaprasad TK, Shetty J (March 2017) Sentiment analysis of product reviews: a review. In: 2017 International conference on inventive communication and computational technologies (ICICCT). IEEE, New York, pp 298–301
5. Tsytsarau M, Palpanas T (2012) Survey on mining subjective data on the web. Data Min Knowl Disc 24(3):478–514
6. Schouten K, Frasincar F (2016) Survey on aspect-level sentiment analysis. IEEE Trans Knowl Data Eng 28(3):813–830
7. Hu M, Liu B (July 2004) Mining opinion features in customer reviews. In: AAAI, vol 4, No 4, pp 755–760
8. Qiu G, Liu B, Bu J, Chen C (July 2009) Expanding domain sentiment lexicon through double propagation. In: IJCAI, vol 9, pp 1199–1204
9. Popescu AM, Etzioni O (2007) Extracting product features and opinions from reviews. In: Natural language processing and text mining. Springer, London, pp 9–28
10. Hu M, Liu B (August 2004) Mining and summarizing customer reviews. In: Proceedings of the tenth ACM SIGKDD international conference on knowledge discovery and data mining. ACM, New York, pp 168–177
11. Titov I, McDonald R (April 2008) Modeling online reviews with multi-grain topic models. In: Proceedings of the 17th international conference on World Wide Web. ACM, New York, pp 111–120
12. Zhuang L, Jing F, Zhu XY (November 2006) Movie review mining and summarization. In: Proceedings of the 15th ACM international conference on Information and knowledge management. ACM, New York, pp. 43–50

13. Marcheggiani D, Täckström O, Esuli A, Sebastiani F (April 2014) Hierarchical multi-label conditional random fields for aspect-oriented opinion mining. In: ECIR pp 273–285
14. Sauper C, Barzilay R (2013) Automatic aggregation by joint modeling of aspects and values. J Artif Intell Res
15. Zhao WX, Jiang J, Yan H, Li X (October 2010) Jointly modeling aspects and opinions with a MaxEnt-LDA hybrid. In: Proceedings of the 2010 conference on empirical methods in natural language processing. Association for Computational Linguistics, pp 56–65
16. Mukherjee A, Liu B (July 2012) Modeling review comments. In: Proceedings of the 50th annual meeting of the association for computational linguistics: long papers, vol 1. Association for Computational Linguistics, pp 320–329
17. Pedregosa F, Varoquaux G, Gramfort A, Michel V, Thirion B, Grisel O, Blondel M, Prettenhofer P, Weiss R, Dubourg V, Vanderplas J (2011) Scikit-learn: machine learning in python. J Mach Learn Res 12:2825–2830
18. McAuley J, Targett C, Shi Q, Van Den Hengel A (August 2015) Image-based recommendations on styles and substitutes. In: Proceedings of the 38th international ACM SIGIR conference on research and development in information retrieval. ACM, New York, pp 43–52
19. He R, McAuley J (April 2016) Ups and downs: modeling the visual evolution of fashion trends with one-class collaborative filtering. In: Proceedings of the 25th international conference on World Wide Web. International World Wide Web Conferences Steering Committee, pp 507–517
20. Bird S (July 2006) NLTK: the natural language toolkit. In: Proceedings of the COLING/ACL on interactive presentation sessions. Association for Computational Linguistics, pp 69–72
21. Kohavi R (August 1995) A study of cross-validation and bootstrap for accuracy estimation and model selection. In: IJCAI, vol 14, No. 2, pp 1137–1145
22. Wall ME, Rechtsteiner A, Rocha LM (2003) Singular value decomposition and principal component analysis. In: A practical approach to microarray data analysis. Springer US, pp 91–109
23. Maaten LVD, Hinton G (2008) Visualizing data using t-SNE. J Mach Learn Res 9:2579–2605
24. Maaten LVD (2017) t-SNE. [online] Available at: https://lvdmaaten.github.io/tsne/. Accessed 31 Aug 2017
25. Hofmann T (2001) Unsupervised learning by probabilistic latent semantic analysis. Mach Learn 42(1):177–196
26. Hofmann T (2000) Learning the similarity of documents: an information-geometric approach to document retrieval and categorization. In: Advances in neural information processing systems, pp 914–920
27. Blei DM (2012) Probabilistic topic models. Commun ACM 55(4):77–84

An Image Deblocking Approach Based on Non-subsampled Shearlet Transform

Vijay Kumar Nath, Hilly Gohain Baruah and Deepika Hazarika

Abstract This paper presents a non-iterative algorithm for removal of blocking artifacts from block discrete cosine transform (BDCT) compressed images using hard thresholding in non-subsampled shearlet transform (NSST) domain. The good denoising property of NSST is utilized here for the removal of blocking artifacts. A relationship between the threshold value and the mean of top left 3×3 values from the quantization table has been shown. This method can determine the hard threshold values adaptive to various BDCT-based compressed images that uses different quantization tables. Experimental results show that the proposed post-processing technique provides better objective and subjective results compared to several existing post-processing techniques.

Keywords Block discrete cosine transform · Non-subsampled shearlet transform · Image deblocking

1 Introduction

Block-based discrete cosine transform (BDCT) is a popular transform among researchers due to its good energy compaction and de-correlation property [1]. It has been considered to be the most significant component in many multimedia compression techniques like JPEG and MPEG as its performance is very near to Karhunen–

V. K. Nath · H. G. Baruah (✉) · D. Hazarika
Department of Electronics & Communication Engineering,
Tezpur University, Tezpur 784028, Assam, India
e-mail: hgbaruah1990@gmail.com

V. K. Nath
e-mail: vknath@tezu.ernet.in

D. Hazarika
e-mail: deepika@tezu.ernet.in

© Springer Nature Singapore Pte Ltd. 2019
A. J. Kulkarni et al. (eds.), *Proceedings of the 2nd International Conference on Data Engineering and Communication Technology*, Advances in Intelligent Systems and Computing 828, https://doi.org/10.1007/978-981-13-1610-4_24

Loeve transform (KLT). But the BDCT-based compression results in visually annoying blocking artifacts at higher compression ratio or low bit rate. The visibility of the blocking artifact is more prominent in homogeneous regions compared to textured regions. Challenge faced by researchers is to reduce these blocking artifacts significantly without tempering the true image details. Many methods have been found in the literature for removal of these artifacts over the past decade either in spatial domain or in transform domain. A 3×3 Gaussian filter was applied across the block boundaries in the earliest and simplest method [16] but it caused blurring around true edges. In [12], the mean squared difference of slope is minimized in order to smooth the block discontinuities. Sun and Cham [18] viewed the degradation caused by compression as additive white Gaussian noise and introduced an image deblocking technique based on fields of experts prior. Averbuch et al. [2] applied the weighted sum on the pixel quartet that is symmetrical around the block boundaries to suppress the block discontinuities. Projection onto convex set-based schemes has higher computational complexity than these methods because of their iterative nature. In [8], Gopinath et al. introduced a deblocking method based on simple wavelet domain soft thresholding. Xu et al. in [20] proposed to suppress the blocking artifacts using discrete Hadamard transform-based scheme by identifying the smooth and coarse regions and adaptively filtering them. Soft thresholding was used by Wu et al. [19] in wavelet domain for reduction of visually annoying blocking artifacts. The soft threshold value in this method is obtained from the difference between magnitude of discrete wavelet transform coefficients of the image blocks and that of the whole image. Kim [10] considered the directional activity of the blocks by directional filter for deblocking. In order to avoid excess smoothing of details, Zhang and Gunturk in [21] first identified the textured regions and the block discontinuities based on which the parameters of bilateral filter are adaptively controlled. Nath et al. in [14] improved the deblocking performance by using a bilateral filter whose parameters were determined adaptively using quantization table which was used during the compression process. Pointwise shape-adaptive DCT (SA-DCT) [6] was able to reduce blocking and ringing artifacts significantly. Chen et al. [4] considered masking system for human visual system (HVS) and adaptive weighing mechanism in transform domain for reducing blocking artifacts from low bit rate transform coded image. Nath et al. in [15] proposed a wavelet domain local Wiener filter-based approach to suppress the block discontinuities. The noise standard deviation is demonstrated to have powerful relationship with the mean of top left 3×3 values from the quantization matrix.

In this work, we present a NSST-based image deblocking method based on hard thresholding of the detail coefficients. The threshold is made adaptive to the mean of top left 3×3 values from the quantization table.

In Sect. 2, we provide the overview of NSST. Section 3 contains the details of proposed method and the experimental results, and discussion has been explained in Sect. 4. The conclusions drawn from the work is summarized in Sect. 5.

2 Non-subsampled Shearlet Transform (NSST)

Wavelet transform is a pioneering multiscale transform for capturing point singularities for both one-dimensional or higher-dimensional signals, but it is unable to capture linear singularities [9]. To deal with this problem, curvelet and contourlet transforms were introduced which allow better sparse representation of higher-dimensional singularities [3, 5]. But the existence of subsampling operation causes lack of shift-invariance property which in turn causes pseudo-Gibbs phenomenon. To overcome this, non-subsampled contourlet transform (NSCT) and NSST were proposed by eliminating downsampling operators which preserved the shift-invariance property. NSST is the multiscale, multidirectional, and shift-invariant version of shearlet transform and it is computationally less complex than NSCT.

NSST is a combination of non-subsampled Laplacian pyramid filter (NSLP) transform and several shearing filters [17]. NSLP decomposes image into low- and high-frequency components, and shearing filters lead to different subbands and different direction shearing coefficients. NSLP is done iteratively and can be expressed as follows [9]

$$\text{NSLP}_{j+1} = A_j f = \left(Ah_j^1 \prod_{k=1}^{j-1} Ah_k^0 \right) f \tag{1}$$

where NSLP_{j+1} is the detail coefficients at scale $j + 1, f$ is the image, and Ah_k^0 and Ah_j^1 are the low-pass and high-pass filters at scale j and k, respectively.

The NSLP procedure for a image f_a^0 of size $N \times N$ and D_j number of direction at a fixed resolution scale j is as follows:

1. Application of NSLP for decomposition of f_a^{j-1} into f_a^j low-pass image and f_d^j high-pass image.
2. Application of FFT on high-pass image f_d^j on a pseudo-polar grid to obtain $\hat{f}_d^j$.
3. Application of directional filters to $\hat{f}_d^j$ to obtain $\{\hat{f}_{d,k}^j\}_{k=1}^{D_j}$.
4. Application of inverse FFT to $\{\hat{f}_{d,k}^j\}_{k=1}^{D_j}$ for obtaining NSST coefficients $\{f_{d,k}^j\}_{k=1}^{D_j}$ in pseudo-polar grid.

3 Proposed Method

Let the degradation in BDCT compressed images due to quantization noise be modeled as some additive noise given by [13]

$$w = x + n \tag{2}$$

where w, x, and n are observed image with BDCT-based compression, original uncompressed image, and noise, respectively. The NSST coefficients of the image w are given by

$$W(p) = X(p) + N(p) \tag{3}$$

where $X(p)$ represents NSST coefficients of original image without compression.

To estimate $X(p)$ in the proposed method, the NSST coefficients of detail subbands are modified by hard thresholding. Hard thresholding either keeps the coefficients or sets the coefficients to zero by comparing it with the threshold. To apply the thresholding in NSST domain, we need an appropriate value of threshold that can reduce the blocking artifacts effectively with preservation of true image details. In order to understand the relationship between NSST domain hard threshold and mean of top left 3×3 values of quantization matrix ($Q_{\text{average}(3 \times 3)}$), the experiments are performed on a variety of test images where the best threshold values that provide highest PSNR are recorded for corresponding values of $Q_{\text{average}(3 \times 3)}$. The measured thresholds are plotted against $Q_{\text{average}(3 \times 3)}$ values, as shown in Fig. 1. It is observed from Fig. 1 that the $Q_{\text{average}(3 \times 3)}$ values are highly correlated to NSST domain hard threshold values.

$$Q_{\text{average}(3 \times 3)} = \frac{1}{9} \sum_{i,j=1}^{3} Q_{i,j}. \tag{4}$$

By fitting the data with a linear polynomial, the following equation has been found,

$$T = \alpha_1 Q_{\text{average}(3 \times 3)} + \alpha_2 \tag{5}$$

where $\alpha_1 = 0.0316$ and $\alpha_2 = 1.436$.

Fig. 1 Scatter plot of measured 'optimal' threshold and the average of top left 3×3 values from quantization matrix with curve fitting

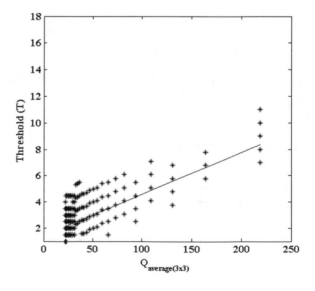

The algorithm can be summarized as follows:

1. Multiscale decomposition of the observed image with blocking artifacts is done using NSST.
2. The NSST detail coefficients are modified using hard thresholding where threshold is computed using (5). The approximation subband is kept untouched.
3. The image is reconstructed by performing inverse NSST on the modified subbands.

4 Experimental Results and Discussion

This section presents the results of the proposed deblocking technique for natural images. The performance of proposed scheme is compared with few existing techniques for Lena and Barbara images. The JPEG image deblocking based on local edge regeneration (LER) [7], deblocking via adaptive bilateral filter (ABF) [14], and deblocking using overcomplete wavelet representation (OWR) [11] is used for comparison in this paper. The deblocking results of [7, 11, 14] were achieved using MATLAB codes given by the original authors. We use NSST with three levels of decompositions here. For objective and subjective performance evaluation, two metrics, peak signal-to-noise ratio (PSNR) and structural similarity index (SSIM), are calculated. It should be noted that the SSIM index is a reliable measure of perceived image quality.

Table 1 presents the comparison of PSNR and SSIM results of proposed method with other well-known image deblocking methods, for image Barbara and Lena respectively. It is observed that the proposed method outperforms the other methods consistently for various levels of JPEG compression.

Table 1 PSNR (in dB) and SSIM comparison between various methods for JPEG image quality factors. (PSNR (in regular) and SSIM (in italics))

Qf	Barbara (512 × 512)				Lena (512 × 512)			
	LER [7]	ABF [14]	OWR [11]	Proposed	LER [7]	ABF [14]	OWR [11]	Proposed
4	24.183	24.393	23.488	**24.504**	27.654	27.828	27.608	**27.857**
	0.7766	*0.7780*	*0.7458*	*0.7916*	*0.8461*	*0.8408*	*0.8300*	*0.8421*
6	24.917	25.263	24.424	**25.318**	29.101	29.547	29.268	**29.632**
	0.8230	*0.8291*	*0.8069*	*0.8422*	*0.8889*	*0.8884*	*0.8748*	*0.8915*
8	25.412	25.885	25.150	**25.895**	29.925	30.613	30.534	**30.720**
	0.8543	*0.8680*	*0.8497*	*0.8788*	*0.9090*	*0.9123*	*0.9134*	*0.9139*
10	25.894	26.381	25.854	**26.412**	30.631	31.424	31.317	**31.571**
	0.8771	*0.8935*	*0.8795*	*0.9048*	*0.9248*	*0.9284*	*0.9295*	*0.9309*
12	26.440	26.855	26.478	**26.916**	31.397	32.054	31.918	**32.186**
	0.8915	*0.9023*	*0.8970*	*0.9198*	*0.9358*	*0.9392*	*0.9394*	*0.9409*

Bold indicates the best results

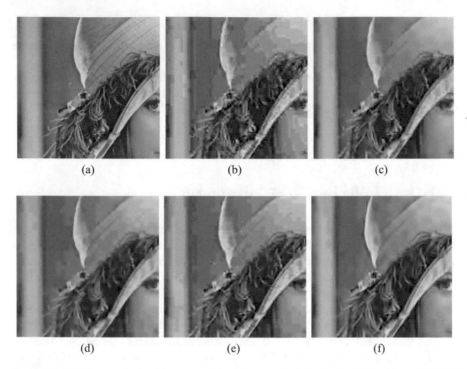

(a)   (b)   (c)

(d)   (e)   (f)

Fig. 2 Deblocking results for a fragment of *Lena* Image, **a** Original uncompressed image **b** JPEG compressed image (QF = 6). **c** LER. **d** ABF. **e** OWR. **f** Proposed method

Table 2 Computation time of different methods in seconds

Method	OWR [11]	LER [7]	ABF [14]	Proposed method
Time (s)	2.250403	9.726564	5.290793	3.897831

Figure 2 depicts the visual performance for a fragment from Lena image. In Fig. 2f, the edges look better and sharper in comparison with Fig. 2c–e. Moreover, Fig. 2f has less blocking artifacts compared to Fig. 2c–e. It can be seen that the artificial block discontinuities are reduced effectively without affecting the true image details.

Table 2 shows the actual computation time of different methods. The computation time of the proposed method is much less than the methods of [7, 14]; however, it is slightly greater than the method described in [11]. But it should be noted that the PSNR and SSIM results of proposed scheme are considerably better than the scheme discussed in [11]. Intel(R)-core(i3)-2120 CPU@3.30 GHz with 2 GB RAM and MATLAB2012b environment has been used for simulation of all the techniques.

The proposed method has also been tested for BDCT compressed images with three different quantization matrices Q1, Q2 and Q3 other than JPEG. The quantization matrices Q1, Q2, and Q3 has been taken from [11, 19]. Since the method discussed in [7] is specifically developed for JPEG compressed images, the perfor-

Table 3 PSNR comparison between [11, 14] and proposed scheme for Q1, Q2 and Q3 tables

Table	Barbara (512 × 512)				Lena (512 × 512)			
	Test image	OWR [11]	ABF [14]	Proposed	Test image	OWR [11]	ABF [14]	Proposed
Q1	25.94	26.17	26.42	**26.57**	30.70	31.56	31.64	**31.85**
Q2	25.59	25.86	26.22	**26.23**	30.09	31.11	31.19	**31.27**
Q3	24.02	24.53	24.80	**24.92**	27.38	28.63	28.66	**28.67**

Bold indicates the best results

mance comparison in Table 3 has been done only with methods [11, 14]. The proposed method performs better than [11, 14] techniques for quantization tables Q1, Q2, and Q3. It should be noted that the methods discussed in [11, 14] are developed for any BDCT compressed images unlike [7].

Except for Lena image at $Q_f = 4$, our technique outperforms all other techniques for most of the BDCT compressed images (both JPEG and non-JPEG) which shows that the deblocking performance is not much affected because of the fitting errors, and Eq. (5) can be considered for most of the images. This method with suitable modifications could also be extended to BDCT-based video compression techniques.

5 Conclusion

This work has presented a simple blocking artifacts suppression approach for BDCT compressed images using hard thresholding in NSST domain. The multiscale, multi-directional, and shift-invariance property of NSST used in the method leads to better representation of image details. The detail NSST coefficients were modified with hard thresholding. The measured 'optimal' threshold is made adaptive to the mean of top left 3 × 3 values from the quantization table. The proposed method with much less execution speed outperforms the other techniques in terms of both subjective and objective measures.

References

1. Ahmad N, Natarajan T, Rao K (1974) Discrete cosine transform. IEEE Trans Comput C–23(1):90–93
2. Averbuch AZ, Schclar A, Donoho DL (2005) Deblocking of block transform compressed images using weighted sums of symmetrically aligned pixels. IEEE Trans Circuits Syst Video Technol 14(12):200–212
3. Candès EJ, Donoho DL (2002) New tight frames of curvelets and optimal representations of objects with piecewise C^2 singularities. Commun Pure Appl Math 57:219–266
4. Chen T, Wu HR, Qiu B (2001) Adaptive postfiltering of transform coefficients for the reduction of blocking artifacts. IEEE Trans Circuits Syst Video Technol 11(5):594–602

5. Do MN, Vetterli M (2005) The contourlet transform: an efficinet directional multiresolution image representation. IEEE Trans Image Process 14(12):2091–2106
6. Foi A, Katkovnik V, Egiazarian K (2007) Pointwise shape adaptive DCT for high quality denoising and deblocking of grayscale and color images. IEEE Trans Image Process 16(5):1395–1411
7. Golestaneh SA, Chandler DM (2014) Algorithm for JPEG artifact reduction via local edge regeneration. J Electron Imaging 23(1):1–13
8. Gopinath RA (1994) Wavelet-based post-processing of low bit rate transform coded images. IEEE Int Conf Image Process 2:913–917
9. Hou B, Zhang X, Bu X, Feng H (2012) SAR image despeckling based on nonsubsampled shearlet transform. IEEE J Sel Top Appl Earth Observations Remote Sens 5(3):809–823
10. Kim NC, Jang IH, Kim DH, Hong WH (1998) Reduction of blocking artifact in block-coded images using wavelet transform. IEEE Trans Circuits Syst Video Technol 8(3):253–257
11. Liew AWC, Yan H (2004) Blocking artifacts suppression in block-coded images using overcomplete wavelet representation. IEEE Trans Circuits Syst Video Technol 14(4):450–461
12. Minami S, Zakhor A (1995) An optimization approach for removing blocking artifacts in transform coding. IEEE Trans Circuit Syst Video Technol 5(2):74–82
13. Nath VK, Hazarika D (2012) Image deblocking in wavelet domain based on local laplace prior. Int J Multimedia Appl 4(1):39–45
14. Nath VK, Hazarika D, Mahanta A (2010) Blocking artifacts reduction using adaptive bilateral filtering. Int Conf Sign Process Commun 6:243–250
15. Nath VK , Hazarika D (2012) Blocking artifacts suppression in wavelet transform domain using local wiener filtering. In: IEEE National conference on emerging trends and applications in computer science (NCETACS) pp 93–97
16. Reeve HC, Lim JS (1984) Reduction of blocking artifacts in image coding. Opt Eng 23(1):34–37
17. Shahdoosti Hamid R, Khayat Omid (2016) Image denoising using sparse representation classification and non-subsampled shearlet transform. Sign Image Video Process
18. Sun D, Cham WK (2007) Postprocessing of low bit-rate block DCT coded images based on a fields of experts prior. IEEE Trans Image Process 16(11):2743–2751
19. Wu S, Yan H, Tan Z (2001) An efficient wavelet-based deblocking algorithm for highly compressed images. IEEE Trans Circuits Syst Video Technol 11(11):1193–1198
20. Xu J, Zheng S, Yang X (2006) Adaptive video blocking artifact removal in discrete Hadamard transform domain. Opt Eng 45(8)
21. Zhang M, Bahadir GK (2009) Compression artifact reduction with adaptive bilateral filtering. Visual Commun Image Process, vol 7257

An Effective Video Surveillance Framework for Ragging/Violence Recognition

Abhishek Sharma, Vikas Tripathi and Durgaprasad Gangodkar

Abstract Ragging is one among the biggest problems in the education system in the current time. This paper proposes an efficient computer vision-based framework to detect ragging/violence during the actual time of occurrence of the event. The framework of the system is based on motion analysis. To determine the motion in the video, the use of optical flow has been made. The code has been trained and tested by a collection of several videos. Three different reference frame delays are considered, and the result analytics are reviewed for effective reference frame selection. Passage of the different frames of the video through the descriptor takes place in order to wrench out feature from the frames. The use of two descriptors HOG and LBP has been made individually with optical flow to generate dataset. Further analysis has been done through Random Forest to validate our results. Our proposed framework is able for effective detection of ragging and normal activities with an accuracy of 88.48%.

Keywords Computer vision · Optical flow · HOG · LBP · Ragging

1 Introduction

Ragging is one of the most common problems that has a deep root in the education society. Since many years, several students have been victim to this malpractice. Although there are many laws against ragging in India, still this practice prevails in most of the institutions of higher studies. The problem is more severe in medical and engineering colleges. Physical violence, mental torture, and rape are some of the darkest sides of ragging. Forcing the juniors to indulge in unhealthy activities or using them to get through their work is another widespread problem in ragging. Ragging can cost life of a student. According to the RTI reply, UGC received 640

A. Sharma (✉) · V. Tripathi · D. Gangodkar
Graphic Era University, Dehradun, Uttarakhand, India
e-mail: abhisheksharma13598@gmail.com

© Springer Nature Singapore Pte Ltd. 2019   239
A. J. Kulkarni et al. (eds.), *Proceedings of the 2nd International Conference on Data Engineering and Communication Technology*, Advances in Intelligent Systems and Computing 828, https://doi.org/10.1007/978-981-13-1610-4_25

and 543 complaints in year 2013 and 2014, respectively. Even an FIR was filed in 9 and 57 cases, respectively, in year 2013 and 2014 [1].

Almost all institutes are well equipped with the CCTV cameras, and almost every part of the campus is covered with these cameras. Most of the time these cameras are used for recording, but they are used after some accident has occurred. So, we came up with the thought, '*Why not use CCTV cameras to their full potential so that they can do something more than just recording?*' If the cameras can detect ragging, then there will be no need for the victim to register complaint and the culprits would be caught and the case will be handled accordingly. In this paper, we proposed a framework that will be self-sufficient to analyze the video and capture ragging. We have used optical flow, HOG, LBP, and other techniques to develop the desired system. The paper is organized as follows: In Sect. 2, we provide the previous work performed on the violence detection in videos. Methodology of the proposed work has been explained in Sect. 3. Section 4 contains the analysis of the result and the last section concludes the paper.

2 Literature Review

The first among those with the ideas to detect violence in video was Nam et al. [2], it gives the idea of recognizing unusual activities in videos using blood detection and fire in the video and analyzing the movement, also by analyzing the sounds in the case of abnormal activity. Cheng et al. [3] took into consideration the gunshots, car-breaking, and explosions in audio. Giannakopoulos et al. [4] also put forward the idea of detecting violent activities using sound. Clarinet al. [5] talks about the framework that uses a Kohonen framework to recognize blood pixels and human skin in the video and detecting motion involved in violent activity with blood in the action involved. Zajdel et al. [6] proposed a framework, which is based on feature extraction from the motion in video and distress audio to detect unusual activity like violence in the videos.

Another work involves that of Gong et al. [7]. He put forward the idea of violence detection using video, sound, and sound effects to detect abusive and unusual activities in the movies. Chen et al. [8] used the feature extraction from motion and collection of dataset in order to detect aggressive behaviors. Lin and Wang [9] used motion in the movie along with the sound training followed by the sound of exploding and blood detection classifier. Giannakopoulos et al. [10] used framework for detecting violent scene in movies on the basis of audio statistics, combined with the statistic of visuals and motion statistics to determine whether a video contains violent scene or not. Some other recent work involves the papers breaking down violence scene by Acar et al. [11] in the year 2015 and violence detection using oriented violent flow by Gao et al. [12] in the year 2016. Finally, to conclude the basic theme of all the papers mentioned, we can say that most ideas are based on the analysis of videos along with the audio. But these systems may fail in the case where there are no audios. Not all CCTV cameras are accompanied by voice recording system.

Generally, the CCTV cameras are used only to record the video. Another important criteria used in many papers is the recognition of blood to detect violence. Since the quantity of blood may not be significant to detect for a camera like that of a CCTV which has low resolution.

3 Methodology

In this section, we have given a detailed explanation of the steps that were followed in getting an efficient result. Firstly, we will discuss the datasets that are used in this experiment, and later, the procedure of the experiment is laid out.

3.1 Datasets

The video dataset is divided into two parts; one is the ragging, and the other is normal activity. The main source of the video is YouTube, and few scenes are from movie 'Hostel of 2011 by Manish Gupta [13].' Although there are a few videos that we shot on our own.

The ragging dataset contains fight scene (pushing, kicking, punching, bullying activities, etc.). The length of each video varies from 30 s to a minute long. In the test set, the total video length is around 7–8 min. Figure 1 shows some scenes of the videos from the ragging activity dataset.

Fig. 1 Images from ragging activities dataset

Fig. 2 Images from normal activities dataset

The normal dataset contains the usual activities in the day-to-day life on the road, institutes, and other public places. The activities involved are walking, talking, sitting, etc. We tried to cover all the normal activities that are observable in the public place. In the test set, the total video length is around 7–8 min. Scenes of some normal activities from normal activity dataset are shown in Fig. 2.

3.2   Proposed Frame Work

The main ideology behind the framework is combining the optical flow with the descriptors such as HOG [14] and LBP. Algorithm 1 shows the steps involved in the implementation of framework where window size refers to reference frame delay of 5, 10, and 15 frames while calculating the optical flow. These steps involve acquiring video, calculating the optical flow, passing the images to a descriptor, classifying the image using random forest, etc.

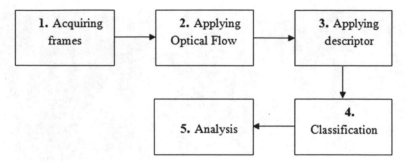

Fig. 3 Flow diagram of proposed framework

Algorithm 1 for Generation of Descriptor

Input　: Video.
Output　: Compute descriptor.

1. Compute frames.
2. Initialize p with 0.
3. Initialize n with window size
4. While p < frames do
　　I.　　If (p % n = 0)
　　　　a.　Calculate optical flow.
　　　　b.　Extract features.
　　　　c.　Store features.
　　　　d.　Increment p by 1.
　　5.　　Dataset generated.

Figure 3 shows the flow diagram of the implementation of the frame work. There are mainly five blocks to implement this framework. Our first step was to read the video files and extract the frames from the video. The video frames are acquired, and the color images are converted into gray scale. The videos are '.mp4' format.

The next step involves applying optical flow on the frames. In order to achieve more accuracy, we have taken variations in applying optical flow. We have taken reference frame delay of 5, 10, and 15. This means that for a frame delay of 5, the current frame will be evaluated with the fifth frame in the video in order to calculate optical flow. Optical flow is nothing but the movement of the pixel in 'x' and 'y' direction. To calculate this 'x' and 'y,' there are two popular methods; one is Horn and Schunck [15] and Lucas–Kanade method. Image in Fig. 4 depicts optical flow in the image. These images were in RGB and were converted into gray scale. The lines in the image depict optical flow.

We have used two well-known descriptors HOG and LBP. Then, an analysis is made among these descriptors. We obtain 1×288 arrays in the case of HOG and 1×348 in the case of LBP. There are two classes; one is named ragging, and the other is normal. So for all the frames, we store values in a file.

Next step involves classifying data. There are almost 75,000 instances in training set with almost equal number of instances for ragging and normal. Similarly, with

Fig. 4 Images with optical flow (lines represent optical flow)

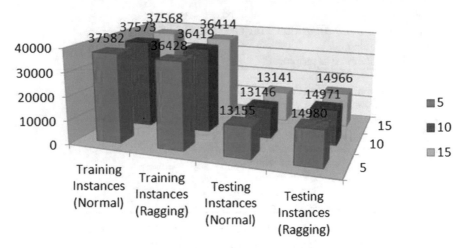

Fig. 5 Graph representing instances of test and train in terms of frames

almost equal number of instances for ragging and normal dataset, there are about 28,000 total instances in testing dataset. The videos are in the format 480 × 854. We have used random forest for classification purpose. Figure 5 depicts the information regarding instances of both train and test for both ragging activity and normal activity classes.

4   Result and Analysis

The framework is coded in MATLAB; training and testing are conducted in Weka. The system on which this framework was developed has 4 GB RAM, Intel core i3, 2.13 GHz processor. The frame resolution chosen is 480 × 854. The videos are in '.mp4' format, and there length varies from 10 s to 2 min. The frame rate is 29 frames/s. Table 1 gives the accuracy in term of percentage.

Table 1 Accuracy with different method

Method	5	10	15
LBP	87.3645	86.7816	88.4805
HOG	75.4256	76.9966	77.3983

Table 2 Detailed result statistics for LBP with 15 frame delay optical flow

Class	True-positive ratio	False-positive ratio	Precision value	Recall value	F-measure	ROC
Normal	0.873	0.104	0.88	0.873	0.876	0.956
Ragging	0.896	0.127	0.889	0.896	0.892	0.956
W. Average	0.885	0.117	0.885	0.885	0.885	0.956

Table 3 Confusion matrix for LBP with 15 frame optical flow

	Ragging	Normal	Total
Ragging	13,416	1564	14,980
Normal	1677	11,478	13,155
Total	15,093	13,042	28,135

In Table 1, values 5, 10, and 15 in first row are referring to the reference frame delay of frames while calculating optical flow. For example, 5 from first row and method LBP in above table imply that the descriptor is LBP with optical flow having reference frame delay of five frames. From table above, we can observe that the best accuracy is obtained by LBP with reference frame delay of 15 frames.

From the Table 2, the value of true-positive ratio, false-positive ratio, and precision gives very promising results. True-positive ratio is the proportion of correctly identified instances. Similarly, we have false-positive ratio that generates a false alarm. Precision value is the refinement in the calculations, recall means the fraction of the relevant instances that are retrieved, and F-measure takes the combination of precision and recall to provide a single value of measurement.

Table 3 represents the confusion matrix. Here, we can see that about 11,478 instances of normal and 13,416 instances of ragging are correctly classified. These results are quite promising as our framework identifies 24,894 instances correctly out of 28,135 instances which is 88.48%.

Figures 6 and 7 are showing threshold model for ragging activities as well as threshold model for normal activities, respectively. The ROC curve is a plot which is plotted in between false-positive rate and true-positive rate for different cutoff points. The points on the ROC curve represent the ratio of true-positive rate and false-positive rate with respect to a particular decision threshold. The ROC curve passes to the upper left corner shows the high sensitivity and specificity which means area under the ROC curve close to 1 represents the high accuracy. The area under both of our model is 0.9958 which shows an excellent test.

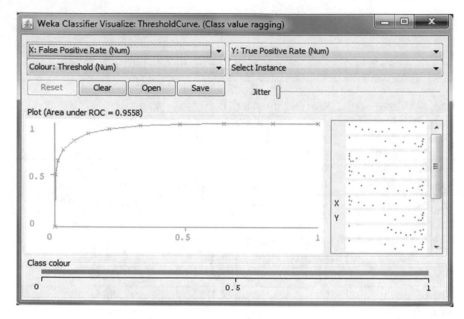

Fig. 6 Threshold model (ragging activity)

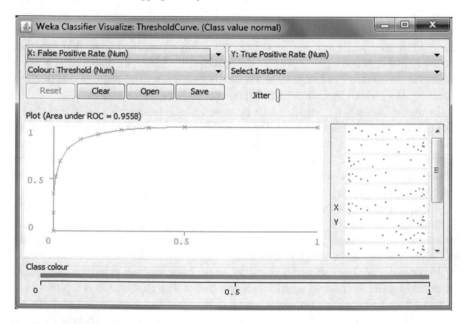

Fig. 7 Threshold model (normal activity)

5 Conclusion

In this paper, a framework for better prevention of ragging using video surveillance in college campus or similar premises has been presented. In particular, this paper discusses the recognition of ragging and normal activity in the concerned area. We have taken into consideration three different cases for optical flow (reference frame delay) and two different types of descriptor (HOG and LBP). Thus, a total of six different cases to observe the accuracy are presented. From the different cases, maximum accuracy in the case of HOG is 77.39% for reference frame delay of 15 frames and in the case for LBP it is 88.48% for reference frame delay of 15 frames. However, this framework is open to improvizations and advancements such as fusion of framework with audio analysis, application of new descriptors or addition of various combinations of optical flow and descriptors.

References

1. Times of India Homepage, http://timesofindia.indiatimes.com/home/education/news/1183-ra gging-cases-66-FIRs-in-two-years/articleshow/47477316.cms. Last accessed 12 Aug 2017
2. Nam J, Alghoniemy M, Tewfik AH (1998) Audio-visual content-based violent scene characterization. In: Proceedings of international conference on image processing, pp 353–357
3. Cheng WH, Chu WT, Wu JL (2003) Semantic context detection based on hierarchical audio models. In: Proceedings of the ACM SIGMM workshop on multimedia information retrieval, pp 109–115
4. Giannakopoulos T, Makris A, Kosmopoulos D, Perantonis S, Theodoridis S (2006) Violence content classification using audio features. Advances in Artificial Intelligence, Lecture Notes in Computer Science, vol 3955, pp 502–507
5. Clarin CT, Ann J, Dionisio Michael M, Echavez T, Naval Jr PC (2005) Detection of movie violence using motion intensity analysis on skin and blood. In: University of the Philippines, pp 150–156
6. Zajdel W, Andringa T, Gavrila DM, Krijnders JD (2007) Audio-video sensor fusion for aggression detection. In: IEEE conference on advanced video and signal based surveillance, pp 200–205
7. Gong Y, Wang W, Jiang S, Huang Q, Gao W (2008) Detecting violent scenes in movies by auditory and visual cues. In: Proceedings of the 9th Pacific rim conference on multimedia, pp 317–326
8. Chen D, Wactlar HD, Chen MY, Hauptmann A (2008) Recognition of aggressive human behaviour using binary local motion descriptors. In: Engineering in medicine and biology society, pp 5238–5241
9. Lin J, Wang W (2009) Weakly-supervised violence detection in movies with audio and video based co-training. In: Proceedings of the 10th pacific rim conference on multimedia, pp 930–935
10. Giannakopoulos T, Makris A, Kosmopoulos D, Perantonis S, Theodoridis S (2010) Audio-visual fusion for detecting violent scenes in videos. In: 6th Hellenic conference on AI SETN Athens Greece, pp 91–100
11. Acar E, Hopfgartner F, Albayrak S (2016) Breaking down violence scene. Neurocomputing 208:225–237
12. Gao Y, Liu H, Sun X, Wang C, Liu Y (2016) Violence detection using oriented violent flow Image and Vision Computing. Image Vis Comput 48:37–41

13. Youtube Homepage. https://www.youtube.com/watch?v=1ktaw31FyiY. Last accessed 21 June 2017
14. Dalal N, Triggs B (2005) Histograms of oriented gradients for human detection. In: International conference on computer vision & pattern recognition (CVPR). IEEE Computer Society, vol 1. San Diego, USA, pp 886–893
15. Horn BKP, Schunck BG (1981) Determining optical flow. In: Artificial intelligence, pp 81–87

DoT: A New Ultra-lightweight SP Network Encryption Design for Resource-Constrained Environment

Jagdish Patil, Gaurav Bansod and Kumar Shashi Kant

Abstract The paper proposes a new lightweight encryption design DoT which supports block length of 64 bits with a key size of 80/128 bits. DoT is the 31-round substitution and permutation-based network. Hardware and software parameters are the main key focused for designing the DoT cipher, and DoT cipher shows the good performance on both hardware and software. The metrics which makes the DoT efficient as compared to other lightweight ciphers are memory size, Gate Equivalent (GEs) and throughput. DoT cipher design results in only 2464 bytes of flash memory, and it needs only 993 GEs for its hardware implementation. While considering the speed of the DoT cipher with block input of 64 bit, it gives 54 Kbps which is very much greater than other existing ciphers. DoT design consists of a single S-box, few shift operators and a couple of XOR gates. We believe that DoT is the compact and smallest cipher till date for GEs, memory requirement and execution time.

Keywords IoT · Lightweight block cipher · Encryption system · WSN

1 Introduction

Nowadays, the humans and machines are connecting through the recent technologies like IoT and WSN. These applications are basically based on the information transfer. When information comes into the picture, then there is a problem of increased hacking and that will be protected by some kind of security. Lightweight cryptography is a recent trend to provide the security for constrained and low-power device applications like wireless sensor node (WSN) and Internet of things (IoT). For such applications, important design parameters are memory size, gate equivalents (GEs) and throughput of the system. Generally, for constrained and low-power devices,

J. Patil (✉) · K. S. Kant
Symbiosis Institute of Technology, Pune 412115, India
e-mail: jagdishpatil108@gmail.com

G. Bansod
Pune Institute of Computer Technology, Pune 411043, India

© Springer Nature Singapore Pte Ltd. 2019
A. J. Kulkarni et al. (eds.), *Proceedings of the 2nd International Conference on Data Engineering and Communication Technology*, Advances in Intelligent Systems and Computing 828, https://doi.org/10.1007/978-981-13-1610-4_26

8-bit controller is preferable for the operation. In case of the WSN and IoT, all the nodes are connecting each other through the network so there are chances of getting hacked that node through the external attack. But the requirement is to protect that node through all possible types of external attack with consideration of memory size and hardware implementation. Lightweight cipher is basically providing the security at node level only. One of the examples is explained with reference to the RFID device. For designing the RFID, 10,000 GEs are required for its hardware implementation; from those, only maximum 2000 GEs are reserved for implantation of security algorithm. Advanced encryption system and data encryption system algorithms use approximately 2500–3600 GEs for implementing their hardware [1]. This data say that AES and DES are not suitable for WSN, RFID and IoT. So to achieve all these parameters, lightweight ciphers are proposed like PRESENT [2], FeW [3], MIDORI [4], BORON, LED, RECTANGLE [5], PICCOLO, TWIN, PICO [6] for giving the security in the low-power devices and memory-constrained environment. These ciphers have GEs around 1300–2500.

This paper proposed the new block cipher called "DoT" which is a SP network and in terms of hardware implementation requires only 993 GEs. In terms of the security, SP network is stronger than a Feistel network. SP network generally uses the long S layer which is only the nonlinear element in the design and will give more security and highest complexity in the data. As far as security concern "DoT" cipher withstands against basic (linear, differential) attacks and also withstands against advanced (biclique) attack. By considering all the above parameters, DoT will be the significant selection for low power and compact memory application.

2 Design Matrices

The first aim to design cipher is to reduce the gate counts so that the cipher should result in small hardware implementation. Moreover, power consumption should also be less as the cipher design is aimed at providing security to battery-powered nodes. Flash memory size of cipher should be very less as targeted processor would be of 8 bit, which generally has very less memory space. The designed cipher should perform efficiently both on hardware and on software platforms. By maintaining all these metrics, throughput of the cipher design also should be competitive. DoT has 8-bit permutation layer which results in the minimum flash memory requirement, whereas other ciphers have 32-bit, 64-bit permutation layers. In DoT cipher, we made a successful attempt to minimize memory requirement by using a bit permutation with high diffusion mechanism. Bit permutation layer is designed in such a way that the bit distribution results in maximum number of active S-box in minimum number of rounds. In DoT cipher design, we used maximum bit permutation, block shuffling and circular shifting which does not require GEs so that we reduced GEs count. Without the use of post-whitening key in the DoT cipher, it achieved more complexity with good throughput. In the cipher design at software level, care is taken to use minimum number of local and global variables which results in less memory requirement. Due to the use of a mesh kind of network in the cipher design and the reuse of limited registers in programming, the power dissipation of DoT cipher is less. By considering all the above aspects, the main aim of the DoT cipher is to provide the high-level encryption to the low-powered, self-battery-operated and memory-constrained devices.

3 Proposed System

DoT is the SP network-based design which has the round function shown in the block diagram (see Fig. 1). DoT cipher supports the input data as of 64 bit, and key data length is 80/128 bit. For 128-bit key, we have different key scheduling algorithms which we will see in the next section. The block shuffle output is given to the 8-bit permutation layer which helps to increase the flash memory in the system. After 8-bit permutation, there is block shuffling and circular shifting which also help to increase the complexity in the design. At the end, the output is again fed to the input of the next round and rotating the same design for 31 rounds. The same analogy is applicable to the key which has been XOR-ed with DoT design at each round. After completing the 31 round of design and the key scheduling, one extra key will be generated called as post-whitening key, but this design does not require the post-whitening key for proving the security.

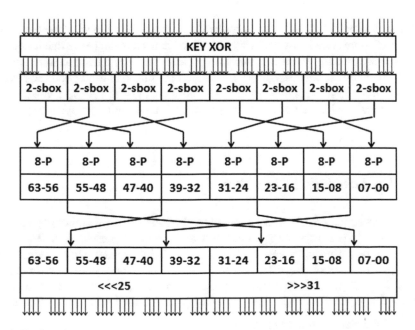

Fig. 1 Design of DoT cipher

Table 1 4 × 4 S-box

X	0	1	2	3	4	5	6	7	8	9	A	B	C	D	E	F
S[x]	3	F	E	1	0	A	5	8	C	4	B	2	9	7	6	D

Table 2 8-bit permutation

i	0	1	2	3	4	5	6	7
P[i]	2	4	6	0	7	1	3	5

4 System Components

4.1 S-Box

In this design, we are using the 4 × 4 S-box which has four input bits and four output bits. The S-box is used to introduce the nonlinearity in the system and also helpful for increasing more complexity in the cipher so that the system should be more secure. There are many design criteria which have been referred from PRESENT [2] and RECTANGLE [5] ciphers. Table 1 indicates the 4 × 4 S-box designed for DoT.

4.2 Permutation Layer

Permutation layer has many advantages like it increases the strength of the cipher and introduces more complexity, and very most important advantages are that it does not require GEs for hardware implementation because there is no need to store the data into the registers. In this design, we introduced the block shuffling and the circular shifting which have been used to increase the number of the active S-box. DoT p layer is mentioned in Table 2. The design criteria of permutation layer are given in paper [1].

4.3 Key Scheduling Algorithm

DoT key scheduling has been encouraged by the PRESENT [2] cipher key scheduling because it is a very strong algorithm designed and presented in PRESENT [9, 10] cipher. The input key is stored in the KEY register which is given as $K_{127}K_{126}K_{125} \ldots K_2K_1K_0$; from that for DoT, we have considered 64 leftmost bits as key $K^i = K_{63}K_{62} \ldots K_2K_1K_0$ as applied to the initial round of cipher [11]. For the next round, the KEY register is updated as per the following key scheduling steps:

1. KEY register is circularly left shifted by 13.

$$KEY \lll 13$$

2. The leftmost 8 bits, i.e. $K_7 K_6 \ldots K_2 K_1 K_0$, are passed through the S-box of DoT.

$$[K_3 K_2 K_1 K_0] \leftarrow S[K_3 K_2 K_1 K_0]$$

$$[K_7 K_6 K_5 K_4] \leftarrow S[K_7 K_6 K_5 K_4]$$

3. Apply the round counter. That is, XOR the round counter RC^i of respective round with the key bits $K_{63} K_{62} K_{61} K_{60} K_{59}$.

$$[K_{63} K_{62} K_{61} K_{60} K_{59}] \leftarrow [K_{63} K_{62} K_{61} K_{60} K_{59}] \oplus RC^i$$

5 Performance of the System

5.1 GEs

In the constrained environment, the main parameter for cipher design is GE. The total gate count for the DoT—128 cipher designs—using serialized architecture is 993 which is very less than other existing ciphers. The detailed comparison is depicted in Fig. 2.

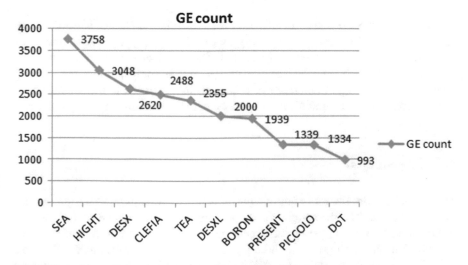

Fig. 2 Shows comparison of GEs of DoT with standard existing ciphers

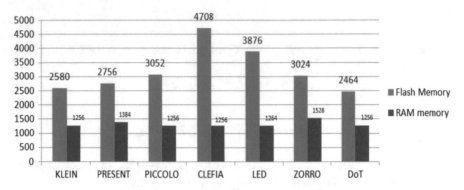

Fig. 3 Comparison of flash memory and RAM of DoT with existing ciphers

Table 3 Comparison of the throughput of other SP network existing ciphers with DoT

Ciphers	Network	Block size	Key size	Execution time (in usec)	Throughput (in Kbps)	No. of cycles
LED	SP	64	128	7092.86	9	85,144
KLEIN	SP	64	96	887.51	72	10,650.1
HUMMINGBIRD-2	SP	16	128	316.51	51	3897.12
PRESENT	SP	64	128	2648.65	24.16	31,783.8
DoT	SP	64	128	1190.25	53.77	14,283

5.2 Memory Requirement

The DoT cipher performance is tested on the ARM 7 LPC2129, 32-bit processor platform, and it gives less flash memory and RAM memory size requirements as compared to existing lightweight ciphers [6]. The flash memory requirement of the DoT cipher is 2408 bytes which is very competitive as compared to other existing ciphers. For analysing the memory of other existing ciphers, the same platform is used. Figure 3 gives the graphical comparison of DoT with other existing ciphers.

5.3 Throughput

Throughput decides the rate of the output. It is one of the important design metrics in cipher design. The throughput of the system can be calculated as the number of the plaintext bits divided by the execution time [9]. Execution time of the DoT is 1190.25 usec, which gives the throughput of the DoT cipher around 53.77 Kbps. Table 3 shows the comparison of the throughput of other SP network existing ciphers with DoT.

6 Performance of the System

6.1 Linear Attack

The linear cryptanalysis is in the form of linear expressions having the plaintext, cipher text and the key bits. In case of DoT cipher, the maximum bias can be calculated from the linear approximation and that is 2^{-2} based on the S-box used in the cipher design [7, 9]. In linear cryptanalysis, we have calculated the probability of bias for "n" rounds. Consider the example of two rounds; from that, we can calculate the minimum number of active S-boxes with the help of linear trails. The total number of active S-boxes from round two is three. Similarly, we can find the active S-boxes for different rounds. For round 6, we got the total number active S-boxes which are 10. The maximum bias for DoT cipher is 2^{-2}; from that for round 6, the total bias is $2^9 \times (2^{-2})^{10} = 2^{-11}$ that can be calculated as per the pilling up lemma principle [10]. So for total round of the DoT cipher, i.e. for 30 rounds, for total 50 min active S-boxes, there are 2^{-51} total bias [10]. Linear attack complexity for this is $\frac{1}{(2^{-51})^2}$ is 2^{102} [10]. From the above analysis, we can conclude that the total number of known plaintext is 2^{102} which is greater than 2^{64}. This shows the DoT cipher is resistible to linear attack with total 30 rounds.

6.2 Differential Attack

Differential cryptanalysis is also one of the basic attacks which are also called as chosen plaintext attack [7]. The difference distribution table gives the minimum number of active trails. Differential trails can be calculated by considering the high probability difference between input and output for each round. The important component used here for nonlinearity is S-box. The differential probability of DoT cipher is $\frac{4}{16} = \frac{1}{4} = 2^{-2}$ [9]. From the above trails, there are two active S-boxes for round 2 and for round 6 there are total 7 active S-boxes. We can say that for round 30 there are 35 minimum active S-boxes. P_d (differential probability) is $(2^{-2})^{35} = 2^{-70}$. Differential attack complexity is given by $N_d = \frac{1}{P_d}$, where $C = 1$ and $P_d = 2^{-70}$ [9]. For DoT, the complexity of the differential attack is 2^{70}. From the above computation, we can say that the differential complexity of the DoT cipher is 2^{70} which are greater than 2^{64} which depicts that DoT is resistible to differential attack with a block length of 64.

6.3 Biclique Attack

Biclique attack is a theoretical attack which can decide the data and computational complexity of any block cipher. Biclique attack is an extension of meet-in-the-middle attack [8]. In this paper, we have shown the DoT cipher is resistible against biclique attack and it gives the maximum data complexity of 2^{36}. DoT is 31-round cipher from which the biclique attack is mounted on round 28–31 with four-dimensional biclique. In this analysis, the most important thing is to select the key bits. Key is selected in such a way that while doing the forward and backward computation no single key should be overlapped. These keys are selected from the key scheduling algorithm which is mentioned in the key scheduling algorithm section. In this paper for DoT cipher, we have selected the two set of keys, i.e. $(K_{40}, K_{39}, K_{38}, K_{37})$ and $(K_{50}, K_{49}, K_{48}, K_{47})$; one is for forward computation indicating with red key, and the other set is for backward computation indicating with blue key. The total computational complexity can be calculated by using the following formula mentioned in the paper [9, 10]. Total computational complexity of the DoT-128 is $2^{127.245}$. This shows that DoT is resistible to biclique attack.

7 Conclusion

The proposed cipher design "DoT" which is based on SP network has maximum number of active S-boxes with minimum number of round. Hardware implementation of the cipher has been done with the help of the 993 GEs with 128-bit key length that are comparatively less than other ciphers and also required very less memory of 2464 bytes. In term of GEs, "DoT" is the smallest design among all SP networks. Block shuffling and permutation of 8 bit make DoT unique. Block shuffling and bit permutation also helped a lot to achieve more throughputs, i.e. 60 Kbps which is high as compared to the existing ciphers. As compared to PRESENT cipher, throughput of the DoT cipher is 250 times more. Security concern is also taken care by implementing the all possible attacks on DoT, but DoT gives very good resistance against the basic and advanced attacks. DoT will give the positive impact in the field of lightweight cryptography and will prove to be crusader in making technologies like IoT feasible.

8 DoT Vectors

Table 4 shows the input and output vectors for two different key inputs which have given the maximum input changes.

Table 4 Input and output vectors for two different key inputs

Plaintext	Key	Cipher text
00000000 00000000	00000000 00000000 00000000 00000000	4707a9269825aaf0
00000000 00000000	FFFFFFFF FFFFFFFF FFFFFFFF FFFFFFFF	ca1d5da75fd01910

References

1. Coppersmith, D (1992) The data encryption standard (DES) and its strength against attacks. IBM Thomas J Watson Research Center technical report RC 18613 (81421)
2. Bogdanov A, Leander G, Knudsen LR, Paar C, Poschmann A, Robshaw MJB, Seurin Y, Vikkelsoe C (2007) PRESENT—an ultra-lightweight block cipher. In: Paillier P, Verbauwhede I (eds) Cryptographic hardware and embedded systems—CHES 2007, vol 4727 in LNCS. Springer Berlin Heidelberg, pp 450–466
3. Kumar M, Pal SK, Panigrahi A (2014) FeW: a lightweight block cipher. Scientific Analysis Group, DRDO, Delhi, INDIA, Department of Mathematics, University of Delhi, India
4. Banik S, Bogdanov A, Isobe T, Shibutani K, Hiwatari H, Akishita T, Regazzoni F Midori: a block cipher for low energy (Extended Version). In: IACR eprint archive
5. Zhang W, Bao Z, Lin D, Rijmen V, Yang B, Verbauwhede I (2014) RECTANGLE: a bit-slice ultra-lightweight block cipher suitable for multiple platforms. Cryptology ePrint Archive, Report 2014/084
6. Bansod G, Pisharoty N, Patil A (2016) BORON: an ultra lightweight and low power encryption design for pervasive computing. Frontiers
7. Heys HM (2002) A tutorial on linear and differential cryptanalysis. http://citeseer.nj.nec.com/443539.html
8. Jeong K, Kang H, Lee, C, Sung, J, Hong S (2012) Biclique cryptanalysis of lightweight block ciphers PRESENT, Piccolo and LED. Cryptology ePrint Archive, Report 2012/621
9. Patil J, Bansod G, Kant KM (2017) LiCi: A new ultra-lightweight block cipher. In: 2017 International conference on emerging trends & innovation in ICT (ICEI) http://doi.org/10.11.9/ETIICT.2017.7977007
10. Matsui M (1994) Linear cryptanalysis method for DES cipher. In: Helleseth T (ed) Advances in cryptology, Proceeding of Eurocrypt'93, LNCS 765. Springer, pp 386–397
11. Poschmann A (2009) Lightweight cryptography—cryptographic engineering for a pervasive World. Number 8 in IT Security. Ph.D. Thesis, Ruhr University Bochum

A Distributed Application to Maximize the Rate of File Sharing in and Across Local Networks

Vishal Naidu, Videet Singhai and Radha Shankarmani

Abstract File sharing is the most important aspect of any network. The measure of quality of the network is done by the rate and accuracy of the data transferred. Any such transferred is initiated by the availability of the file on a server that is within reach of the network admin (in case of a compliance restricted network) and every other computer (in an unrestrained network) in the network. The next key factor determining the rate of access to the file is the available bandwidth. The solution that provides maximum performance in sharing files in LAN would be a distributed application that manages file sharing in a way that the required file, found on one or more computers, is shared in pieces via distinct lines with independent bandwidth. Thus, requiring the receivers to download fragments from different devices and stitch them together to get the original file.

Keywords LAN · Torrent · File sharing · Distributed system
Distributed application · Latency · Download

1 Introduction

The time it takes to share a file in a network to different computers is a measure of performance of the network. This can be done by reading the LAN network statistics for the packet rate and the latency to each system from every other system. Now, one of the most common problems faced by the administrators and the users of a network is the sharing of a particular file [1, 2], in a case wherein the file is to be shared to every computer in the network. For instance, file sharing in the local

V. Naidu · V. Singhai · R. Shankarmani (✉)
Sardar Patel Institute of Technology, Andheri, Mumbai, India
e-mail: radha_shankarmani@spit.ac.in

V. Naidu
e-mail: naiduvishal13@gmail.com

V. Singhai
e-mail: videetssinghai@gmail.com

© Springer Nature Singapore Pte Ltd. 2019
A. J. Kulkarni et al. (eds.), *Proceedings of the 2nd International Conference on Data Engineering and Communication Technology*, Advances in Intelligent Systems and Computing 828, https://doi.org/10.1007/978-981-13-1610-4_27

area network in colleges. Another instance would be to download Linux on every computer in a laboratory for a kernel programming practice/teaching session. The most conventional and straightforward approach to this problem would be to do this in a time frame where the laboratory is reserved only for installing Linux on the machines. The first stepping stone in this path is to make the installation image available on every computer in the laboratory.

2 Problem Statement

The current problem is to transfer a file from one computer to N other computers in the same local area network, in the fastest possible way [3, 4]. The following problem can be overcome by using one of the proposed solutions:

Download file on one computer and use one of the following methods:

(A) Smart pairing (Progressive Pairing)
(B) Split and seed.

3 Proposed Solution

3.1 Architecture of the Proposed System

1. **Archived Storage**

The archived storage is the storage location where all the downloaded files are archived and split, ready for supply. This is also the place where all the fragments and their metadata are stored, so that whenever a new request arrives, there is a query to search for the exact fragment, for the file signature specified (Fig. 1).

2. **Application Layer**

This layer is the controller of the application. Our main algorithm of sending and receiving data is implemented in this layer. Similarly, it stitches and unarchives the received fragments into a single file, which is the desired file.

3. **Fragmentation Layer**

This layer consists of seven-zip tool which we have used to archive and split. By the comparative analysis of [5], seven-zip is the most relevant tool for our application.

4. **Local Network Layer**

This layer keeps an account of all the computers up in the network. This layer ensures the network connectivity and configuration of all the computers connected to the LAN. In a network, any node can go down any moment; whatever happens, the

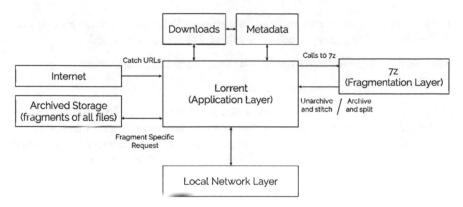

Fig. 1 Elements in the proposed application architecture and their dependencies among each other

download must not get interrupted. Our local network layer uses a timeout mechanism to check whether the request node is up or not. While receiving the receiving node sends a request to the other nodes if this request is not acknowledged within a given time interval, the receiver concludes that the node is down and stops sending the request.

3.2 Proposed Methodology

The proposed solution is in the form of a desktop application that has a background service paired with a browser extension. Every file that is downloaded on a computer is uniquely marked using a hash of the file's download address on the Internet and its signature. A public table is maintained for all the downloaded files that are available on the computer. This table will be visible to all the other computers running this application on the same network and on the same port.

Now, any other user requiring to download a file from the Internet would first go through a constraint check before the browser downloads it, thanks to the extension. This check would hit all the clients on the network and would prioritize them based on the factors such as file availability and n. of downloaders latched to that seeder.

Distributed download initiation algorithm (steps 1–4 from the '**common for both**' pseudocode):

1. Browser extension grabs the download URL and metadata obtained from the site
2. If (file already exists in local download storage)

   a. Open folder containing the file

   //pSeeders is a list of all computers in LAN
3. Look in the ARP table to enumerate all potential seeders

4. Send a request to all seeders, containing details to the computers in pSeeders, in the format: (Req_comp's_addr, Download_URL, File_signature)

From here on, there are two ways to download the file from the seeder:

1. **Smart pairing (or Progressive pairing)**

In this method, there may be N seeders and M downloaders, where M > N. Now, N seeders can share the file fastest to N downloaders, because it is a 1:1 transfer. Smart pairing or progressive pairing would set up a chain reaction in the network where the files would be progressively shared to that neighbor which has a pending request following an FCFS pattern, wherein each share, N increases to 2N and thus the next transfer would lead to a share size of 2N.

The algorithm (Request end):

5. Receive responses from all the computers that are not currently engaged in file transfers themselves
6. Sort the responses based on their latencies
7. Request file from the computer with the lowest pair-up latency
8. If (No reply contains complete affirmation)

    a. Set timeout
    b. At the end of timeout, send requests to those computers with partial affirmation (i.e., engaged) and go to step (9)

9. Else: Request file to addr of first complete affirmation and begin transfer on OK

The algorithm (Seeder end):

1. Background service that waits for requests from other computers in LAN
2. Upon request attained, scan the download history to find out if the signature specified matches with any file present in the storage
3. If (present)

    a. If (free): Reply with a response to the computer that requested the file
    b. Else: Reply with partial affirmation (Indicating current status)

4. Else: No reply.

2. **Split and seed**

The second method, split and seed is a variation of the approach 2A specified in the problem statement. But, here, the file on the seeder is split into X fragments, where X is determined using the factors:

(1) No. of downloaders demanding the file
(2) Size of the file
(3) File signature (MD5)
(4) Fragment limit.

Keeping in mind that bandwidth is spent when one seeder–downloader pair is in active sharing, N seeders share those fragments that are unavailable in the downloaders. This increases the bandwidth usage by boosting the download speed by increasing the fragment availability across more than one seeder. Split and seed also finds its advantages in topologies other than star.

The algorithm (Request end) (Steps 1–4 from the '**common for both**' pseudocode):

5. While (Response timeout, t, ends)
    a. Filter out all computers by the following factors:
        i. Computers that have the entire file
        ii. Computers that have M fragments
            (Where M $\leq$ n(Fragments based on config.ini))
        iii. Computers that don't have the file
6. If (nSeeders == 0): Go to step (6)
    // mSeeders is a list of all confirmed seeders
7. If (mSeeders == 1)
    Request the entire file in the format:
    (Req_comp_addr, –1(Fragment_ID), Download_URL, File_signature)
    // (–1) is to indicate that the entire file is targeted instead of a fragment
    /* fragList is a map of (computers to fragments).
    * e.g. ((0, [0,1,2,3]), (1, [0,1,2,3]), (2, [2]))
    * This indicates that comps 0 and 1 have all frags, and comp 2, only has frag_ID 2
    */
    // downMap is a map that stores download states of each fragment
8. If (mSeeders > 1)
    a. For (each frag in downMap)
        If frag.downloadStatus == Not Downloaded
            i. Map 1 fragment to 1 computer, From mSeeders
            ii. Request corresponding fragment to each computer
9. Wait till the current downloads complete
10. On download completion of frag_ID x: Update downMap
11. If (fragments in downMap left): Go to step 10
12. 7z command to stitch the 10 fragments downloaded and unarchive that file
13. Save the stitched file in Downloads folder
    // Archived storage is the storage where all the files
    Splitter background service (Seeder end)
    For each file in archived storage
    If (File not an archive and size > Threshold):

1. 7z command to archive file
    (To standardize the transfer format for all types of files, e.g., .txt, .mp4)
2. 7z command to split into N fragments (Where N is an experimental constant)
    Refer config.ini for fragmentation rules
3. Associate a unique ID to each frag, create a map, and store as fragList

The above algorithm is explained in simple terms with the help of flowchart (Fig. 2):

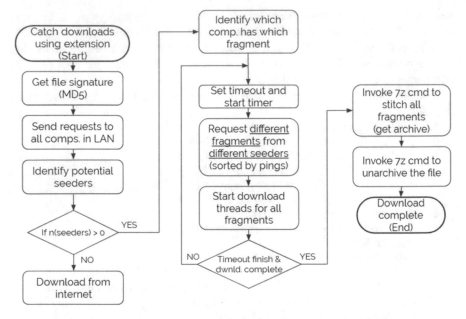

Fig. 2 The main stages and protocols involved in the distributed download process at the receiving computer's end

//**Raw** file is one that has not been archived and split and still resides in downloads
   The algorithm (Seeder end):

1. Background service that waits for requests from other computers in LAN
2. Upon request attained, check if the signature specified matches with any file present in the storage
3. If (File absent in downloads): No reply
4. Match download request format
   (Req_comp_addr, -1 (Fragment_ID), Download_URL, File_signature)
5. If (Fragment_ID ==-1)

   a. Start direct file transfer
   b. If transfer complete then Declare supply complete

6. If (Fragment_ID !=-1)

   a. Find fragment in temporary storage using the following:
      (File signature, Download URL, Fragment_ID)
   b. If fragment match found
      i.   Start fragment transfer
      ii. If transfer complete then Declare supply complete
   c. Else
      i.   Declare supply complete and Go to step 2.

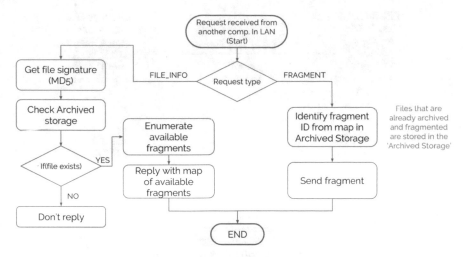

Fig. 3 The sequence of steps involved in communicating the availability of the target file, transferring the fragments from the seeder end to the downloader

The above algorithm is explained in simple terms with the help of a flowchart (Fig. 3):

4 Advantages

4.1 Specific Advantages

There are scenarios that help maximize the advantages that application can provide.

Mesh topology:

A mesh topology is a case where every computer in the LAN is connected to every other computer, not via a centralized communicator, but by a peer-to-peer approach. This LAN type proves to be of much greater advantage [6], particularly because there is no single line where the bandwidth is restricted by the capacity of the switch and the wire used to connect it to the computer. Moreover, traffic load will be handled thoroughly in mesh [7, 8]. Even if one node fails, the other nodes continue to seed the files. This is a huge advantage in mesh topology.

A disadvantage the centralized star topology faces is the bandwidth limit specified in the above paragraph. In such a scenario, progressive pairing and download will enjoy greater performance over the split and share method, for there is no computation time involved in splitting or unarchiving and that the data transfer rates are fairly similar in both cases (Figs. 4 and 5).

4.2 Overall Advantages

The following are some of the overall advantages of using the proposed model.

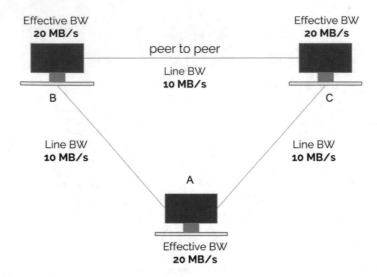

Fig. 4 Figure indicates that the star topology depends on the switch to supply adequate bandwidth for data transfer across one computer to another on the LAN [6]

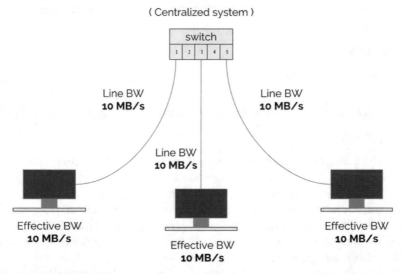

Fig. 5 Figure shows that an absence of centralized control and the vigorous interconnection between every node allows a much greater freedom to be exercised on the bandwidth available to the computer. Thus, boosting download speed if one were to download two fragments of the same file from two separate computers

1. Increasing the data sharing rate by a proportion equivalent to the number of seeders engaged in the transfer

2. Sharing speed would be consistent and independent of the internet connection speed
3. Lower bandwidth usage, leading to faster downloads from other sources
4. Boosted availability of files
5. No need of centralized file server.

5 Conclusion

Thus, we have presented an efficient solution to solve the data sharing problem faced in a LAN and in a limited Internet bandwidth. Also provided a system to manage and organize the downloads. Suggested the automation of the process by catching downloads from the browser using a browser extension and checking neighboring clients to download from or to declare the current download as the first instance.

6 Future Work

The future works might include the following solutions: To adapt the model in topologies other than the most efficient one, mesh network. So that the consumers of the model won't be limited to just organizations making use of the mesh network. The next step might be to solve the bandwidth exclusiveness problem in a centralized network, i.e., the bandwidth available to a particular device in a centralized network is always associated with the line that connects the switch to the device. Further works may make use of the same model to widen the network by connecting more than one LAN. Although there might arise latency issues when transferring from one LAN to another, the transfer rate would however be advantageous over simple external downloads.

References

1. Zhu D et al (2005) Using cooperative multiple paths to reduce file download latency in cellular data networks. Paper presented at IEEE global telecommunications conference
2. Funasaka J et al (2005) Implementation issues of parallel downloading methods for a proxy system. Paper presented at 25th IEEE international conference on distributed computing systems workshops
3. Gong T (2005) The use of multiple proxies for fast downloads. Paper presented at Canadian Conference on electrical and computer engineering, IEEE
4. Kato T et al (2012) An efficient content distribution method using overhead and file piecing in wireless LAN. Paper presented at 9th Asia-Pacific symposium on information and telecommunication technologies
5. Singh R (2016) The survey of different data compression tools and techniques. Int J Multi Res Technol 1(1)

6. Meador B (2008) A survey of computer network topology and analysis examples. Paper presented at Washington University
7. West MA (2002) Use of performance enhancing proxies in mesh topologies. IEEE Savoy Place, London WC2R OBL, UK (2002)
8. Farej ZK et al (2015) Performance comparison among topologies for large scale WSN based IEEE 802.15.4 Standard. Int J Comput Appl 124(6)

Obstacle Detection for Auto-Driving Using Convolutional Neural Network

Pritam Kore and Suchitra Khoje

Abstract This paper details about a framework for preceding obstacle detection for autonomous vehicles or driver assistance systems (DAS). Detection and tracking surrounding moving obstacles such as vehicles and pedestrians are crucial for the safety of people and autonomous vehicles. The proposed system uses four main stages: input, feature extraction, obstacle detection, and output of system. The system has input in video format which will be converted into frames. Feature extraction stage extracts LBP+LTP feature and Gabor feature separately. With the help of training set and extracted features, CNN will detect the obstacle in the frame. The system with LBP+LTP feature in cooperation with CNN gives 86.73% detection rate and 4% false alarm rate and the system with Gabor feature in cooperation with CNN gives 86.21% detection rate and 0.27% false alarm rate. From the proposed system, we can conclude that computer vision in combination with deep learning has the potential to bring about a relatively inexpensive, robust solution to autonomous driving.

Keywords Obstacle detection · Local binary pattern · CNN

1 Introduction

Every year a lot of people die in road accidents involving more than one vehicles involving in it. As a result in recent times, there has been a large boost in the research giving scope to advanced assistance for the drivers which help a vehicle to be driven more autonomously. In such research, a lot of attention is provided to design a system that will inform the driver in any means regarding any possible collisions and the

P. Kore (✉) · S. Khoje
Department of Electronics and Telecommunication, MAEERs MIT College of Engineering,
Pune, Maharashtra, India
e-mail: pritamk555@gmail.com

S. Khoje
e-mail: suchiamol08@gmail.com

© Springer Nature Singapore Pte Ltd. 2019
A. J. Kulkarni et al. (eds.), *Proceedings of the 2nd International Conference on Data
Engineering and Communication Technology*, Advances in Intelligent Systems
and Computing 828, https://doi.org/10.1007/978-981-13-1610-4_28

available obstacles in the vicinity to the vehicle. This paper details about a possible way of how the vehicles can be detected by using a mounted camera on the vehicle itself [1, 2].

A stationary camera makes it simpler to detect a moving object. Background subtraction is the widely used concept for the detection of obstacles. Consequently, the task becomes more challenging if the camera is moving. The camera motion along with structure and the motion of objects moving independently is used to create image motion [3–5].

The proposed system has been used to detect obstacles in the road scenario. For this input, video is converted into frames. These frames are used to train CNN to detect obstacles from it. Various images of obstacle like vehicles or pedestrians are used to train CNN. The system will take one frame and then convert that into image patches. Each image patch is then matched with the training dataset, if the match found, it is located as the obstacle in that frame. Similarly, process continues for all the frames. This gives general idea about proposed system.

2 Related Work

In the literature computer vision, LiDAR and RADAR are the various modes discussed for the detection of obstacles. RADAR technology has applications like adaptive cruise control (ACC) and side impact warning assist [6]. Unlike other methods, RADAR works accurately in critical weather conditions.

LiDAR is used in the autonomous vehicles for the detection of the obstacles. It is also used in applications like ACC to perform the same task. Weather and the illuminating conditions make LiDAR more sensitive. Also, it is a costlier option compared to computer vision techniques and RADAR [6, 7].

Deciding camera position in case of on-road vehicle detection is a challenging task. Lin et al. [8] state a possible solution by mounting the camera on the vehicle roof in order to capture blind spot area. For the tracking of moving objects, the most commonly used technique is Kalman filtering. It has a feature of detecting vehicles traveling parallel to the subject vehicle. Wang et al. [9] use Kinect depth camera for detecting the obstacles.

Nowadays, it has become a need to consider the robust features of vehicles while detecting them. Some of these features are Haar-like features. Also, histogram of oriented gradient makes up the list of such features. These two features work in combination with each other [10]. Based on the symmetry feature, the vehicles are detected by using AdaBoost method of classification [11]. The front and rear vehicles are detected by using the combination of AdaBoost and Haar-like feature.

A novel deep learning approach is established by Nguyen et al. [12] for detection, classification, and tracking of real-time on-road obstacles. For detection, they have used UV Disparity algorithm.

3 Proposed Methodology

By doing classification process on different patches or windows or regions of image, one could perform detection of object from image. Image patch which shows the object of interest it also locates the position of that object in the image. One of the methods is sliding a patch or window over the image to cover all the locations from the image and apply classification for all the patches from the image, as shown in Fig. 1. From this process, one can get the classified object and location of object in the image.

In proposed system, we first train the CNN with various obstacle images, i.e., vehicle or pedestrians. We apply videos as input to our system, after that convert the input video into frames. Now CNN will check all image patches from each frame in that video with the training database. If the match found, then the location of that matched image patch is detected as obstacle in that respective frame. Similarly, check obstacles from all the frames. This is the general proposed system.

3.1 Training Dataset Preparation

In the proposed system, five videos have been taken as input from real-time on-road scenario. From these videos, the obstacles images have been cropped. The CNN is trained using these obstacle images. Initially, one video is chosen and it is converted into frames. In the proposed system, the videos are chosen randomly and thus each video has different frame rates per second. From all the frames of video, the image patches containing various obstacles are removed. In the proposed system, first frame is taken and the image patches are cropped in such a way that all the locations in the frame are covered. The image patches are selected by a fixed window. This window will slide over all the locations in frame. Then all the image patches are cropped and saved in a separate folder.

In proposed system, window size of 32 pixels is used. Slide this window over the frame with step size of 16. Like this, window will cover all the locations in a frame.

Fig. 1 Sub-windows formed by sliding different sized patches

Repeat the process for all the frames in a video. Similarly, take obstacle images from other videos.

3.2 Feature Extraction

In the proposed system, local binary pattern method (LBP), local ternary pattern (LTP), and Gabor filter are used to extract the texture features from the given input image.

(1) Local binary pattern (LBP): LBP is an image descriptor whose purpose is to compute binary patterns for the local structure of an image. Maybe the most essential property of the LBP descriptor in genuine applications is its robustness to monotonic grayscale changes created, for instance, by intensity varieties. Another imperative property is its computational effortlessness, which makes it conceivable to break down pictures in testing real-time settings.

The construction of the LBP features for given image is shown in Fig. 2. This shows the 3*3 pixel array. To construct the LBP code for respective image, LBP operator focuses on central pixel for reference. The grayscale values of neighborhood 8 pixels are used to generate respective binary value. If the neighborhood pixel value is greater than or equal to central pixel value, then LBP operator will interpret it as a '1'. And if the neighborhood pixel value is less than central pixel value, then LBP operator will interpret it as a '0'. As shown in Fig. 2, computed code will be 11100101. The same code will be computed for whole image by following same procedure. The computed LBP code will be used to form a global histogram of by using locally obtained histogram [13].

Given a pixel in the image, an LBP code is computed by comparing it with its neighbors

$$LBP_{P,R} = \sum_{p=0}^{p-1} s(g_p - g_c)$$
$$s(x) = 1, \quad x \geq 0$$
$$s(x) = 0, \quad x \leq 0 \tag{1}$$

where g is the gray value of the central pixel, g_p is the value of its neighbors, P is the total number of involved neighbors, and R is the radius of the neighborhood. Suppose the coordinate of g_p is (0, 0), then the coordinates of ($R\cos(2\pi = P)$, $R\sin(2\pi = P)$).

Fig. 2 Input array and respective LBP code

15	12	25
19	11	8
10	13	7

1	1	1
1		0
0	1	0

The gray values of neighbors that are not in the image grids can be estimated by interpolation. Suppose the image is of size $I*J$. After the LBP pattern of each pixel is identified, a histogram is built to represent the texture image.

$$H(k) = \sum_{i=1}^{I} \sum_{j=1}^{J} f\left(LBP_{P,R}(i, j), k\right), k \in [0, k]$$

$$f(x) = 1, x = y$$

$$f(x) = 0, \text{otherwise} \tag{2}$$

(2) Local ternary pattern(LTP): However, the LBP is sensitive to random noise in uniform regions or monotonic gray-level changes. Therefore, modifying the two-value encoding in the LBP to have three-value encoding instead, as follows:

$$LTP_{N,R}(z_c) = \sum_{n=1}^{N} f_{ltp}(z_n, z_c, \beta)2^n$$

$$f_{ltp}(z_n, z_c, \beta) = 1, \quad z_n \geq z_c + \beta$$

$$f_{ltp}(z_n, z_c, \beta) = 0, \quad |z_n - z_c| < \beta$$

$$f_{ltp}(z_n, z_c, \beta) = -1, \quad z_n \leq z_c - \beta \tag{3}$$

where β is the threshold, z_c and z_n are the center pixel, and nth neighbor pixel, respectively, and R is the radius of the neighborhood. The LTP results are divided into two LBP channels (LTP upper and LTP lower), as shown in Fig. 3. LTP upper is formed by replacing '−1' by '0' and LTP lower is formed by replacing '1' by '0' and '−1' by '1'.

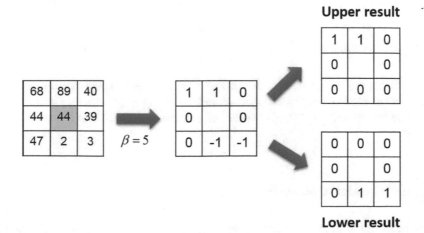

Fig. 3 Construction of local ternary pattern

(3) Gabor filter: The Gabor filter is a linear filter used for texture analysis, which means that it basically analyses whether there is any specific frequency content in the image in specific directions in a localized region around the point or region of analysis. A Gabor filter is obtained by modulating a sinusoid with a Gaussian. For the case of one-dimensional (1D) signals, a 1D sinusoid is modulated with a Gaussian. This filter will therefore respond to some frequency, but only in a localized part of the signal. The Gabor filter can be expressed as:

$$g(x, y, \theta, \emptyset) = \exp\left(-\frac{x^2 + y^2}{\sigma^2}\right)\exp(2\pi\theta(x\cos\emptyset + y\sin\emptyset)) \qquad (4)$$

It has been shown that σ, the standard deviation of the Gaussian kernel depends upon the spatial frequency to measured, i.e., θ [14].

3.3 Training Dataset

To prepare training dataset, LBP, LTP, and Gabor feature of all the images of obstacles are extracted separately. Table all the extracted features of obstacle in an Excel Sheet. LBP and LTP features are combined and created a feature vector containing both features to get better results. A feature vector for Gabor feature is created separately. So one feature vector which contains LBP+LTP features and second feature vector contains Gabor feature. These two feature vectors for each video are formed separately. These will be training set for respective video and repeat the process for all videos and make separate training dataset for each video. So, when any of the video is taken as input then respective training set will be selected according to the selected video.

3.4 Flowchart

Flowchart gives the detail steps for proposed system. All explanation given till in the paper is arranged stepwise for detail understanding of procedure. Form input as video to the output as detected obstacle, all procedure is covered in flowchart given in Fig. 4.

4 Experimental Results

The proposed system uses convolutional neural network for obstacle detection in the images. In this method, CNN is trained manually with all possible images of

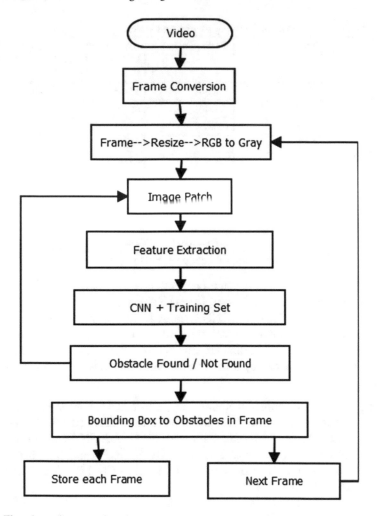

Fig. 4 Flowchart of proposed work

obstacle like vehicle or pedestrian from videos which are given as input. These images of obstacles are taken from frames of input videos and are given to CNN as training images. This methodology has been implemented in MATLAB R2013a on a laptop having 4 GB RAM and 2.5 GHz INTEL CORE i3 Processor.

To finalize the proposed system, two approaches are tested, namely

(1) Designing CNN with LBP+LTP feature.
(2) Designing CNN with Gabor feature.

The performance of both designed systems has been evaluated by detection rate and false alarm rate accuracy, which is defined as below,

Table 1 Training, testing set, and accuracy of proposed systems

S. no.	Training images	Testing images	Accuracy of LBP+LTP+CNN (%)	Accuracy of Gabor+CNN (%)
1	2485	106,547	85.44	85.02
2	3808	71,961	86.76	85.95
3	815	100,037	85.95	86.11
4	1993	73,997	84.75	83.88
5	684	54,467	86.08	85.78

$$DR(\%) = \frac{\text{No. of detected Obstacles}}{\text{No. of real Obstacles}} * 100 \tag{5}$$

$$FA(\%) = \frac{\text{No. of false detection}}{\text{Detected Obstacles + False detection}} * 100 \tag{6}$$

After frame conversion and patch conversion, each image patch undergoes to feature extraction stage. In that stage, LBP+LTP and Gabor features of that image are extracted and applied to CNN. CNN is trained with training dataset. It will check for the obstacles in the image. If match found, then CNN will show it as obstacle at output. For five videos, detection accuracy is near about same by using proposed method. The detection accuracy with these proposed methods for each video is given in Table 1.

- Comparison of two proposed methods: The performance of proposed two methods is comparatively same. Maximum detection rate of LBP+LTP+CNN method is 86.73%, whereas maximum detection rate of the Gabor+CNN method is 86.21%. But the false alarm rate for LBP+LTP+CNN method is much higher than the Gabor+CNN method. False alarm rate of LBP+LTP+CNN is 4%, and false alarm rate of Gabor+CNN is 0.27%.

  CNN will take same time for both the methods to run single frame. LBP+LTP+CNN method take 20.26 s to run single frame and Gabor+CNN method take 20.15 s to rum single frame.

Figure 5 shows the detected vehicles from respective images by using LBP and LTP features with help of CNN. Figure 6 shows the detected vehicles from respective images by using Gabor feature with help of CNN.

Figure 7 shows the detection rates and false alarm rates for various obstacle detection systems and proposed systems. Different detection systems have different detection accuracy. Many systems have detection accuracy greater that proposed systems but all these systems works on predefined datasets and proposed systems works on manually created datasets.

Fig. 5 Detected vehicles with LBP+LTP+CNN system

Fig. 6 Detected vehicles with Gabor+CNN system

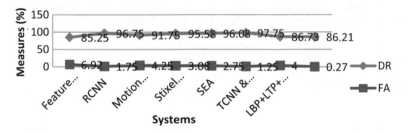

Fig. 7 Performance of proposed systems is compared with other detection systems

5   Conclusion and Future Scope

A proposed framework has been introduced to robustly detect the obstacles on-road in real time. The proposed methodology with LBP and LTP features have detection rate 86.73% and false alarm rate 4%. On the other hand, the proposed methodology with Gabor features has detection rate 86.21% and false alarm rate 0.27%. The proposed system takes large amount of time to run on MATLAB because it checks the every image patch in all frames in a video. So with the help of deep learning combination with GPU will reduce the run time. Classification of obstacles, like vehicles or pedestrians or other, will be the future scope for the proposed system. In future work, we can also add the traffic sign detection and lane detection.

References

1. Nguyen VD, Nguyen TT, Nguyen DD, Lee SJ, Jeon JW (2013) A fast evolutionary algorithm for real-time vehicle detection. IEEE Trans Veh Technol 62(6)
2. Keller CG, Enzweiler M, Rohrbach M, Llorca DF (2011) The benefits of dense stereo for pedestrian detection. IEEE Trans Intell Trans Syst 12(4)
3. Yu H, Hong R, Huang XL, Wang Z (2013) Obstacle detection with deep convolutional neural network. In: Sixth international symposium on computational intelligence and design
4. Huval B, Wang T, Tandon S, Kiske J, Song W, Pazhayampallil J (2015) An empirical evaluation of deep learning on highway driving
5. Erbs F, Barth A, Franke U (2011) Moving vehicle detection by optimal segmentation of the dynamic stixel World. In: 2011 IEEE Intelligent Vehicles Symposium (IV) Baden-Baden, Germany, 5–9 June 2011
6. Mao X, Inoue D, Kato S, Kagami M (2012) Amplitude-modulated laser radar for range and speed measurement in car applications. IEEE Trans Intell Transp Syst 13(1):408–413
7. Sato S, Hashimoto M, Takita M, Takagi K, Ogawa T (2010) Multilayer lidar-based pedestrian tracking in urban environments. In: Proceeding of IEEE IV, pp 849–854
8. Cortes C, Vapnik V (1995) Support vector networks. Mach Learn 20(3), 273–297
9. Wang T, Bu L, Huang Z (2015) A new method for obstacle detection based on Kinect depth image. IEEE
10. Nuevo J, Parra I, Sjoberg J, Bergasa L (2010) Estimating surrounding vehicles' pose using computer vision. In: Proceeding of 13th ITSC, pp 1863–1868
11. Liu T, Zheng N, Zhao L, Cheng H (2005) Learning based symmetric features selection for vehicle detection. In: Proceeding of IEEE Intelligence Vehicle Symposium, pp 124–129
12. Nguyen VD, Nguyen HV, Tran DT, Lee S, Jeon J (2016) Learning framework for robust obstacle detection, recognition, and tracking. IEEE Trans Intell Transp Syst
13. Guo Z, Zhang L, Zhang D (2010) A completed modeling of local binary pattern operator for texture classification
14. Prasad VSN Gabor filter visualization

Leakage Power Improvement in SRAM Cell with Clamping Diode Using Reverse Body Bias Technique

Chusen Duari and Shilpi Birla

Abstract Leakage has been the main issue in SRAM design due to the scaling of CMOS devices. In this paper, a novel technique has been proposed for 8T SRAM to reduce leakage current at 45-nm technology node. Here, we have used three different reverse body biasing techniques to reduce the leakage power. The three techniques used are: clamping of NMOS diode, clamping of PMOS diode, and clamping of NMOS and PMOS diode. Out of the three techniques, leakage current in 8T SRAM was minimum using PMOS clamping diode as compared to other proposed techniques. The supply voltage has been varied from 0.5 to 0.85 V. The leakage current has been reduced by both NMOS and PMOS clamping techniques, but the results for PMOS and NMOS were high as compared to two but lower than that of SRAM cell. The leakage current improved by 3.3x using PMOS technique and 2.7x using NMOS technique at supply voltage of 0.5V. The static power has also been calculated and compared for the three techniques.

Keywords SRAM · Leakage current · Clamping diode · Reverse body bias

1 Introduction

Static random-access memory is the most popular candidate for embedded memory in various ultra-low-power designs like PDAs, cell phones, biomedical devices, and other handheld devices which occupy a major area within the die. Low-leakage SRAM is the primary concern of the battery-operated portable devices as they are responsible for large percentage of total power consumption. The design of SRAMs for ultra-low-voltage application remains significantly challenging because of the

C. Duari (✉) · S. Birla
Department of Electronics & Communication Engineering, Manipal University Jaipur, Jaipur 303007, India
e-mail: chusen.duari@jaipur.manipal.edu

S. Birla
e-mail: shilpi.birla@jaipur.manipal.edu

© Springer Nature Singapore Pte Ltd. 2019
A. J. Kulkarni et al. (eds.), *Proceedings of the 2nd International Conference on Data Engineering and Communication Technology*, Advances in Intelligent Systems and Computing 828, https://doi.org/10.1007/978-981-13-1610-4_29

limitations such as high sensitivity to process–voltage–temperature variations, lower cell stability, lower voltage margin, and leakage current. Various SRAM cells have been proposed in the past for the reduction of leakage current. Single-port SRAM which is conventional six-transistor architecture has the read destruction problem which degrades the stability at low-power mode [1]. Dual-port memories with independent read and write bit lines can improve the stability [2].

Dual-threshold voltage with forward body bias technique can reduce the gate leakage current and total power [3]. Auto-back gate-control MTCMOS technique for the reduction of leakage power in SRAM cells is reported in [4]. SRAM cells designed with separated cell nodes from read bit lines have been used to improve cell stability and become a popular choice. [5]. A novel 8T SRAM cell with increased data stability is discussed, and ultra-low-voltage operation is reported [6, 7].

In our work, we have implemented three different leakage minimization techniques to improve the performance of 8T SRAM cell which can be used for ultra-low-power application in sub-nanometer regime. We have implemented reverse body bias technique by using one extra transistor to reduce the leakage current. This design uses dedicated read and write bit line which avoids read disturb. Here, we have focused mainly on the leakage current. The circuits are simulated at various supply voltages ranging from 0.85 to 0.5 V at 45-nm technology node.

This paper is organized as follows: Section 2 gives a review of transistor leakage mechanism. Section 3 discusses various leakage reduction techniques. In Sect. 4, we explain about 8T SRAM. The proposed techniques have been discussed in Sect. 5. Analysis and simulation result has been reported in Sect. 6.

2 Transistor Leakage Mechanism

The main components of leakage current in a MOS transistor can be summarized as follows:

A. Reverse-Bias Junction Leakage Current

Junction leakage current flows from the source to substrate or drain to the substrate through the reverse-biased diodes when the transistor is off [8].

B. Subthreshold (weak inversion) leakage current

This is the drain-to-source current when the transistor is operating in the weak inversion region, when the voltage between gate and source (V_{gs}) is below the transistor threshold voltage (V_{th}). The subthreshold leakage current can be expressed as [8]

$$I_{sub} = \mu_0 C_{ox} \frac{W_{eff}}{L_{eff}} V_T^2 e^{1.8} \exp\left(\frac{V_{gs} - V_{th}}{n V_T}\right) \cdot \left(1 - \exp\left(\frac{-V_{ds}}{V_T}\right)\right)$$

where μ_0 is the carrier mobility, C_{ox} is the gate oxide capacitance per unit area, W_{eff} and L_{eff} denote the transistor effective width and length, $V_T = kT/q$ is the thermal voltage at temperature T, 'n' is the subthreshold swing coefficient of the transistor, V_{gs} is the gate-to-source voltage of the transistor; V_{th} is the threshold voltage, and V_{ds} is the drain-to-source voltage of the transistor [8].

C. Oxide tunneling current

Gate leakage current flows from the gate terminal through the insulating oxide layer to the substrate. Direct tunneling current is significant for low oxide thickness. The gate leakage of a NMOS device is larger than the PMOS device typically by one order of magnitude with identical oxide thickness and Vdd [8].

D. The Gate-Induced Drain Leakage (GIDL) current

GIDL current occurs due to the strong electric field effect along the drain junction of MOS transistors. It becomes more prominent when drain-to-body voltage and drain-to-gate voltage is higher.

3 Leakage Reduction Techniques

A. Reverse Body Bias Technique:

Body biasing sets the substrate voltage of transistors at different levels. Controlling the body bias voltage has a strong effect on leakage current of a particular transistor. Opposite bias voltages are applied to PMOS and NMOS devices consistently in order to get a symmetric effect on both devices. Usually, in zero-bias schemes the bulk of NMOS is grounded and that of PMOS devices is connected to Vdd. The threshold voltage of a MOS transistor changes with change in bias voltage as governed by equation [9]

$$V_{th} = V_{th0} + \gamma.(\sqrt{|2\emptyset_F - V_{BS}|} - \sqrt{2\emptyset_F}$$

Reverse body bias reduces the subthreshold leakage by increasing the substrate voltage. Reverse substrate voltage increases the depletion layer width which results in the increase of threshold voltage. Therefore, higher gate voltage is required to create an inversion layer. Therefore, an optimized reverse body bias can reduce the total leakage current [10].

B. Dual-V_{th} Technology

In dual-V_{th} technique, dual-V_{th} devices switch between low V_{th} and high V_{th}. The low V_{th} can be attained by applying forward body bias, and high V_{th} devices can be formed with zero body steps. This technique improves the short-channel effects [11]. Dual-V_{th} can also be achieved by creating two different oxide layer thicknesses for different transistors. In SRAM bit cell, the gate oxide thicknesses of NMOS access transistor and pull-down transistors are increased, which increases threshold voltage, thereby decreasing the subthreshold leakage current.

C. VTCMOS Technique

Variable threshold CMOS (VTCMOS) technique employs the concept of controlling the threshold voltage and hence the leakage current, by using body biasing technique. Self-body biasing transistors are used to achieve different threshold voltages. A nearly zero substrate bias is used in active mode which enhances the speed of the device, while in standby mode reverse substrate biasing is used to minimize leakage current. However, some additional circuit elements are required for substrate biasing, which trade off the chip area [11].

D. Clamping Diode

Leakage current in SRAM cell can be reduced significantly by connecting a high-threshold voltage NMOS transistor in between source line of SRAM cell and ground terminals. During the active mode, this NMOS will turn on as the gate voltage is high and the resistance is small. Consequently, the source node will be grounded and the SRAM cell functions in the traditional manner. In the sleep mode, NMOS is turned off, and the source will be raised to higher voltage, which in turn reduces the gate and subthreshold leakage currents. A clamping PMOS diode is used along with the NMOS in parallel to resolve the issue of floating voltage by raising the voltage of the source line to certain voltage which depends on the threshold voltage and size of the PMOS diode [12]. Source biasing scheme in association with clamping diode also reduces the subthreshold and gate leakage current [13].

E. Stacking Technique

When one or more transistors are turned off in a stack of series-connected transistor, the subthreshold leakage current flowing through the stack is reduced. The voltage at the intermediate node becomes positive because of this stacking effect which results in negative gate-to-source voltage, thus decreasing subthreshold current exponentially. Also the stacking of transistors results in decrease of drain-to-source voltage, which results in reduction of drain-induced barrier lowering current and the subthreshold leakage current. The bulk-to-source voltage of NMOS also decreases, which in turn increases threshold voltage, hence reducing leakage.

F. Power-Gating Technique

In power-gating technique, the leakage current is reduced by external header and footer transistors which block the path between supply voltage and ground when the device is in idle state. In SRAM bit cell, additional NMOS transistor is used in the leakage path between power supply and ground line which gates the supply line by switching 'ON' the transistor in active mode and switching 'OFF' in the inactive mode. Similarly, extra PMOS is used in between power supply and source line of the cell [10].

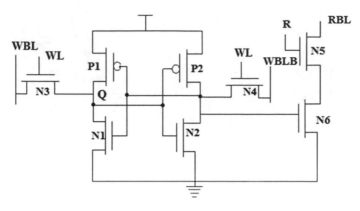

Fig. 1 Circuit diagram of 8T SRAM cell [1]

4 Existing 8T SRAM Cell Design

To overcome the read destruction problem dedicated write and read bit lines are used with addition of two stacked transistors in the basic 6T SRAM cell which is shown in Fig. 1. Though area overhead is a trade-off in 8T SRAM, it operates efficiently than 6T SRAM at lower supply voltage [1, 11]. The 'Read' operation of 8T SRAM cell in Fig. 1 can be performed by pre-charging 'RBL' and asserting the control signal 'R' with logic '1', while word line 'WL' grounded. During 'Write' operation, 'WL' is kept at logic '1' which turns 'ON' the data access transistors N3 and N4. The data to be written is provided through bit line 'WBL' and its complement 'WBLB'. During standby mode, the data access transistors are turned off by de-asserting 'WL' and 'R' and keeping 'WBL' and 'WBLB' at logic '1'. Two separate ports are used for simultaneous 'Read' and 'Write' operation.

5 Proposed Technique

In the proposed work, three different techniques have been implemented to reduce leakage current in 8T SRAM Cell. The circuit shown in Fig. 2a employs one extra NMOS, in Fig. 2b employs one extra PMOS transistor, and in Fig. 2c employs an extra pair of NMOS–PMOS transistors in between source line of SRAM cell and ground. In Fig. 2a, the extra NMOS during 'Read' and 'Write' operation will be in 'ON' state and circuit will behave in a similar fashion as the conventional 8T-SRAM cell shown in Fig. 1. In standby mode, the NMOS transistor will be cut off which in turn reduces the leakage. In the circuit shown in Fig. 2b, the extra PMOS will be kept on by shorting the gate terminal to ground during normal operation and kept off during standby mode. The circuit shown in Fig. 2c uses two extra transistors. During 'Read' & 'Write', these two transistors will be in 'ON' state, and during standby

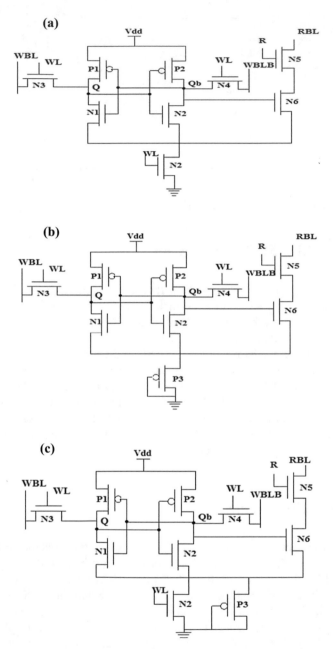

Fig. 2 **a** 8T SRAM with extra NMOS clamping diode. **b** 8T SRAM with extra PMOS clamping diode. **c** 8T SRAM with extra PMOS–NMOS clamping diode

mode, they will be kept 'OFF' to reduce leakage. In all these circuits, two separate bit lines are used for read and write operations. Reading operation is performed by pre-charging the read bit line (RBL) to Vdd and then raising the read control line (R) to Vdd. If originally Q stores '0', then Qb stores '1'; when the read control line is chosen, the two transistors in the read channel will conduct to ground through the extra transistor. Writing operation is performed by raising write word line (WL) to Vdd.

This will turn the access transistors on, and consequently, the desired bit will be stored in the memory cell. During hold state, the two access transistors are kept off by grounding the word line 'WL' and read control signal 'R'. Since the extra transistor is inserted between the source line of the cell and ground which is reversed body-biased, it reduces the leakage current of the memory cell.

6 Simulation Results

The total leakage current for the three different SRAM cells at various supply voltages has been evaluated using HSPICE simulation tool, and the result is shown in Table 1. The total static power of three different implemented cells is shown in Table 2.

The comparison of leakage power of three different implemented techniques is shown in graph in Fig. 3. The leakage current improved by 3.3x using PMOS technique and 2.7x using NMOS technique at supply voltage of 0.5V. All the techniques show a considerable reduction of leakage current. Among the three techniques, the cell with PMOS clamping diode has the least leakage current as well as power. Simulation of 8T SRAM as shown in Fig. 1 has also been carried out at 45-nm technology and compared with the three leakage reduction techniques. The leakage current of 8T SRAM was high as compared to all the reduction techniques used in the paper.

Table 1 Comparative analysis of leakage current at different supply voltages

Supply voltage	Leakage current in nA			
	8T_SRAM	8T_SRAM NMOS	8T_SRAM PMOS	8T_SRAM NMOS PMOS
0.85	821.1	4.48	2.86	172.25
0.8	280	3.43	1.95	78.23
0.75	105.9	1.96	1.25	53.94
0.7	35.66	1.36	0.45	36.97
0.65	12.98	0.60	0.32	26.42
0.6	4.73	0.28	0.24	19.22
0.55	1.72	0.26	0.14	11.38
0.5	0.61	0.22	0.18	7.53

Table 2 Comparative analysis of static power at different voltages

Supply voltage	Static power in 'HOLD' state (W)			
	8T_SRAM	8T_SRAM NMOS	8T_SRAM PMOS	8T_SRAM NMOS PMOS
0.85	4.49u	5.26n	5.12n	96.47n
0.8	2.76u	4.89n	2.36n	63.23n
0.75	1.62u	2.54 n	1.76n	41.13n
0.7	918.61n	2.46n	1.03n	26.54n
0.65	504.47n	1.18n	0.62n	17n
0.6	270.30n	1.25n	0.35n	10.8n
0.55	141.73n	0.55n	0.27n	6.8n
0.5	72.79n	0.46n	0.23n	4.24n

Fig. 3 Comparison of 'HOLD' static power at different supply voltages

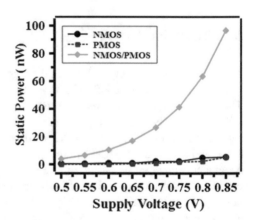

7 Conclusions

In this paper, three different techniques are implemented in an existing 8T dual-port SRAM cell to minimize leakage current and compared their results at various supply voltages at 45-nm technology. The total static power dissipations at different supply voltages have also been evaluated. All the techniques show reduction of leakage power among which the cell with PMOS clamping diode shows best results at all supply voltages as shown in Tables 1 and 2. Though PMOS transistor may increase the cell area, the increased area can be reduced by decreasing the width of the transistor. The circuit can be useful for low-voltage low-power application. This work will be further extended for finding stability of the SRAM by calculating static noise margin.

References

1. Zhang K-J, Chen K, Pan W-T, Ma P-J (2010) A research of dual-port SRAM cell using 8T. In: 2010 10th IEEE International Conference on Solid-State and Integrated Circuit Technology (ICSICT). IEEE, pp 2040–2042
2. Wen L, Li Z, Li Y (2013) Single-ended, robust 8T SRAM cell for low-voltage operation. Microelectron J 44(8):718–728
3. Razavipour G, Afzali-Kusha A, Pedram M (2009) Design and analysis of two low-power SRAM cell structures. IEEE Trans Very Large Scale Integr (VLSI) Syst 17(10):1551–1555
4. Nii K, Makino H, Tujihashi Y, Morishima C, Hayakawa Y, Nunogami H, Arakawa T, Hamano H (1998) A low power SRAM using auto-backgate-controlled MT-CMOS. In: Proceedings of 1998 international symposium on low power electronics and design. IEEE, pp 293–298
5. Kim, T-H, Liu J, Keane J, Kim CH (2007) A high-density subthreshold SRAM with data-independent bitline leakage and virtual ground replica scheme. In: 2007 IEEE International Solid-State Circuits Conference, ISSCC 2007. Digest of Technical Papers. IEEE, pp 330–606
6. Kushwah CB, Vishvakarma SK (2016) A single-ended with dynamic feedback control 8T subthreshold SRAM cell. IEEE Trans Very Large Scale Integr VLSI Syst 24(1):373–377
7. Pal S, Islam A (2016) Variation tolerant differential 8T SRAM cell for ultralow power applications. IEEE Trans Comput Aided Des Integr Circuits Syst 35(4):549–558
8. Roy K, Mukhopadhyay S, Mahmoodi-Meimand H (2003) Leakage current mechanisms and leakage reduction techniques in deep-submicrometer CMOS circuits. Proc IEEE 91(2):305–327
9. Manuzzato A, Campi F, Rossi D, Liberali V, Pandini D (2013) Exploiting body biasing for leakage reduction: a case study. In: 2013 IEEE computer society annual symposium on VLSI (ISVLSI). IEEE, pp 133–138
10. Bikki P, Karuppanan P (2017) SRAM cell leakage control techniques for ultra low power application: a survey. Circuits Syst 8(02)
11. Lorenzo R, Chaudhury S (2017) Review of circuit level leakage minimization techniques in CMOS VLSI circuits. IETE Tech Rev 34(2):165–187
12. Zhang L, Chen W, Mao L-F, Zheng J (2012) Integrated SRAM compiler with clamping diode to reduce leakage and dynamic power in nano-CMOS process. Micro Nano Lett 7(2):171–173
13. Wu C, Zhang L-J, Wang Y, Zheng J-B (2010) SRAM power optimization with a novel circuit and architectural level technique. In: 2010 10th IEEE international conference on solid-state and integrated circuit technology (ICSICT). IEEE, pp 687–689

An Investigation of Burr Formation and Cutting Parameter Optimization in Micro-drilling of Brass C-360 Using Image Processing

Shashank Pansari, Ansu Mathew and Aniket Nargundkar

Abstract A lot of research has been done in area of conventional micro-drilling but the measurement techniques for measuring burr size and holes circularity were either very expensive or inaccurate. This paper attempts to investigate role of input parameters like the spindle speed and the feed rate on burr height and burr thickness at hole exit for Brass C-360, which is a widely used material in micro-fabrication. The measurements were taken from scanning electron microscope (SEM) images of micro-drilled holes using image processing which makes the measurement simple, fast, and accurate. All the experiments were conducted using response surface methodology to develop second-order polynomial models for burr thickness and height of burr. Optimization by using multi-objective genetic algorithm and cohort intelligence algorithm using MATLAB is done to generate optimum output results. All the micro-hole SEM images were analyzed to detect the types of burrs formed during different experiments.

Keywords Burr · Cohort intelligence · Genetic algorithm · Image processing
Micro-drilling · Response surface methodology

1 Introduction

Mechanical micro-drilling is one of the most widely used methods among many micro-hole making methods because it has less dependency on the material properties. Micro-drilling has a wide variety of applications for example in PCB circuits,

S. Pansari · A. Mathew
Manipal Institute of Technology, Manipal Academy of Higher Education, Manipal, India
e-mail: shashank.pansari@manipal.edu

A. Mathew
e-mail: ansu.mathew@manipal.edu

A. Nargundkar (✉)
Symbiosis Institute of Technology, Symbiosis International University, Pune, India
e-mail: aniket.nargundkar@sitpune.edu.in

© Springer Nature Singapore Pte Ltd. 2019
A. J. Kulkarni et al. (eds.), *Proceedings of the 2nd International Conference on Data Engineering and Communication Technology*, Advances in Intelligent Systems and Computing 828, https://doi.org/10.1007/978-981-13-1610-4_30

microprocessors, in automotive industry for making micro-holes in fuel injectors, in making fasteners such as micro-jacks and micro-pins.

It is very difficult to obtain good surface finish, reduced burr height, etc., in micro-drilling. Various factors such as tool diameter, spindle speed, tool helix angle, twist angle, feed rate, and material control the hole quality, and thus, they have to be chosen very carefully. In microholes, burr formation is very problematic as it causes deterioration of surface quality which reduces product durability and precision, assembly problems, wear and tear on the surface, etc. In macro-level, de-burring is easy but in micro-holes rarely it is possible as in micro-drilled holes; it is difficult because of poor accessibility of burr area and tight tolerance. In the past numerous experimental and theoretical studies have been conducted for studying burr formation in drilling [1–4].

This paper deals with investigation of burr formation with respect to cutting parameters like spindle speed and feed rate by measuring burr size accurately by using image processing in MATLAB and optimizing process parameters for minimize burr size that indicates reducing its thickness and its projection at the exit which is also known as burr height. Combining the input parameters in optimum way for the burr formation in the machining of Brass C-360 was determined by applying the response surface methodology optimization technique [5], GA [6] and CI algorithm [7]. The design of experiments was utilized to create a model of drilling process.

2 Experimental

2.1 Material

In present work, workpiece material is selected specifically which is commercially used in micro-fabrication industries, i.e., brass C360. It is a combination of copper and zinc as shown in Table 1, having the highest machinability among all other copper alloys, is the standard against which all the others are compared to C360 brass, known for its strength and resistance to corrosion with properties closely resembling that of steel, and is one of the most popular copper alloys used when it comes to micro-machining.

C360 brass has lead in it which makes it self-lubricating and can be machined accurately and precisely. C360 brass has a natural behavior of forming a thin protective "patina," which is rust proof when exposed to the atmospheric conditions [8].

Table 1 Composition of brass C-360 (Mass fraction, %)

Copper	Lead	Iron
61.5 nominal	2.5 minimum	0.35 maximum
Copper	Lead	Iron

Table 2 Machining parameters

Symbol	Factors	Low level	High level
A	Cutting speed (rpm)	1000	2500
B	Feed rate (mm/min)	1	4

2.2 Experimental Design

Response surface methodology (RSM), one of the popular and powerful experiment designing methods, is chosen in this work. It is a combination of mathematical and statistical techniques which models and analyzes problems in which a response, i.e., the output parameter is influenced by several input variables and its main objective is to optimize this response [9]. As there are only two input parameters are taken into consideration, central composite design (CCD) approach is used and it is generating thirteen experiments for each one of the four drills of diameter 0.5, 0.6, 0.8, and 0.9 mm, the upper and lower bounds of the experiment are shown in Table 2.

2.3 Experimental Procedure

For conducting experiments, a multiprocess machine tool is used, which is a highly precise and integrated multiprocess machine for micro-machining developed in NUS [10] and manufactured by Mikrotool Pte. Ltd. as shown in Fig. 1. All the tools used in this work are of HSS (high-speed steel) straight flute micro-drills having 65 HRC hardness value, and its cutting speed can go up to 50 m per min. Workpiece samples of dimension 1.5 cm by 1.5 cm by 4 mm are used for each drill size.

Measurement

For the measurement of burr height and burr thickness at the exit of through hole of the same, several methods were used in past such as contact method, SEM images, optical

Fig. 1 Micro-tools integrated multiprocess machine tool DT110

Table 3 Experimental results for BH (burr height) and BT (burr thickness) in micron using image processing

S. No.	Speed (RPM)	Feed (mm/min)	0.5 mm Diameter		0.6 mm Diameter		0.8 mm Diameter		0.9 mm Diameter	
			BH	BT	BH	BT	BH	BT	BH	BT
1	2280.33	3.56	152.82	22.23	257.97	31.37	260.12	26.11	395	32.73
2	2280.33	1.43	107.34	20.32	234.52	29.82	305.3	30.50	463.67	33.83
3	1750	2.5	100.04	13.20	214.32	23.57	248.17	24.82	354.9	21.56
4	1750	4	162.49	27.57	307.85	27.31	258.25	25.80	378.5	34.11
5	1750	2.5	100.04	13.20	214.32	23.57	246.08	24.82	354.9	21.56
6	1750	2.5	100.04	13.20	214.32	23.57	250.34	24.82	354.9	21.56
7	1750	2.5	100.04	13.20	214.32	23.57	244.31	24.82	354.9	21.56
8	1750	1	160.28	24.46	260	19.5	256.22	25.59	353	28.33
9	2500	2.5	173.55	21.35	225.32	27.113	322.16	32.22	380.256	32.31
10	1000	2.5	138.15	22.03	192.81	21.67	203.00	20.31	361	23.88
11	1750	2.5	100.04	13.20	214.32	23.57	252.15	24.8	354.9	21.56
12	1219.67	1.43	129.16	24.96	257.5	27.11	213.01	21.35	353.06	33.83
13	1219.67	3.56	159.16	23.72	256.13	26.71	270.19	27.14	302.32	36.58

Fig. 2 SEM image of
0.5 mm hole

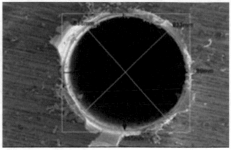

Fig. 3 Flowchart of
MATLAB process

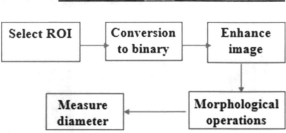

microscope method, and coordinate measuring machine [11, 12]. In the present study, for measurement of the burr height (exit) of the drilled holes, a different method is developed which unlike other methods which choose 5–10 points on the periphery of burr and take the average of height and the same for thickness, image processing is been used. Such methods are not accurate so to get the average of the whole surface of burr, SEM images were taken from two sides, perpendicular and exactly parallel to hole axis for each hole on every sample.

Burr Thickness

The sample image of drilled holes obtained from SEM is shown in Fig. 2. The thickness of the hole is identified using MATLAB and methodology I shown in Fig. 3. The region of interest is extracted from the given image and thresh holding is performed to convert the image into binary. The threshold level used is 0.25 and the output of thresholding operation is shown in Fig. 4. The spatial domain technique averaging is performed in image shown in Fig. 4 for removing noise. It is followed by morphological closing operation for removing the insignificant holes. The closing operation is defined as dilation followed by erosion. Output of the closing operation is shown in Fig. 5. For obtaining the size of hole, image obtained after performing the closing operation is given to the function "region props." Since the image is not a perfect circle, the average minimum diameter and the average maximum diameter are extracted using region props parameters. The burr thickness can be computed as:

$$\text{Burr Thickness} = \frac{D_{\max} - D_{\min}}{2} \tag{1}$$

Fig. 4 After thresholding

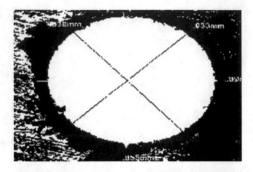

Fig. 5 Binary image

Fig. 6 Re-oriented image of burr and binary image

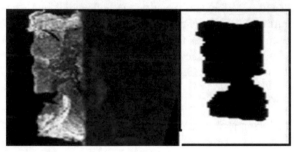

Here,

D_{max} average outside diameter
D_{min} average inside diameter.

Burr Height

For identifying the burr height, hole side view images from two sides are obtained from SEM are again given to MATLAB and procedure followed is shown in Fig. 7. It is rotated for easy processing and region of interest is extracted. Output image after selecting region of interest is shown in Fig. 6. The preprocessing steps performed for burr height are same as of burr thickness. Output of the closing operation is shown Fig. 6. A counter is used to keep track of the number of black pixels in each row for

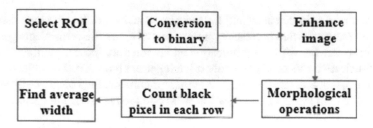

Fig. 7 Flowchart of MATLAB process

Table 4 GA optimized predicted burr height and burr thickness

Tool diameter (mm)	Optimized speed (RPM)	Optimized feed rate (mm/min)	Predicted burr height (μ)	Confirmed burr height (μ)	Predicted thickness (μ)	Confirmed burr thickness (μ)
0.5	1805	2.36	99.46	99.50	13.15	13.15
0.6	1388	1.85	219.1	218.77	23.11	22.76
0.8	1214	1.8	249	252.1	29.56	27.88
0.9	1615	2.64	357.7	351.28	31.49	32.1

calculating the average width of burr projection at hole exit. Finally, a total number of black pixels were computed and average height is determined.

3 Data Analysis

Once the experimental results are obtained after using image processing, from SEM images side view and top view of burr at the hole exit as shown in Table 4, a data analyzer has to be used to generate regression models which shows the relation between output function and input parameter. In this work, response surface analyzer is used for analysis of the influence of input cutting parameters on responses. ANOVA is used for getting most significant factor and equations, i.e., Eqs. 2–9 are framed based on response surface regression. It is given as objective function in multi-objective GA solver and fed into CI algorithm in MATLAB R2010a. Pareto fronts by GA and convergence charts by CI algorithm are obtained, and finally, confirmation experiments are performed for checking the validity of the optimal parameters.

3.1 Response Surface Analyzer

Two responses are entered in Minitab 16 worksheet of experiments. Then, to analyze the responses, response surface analyzer is used. Here, quadratic regression equation in terms of spindle speed and feed rate are considered. Regression equation is generated using response surface regression analysis with the coefficients generated by the analyzer. For further optimization and prediction, these equations are given in multi-objective GA solver and cohort intelligence algorithm. The objective functions for 0.5, 0.6, 0.8, and 0.9 mm holes are for minimized burr height (BH) and burr thickness (BT) in terms of speed(X1) in RPM and feed rate (X2) in mm/s are given as follows:

$$\text{Min BH (0.5 mm)} = 420.94 - 0.234 * X_1 - 99.91 * X_2 + 6.55 \times 10^{-5} * X_1^2$$
$$+ 22.152 * X_2^2 \tag{2}$$

$$\text{Min BT(0.5mm)} = 90.57 - 0.049 * X_1 - 27.12 * X_2$$
$$+ 1.32 \times 10^{-5} * X_1^2 + 5.54 * X_2^2 \tag{3}$$

$$\text{Min BH(0.6 mm)} = 369.67 - 0.028 * X_1 - 156.79 * X_2$$
$$+ 6.64 \times 10^{-6} * X_1^2 + 23.162 * X_2^2 \tag{4}$$

$$\text{Min BT(0.6mm)} = 35.34 - 0.019 * X_1 - 0.59 * X_2$$
$$+ 6.44 * 10^{-6} * X_1^2 + 0.51 * X_2^2 \tag{5}$$

$$\text{Min BH(0.8mm)} = 106.116 - 0.13 * X_1 - 6.62 * X_2$$
$$+ 1.49 * 10^{-6} * X_1^2 + 4.75 * X_2^2 \tag{6}$$

$$\text{Min BT(0.8mm)} = 59.79 - 0.024 * X_1 - 11.3 * X_2 - 7.78 * 10^{-6} * X_1^2$$
$$+ 2.18 * X_2^2 \tag{7}$$

$$\text{Min BH(0.9mm)} = 450.7 - 0.09 * X_1 - 38.48 * X_2 + 2.34 * 10^{-5} * X_1^2$$
$$+ 5.03 * X_2^2 \tag{8}$$

$$\text{Min BT(0.9mm)} = 80.07 - 0.040 * X_1 - 14.81 * X_2$$
$$+ 1.516 * 10^{-5} * X_1^2 + 4.65 * X_2^2 \tag{9}$$

Subject to constraints:

$$1000 \le X_1 \le 2500, 1 \le X_2 \le 4$$

3.2 Genetic Algorithm (GA)

Genetic algorithm works equally well in either continuous or discrete search space. It is a heuristic technique inspired by the natural biological evolutionary process comprising of selection, crossover, mutation, etc. The evolution starts with a population of randomly generated individuals in first generation. In each generation, the fitness of every individual in the population is evaluated, compared with the best value, and modified (recombined and possibly randomly mutated), if required, to form a new population [13].

The new population is then used in the next iteration of the algorithm. The algorithm terminates, when either a maximum number of generations has been produced or a satisfactory fitness level has been reached for the population. The fitness function of a GA is defined first. Thereafter, the GA proceeds to initialize a population of solutions randomly and then improves it through repetitive application of selection, crossover, and mutation operators. This generational process is repeated until a termination condition is reached. In multi-objective GA solver, tournament selection is available as a selection procedure. The crossover operator of GA may exploit structures of good solutions with respect to different objectives to create new non-dominated solutions in unexplored parts of the Pareto front. In addition, most multi-objective GA does not require the user to prioritize, scale, or weigh objectives. Therefore, GA has been the most popular heuristic approach to multi-objective design and optimization problems, and hence it is adopted here.

3.3 Cohort Intelligence (CI)

In this algorithm, behavior of self-supervised learning is the inspiration of the candidates in a group which is also referred to cohort in this particular algorithm. This unit is a group of candidates who compete and learn with each other in order to reach some individual goal which is intrinsically common to all of them. While one is working in such environment, every candidate attempts to improve its own behavior through observation of other candidate's behavior.

The individual candidate may follow a certain behavior in the group which according to itself may result in improvement in its own behavior. As the behavior is dependent on certain qualities, when followed by the candidate, it actually goes for achieving those qualities. This makes the evolution of cohort behavior of each candidate. If after considerable number of attempts of learning from one another, the candidate's behavior ceases to improve as it reaches saturation then it becomes indistinguishable between behaviors of two individuals. The iteration finally could be expected to become successful when after significant number of tries the performance reaches same saturation point [14]. In this work, CI is applied for single objective function individually.

4 Results and Discussions

The experimental results as shown in Table 3 are used by response surface analyzer to generate regression Eqs. 2–9 which were used as an input to multi-objective GA solver and CI algorithm.

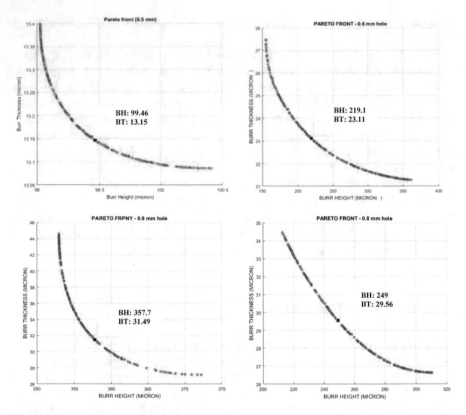

Fig. 8 Optimized results are shown on Pareto front generated from multi-objective GA solver for 0.5, 0.6, 0.8, 0.9 mm diameter tools (clockwise from top left)

4.1 Multi-objective GA Optimization Results

The parameters for GA which was identified as important by literature survey and by trial and error methods are initial population size, crossover function and fraction and mutation function.

Based on number of trials taken, it is observed as initial population size increases no. of solutions in Pareto solution increases. Hence, after taking trials with 30, 50, and 100 the highest value, i.e., 100 is selected. Crossover fraction 0.8 is giving best results. Mutation function "adaptive feasible" gives the best results, and hence, it is selected. As it is a multi-objective optimization, it selects the intersection of both minimization functions in order to get a compromised minimum value to keep Burr size to a minimum level. This can be observed from the Pareto fronts generated as shown in Fig. 8. The predicted and confirmed results are shown in Table 5.

Table 5 Consolidated results for burr thickness and burr height using CI variations

Tool diameter (mm)	Optimized parameters							
	Cohort intelligence							
	Speed (RPM)	Feed (mm/min)	Burr thickness (μ)		Speed (RPM)	Feed (mm/min)	Burr height (μ)	
			Predicted	Confirmed			Predicted	Confirmed
0.5	1856	2.45	11.9	12.12	1786	2.25	99.289	102.3
0.6	1516	1.15	21.33	23.17	1000	3.53	113.593	119.16
0.8	1524	2.59	26.63	27.32	1000	1	252.586	250.68
0.9	1315	1.59	41.96	38.25	1923	3.82	290.637	312.67

4.2 Pareto Fronts for Multi-objective Function for Min. Burr Height and Min. Burr Thickness

See Fig. 8.

4.3 Cohort Intelligence Optimized Results

CI algorithm is used for both the objective functions burr height and burr thickness individually. The parameters reduction factor "r" is 0.9 and no. of candidates "c" is 150 for different variants of CI that has been used in coding. Table 5 shows the optimum solution for burr height and burr thickness.

The solution convergence plots for the objective functions burr height and burr thickness with "Roulette Wheel" approach are presented in Figs. 9 and 10, respectively, which show the whole development of the cohort by refining individual performances. Initially, the comportment of the individual candidates was discernable. As the reduction in sampling interval takes place and evolution of group happens, behavior of the candidate's performance ceases to show significant improvement. At a specific point, it is challenging to decide between the behaviors of the candidates and it is the final saturation.

4.4 Plots for Objective Function Burr Height According to Roulette Wheel Approach

See Fig. 9.

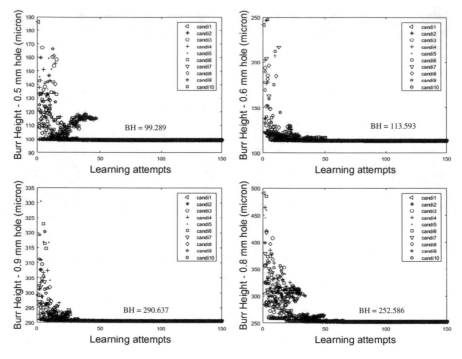

Fig. 9 Optimized burr height from CI for 0.5, 0.6, 0.8, 0.9 mm dia. tools (clockwise from top left)

4.5 Plots for Objective Function Burr Thickness According to Roulette Wheel Approach

See Fig. 10.

5 Burr Analysis

Different shapes of burrs were obtained after experiments for different speed and feed values. Broadly, there are three types of burrs, i.e., crown type, transient, and uniform burrs. All of them are shown in Fig. 11. It has been observed that crown type burrs are formed at very low feed rate and high speed, while transient burrs are obtained at higher feed rate and higher speed. The uniform burrs are not solely dependent on feed rate and speed but tool diameter and tool type also, thus we calculated optimum values of feed rate and speed for different tools and results shows wide range of optimum values.

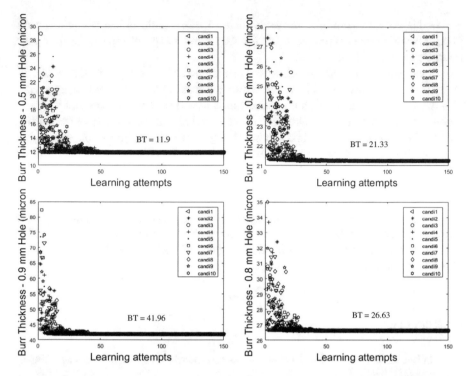

Fig. 10 Optimized burr thickness from CI for 0.5, 0.6, 0.8, 0.9 mm dia. tools (clockwise from top left)

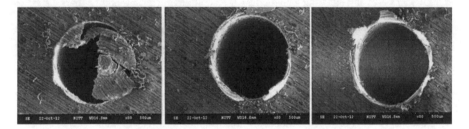

Fig. 11 Crown type burr, uniform type burr, transient type burr (left to right)

6 Conclusion

In the present paper for measuring burr height and burr thickness, a new micro-measurement approach is developed, i.e., through image processing of SEM images which proved more accurate and also avoided any kind of contact measurement which in a way is very reliable as burrs are not disturbed by contact. It is a cheaper and reliable method as instead of just choosing few point whole burr size average is detected.

In this work, two optimization tools are used; one is multi-objective optimization, i.e., GA which gives a unique value for feed rate and spindle speed for minimum burr height and burr thickness.

In the second method, i.e., CI which is a single objective optimization tool with very wide variations available, out of which we chose only roulette wheel approach. The results obtained for each objective function is different for each tool diameter which gives a flexibility for choosing correct feed rate and spindle speed according to requirement, i.e., either minimum burr height or minimum burr thickness.

It is concluded that when selective output parameter is required to be optimized then CI gives better and best minimum value of burr height and burr thickness than GA while when both the output parameters are to be minimized together than GA gives suitable results.

Finally, it is observed during the experimentation and data analysis that in future further research can be done in area of conventional micro-drilling at very high speeds as well as by varying tool twist angles and different tool materials to get further minimized burrs.

References

1. Shikata H, DeVries MF, Wu SM (1980) An experimental investigation of sheet metal drilling. Ann CIRP 29(1):85–88
2. Miyake T et al (1991) Study on mechanism of burr formation in drilling. J Jpn Soc Precis Eng 57(3):485–490 (in Japanese)
3. Takeyama H et al (1993) Study on oscillatory drilling aiming at prevention of burr. J Jpn Soc Precis Eng 59(10):1719–1724 (in Japanese)
4. Guo Y, Dornfeld DA (2000) Finite element modeling of drilling burr formation process in drilling 304 Stainless Steel. Trans ASME J Manuf Sci Eng 122(4):612–619
5. Pansari S et al (2013) Experimental investigation of micro drilling of C360 brass. In: Proceedings of 1st national conference on micro and nano fabrication (Mnf—2013), CMTI, Bangalore, India
6. Saravanan M et al (2012) Multi objective optimization of drilling parameters using genetic algorithm. Procedia Eng 38:197–207
7. Kulkarni AJ et al (2016) Application of the cohort-intelligence optimization method to three selected combinatorial optimization problems. Eur J Oper Res 250(2):427–447
8. Bolton W (1989) Copper. In: Newnes engineering materials pocket book, pp 84–114
9. Donglai W et al (2008) Optimization and tolerance prediction of sheet metal forming process using response surface model. Comput Mater Sci 42:228–233
10. Asad ABMA et al (2007) Tool-based micro-machining. J Mater Process Technol 192–193:204–211
11. Rajmohan T et al (2012) Optimization of machining parameters in drilling hybrid aluminum metal matrix composites. Trans Nonferrous Met Soc China 22:1286–1297
12. Ali SHR (2012) Review article advanced nano-measuring techniques for surface characterization. ISRN Opt 2012(859353):23
13. Chen X et al (2016) Optimization of transmission system design based on genetic algorithm. Adv Mech Eng 8(5):1–10
14. Kulkarni et al (2013) Cohort intelligence: a self-supervised learning behavior. In: Proceedings of the IEEE conference on systems, man and cybernetics 2013, pp 1396–1400 6

Effect of Training Sample and Network Characteristics in Neural Network-Based Real Property Value Prediction

Sayali Sandbhor and N. B. Chaphalkar

Abstract Property value is its current worth in the market which needs to be ascertained for buying, selling, mortgage, tax calculation, etc. Neural network uses past data to predict future data, making it suitable for predicting value of property based on previous valuation reports. Study involves identification of variables affecting property value, data collection, data standardization, and application of neural network for predicting value of property for Pune City, India. More than 3000 sale instances have been recorded and treated as input data. Neural network is applied for value prediction. A well-generalized network is always beneficial for better predictability. A robust network is found out by testing various combinations of network parameters and dataset. Results of prediction have shown varying percentage of rate of prediction. Results obtained by change in training sample, training sample size, network characteristics like number of runs of training, and the combinations of cases given for training, testing, and cross-validation are compared. The process ensures better generalization resulting in better predictability.

Keywords Valuation · Real property · Prediction · Neural networks
Sample size

1 Introduction

Price of a property is one of the most important criteria for users in making their property sell, purchase, and related decisions [1]. The rapid growth in housing prices around the world since the late 1990s has motivated a growing number of studies to examine the variation in housing price dynamics across cities or regions [2].

S. Sandbhor (✉)
Civil Engineering Department, SIT, SIU, Pune, India
e-mail: sayali.sandbhor@sitpune.edu.in

N. B. Chaphalkar
Civil Engineering Department, RSCOE, Pimpri-Chinchwad, India
e-mail: nitin_chaphalkar@yahoo.co.in

© Springer Nature Singapore Pte Ltd. 2019 303
A. J. Kulkarni et al. (eds.), *Proceedings of the 2nd International Conference on Data Engineering and Communication Technology*, Advances in Intelligent Systems and Computing 828, https://doi.org/10.1007/978-981-13-1610-4_31

Being one of the most volatile sectors of nation's economy, the behavior of housing prices has been attracting considerable research attention. Careful attention needs to be given to the dynamics of the factors affecting property value and mapping the same using non-conventional techniques. The conventional methods of value estimation are time-consuming and cumbersome which alarms the use of automated valuation models which would not only guide the appraiser and predict value but also save time and provide guideline for further assessment. Many studies have used artificial intelligence techniques to predict value of property for a specific study region and have found varied results. Indian housing market has observed fluctuations in property rates which are attributed to movements at macroeconomic level. Value of property, on the other hand, depends primarily on property characteristics. Very less research has been done in the field of property value prediction in India. This necessitates the purpose of the study of implementing neural network for predicting property value for which Pune city has been considered as a case study. As Neural Network (NN) is data sensitive, it thus becomes difficult to prepare a robust model and requires number of runs with various combinations to find the best-suited model.

1.1 Property Valuation Using NN

Valuation is a process of finding the current worth of the property, and methods like sales comparison method, rental method, land, and building method, profit method have traditionally been used to predict the property value. An accurate prediction of real property price is important to all the stakeholders like prospective owners, developers, investors, appraisers, tax assessors, and other real estate market participants. Hence, it is required to prepare a prediction model which would take into consideration the effect of various dynamic factors on property value. Various soft computing techniques with higher data handling capabilities have facilitated the study of complicated relationship between property price and the affecting factors. This study tests the applicability of neural network to value prediction for properties in a specific region and tries to assess the effect of change in network characteristics on the results of prediction.

From literature studied, NN's performance is observed to be better than other prediction tools as it can learn and recognize complicated patterns and it can easily be applied with little statistical knowledge of the dataset [3–5]. A study employed backpropagation NN to produce four different housing price models by varying the contributing variables and compared their performance for Hong Kong [6]. Varying results for these models showed the effect of relevancy of the variables on the predictability of the model. Another study employed recurrent NN to produce housing price models in Kaohsiung City and found NN to generate acceptable forecasting error [7]. Inconsistent results obtained in a study have been attributed to the size of the sample, variable selection [8]. It is observed that optimal NN models depend upon specific datasets and variables involved. Large dataset augments the performance of the model but selection of variables is of paramount importance in the development of NN models in property rate forecasting. Hence, the fluctuation in the NN performance can be accredited to parameter selection which plays important role in deciding the predictive power of the model. The results of NN model, thus, depend on reliability of the data, spatial coverage of the data, variable selection, and model parameters [9]. Present study focuses mainly on the impact of the selection of training data, training sample size, number of runs, variations, sequence of cases, etc., on the value predictability for a given network.

2   Data Preprocessing

Indian property market is mainly driven by the predominant economy of the nation and in turn the rates of properties across various regions in the country which are derived from the macroeconomic drivers. Value of existing property is dependent on property characteristics and is a certain fraction of the standard rate of any newly constructed property. Prevalent market rates in addition to fraction of property characteristics give the property value for an existing structure. It is, thus, imperative to select variables that represent most of the characteristics. Nineteen variables have been identified from the literature and validated from experts. Data in terms of valuation reports that contain property details along with its current worth in the market has been obtained from various practicing valuers in and around Pune City.

2.1   Application of Principal Component Analysis (PCA)

PCA is a multivariate analysis technique which is used to reduce data, and it seeks to construct a new set of variables, called principal components, that are less numerous than the original data [10]. These components adequately summarize the information contained in the original variables. Statistical Package for Social Sciences (SPSS) version 21.0 has been used for implementing PCA with an aim to identify latent construct underlying identified variables. It has been observed that the nineteen originally identified variables could be grouped under 7 distinct factors or Principal Components (PCs) [11]. There are total of 8 most important variables as observed from PCA results that contribute to first two PCs. There is a separate set of 7 variables that have highest correlation with their respective PCs, thus are supposed to represent the data better (Table 1). These variables have been chosen from each of the 7 groups of variables obtained from application of PCA. Separate datasheets for 19 variables, 8 variables, and 7 variables have been prepared from the obtained data for assessing the effect of data on prediction error generated by network.

Table 1 Variable sets obtained from PCA

All original 19 variables

1. Saleable area in sqft	2. Future scope of development	3. Nature of workmanship
4. State of property/structural condition	5. Legal aspect of property	6. Age of building
7. Shape and margin of plot	8. Locality	9. Prospect (View)
10. Access road width	11. Construction specifications	12. Parking space
13. Water supply	14. Lift with generator backup	15. Amenities by builder
16. Nearness to facilities	17. Nearness to nuisance	18. Local transport facilities
19. Intercity transport facilities		

8 variables (from PC1 and PC2)

1. Saleable area in sqft	2. Nature of workmanship	3. Legal aspect of property
4. Locality	5. Prospect (View)	6. Lift with generator backup
7. Amenities by builder	8. Local transport facilities	

7 variables (one from each PC)

1. Saleable area in sqft	2. Future scope of development	3. Nature of workmanship
4. Age of building	5. Parking space	6. Nearness to facilities
7. Nearness to nuisance		

2.2 Details of Data

Obtained Data is in the form of property valuation reports collected from various valuers in Pune City and range from year 2006 to 2016. A total of 3094 valuation reports have been collected. As neural network processes numeric data, available data is required to be encoded to numeric values. Thus, obtained data in raw format is converted by following a standard conversion table for transferring the qualitative information into quantitative format. For example, availability of water supply is represented by numeric value 1 for continuous supply and 2 for intermittent supply. Likewise, all the variables have been coded and the data is converted as per the case details. Outlier is a part of dataset that appears to be inconsistent with the remainder of the available set of data [12]. Application of NN demands a good quality data that can form a good training set for the network to understand the hidden relations between the study measurements under consideration. Median method of outlier detection is selected for data reduction. Final dataset is prepared after removing outliers obtained by median method.

2.3 Data Transformation

In addition to the necessity of encoding categorical data, NN training is usually more efficient when numeric data is normalized so as to have similar magnitudes. Normalization is a preprocessing step used to rescale attribute values to fit in a specific range. Data normalization methods enable to bring all of the variables into proportion with one another [13]. Proper data normalization speeds up convergence of NN results [14]. Normalization techniques enhance the reliability of trained feedforward backpropagation NN [15]. Here, min–max method is used as it is beneficial in case of a fixed interval rescaling. It is based on maximum and minimum values of the variable under consideration, and it preserves the basic relations shared by original data. Outlier detection and normalization are carried out separately for every set of data.

3 Application of NN and Parameter Selection for Value Prediction

NN is a simplified model of the biological nervous system and follows the kind of computing performed by a human brain. It is a data processing system that consists of a large number of simple highly interconnected processing elements called artificial neurons. These are placed in an architecture that is inspired by the structure of the brain. NN learns from examples and therefore can be trained with known examples of a problem for acquiring knowledge about it. Once properly trained, the network can be effectively used in solving unknown or untrained instances of the problem. Neural network, with its remarkable ability to derive meaning from complicated or imprecise data, can be used to extract patterns and detect trends that are too complex. It can thus be used to provide projections, given new situations of interest. NeuroSolutions version 7.0 has been used to carry out network runs for present study.

It is observed from the literature that function approximation and pattern recognition find the underlying relationships from a given finite data. It is also observed that multilayer feedforward network can be easily trained for nonlinear regression or pattern recognition [16]. Training algorithm, Levenberg–Marquardt, is best suited for approximation and pattern recognition problem [6]. Fastest transfer function is tanh and is the default transfer function for feedforward network. Thus, a multilayer feedforward network with training algorithm Levenberg–Marquardt and transfer function tanh is chosen for the study.

For selecting the percentage-wise division of cases given for training, testing and Cross-Validation (CV), three standard combinations as shown below have been tested. Combination 2 is more frequently used for prediction models. One combination on either side of the standard combination is considered, and a comparative study is conducted to find the best-suited combination.

Table 2 Comparison of percentage prediction errors for ten cases

Prediction case	% error for combination 1	% error for combination 2	% error for combination 3
Case 1	−0.0618	21.7120	11.0200
Case 2	−27.8241	−33.4770	−16.7160
Case 3	9.9976	39.8230	9.5990
Case 4	53.6651	7.8760	−35.8690
Case 5	19.9472	−11.8850	23.6480
Case 6	48.0262	−67.3900	0.2600
Case 7	47.6524	−54.1580	−27.0310
Case 8	3.7920	−10.1720	14.3110
Case 9	14.5383	−16.0830	38.5030
Case 10	−22.6305	−60.2880	12.3680

1. Combination 1—60% training, 20% testing, 20% validation
2. Combination 2—70% training, 15% testing, 15% validation
3. Combination 3—80% training, 10% testing, 10% validation.

Dataset comprises of a total of 2808 cases obtained after outlier removal, making it 1684 training cases for combination 1, 1966 cases for training in combination 2, and 2248 cases in combination 3. Network is seen to generate good results for one hidden layer when compared with multiple hidden layers. Also, with change in number of Processing Elements (PEs), it is observed that a total number of 5 PEs give best result. For the network to generate predicted value, certain number of cases needs to be given for prediction. Here, ten cases are given for prediction in each combination's run. Predicted value is compared with actual value to obtain percentage prediction error. Table 2 shows percentage prediction error when compared with actual value of the property under consideration. Combination 1 has five cases, combination 2 has four cases, and combination 3 has six cases with less than 20% prediction error. Table 3 shows that combination 1 and combination 3 give improved correlation coefficient of 0.845 and 0.862, respectively. Also, combination 1 shows least values for RMSE and other types of errors generated by the network as compared with other values in Table 3. It is observed that combination 3 gives improved results as the training is carried out with more cases. Combination 1, with 60% training cases, 20% CV cases, and 20% testing cases, is considered to be a better choice than combination 2 and combination 3 and thus is selected for further analysis.

Next, test is carried out to check the effect of number of runs and number of variations on the predictability of the network. Networks are tested for varied combinations of number of runs and variations, a few of which have been shown in Table 4. Network runs are conducted for every combination. Corresponding R-value and percentage prediction error are compared. It is observed that training run with 4 runs and 4 variations with 1000 epochs gives better results. Table 4 exhibits the performance of network for varied combinations and corresponding percentage prediction error for 10 cases considered for prediction. It is seen that combination with 4 runs and

Table 3 Comparison of network performance

Performance	Combination 1	Combination 2	Combination 3
RMSE	0.0895	0.1104	0.0908
MAE	0.0533	0.0554	0.0538
Min Abs Error	0.000096	0.00011	0.00045
Max Abs Error	0.5116	1.4210	0.6766
R	0.8450	0.8274	0.8620
Score	89.5077	88.3592	90.3335

4 variations gives a relatively high value of R and maximum number of cases with <20% prediction error.

4 Selection of Best-Suited Training Sample

In order to obtain a generalized network, it is necessary to take number of training runs and check accuracy to avoid unnecessary errors. After removal of outliers by median method, a total of 2818 cases remain for trial. Out of these, 10 cases are reserved for prediction analysis. Remaining 2808 cases are divided into training sample, CV sample, and testing sample as 60%, 20% and 20%, respectively. Selection of the training, testing, and CV sample can be varied by changing the sequence, for example, selecting first 60% cases in the dataset as training for one run, selecting last 60% cases for training for other training run, i.e., cycling samples between the training set, validation set, and test set with fixed production set. To select the combination sequence, i.e., sequence of cases to be given for the run, 3 factorial combinations, i.e., a total of 6 combinations of cases with different sequence of training sample, testing sample, and CV sample are fed to NeuroSolutions version 7.0, and their results are compared.

The prediction results are converted into percentage error in prediction for each case given for prediction. More the percentage error lesser is the accuracy of that sequence. Table 5 gives combination details along with their respective R-values and percentage prediction error cases falling within 10%, 20%, and above 40% for the dataset with 19 original variables. The original dataset is divided into two parts, to ensure that a completely independent training set and independent prediction set is preserved. Same procedure is carried out for datasets prepared for 8 variables and 7 variables. Table 6 gives combination details along with their respective R-values and percentage prediction error cases falling within 10%, 20%, and above 40% for the dataset with 8 variables represented by first two principal components, and Table 7 shows the results of NN run for 7 variables dataset.

It is seen that in the case of dataset of 19 variables, combination 2 with training, testing, and CV as the sequence gives optimum result in terms of R-value and number of cases falling within 10% and 20% prediction error. For 8 variables dataset,

Table 4 Impact of change in number of runs and variations on prediction

Combination	1 run 1 var	1 run 2 var	2 runs 1 var	2 runs 2 var	2 runs 3 var	3 runs 1 var	3 runs 2 var	4 runs 4 var
RMSE	0.1280	0.1280	0.1280	0.1280	0.1197	0.1280	0.1262	0.0904
MAE	0.0617	0.0588	0.0616	0.0613	0.0585	0.0617	0.0631	0.0572
Min Abs Error	0.00055	0.00013	0.00055	0.00022	0.00028	0.00055	0.000075	0.00018
Max Abs Error	1.4102	1.5015	1.4102	1.4508	1.3345	1.4102	1.3759	0.5161
R	0.8283	0.8120	0.8283	0.8323	0.8286	0.8283	0.8220	0.8499
Score	88.1905	87.3835	88.1905	88.3901	89.3043	88.1905	87.8917	89.7281
% Prediction error								
Case 1	11.5476	−18.4865	11.5476	0.9635	−17.6553	11.5476	−2.9914	−35.7421
Case 2	−31.4815	−35.2161	−31.4815	−7.3223	−28.9793	−31.4815	−7.5851	−10.1662
Case 3	14.2600	7.9890	14.2600	10.3082	18.4656	14.2600	1.0571	1.2296
Case 4	−37.8742	−76.5935	−37.8742	−73.5279	−10.1023	−37.8743	−37.9394	−13.6515
Case 5	15.8252	23.1312	15.8252	13.6448	10.0992	15.8252	16.2214	32.3523
Case 6	48.8725	46.4902	48.8725	22.5538	27.3703	48.8725	47.1145	−26.7394
Case 7	50.1815	46.3018	50.1815	47.1559	51.5995	50.1814	41.2918	42.0961
Case 8	12.2298	10.5975	12.2298	28.1722	22.1449	12.2298	29.9853	26.8038
Case 9	10.6267	2.9845	10.6267	13.9351	6.1083	10.6267	9.3602	−12.972
Case 10	−10.4072	5.0441	−10.4072	10.0430	12.8081	−10.4072	20.2554	12.3056

Table 5 Comparison of prediction output for changing sequence of cases for 19 variables dataset

Combination number	1st Part	2nd Part	3rd Part	R-value	No. of production cases with <20% error	No. of production cases with error >40%	No. of production cases with <10% error
Combination 1	Training	CV	Testing	0.8482	2	5	1
Combination 2	Training	Testing	CV	0.8499	5	1	1
Combination 3	Testing	Training	CV	0.8213	5	3	3
Combination 4	Testing	CV	Training	0.8148	2	4	1
Combination 5	CV	Training	Testing	0.8444	3	3	1
Combination 6	CV	Testing	Training	0.8315	3	2	2

Table 6 Comparison of prediction output for changing sequence of cases for 8 variables dataset

Combination number	1st Part	2nd Part	3rd Part	R-value	No. of production cases with <20% error	No. of production cases with error >40%	No. of production cases with <10% error
Combination 1	Training	CV	Testing	0.8242	7	2	4
Combination 2	Training	Testing	CV	0.8439	6	1	5
Combination 3	Testing	Training	CV	0.8266	8	0	6
Combination 4	Testing	CV	Training	0.8260	7	3	4
Combination 5	CV	Training	Testing	0.8245	4	3	3
Combination 6	CV	Testing	Training	0.8362	4	3	1

Table 7 Comparison of prediction output for changing sequence of cases for 7 variables dataset

Combination number	1st Part	2nd Part	3rd Part	R-value	No. of production cases with <20% error	No. of production cases with error >40%	No. of production cases with <10% error
Combination 1	Training	CV	Testing	0.7387	7	2	4
Combination 2	Training	Testing	CV	0.8308	6	1	4
Combination 3	Testing	Training	CV	0.7606	7	3	6
Combination 4	Testing	CV	Training	0.8301	8	2	5
Combination 5	CV	Training	Testing	0.8139	5	3	5
Combination 6	CV	Testing	Training	0.8258	6	0	5

combination 3 with testing, training, and CV as the sequence gives optimum result as compared with other five combinations. In case of 7 variables dataset, combination 4 gives better results as compared with other combinations as it has 5 cases with <10% error, 8 cases with <20% error, and only 2 cases in >40% category.

5 Conclusion

It is observed that the output of NN prediction is data sensitive. From number of runs, it is seen that the prediction results give varied percentage errors for even a slight change in the network characteristics which imply that the network is data sensitive as well as its performance depends upon the chosen parameters. Combination 2 is observed to be the best-suited combination in case of data for 19 variables as it gives comparatively high R-value with 10% cases falling within 10% prediction error and a total of 50% cases within 20% prediction error. Combination 3 is the optimum choice in case of data for 8 variables as it represents 60% of the cases within 10% prediction error and a total of 80% cases within 20% prediction error. Dataset with 7 variables shows 50% cases within 10% error and 80% within 20% error in prediction.

It can be commented that NN behaves better in response to reduced dataset, i.e., dataset of 7 and 8 variables performs better than extensive dataset of 19 variables. Combination of division of dataset into training, testing, and CV, i.e., 60%, 20% and 20%, respectively, works well with this application. Network fails to generalize in less number of training runs selected for a certain dataset. More the number of runs more is the understanding of the network, and the prediction rate gets improved. The best-suited model for the prediction of property value for the present case study is the one with 60% cases for training, 20% cases for CV, and 20% cases with training, testing, and CV as the optimum sequence in case of 19 variables data. Likewise, network converges in case of 8 variable data at sequence of cases as testing, training, and CV. Similarly, dataset of 7 variables gives best results for testing, CV, training sequence of cases selected from the database.

Ascertaining the best-suited network for a particular problem is very important to arrive at good prediction results. The outcome of present study is further used for predicting value of real property, and it is observed to give better results for prediction model than any other combination. Similar study can be carried out for determining best combination of network parameters to get a good generalized NN model with improved predictability. This study gives guidelines as to how to proceed with model preparation when dealing with neural network prediction tool. Any change in the dataset hampers the model prediction. Thus, prior to finding best-suited model parameters, it is required to prepare the dataset meticulously. Any further improvement in the dataset, to improve the model prediction in future, should be accompanied every time with model parameters determination.

References

1. Jim CY, Chen WY (2009) Value of scenic views: hedonic assessment of private housing in Hong Kong. Landscape Urban Plann. 91:226–234
2. Wang S, Chan SH, Xu B (2012) The estimation and determinants of the price elasticity of housing supply: evidence from China. J. Real Estate Res. 34(3):311–344
3. Tay D, Ho D (1991) Artificial intelligence and the mass appraisal of residential apartments. J. Property Valuation Investment 10:525–539
4. Do A, Grudnitski G (1992) A neural network approach to residential property appraisal. The Real Estate Appraiser, pp 38–45
5. Nguyen N, Al C (2001) Predicting housing value: a comparison of multiple regression analysis and artificial neural networks. J Real Estate Res 22(3):313–336
6. Ge Xin J, Runeson G (2004) Modeling property prices using neural network model for Hong Kong. Int Real Estate Rev 7(1):121–138
7. Pi-ying Lai (2011) Analysis of the mass appraisal model by using artificial neural network in kaohsiung city. J Mod Account Auditing 7(10):1080–1089
8. Zurada J, Levitan AS, Guan J (2006) Nonconventional approaches to property value assessment. J Appl Bus Res 22(3)
9. Chaphalkar NB, Sandbhor S (2013) Use of artificial intelligence in real property valuation. IJET 5(3)
10. Krishnakumar J, Nagar AL (2008) On exact statistical properties of multidimensional indices based on principal components, factor analysis, MIMIC and structural equation models. Social Indic Res 86(3):481–496
11. Sandbhor S, Chaphalkar NB (2017) Determining attributes of Indian real property valuation using principal component analysis. J Eng Technol 6(2):483–495
12. Johnson R (1992) Applied multivariate statistical analysis. Prentice Hall
13. Poucke SV, Zhang Z, Roest M, Vukicevic M, Beran M, Lauwereins B, Zheng MH, Henskens Y, Lancé FM, Marcus A (2016) Normalization methods in time series of platelet function assays, A SQUIRE compliant study. Medicine (Baltimore) 95(28):e4188
14. Jin J, Li M, Jin L (2015) Data normalization to accelerate training for linear neural net to predict tropical cyclone tracks. Math Probl Eng 2015(931629):8
15. Jayalakshmi T, Santhakumaran A (2011) Statistical normalization and back propagation for classification. Int J Comput Theory Eng 3(1):89–93
16. Limsombunchai GC, Lee M (2004) House price prediction: hedonic price model vs. artificial neural network. Am J Appl Sci 1(3):193–201

Comparative Analysis of Hybridized C-Means and Fuzzy Firefly Algorithms with Application to Image Segmentation

Anurag Pant, Sai Srujan Chinta and B. K. Tripathy

Abstract In this paper, we combine two famous fuzzy data clustering algorithms called fuzzy C-means and intuitionistic fuzzy C-means with a metaheuristic called fuzzy firefly algorithm. The resultant hybrid clustering algorithms (FCMFFA and IFCMFFA) are used for image segmentation. We compare the performance of the proposed algorithms with FCM, IFCM, FCMFA (fuzzy C-means fused with firefly algorithm), and IFCMFA (intuitionistic fuzzy C-means fused with firefly algorithm). The centroid values returned by firefly algorithm and fuzzy firefly algorithm are compared. Two performance indices, namely Davies–Bouldin (DB) index and Dunn index, have also been used to judge the quality of the clustering output. Different types of images have been used for the empirical analysis. Our experimental results prove that the proposed clustering algorithms outperform the existing contemporary clustering algorithms.

Keywords Data clustering · Fuzzy firefly · Intuitionistic fuzzy C-means
Fuzzy C-means · DB index

1 Introduction

Image segmentation can be defined as the process of partitioning an image into several non-overlapping meaningful homogeneous regions. Data clustering algorithms have been used extensively in the past to segment images. In our previous work [1–4], we have shown that fusing clustering algorithms such as fuzzy C-means [5]

A. Pant (✉) · S. S. Chinta · B. K. Tripathy
School of Computing Science and Engineering, VIT University, Vellore
632014, Tamil Nadu, India
e-mail: anurag.pant2014@vit.ac.in

S. S. Chinta
e-mail: chintasai.srujan2014@vit.ac.in

B. K. Tripathy
e-mail: tripathybk@vit.ac.in

© Springer Nature Singapore Pte Ltd. 2019   315
A. J. Kulkarni et al. (eds.), *Proceedings of the 2nd International Conference on Data Engineering and Communication Technology*, Advances in Intelligent Systems and Computing 828, https://doi.org/10.1007/978-981-13-1610-4_32

and intuitionistic fuzzy C-means [6] with metaheuristics like firefly algorithm [7] improves the quality of the clustering output, and consequently, the segmentation quality also improves. This is primarily because of the fact that all the data clustering algorithms heavily depend on the random initialization of the centroids. This dependency severely degrades the stability of the output produced by these data clustering algorithms. We had eliminated this shortcoming of the clustering algorithms by fusing them with firefly algorithm which returns the near-optimal centroids to the clustering algorithms. However, in the firefly algorithm, the movement of the fireflies is heavily influenced by the local optima, so much so that the global optima has no effect on the movement of the fireflies. This leads to decreased exploration of the solution space which in turn results in sub-optimal results. In this paper, we replace firefly algorithm by the fuzzy firefly algorithm [8], wherein the movement of each firefly is influenced not only by the global optima but by a set of fireflies which glow with a brighter intensity. The rest of the paper is structured as follows: Sect. 2 consists of a brief explanation of FCM algorithm. IFCM algorithm is explained in Sect. 3. We describe the firefly algorithm in detail in the fourth section. Section 5 consists of a brief summary of fuzzy firefly algorithm followed by our proposed clustering algorithms in Sect. 6. Our experimental results have been displayed in Sect. 7 followed by the conclusion in Sect. 8.

2 Fuzzy C-Means Algorithm (FCM)

Fuzzy C-means algorithm is based on the concept of fuzzy sets [9]. In FCM algorithm, random cluster centroids are initiated. The distance d_{ik} between every cluster center i and every pixel of the image k is computed using some distance measure such as the Euclidean distance. The membership matrix is computed as follows:

$$\mu_{ik} = \frac{1}{\sum_{j=1}^{c} \left(\frac{d_{ik}}{d_{jk}} \right)^{\frac{2}{m-1}}} \tag{1}$$

Here, c is the total number of clusters and m is called the fuzzifier. In this paper, the value of m is taken as 2. The cluster centers are calculated using the following formula:

$$v_i = \frac{\sum_{j=1}^{N} (\mu_{ij})^m x_j}{\sum_{j=1}^{N} (\mu_{ij})^m} \tag{2}$$

Here, N represents the total number of data points and x_j refers to the jth data point. Computing the membership matrix and updating the cluster centers constitutes a single iteration. The distances between every pixel and every cluster center are re-computed at the beginning of the next iteration. The major contribution of this paper [3] was that it provided a method to overcome the limitations faced by the infinite solution space. This was achieved by transforming the original problem to the minimization of the objective function J given by:

$$J = \sum_{i=1}^{c} \sum_{k=1}^{n} (\mu_{ik})^m d^2(x_k, v_i) \tag{3}$$

3 Intuitionistic Fuzzy C-Means (IFCM)

The main difference between FCM and IFCM is the concept of non-membership of data. IFCM is based on the intuitionistic fuzzy set model [10]. In IFCM, the hesitation degree is calculated and added to the membership matrix which further enhances the clustering process. In this paper, the non-membership values are calculated using Yager's intuitionistic fuzzy complement [11]:

$$f(x) = (1 - x^\alpha)^{\frac{1}{\alpha}} \tag{4}$$

α value is taken as 2. The hesitation degree of data point x in cluster center A is given as:

$$\pi_A(x) = 1 - \mu_A(x) - (1 - \mu_A(x)^\alpha)^{\frac{1}{\alpha}} \tag{5}$$

In Eq. (5), the value of $\mu_A(x)$ is calculated using Eq. (1). The modified membership matrix represented as μ' is given by:

$$\mu'_A(x) = \mu_A(x) + \pi_A(x) \tag{6}$$

The cluster centers are updated using Eq. (2), except that μ' is used instead of μ. Furthermore, the cost function changes to the following equation:

$$J = \sum_{i=1}^{c} \sum_{k=1}^{n} u'_{ik}{}^m d(x_k, v_i)^2 + \sum_{i=1}^{c} \pi'_i e^{1-\pi'_i} \tag{7}$$

4 Firefly Algorithm

A nature-inspired algorithm based on the mating behavior of fireflies in real life, the firefly algorithm [12] uses the brightness of each firefly to determine its attractiveness (direct proportion). The attractiveness between two fireflies decreases as the distance between them increases. The objective function can be used to determine the light intensity of the firefly, which in turn affects its brightness and therefore, has a direct correlation with its attractiveness. The attractiveness of the firefly can be defined by:

$$\beta(r) = \beta_0 e^{-\gamma r^2} \tag{8}$$

where β_0 denotes the initial attractiveness when $r = 0$, γ is the light absorption coefficient at the source, and r denotes the distance between the two fireflies. The brighter fireflies attract the less brighter fireflies toward them. In the firefly algorithm, in one iteration, only the brightest firefly influences its neighbors and attracts them toward itself. The movement of the firefly i toward the more attractive firefly j can be determined by:

$$x_i = x_i + \beta_0 e^{-\gamma r_{i,j}^2}(x_i - x_j) + \alpha\left(\text{rand} - \frac{1}{2}\right) \tag{9}$$

where α denotes the randomization parameter, and 'rand' is a function used to generate random numbers in the interval [0, 1].

5 Fuzzy Firefly Algorithm

In the fuzzy firefly algorithm proposed in 2014 [8], the aim is to increase the area of exploration by each firefly and to decrease the total number of iteration. In order to do this, k-brighter fireflies are selected in each iteration which can exert their influence over the other less brighter fireflies, where k is a user-set parameter depending on the complexity of the problem and the swam population. If h is one of the k-brighter fireflies in each iteration, $f(p_h)$ be the fitness of the firefly h and $f(p_g)$ be the fitness of the local optimum firefly, then the attractiveness $\psi(h)$ of the firefly h is given by:

$$\psi(h) = \frac{1}{\left(\frac{f(p_h) - f(p_g)}{\beta}\right)} \tag{10}$$

where β is defined as:

$$\beta = \frac{f(p_g)}{l} \tag{11}$$

where l is a user-specified parameter. The movement of a firefly i toward firefly h, one of the k-brighter fireflies, is given by:

$$X_i = x_i + \left(\beta_0 e^{-\gamma r_{ij}^2}(x_j - x_i) + \sum_{h=1}^{k} \psi(h)\beta_0 e^{-\gamma r_{ij}^2}(x_h - x_i)\right) \times \alpha\left(\text{rand} - \frac{1}{2}\right) \tag{12}$$

6 Methodology

The main intuition behind combining FCM and IFCM with firefly and fuzzy firefly algorithms is to eliminate their dependencies on the initial random cluster centers. Throughout this paper, the cluster centers and data points are taken as the grayscale values of the pixels. Each firefly represents a set of randomly initialized cluster centers. The objective functions of FCM and IFCM are used to compute the intensities of the fireflies. Thus, a firefly with brighter intensity is basically a firefly which returns low cost when its cluster center values and membership matrix values are plugged into the objective function of the corresponding data clustering algorithm. Upon convergence, the values of the best firefly are returned to the clustering algorithm. Therefore, if n segments of the input image are desired, each firefly will represent n randomly initialized grayscale values. The problem with firefly algorithm is that all the fireflies with lower intensities move toward the brightest firefly. This reduces the scope for exploration of the solution space. In fuzzy firefly algorithm, this problem is resolved by taking into account the effect of the k-brighter fireflies instead of just the brightest one. Our experimental results prove that this boosts the performance of the clustering algorithms.

7 Results

We have used two performance indices: DB index and Dunn index. DB index is the ratio of intra-cluster distance to inter-cluster distance. It gives a measure of how similar two different clusters are. Dunn index gives a measure of how similar two objects belonging to the same cluster are. Thus, low DB index and high Dunn index imply good clustering quality.

$$\text{DB} = \frac{1}{c} \sum_{i=1}^{c} \max_{k \neq i} \left\{ \frac{S(v_i) + S(v_k)}{d(v_i, v_k)} \right\} \quad \text{for} \quad 1 < i, k < c \tag{13}$$

$$\text{Dunn} = \min_{i} \left\{ \min_{k \neq i} \left\{ \frac{d(v_i, v_k)}{\max(S(v))} \right\} \right\} \quad \text{for} \quad 1 < k, i, l < c \tag{14}$$

To evaluate the centroid values provided by the firefly and the fuzzy firefly, we have used another index named 'Error', which evaluates the proximity of the final centroid values obtained by firefly and fuzzy firefly metaheuristic to the final centroid values obtained by the clustering algorithm.

7.1   Brain MRI Segmentation

It can be clearly seen in Fig. 1 that the algorithm manages to successfully segment the brain and its tumor into separate clusters. It can be observed in Table 1 that the FCMFA and IFCMFA require lesser iterations than FCM and IFCM to find the final cluster values. The fuzzy firefly implementation of both algorithms manages to further improve the results of the original firefly metaheuristic and reduces both the error and the number of iterations by a greater margin. The changes in both the indices are negligible among the normal, firefly, and fuzzy firefly implementation.

7.2   Rice Image Segmentation

The results in Fig. 2 and Table 2 are observed to be similar to the previous brain MRI image with both the firefly and the fuzzy firefly implementations successfully reducing the number of iterations. The fuzzy firefly implementation is the most efficient, as can be evidently observed by the least amount of iterations and the least error value generated, as compared to the firefly and normal implementation.

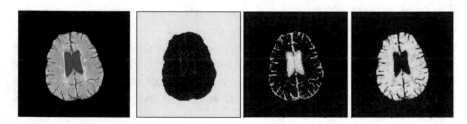

Fig. 1 Brain MRI input image, i.e., the first image on the left is segmented into different clusters represented by the yellow color in the other three images by FCMFFA

Table 1 DB and Dunn index values for brain MRI image

Centroids	2				3			
Algorithm	Iterations	DB	Dunn	Error	Iterations	DB	Dunn	Error
FCM	9	1.3645	1.1384	–	12	2.3424	0.5358	–
FCMFA	5	1.3645	1.1384	26.5	9	2.3420	0.5356	16.5
FCMFFA	3	1.3645	1.1384	0.5	5	2.3424	0.5358	3.0
IFCM	8	1.3893	1.1277	–	17	2.5478	0.5459	–
IFCMFA	6	1.3892	1.1277	25.5	12	2.5479	0.5459	22.0
IFCMFFA	4	1.3892	1.1277	3.5	9	2.5470	0.5462	19.5

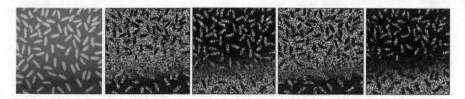

Fig. 2 Rice input image, i.e., the first image on the left is segmented into different clusters represented by the yellow color in the other four images by FCMFFA

Table 2 DB and Dunn index values for Rice image

Centroids	3				4			
Algorithm	Iterations	DB	Dunn	Error	Iterations	DB	Dunn	Error
FCM	16	2.5042	0.6098	–	20	3.7913	0.3930	–
FCMFA	12	2.5016	0.6109	31.2	18	3.7934	0.3931	59.4
FCMFFA	8	2.5018	0.6108	1.1	10	3.7904	0.3930	24.6
IFCM	19	2.5043	0.6303	–	21	4.1639	0.3729	–
IFCMFA	13	2.5044	0.6302	33.1	14	4.1665	0.3730	42.7
IFCMFFA	6	2.5038	0.6305	14.5	12	4.1664	0.3730	20.3

Fig. 3 Lena input image, i.e., the first image on the left is segmented into different clusters represented by the yellow color in the other four images by FCMFFA

7.3 Lena

It can be observed in Fig. 3 and Table 3 that the image is successfully segmented into different clusters. The results follow the same pattern as can be seen in the other two images with the fuzzy firefly implementation giving us the best results (least iterations and least error value).

Table 3 DB and Dunn index values for Lena image

Centroids	3				4			
Algorithm	Iterations	DB	Dunn	Error	Iterations	DB	Dunn	Error
FCM	27	2.0580	0.7280	–	30	2.7431	0.4571	–
FCMFA	24	2.0583	0.7278	26.4	23	2.7431	0.4571	54.0
FCMFFA	14	2.0583	0.7278	21.6	8	2.7433	0.4573	19.1
IFCM	21	2.2991	0.6842	–	23	2.8686	0.4484	–
IFCMFA	17	2.2992	0.6844	46.9	17	2.8687	0.4484	34.4
IFCMFFA	15	2.2991	0.6842	16.9	9	2.8683	0.4484	22.2

8 Conclusion

FCMFFA and IFCMFFA retain the advantages of FCM, IFCM, FCMFA, and IFCMFA and further enhance the performance by drastically reducing the number of iterations required for convergence. This is made possible due to fuzzy firefly algorithm which not only supplies IFCM with a near-optimal solution for it to work with instead of random membership matrix and cluster centers but also increases the coverage of the solution space. By doing so, we are eliminating the problem of being stuck at the local optima. Our experimental results prove that the cluster centers returned by fuzzy firefly algorithm are closer to the actual cluster centers than those returned by firefly algorithm. Our future works include using fuzzy firefly algorithm to find the optimal fuzzification parameter and extending these hybrid clustering algorithms to big data clustering.

References

1. Abhay J, Srujan C, Tripathy BK (2017) Stabilizing rough sets based clustering algorithms using firefly algorithm over image datasets. In: Information and communication technology for intelligent systems (ICTIS 2017)—vol 2, pp 325–332
2. Tripathy BK, Namdev A (2016) Scalable rough C-means clustering using firefly algorithm. Int J Comput Sci Bus Inf 16(2)
3. Srujan C, Abhay J, Tripathy BK (April 2017) Image segmentation using hybridized firefly algorithm and intuitionistic fuzzy C-means. In: Presented at 1st international conference on smart systems, innovations and computing, vol 1. Manipal University, Jaipur, India
4. Srujan C, Tripathy BK, Govindarajulu K (2017) Kernelised intuitionistic fuzzy C-means algorithms fused with Firefly Algorithm for Image Segmentation. In: Presented at IEEE international conference on microelectronics devices, circuits and systems (ICMDCS 2017), vol 1. VIT University, Vellore, India
5. Ruspini EH (1969) A new approach to clustering. Inf Control 15(1):22–32
6. Chaira T (2011) A novel intuitionistic fuzzy C means clustering algorithm and its application to medical images. Appl Soft Comput 11(2):1711–1717
7. Yang X-S (2009) Firefly algorithms for multimodal optimization. In: International symposium on stochastic algorithms. Springer, Berlin

8. Hassanzadeh T, Kanan HR (2014) Fuzzy FA: a modified firefly algorithm. Appl Artif Intell 28(1):47–65
9. Zadeh LA (1965) Fuzzy sets. Inf Control 8(3):338–353
10. Atanassov KT (1986) Intuitionistic fuzzy sets. Fuzzy Sets Syst 20(1):87–96
11. Yager RR (1980) On the measures of fuzziness and negation part II lattices. Inf Control 44:236–260
12. Yang X (2010) Firefly algorithm, stochastic test functions and design optimization. In: Proceedings of IJBIC, pp 78–84

Sanction Enforcement for Norm Violation in Multi-agent Systems: A Cafe Case Study

Harjot Kaur and Aastha Sharma

Abstract Norms are established in an organization for its smooth functioning and well being. In multi-agent systems, norms are expectations of an agent about the behavior of other agents. It is norm violation that results in sanctions enforcement in an organization. The work presented in this paper primarily investigates various types of sanction enforcement dynamics existing in multi-agent systems in the form of a threefold task. First, it presents life-cycle model of sanctions in multi-agent systems which are deduced from norms' life-cycle model. Second, it also helps researchers to investigate various types of circumstances subject to whom norms are violated and sanctions are enforced in a multi-agent system. Third, the proposed life-cycle model of sanctions and sanction enforcement dynamics are experimentally assessed by modeling a cafe (with norms and sanctions) in the form of a multi-agent system.

Keywords Multi-agent systems · Norms · Sanctions · Violation · Enforcement

1 Introduction

Multi-agent systems are collection of *heterogeneous software agents* [25], which are capable of acting in an *autonomous* manner. The software agents present in multi-agent systems act on the behalf of human agents; hence, they should also possess a mental model of norms [17], which they are suppose to follow.

Norm [23] can be defined as

an authoritative standard or a principle of right action binding upon the members of a group and serving to guide, control, or regulate proper and acceptable behavior.

H. Kaur (✉) · A. Sharma
Department of Computer Science & Engineering,
Guru Nanak Dev University Regional Campus, Gurdaspur 143521, Punjab, India
e-mail: harjotkaursohal@rediffmail.com

A. Sharma
e-mail: aasthasharma17@ymail.com

© Springer Nature Singapore Pte Ltd. 2019   325
A. J. Kulkarni et al. (eds.), *Proceedings of the 2nd International Conference on Data Engineering and Communication Technology*, Advances in Intelligent Systems and Computing 828, https://doi.org/10.1007/978-981-13-1610-4_33

Norms are established in an organization for its well being. It is well proved that deviations occur in organizations [1] whenever norms are violated. Many distinct types of norms are defined in organizations besides *sanctions*, corresponding to norm violation at different stages of operations in it. This structure of norms is normally utilized in organizations [6] in order to ensure discipline and earn more profitability. The fact that only norms [3] cannot be used to ensure rules and regulations in an organization is proved time and again. Therefore, for ensuring norm abidance in an organization, sanctions are required.

In multi-agent systems, *norms* [22] *are expectations of an agent about the behavior of other member agents residing in there*. Norms play a central role in various social phenomena of a multi-agent system such as coordination, cooperation, decision-making [12]. Every multi-agent system has its own norms that are to be followed for its smooth working. In any case, whenever these norms are violated, then it becomes necessary to give punishment to the norm-violating member agent. That punishment [5] can be given in the form of *sanctions*. The norm-violator agent is sanctioned according to the norm violated by it.

1.1 Norms

Norms [12] are social rules and standards. Standards can be unequivocal, (e.g., laws) or verifiable, (for instance, codes of considerate conduct). Standards can be hard to recognize in the light of the fact that they are so profoundly ingrained in individuals [23] from a given society.

1.1.1 Norms' Life Cycle

Norms in an organization go through the set of phases [3, 11] or *a life cycle*. A lot of research work [3, 6, 11, 22] has already been carried on norms' life cycle [20] which comprises of the following four phases. Each stage of norm life cycle has been associated with various simulation mechanisms, which can be used to implement it. These phases are:

- **Norm Creation**: This phase is involved with creation of norms in an organization in order to ensure order in the same.
- **Norm Spreading**: This phase guarantees content spreading related to various norms by conveying already created norms to various members of an organization, for effective norm adherence.
- **Norm Enforcement**: This phase is responsible for ascertaining norm enforcement in an organization, after establishment and spreading of norms to various agents in an organization, so that they can be abided in an effectual manner.
- **Norm Emergence**: This phase is primarily responsible for the emergence of new norms if currently existing norms in an organization are not fulfilling its norm specifications.

The life cycle of norms outlines the process of initialization and perpetuation of norms in an organization in an effective manner. It is basically because of norm violation [21] in norm enforcement phase, need of sanctions arises. The severity of sanction enforced against the norm violation is according to the type of the norm violated by the norm violator.

1.1.2 Types of Norms

Various researchers have categorized norms into different categories according to their use and adherence in organizations. According to Tuomela [22], *societal norms* in can be categorized as *social norms* [4], *rule norms (private norms)*, *legal norms*, *triangular norms* [15] *moral norms*, and *prudential norms*. Whenever, these norms are violated by members of a group or a society, various sanctions are imposed on them as a penalty or punishment.

1.2 Sanctions

In generic terms, sanction [7] is an official approval for a particular action to be performed in a correct manner. But in terms of norm violation it is defined as,

> a kind of punishment or penalty for disobeying a norm.

Sanction establishment is necessary in an organization, in order to enforce a *sufficient punishment* in case of norm violation. In other words, one can say that it is norms that give rise to sanctions but not vice-versa. Sanctions like norms also possess a life cycle. A sanctions' life cycle has also been proposed in this paper, which has been deduced from norms' life cycle. This life cycle clearly defines all the stages of a sanction [8] including the requirement of sanction creation. Once sanction has been created, how it can be enforced and from where it has to be reinforced if it has been disobeyed is demonstrated by sanctions' life cycle. The sanctions' life cycle consists of the following four phases:

- **Sanction Need**: This phase justifies the necessity of sanction creation in norm enforcement phase because of norm violation.
- **Sanction Enforcement**: Sanction enforcement is performed to penalize or punish norm-violator agent(s) who do not fulfill their commitments in an organization. It is norm violation that generally gives rise to sanction enforcement.
- **Sanction Obedience**: Sanction obedience phase specifies the roadmap which can be adopted to assure successful sanction abidance in an organization.
- **Sanction Reinforcement**: After actualization of new norms, new sanctions also need to be actualized and enforced. Also, it can lead to alteration or reinforcement of already existing sanctions corresponding to newly created norms. Hence, this phase is termed as sanction reinforcement.

1.2.1 Types of Sanctions

According to the type of norms which are violated in an organization/multi-agent system and style in which sanction is enforced [8, 21], sanctions can be classified into the following categories:

- **Implicit sanctions**: These are the types of sanctions that are *autonomously* decided by agents. They are not publicly known, and agents have to discover whether or not they have been sanctioned (for instance, by noticing that others do not communicate with them anymore). They can be further classified as follows:

  - *Personal Sanctions*: These are the sanctions that are used for individual purposes.
  - *Selective Sanctions*: These sanctions are selective in nature.

- **Explicit Sanctions**: These sanctions are publicly known (at least among the interacting agents). They can be further classified as follows:

  - *Diplomatic Sanctions*: These sanctions are implemented by organizations that trigger removal and reduction of diplomatic ties.
  - *International Sanctions*: These are the actions taken by organizations against other ones for political reasons.
  - *Economic Sanctions*: These are commercial or financial penalties applied by one or more organizations against a targeted group.

In this paper, what are various types of sanctions which can enforced and when they are enforced, corresponding to norm violation in an organization, are studied and analyzed. This is performed by modeling and designing a cafe case study with cafe being an organization with certain set of norms [9] and visitors to the cafe being modeled as agents. And violation of these norms causes certain sanctions to be enforced on the cafe visitors who violate the same.

2 Background Work

This section provides a detailed description of existing work in the area of norms and sanctions in multi-agent systems.

The norm mechanism for open multi-agent systems has been studied by Hamid et al. [14]. They have studied the impact of norms establishment on behavior of agents in a multi-agent system. They have stated that norms are no longer created during organization's configuration. Rather, they emerge during agent interaction. They have also studied run-time allocation of norms in this work.

Mahmoud et al. [18] have studied that the norm mining approach can be followed alongside the concept of community norms, in order to ascertain that whether a particular agent is following the code of conduct or not. Various norm constituents which help them in their reinforcement also have been pinpointed in this work, and the norm identification process varies with the state of affairs of an organization.

The adaptive sanctioning mechanism has been studied by Centeno et al. [8]. It provides continuous revision of norms or protocols which further are dependent on the agent interaction in an organization's environment. The dynamic environment is contemplated by authors for sanction enforcement in an organization, although it is responsible for some flaws in the same as well.

The normative run-time reasoning agents in multi-agent systems have been studied by Balke et al. [2] in their work. In this study, they have considered the dynamic behavior alteration of agents besides norm modification.

A framework for norm aware multi-agent systems has been proposed by Dybalova et al. [10]. It is 2OPL (Organizational Programming Language) which is used by authors to design a norm identification framework in which norms identification is performed in a dynamic manner. The presented framework is used to ascertain whether norms established by normative organization have been followed by agents or not.

Social norms which originate naturally within human societies are considered by Hexmoor et al. [16]. This work reviews usage of norms in a society by using multi-agent systems. They also state that norm establishment in a small organization is relatively easy as compared to complex environment where norm establishment process is found to be extremely difficult.

An automated approach to handle social norms in a normative organization has been described by Neruda et al. [19]. Here, norms establishment and its sufficient enforcement are ensured by using normative organization. But, they have not considered the concept of sanctions in this work. The automated reasoning established in this paper only works for very large organizations, which can be presented in the form of multi-agent systems.

The clarity of definition of norms in normative environment is stated by Balke et al. [3]. This work also describes about nature of norms. According to this work, agent interaction is an essential feature of norms and this aspect is highlighted as well.

All the above-surveyed works have considered *norm enforcement* in static as well as *dynamic environments*, but none of them have considered *sanction enforcement* as a result of norm violation. And, it is also not mentioned in any of these works that which protocols or structures should be followed for sanction enforcement in multi-agent systems. This paper broadens the perspective of readers on sanction enforcement in multi-agent systems by proposing various types of sanctions and life cycle of sanctions. It precisely investigates why, where, and how sanctions are enforced and exercised on various member agents in artificial agent societies? The proposed work will be validated by proposing a case study of a cafe which is modeled in the form of a multi-agent system, with norms being enforced as a part of cafe in the form of a norm base. This work primarily investigates how various agents who will be violating norms in the cafe will be sanctioned by various sanctions?

3 Sanction Enforcement—A Cafe Case Study

The case study of a cafe is considered in the form of a multi-agent system. In the cafe, *groups* of similar natured agents are formed who visit cafe, and eat or drink there. The cafe has certain rules and regulations in the form of norms [9] that must be followed by the visitors of the cafe. If the norms of the cafe are not followed, then various sanctions are enforced. The norms of the cafe can be categorized as social norms, legal norms, and private norms, where

- *social norms* are the rules which ensure the safety of other agents sitting next to the norm-abiding agent, i.e.,

    - no smoking inside the cafe;
    - no drinking inside the cafe;
    - not to stare at anyone in the cafe;

- *legal norms* are the rules written under law and are used to protect the cafe's property and infrastructure, i.e.,

    - not to steal anything from the cafe;
    - not to harm infrastructure of the cafe;

- private norms are the rules that are defined by an organization (cafe) itself. These are the norms that are defined according to the organizational benefits of the cafe, i.e.,

    - not to misbehave with any of the cafe's staff;
    - not getting involved in any kind of dispute in the cafe.

If the norms defined for the cafe are violated by its visitors, the following sanctions are enforced upon them; all these sanctions [13] are implicit sanctions in nature, i.e.,

- a person will be warned first, and if person ignores the warning, then
- a person could be fined;
- a person could be jailed;
- a person could be thrown out from the cafe;
- a person could be debarred from entering the cafe.

Various agents who are present in the cafe environment in the form of multi-agent system can be categorized as follows:

1. *visitor agents*—various agents which are visiting the cafe and are supposed to abide the norms formulated in the cafe.
2. *analyzer agents*—these are agents residing in the cafe, which are a part of cafe's management. They will be observing all the agents present as well as visiting in the cafe, in order to check whether they are abiding the norms formulated by the cafe's management. If visitor agents are not abiding the norms of the cafe, they will be sanctioned using various sanctions.

3.1 Working Algorithm for Sanction Enforcement

The cafe is modeled as a multi-agent system comprising of various heterogeneous agents, i.e., *analyzer agents* and *visitor agents*. The sanction enforcement mechanism in cafe will check which agents in the cafe are following the norms [20] and which are violating them. It will also make sure that the norm-violator agent gets penalized by the sanction. This sanction enforcement mechanism works according to the sanction enforcement algorithm.

Algorithm 1 Sanction Enforcement Algorithm

1: Start
2: Initialize MAS.   ▷ Create a multi-agent system
3: MAS_i = Number_of_Agents   ▷ It is a set of agents simulating a caf
4: *Counter* = 0   ▷ It is used to count the number of agents in a MAS
5: *Threshold* = 5   ▷ It is maximum number of times an agent can violate a norm
6: *Count_th* = 0   ▷ It is used to count the threshold value
7: Start the simulation and initialize the analyzer agent
8: **for** *Counter* = 0 to Number_of_Agents **do**   ▷ Analysis of visitor agents is done
9:   **if** norm _violated(Agent) **then**   ▷ Agent violates any of the norms
10:     **if** Count_th >= Threshold **then**
11:       Apply appropriate sanction   ▷ Sanction is enforced
12:     **end if**
13:     Count_th = Count_th +1
14:   **end if**
15: **end for**
16: End

According to the working algorithm for sanction enforcement, a cafe in the form of a multi-agent system with a set of agents (analyzer agents and visitor agents) and norms is initialized. After initialization, an analyzer agent observes each and every agent for its cafe's norms' abidance or violation. For this, a variable counter is used to count the number of agents residing in a cafe. There is a constant called norm threshold, which is maximum number of times, an agent can violate a norm. The cafe simulation is activated and analyzer agent starts observing various visitor agents in a cafe. And, as long as, visitor agents are abiding the norms, there is no problem.

But, whenever, there is a norm violation performed by any of the visitor agents, first, it is checked that whether norm violation counter (count_th) for this particular agent has reached norm threshold or not. If yes, then the agent is sanctioned with an appropriate sanction(s) corresponding to the norm(s) violated by it. But, if norm violation counter has not reached norm threshold yet, the visitor agent is again given a chance to improve his behavior. To count the number of agents visiting the cafe, a variable counter is used. It is initialized with zero, and it ranges till number of agents existing in cafe multi-agent system.

3.2   Simulation of Cafe Case Study in NetLogo

The cafe is simulated by modeling a multi-agent system in NetLogo [24], a multi-agent programming environment, in which population of various member agents participating in it is initialized. Various norms are formulated for the cafe in the form of a multi-agent system and stored in norm base, which are supposed to be followed by agents inside the cafe. There are some agents who are following the norms defined in norm base of the cafe while some are violating them.

The cafe is modeled and designed in the form of multi-agent system using Net-Logo's artificial world, with visitor agents (in green color) and analyzer agent (in purple color) being active objects in it (as shown in Fig. 1). The population of visitor agent in a cafe can be controlled by using No._of_Agents slider. The artificial world designed for modeling a cafe also comprises of norm base and sanction base files for storing various norms modeled for the cafe and appropriate sanctions formulated corresponding to them, respectively.

Various monitors (private_norms_violating_agent, private_sanctioned_agents, etc.) are also present along with the cafe's artificial world for reporting the number of agents which are violating various types of norms and sanctions which are enforced on them. The count of population of agents who are violating various norms and appropriate sanction enforcement against them is also shown in graphical mode using norms and sanctions plot. The *analyzer agent* present in a cafe environment first checks each and every *visitor agent* that whether it is abiding the norms of the cafe or not? If the visitor agents are abiding the norms, then it allows them to sit in the cafe. But on the other hand, the agents that are violating the norms are sanctioned accordingly.

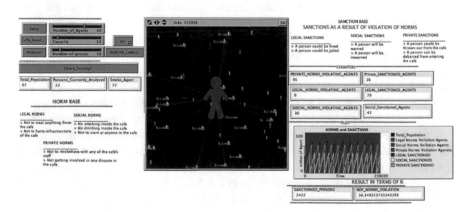

Fig. 1 Simulation of cafe in NetLogo

3.3 Experimental Analysis for Sanction Enforcement

The cafe in the form of multi-agent system (shown in Fig. 1) has various types of norms formulated in it in the form of *legal, social* and *private norms*. The visitor agents who violate *social norms* are sanctioned firstly with *social sanctions*, but if they do not acknowledge the social sanctions, they can be penalized further with legal sanctions. But, the visitor agents who violate *legal norms* are always penalized with *legal sanctions*.

The visitor agents are also penalized with *social and private sanctions*, if they violate *private norms*. The analysis of norms and sanctions within the cafe as shown in its model expresses in detail the relationship that exists between various visitor agents and cafe organization committee (analyzer agent). The positive relationship results in abiding of the norms associated with the cafe and contrariwise results in sanction enforcement against various agents in a cafe.

3.3.1 Results and Interpretations

In the above-modeled multi-agent system for a cafe case study, the visitor agents which are violating various types of norms in a cafe (present in norm base) are sanctioned using various sanctions mentioned in the sanction base of the cafe. The visitor agents, whether they are following the norms or not, are analyzed in the form of groups. The results of norm violation for different types of norms formulated in a cafe are shown in Fig. 2. It can be observed that among the agent groups which are formed of visitor agents, the norms which are violated the most in the cafe are social norms, after that, it is private norms, and legal norms are leastways violated.

It is also observed that visitor agents who violate social, legal, or private norms of the cafe are sanctioned with social, legal, or private sanctions, respectively. But,

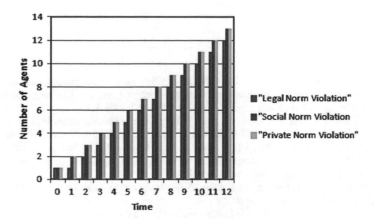

Fig. 2 Various types of norm violation in a cafe multi-agent system

maximum sanctioning in the cafe is done with social sanctions as they are more related to warning or ultimatum. Only after the agent does not abides to a social sanction or repeat the norm violation process, it will be sanctioned with legal or private sanctions; i.e., agent will be fined or debarred. Hence, severity of penalty totally depends upon the nature of norm violation that has been done.

4 Conclusions and Future Research Directions

In this paper, we have primarily tried to investigate norm violation and sanction enforcement dynamics in multi-agent systems. It is primarily investigated and analyzed in this work that why, where, and how sanctions are enforced in multi-agent systems. This is accomplished by first modeling and outlining a sanction life cycle along with various types of sanctions which can be enforced corresponding to a variation of norms which are violated in a multi-agent system. Then, this sanction life cycle is experimentally assessed and validated by presenting a case study of cafe modeled in NetLogo. It is also concluded experimentally that among the type of norms formulated, social norms are the one which is violated the most, and also, it is social sanctions in majority, which are used to sanction agents in case they violate any type of norms (be it social, legal, or private norms). The study of sanction enforcement dynamics presented in this paper is applicable to all the organizations, where norms are created for their streamlined operations, and there are fair chances of norm abidance, violation, and sanction enforcement as well.

Our work has considered only social, legal, and private norms for sanction enforcement in multi-agent systems. But apart from these norms, sanction enforcement can be studied and analyzed corresponding to prudential and moral norms also. We leave this work as one of the future directions which can be pursued in sanction enforcement dynamics in multi-agent systems.

References

1. Astafanoaei L, Boer FSD, Dastani M (2009) Rewriting agent societies strategically. In: 2009 IEEE/WIC/ACM International conference on web intelligence and intelligent agent technology, Milan, Italy, pp 441–444. https://doi.org/10.1109/WI-IAT.2009.321
2. Balke T, Vos de M, Padget J, Traskas D (2011) Normative run-time reasoning for institutionally-situated BDI agents. In: 2011 IEEE/WIC/ACM International conferences on web intelligence and intelligent agent technology, Lyon, pp 1–4. https://doi.org/10.1109/WI-IAT.2011.49
3. Balke T, Pereira C da C, Dignum F (2013) Norms in MAS: definitions and related concepts. In: Normative multi-agent systems. Dagstuhl Publishing
4. Bicchieri C (2006) The grammar of society: the nature and dynamics of social norms. Cambridge University Press, New York
5. Boella G, van der Torre L (2003) Local policies for the control of virtual Communities. In: International conference on web intelligence, pp 161–167. https://doi.org/10.1109/WI.2003.1241188

6. Boella G, van der Torre L (2003) Norm governed multi-agent systems: the delegation of control to autonomous agents. In: IEEE/WIC International conference on intelligent agent technology, pp 329–335. https://doi.org/10.1109/IAT.2003.1241092
7. Cardoso HL, Oliveira E (2009) Adaptive deterrence sanctions in a normative framework. In: 2009 IEEE/WIC/ACM International conference on intelligent agent technology, Milan, Italy, pp 36–43. https://doi.org/10.1109/WI-IAT.2009.123
8. Centeno R, Billhardt H, Hermoso R (2011) An adaptive sanctioning mechanism for open multi-agent systems regulated by norms. In: IEEE 23rd International conference on tools with artificial intelligence, pp 523–530. https://doi.org/10.1109/ICTAI.2011.85
9. Dechesne F, Tosto G, Dignum V, Dignum F (2013) No smoking here: values, norms and culture in multi-agent systems. Artif Intell Law 21:79–107
10. Dybalova D, Testerink B, Dastani M, Logan B (2014) A framework for programming norm-aware multi-agent systems. In: Balke T, Dignum F, van Riemsdijk M, Chopra A (eds) Coordination, organizations, institutions, and norms in agent systems IX COIN 2013. Lecture Notes in Computer Science, vol 8386, pp 364–380
11. Elsenbroich C, Gilbert N (2014) Modelling norms. Springer Netherlands, Dordrecht
12. Gabbai J, Yin H, Wright W, Allinson N (2005) Self-organization, emergence and multi-agent systems. In: 2005 IEEE International conference on neural networks and brain, Bejing, pp 1858–1863. https://doi.org/10.1109/ICNNB.2005.161498
13. Gopalan S (2007) Alternative sanctions and social norms in international law: the case of Abu Ghraib. Michigan State Law Rev pp 785–839
14. Hamid ANH, Ahmad MS, Ahmad A, Mahmoud MA, Yusoff MZM, Mustapha A (2014) Trusting norms in normative multi-agent system. In: 6th International conference on information technology and multimedia, Putrajaya, pp 217–222. https://doi.org/10.1109/ICIMU.2014.7066633
15. Hernndez P, Cubillo S, Torres-blanc C (2015) On t-norms for type-2 fuzzy sets. IEEE Trans Fuzzy Syst 23(4):1155–1163
16. Hexmoor H, Venkata SG, Hayes D (2006) Modeling social norms in multi-agent systems. J Exp Theor Artifi Intell 18(1):49–71
17. Kaur H, Kahlon KS (2015) Dynamics of artificial agent societies: a survey and agent migration perspective. AI Commun 28:511–537
18. Mahmoud MA, Ahmad MS, Ahmad A, Mohd Yusoff MZ, Mustapha A (2012) A norms mining approach to norms detection in multi-agent systems. In: 2012 International conference on computer and information science (ICCIS), pp 458–463. https://doi.org/10.1109/ICCISci.2012.6297289
19. Neruda R, Kazk O (2012) Formal social norms and their enforcement in computational MAS by automated reasoning. Int J Comp Sci IAENG 39(1):80–87
20. Oliveira A, Girardi R (2016) An analysis of norm processes in normative multi-agent systems. In: International conference on web intelligence workshops, pp 68–71. https://doi.org/10.1109/WIW.2016.6
21. Robert MR, Corts MI, Campos GAL, Friere ESS (2015) A sanction application mechanism considering commitment levels in hierarchical organizations. In: 2015 Latin American computing conference (CLEI), pp 1–11. https://doi.org/10.1109/CLEI.2015.7359468
22. Savarimuthu BTR, Cranefield S (2009) A categorization of simulation works on norms. In: Boells G, Noriega P, Pigozzi G, Verhagen H (eds) Dagstuhl seminar proceedings, vol 09121. Dagstuhl, Germany
23. Vanhe L, Aldewereld H, Dignum F (2011) Implementing norms? In: 2011 IEEE/WIC/ACM International joint conferences on web intelligence and intelligent agent technology, Lyon, pp 13–16. https://doi.org/10.1109/WI-IAT.2011.184
24. Wilensky U, Stroup W, Hubnet (1999) Center for connected learning and computer-based modeling. Northwestern University. Evanston, IL http://ccl.northwestern.edu/netlogo/hubnet.html
25. Woolridge M, Jennings R (1995) Intelligent agents: theory and practice. Knowl Eng Rev 10(2):115–152

Interdigitated Electrodes and Microcantilever Devices for Sensitive Gas Sensor Using Tungsten Trioxide

Chetan Kamble and M. S. Panse

Abstract Microdevices such as interdigitated electrodes (IDEs) and piezoresistive-based microcantilevers are the most prominent gas sensor devices to miniaturize the sensing device and enhance the sensing properties. This paper presents the design and development of these microdevices for gas sensing applications. Tungsten trioxide (WO_3) is one of the well-studied materials from n-type semiconducting metal oxides which has already been used for gas sensing. A combination of a sensitive microdevice with high sensing property of WO_3 can be used for low-level gas detection. COMSOL Multiphysics software is used to determine optimal design conditions to get the highest sensitivity from the microdevices. IDEs and microcantilever are fabricated using optimized dimensions. SEM and HRXRD of deposited WO_3 film have been acquired.

Keywords IDEs · Gas sensing · WO_3 · Tungsten trioxide · Microcantilever

1 Introduction

Gas sensors are used for the detection of various explosives, toxic and flammable gases; such type of gas sensors are mostly used in the industrial and commercial purposes for pollution control and human safety purposes. Gas sensors can also be useful for diagnosis of diseases by exhale breath analysis. Existing methods such as gas chromatography, mass spectroscopy can be used for gas detection, but some are expensive, bulky, have poor response-recovery time and less resolution. Hence, they need to be replaced with portable, cheap, quick response and low concentration level sensor [1]. In recent years, metal oxide nanostructure-based chemoresistive sensors are popular as a future technology. Metal oxide-based gas sensing is of low cost, high stability, good sensitivity, good selectivity and has short response-recovery time. Till date, several n-type semiconducting metal oxides such as zinc oxide (ZnO),

C. Kamble (✉) · M. S. Panse
Electronics Engineering, Veermata Jijabai Technological Institute, Mumbai, India
e-mail: chetanpkamble@gmail.com

© Springer Nature Singapore Pte Ltd. 2019
A. J. Kulkarni et al. (eds.), *Proceedings of the 2nd International Conference on Data Engineering and Communication Technology*, Advances in Intelligent Systems and Computing 828, https://doi.org/10.1007/978-981-13-1610-4_34

tin dioxide (SnO_2), titanium dioxide (TiO_2), iron oxide (Fe_2O_3) and tungsten trioxide (WO_3) have been reported as successful gas sensing materials for detecting a variety of toxic and inflammable gases. Among these metal oxides, WO_3 nanostructured is one of the well-studied material for smart windows, pigment in ceramics and paints, photocatalytic water splitting, photoelectrochemical cell and gas sensing applications [2, 3]. WO_3 has been considered as potential candidate for pollutant gas sensing like NO_2, SO_2, CO, LPG, NH_3 and H_2S. WO_3 also shows sensitive towards acetone vapour. However, WO_3 has relatively less sensitivity and selectivity for the low level of gas sensing use; also it has high energy consumption. Therefore, WO_3 can be deposited on microdevices for improving sensitivity and selectivity, and it operates at less power. Microdevices such as microcantilevers or IDEs are more sensitive, and it can respond to small changes, so they are more promising for the low level of gas sensing. In this work, piezoresistive microcantilever and IDEs are simulated and fabricated with optimized geometry and materials.

2 Experimental

2.1 Preparation of WO₃ Film

High purity with nanocrystalline size WO_3 powder was procured from pallav chemicals, Mumbai, India. These WO_3 nanoparticles were mixed with organic solvent and binder at room temperature, and it was stirred for 60 min to form a homogeneous viscous liquid paste, which was then drop deposited using micropipette on a silicon substrate and annealed at 500 °C for 2 h in N_2 ambient for the preparation of WO_3 thick film.

2.2 Theoretical Considerations for Microdevices

In both the microdevice sensors, sensing material will be deposited on top of microcantilever and electrodes of IDEs devices. Target gas interacts with the sensing material to cause deflection in the cantilever and change in current in the IDEs sensor. Change in deflection and change in current has a correlation with sensed target gases in microcantilever and IDEs sensor, respectively.

2.2.1 Microcantilever Sensor

Microcantilevers are one of the basic micromechanical structures. Their micrometre and nanometre sizes accounts for its extremely high sensitivity. Microcantilevers are called microcantilever-based sensors when their surfaces are functionalized to

Fig. 1 Geometry of
piezoresistive
microcantilever gas sensor

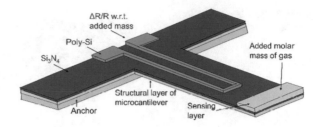

detect specific molecules. Deflection gives the relation between gas molecules which interacts with sensing layer of the cantilever. As WO_3 has good selectivity towards acetone, we considered acetone as target gas for the calculation. In this work, piezoresistive microcantilever with WO_3 as sensing material for acetone vapour detection is simulated using COMSOL Multiphysics 5.1.

By using Stoney's formula, the displacement in microcantilever can be derived as a function of the generated surface stress [4, 5]. In static mode, the cantilever's deflection (Δz) is given by,

$$\Delta z = \frac{3(1-v)L^2}{Et^2}(\sigma_1 - \sigma_2) \tag{1}$$

where L and t are the length and thickness of the cantilever beam, E is young's modulus of the cantilever material, $(\sigma_1 - \sigma_2)$ is generated differential surface stress, and v is Poisson's ratio.

The deflected cantilever produces change in resistance to the piezoresistor; hence, the change in resistance with respect to original resistance is given by,

$$\frac{\Delta R}{R} = \beta \frac{3\pi_1(1-v)}{t}(\sigma_1 - \sigma_2) \tag{2}$$

where π_1 is the longitudinal piezoresistive coefficient of silicon at the operating temperature, at a given doping and its value is 71.8×10^{-9}, β is correction factor between 0 and 1. By equating both Eqs. (1) and (2),

$$\frac{\Delta R}{R} = \frac{Et\beta\pi_1\Delta z}{L^2} \tag{3}$$

Therefore, by using deflection value Δz resulting resistance change ($\Delta R/R$) of the piezoresistive microcantilever can be determined (Fig. 1).

2.2.2 Interdigitated Electrodes (IDEs) Sensor

Interdigitated electrodes are composed of two interdigitated electrodes called anode and cathode as shown in Fig. 2. The sensing material is deposited over interdigitated electrodes, which form the electrical connections through which the relative resis-

Fig. 2 Geometry of IDEs device

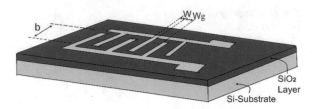

tance change is measured [6]. These IDEs sensors offer several advantages, such as working with small sensing layer, enhanced sensitivity and detection limits, operate at less power. After deposition of WO_3 on the IDEs, the relative change in resistance can be measured before and after target gas exposed on the device.

In electrochemical, when the potential difference between two interdigitated electrodes is sufficiently high, an electrochemical recycling was observed on electrode spacing, where reduction and oxidation occur at adjacent electrodes. As a result, limiting current can be measured at the electrodes. Limiting current can only be maximized by maximizing f (dimensionless current at each electrode) which is dependent on an electrodes' geometry [7, 8].

$$f = I/(m\,b\,n\,F\,C\,D) \tag{4}$$

Limiting current response of both anode and cathode is described by,

$$I = m\,b\,n\,F\,C\,D\left\{0.637\ln\left[2.55(1 + W/W_g)\right] - \frac{0.19}{(1 + W/W_g)^2}\right\} \tag{5}$$

where m is no. of fingers, b is length of fingers, F is Faraday's constant, so nF is the charge transferred per mole of analyte reacted, $W_a = W_c = W$ is width of anode and cathode, W_g is width of gap between fingers, C is the bulk analyte concentration, D is the analytes' diffusion coefficient.

The study of IDEs with different geometries shows that smaller gap between anode and cathode electrodes increase the cycling efficiency because the time for reaching the recycling equilibrium is shorter. Thus, by decreasing the gap between electrodes can minimize the device response time also increases the overall current in a redox reaction. From Eq. (5), it shows that the current signal of the IDEs structure depends on the active area, gap, width and numbers of the fingers [9].

3 Morphological and Structural Analysis of WO₃ Film

The surface morphology of the WO_3 film was studied using EVO Scanning Electron Microscope (SEM). The SEM of the drop deposited WO_3 film on the silicon substrate is as shown in Fig. 3a, which indicates film is in porous structures. This porous film can interact with more number of gas molecules due to its more surface area.

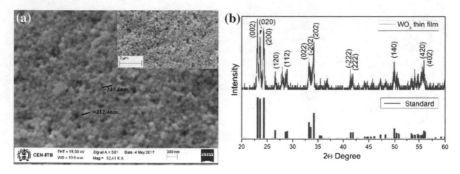

Fig. 3 a SEM of WO₃ film, b HRXRD of WO₃ film

The structural analysis of WO_3 film was characterized by High Resolution X-ray Diffraction (HRXRD). The HRXRD patterns of the WO_3 thin film (Fig. 3b) indicate that the film match well with JCPDS data file 43-1035 confirming the monoclinic crystal structure of WO_3.

4   Design and Fabrication of Microcantilever

In the simulation, the silicon material is selected as a structural layer for the microcantilever with dimensions of $200 \times 50 \times 0.5$ μm. Silicon nitride layer of 0.1 μm deposited above silicon is used as an insulating layer. Polysilicon material is used as a piezoresistive layer. WO_3 sensing material is put at the free end of the microcantilever as shown in Fig. 1.

Calculation for the conversion of the acetone molecules from ppm to mg/m³ is done to determine the weight of adsorbed acetone on sensing layer [10]. Conversion of ppm to mg/m³ is done using ideal gas law; hence, the number of moles for acetone gas is $4.0896 \times 10^{-5} \times 1$ ppm. But, molecular weight of acetone is 58.08 kg/kmol, so calculated weight for 1 ppm acetone is 2.3752×10^{-6} kg/m³ at NTP. When this force is given to the sensing region of the cantilever as shown in Fig. 1, which is considered as adsorption of acetone with sensing region, it causes a change in mass to deflect the cantilever in the direction of gravity. So, the relation between adsorbed target gas molecules with cantilever's deflection can be found.

The simulation of microcantilever sensor expressed, microcantilever works effectively only when the piezoresistive layer is placed close to the region of maximum stress. Figure 4a shows that when the length of the piezoresistor is varied by keeping other parameters constant, more deflection is obtained for minimum piezoresistor length. The sensing area's length from the free end of the cantilever is inversely proportional to the cantilever's deflection as shown in Fig. 4b. At the optimized value of the cantilever's geometry, if acetone concentration is increased the deflection in cantilever increases linearly as shown in Fig. 5a. Cantilever's $\Delta R/R$ for applied ace-

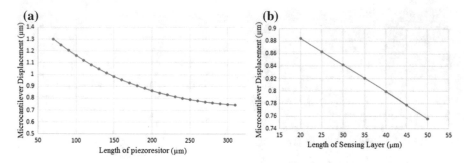

Fig. 4 **a** Piezoresistor length versus deflection of microcantilever and **b** length of sensing layer versus deflection of microcantilever

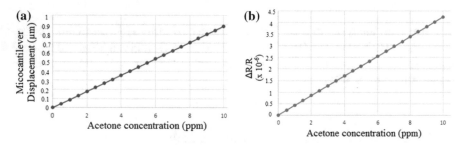

Fig. 5 **a** Total displacement of cantilever versus acetone concentration and **b** relative change in resistance of piezoresistive microcantilever versus acetone concentration

tone load was found by measuring the change in resistivity when a load was applied to the cantilever with respect to no load condition as shown in Fig. 5b.

In the fabrication of microcantilever, surface micromachining technique was used. Dry oxidation was used to form SiO_2 sacrificial layer on silicon substrate. Si_3N_4 was deposited as an insulating layer in between silicon and polysilicon layers. Si_3N_4, silicon and polysilicon were deposited using HWCVD. To release the microcantilever, SiO_2 layer was removed using BHF etching. Extra care should be taken during releasing of the cantilever to avoid stiction problem. Fabricated structure of microcantilever device is shown in Fig. 6.

5 Design and Fabrication of IDEs Device

In the simulation of IDEs (Fig. 7a), fingers' length, width and thickness are equal to 200, 10 and 0.5 μm, respectively, and the gap between fingers is 2 μm and 40 numbers of fingers are used. Gold is selected for electrodes material. In physics, electrostatics is added. An electric potential is added for giving +1 V to one electrode and ground to another. IDEs show more effective surface area by increasing length of the fingers,

Fig. 6 Fabricated structure of piezoresistive microcantilever

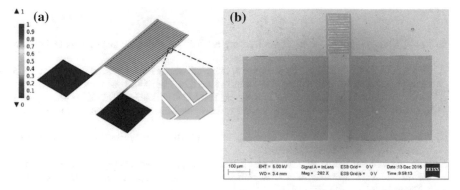

Fig. 7 a Simulation of potential distribution in IDEs and **b** fabricated structure of IDEs device

increasing the width of the fingers, reducing the gap between the fingers, slightly increase in fingers thickness and by increasing number of fingers.

In the fabrication of IDEs device, SiO_2 layer is deposited on silicon substrate using dry oxidation which acts as an insulating layer. UV exposure, patterning and development were done. Thermal evaporation was used for deposition of gold material. Finally, lift-off process is used to remove remaining photoresist. SEM image of the fabricated IDEs device is shown in Fig. 7b. Deposition of WO_3 thin film on microdevices and its gas sensing are under progress.

6 Conclusions

The surface morphology of deposited WO_3 film was studied using SEM. HRXRD of the film shows monoclinic crystal structure, which is good for gas sensing applications. Simulation and optimization of microdevices are done. Microcantilever's simulation shows that piezoresistor and sensing area affected on microcantilevers' deflection. Simulation of IDEs device shows that IDEs limiting current response depends on finger's length, width, thickness, numbers and the gap between them. We have successfully fabricated IDEs and piezoresistive microcantilever devices which can be used in low-level gas sensing functionalized with suitable metal oxides. This work provides a novel approach to develop low cost, portable and quick in response gas sensor devices.

Acknowledgements We acknowledge the support of the Indian Nanoelectronics Users Program at the Indian Institute of Technology (IIT), Bombay. Authors would like to thank Shrivatsa Bhat, for his help in fabrication of microcantilevers.

References

1. Liu X, Cheng S, Liu H, Hu S, Zhang D, Ning H (2012) A survey on gas sensing technology. Sensors 12(7):9635–9665
2. Gao P, Ji H, Zhou Y, Li X (2012) Selective acetone gas sensors using porous WO_3–Cr_2O_3 thin films prepared by sol–gel method. Thin Solid Films 520(7):3100–3106
3. Shi J, Hu G, Sun Y, Geng M, Wu J, Liu Y, Ge M, Tao J, Cao M, Dai N (2011) WO_3 nanocrystals: synthesis and application in highly sensitive detection of acetone. Sens Actuators B Chem 156(2):820–824
4. Boisen A, Dohn S, Keller SS, Schmid S, Tenje M (2011) Cantilever-like micromechanical sensors. Rep Prog Phys 74(3):036101
5. Patil SJ, Duragkar N, Rao VR (2014) An ultra-sensitive piezoresistive polymer nano-composite microcantilever sensor electronic nose platform for explosive vapor detection. Sens Actuators B Chem 192:444–451
6. Kalita H, Palaparthy VS, Baghini MS, Aslam M (2016) Graphene quantum dot soil moisture sensor. Sens Actuators B Chem 233:582–590
7. Liu F, Kolesov G, Parkinson BA (2014) Time of flight electrochemistry: diffusion coefficient measurements using interdigitated array (IDA) electrodes. J Electrochem Soc 161(13):H3015–H3019
8. Aoki K, Morita M, Niwa O, Tabei H (1988) Quantitative analysis of reversible diffusion-controlled currents of redox soluble species at interdigitated array electrodes under steady-state conditions. J Electroanal Chem Interfacial Electrochem 256(2):269–282
9. Min J, Baeumner AJ (2004) Characterization and optimization of interdigitated ultramicroelectrode arrays as electrochemical biosensor transducers. Electroanalysis 16(9):724–729
10. Subhashini S, Juliet AV (2015) Analytical investigations involved in a microcantilever for gas detection. IJEECC 2(1):301–305

Emotion Recognition from Sensory and Bio-Signals: A Survey

Kevin Vora, Shashvat Shah, Harshad Harsoda, Jeel Sheth,
Seema Agarwal, Ankit Thakkar and Sapan H. Mankad

Abstract Emotion detection is crucial in several applications including voice-based automated call centers and machine-to-human or machine-to-machine conversation systems. This area of research is still in its infancy stage. In this article, we present an overview of emotion recognition from various types of sensory and bio-signals and provide a review of existing literature.

Keywords Emotion recognition · Affective computing · Speech · EEG · Facial animation parameters

1 Introduction

Recognizing emotions using machines is a very challenging task. Fields such as human–computer interaction (HCI), psychology, affective and cognitive computing highly depend on emotion recognition [1]. Many attempts have been made for

K. Vora · S. Shah · H. Harsoda (✉) · J. Sheth · S. Agarwal · A. Thakkar · S. H. Mankad
Department of Information Technology, Institute of Technology, Nirma University,
Ahmedabad 382481, Gujarat, India
e-mail: 15BIT012@nirmauni.ac.in

K. Vora
e-mail: 15BIT066@nirmauni.ac.in

S. Shah
e-mail: 15BIT054@nirmauni.ac.in

J. Sheth
e-mail: 15BIT058@nirmauni.ac.in

S. Agarwal
e-mail: 15BIT052@nirmauni.ac.in

A. Thakkar
e-mail: ankit.thakkar@nirmauni.ac.in

S. H. Mankad
e-mail: sapanmankad@nirmauni.ac.in

© Springer Nature Singapore Pte Ltd. 2019
A. J. Kulkarni et al. (eds.), *Proceedings of the 2nd International Conference on Data Engineering and Communication Technology*, Advances in Intelligent Systems and Computing 828, https://doi.org/10.1007/978-981-13-1610-4_35

modeling emotions in the literature using various sensory and bio-signals such as electrocardiogram (ECG) [2], electroencephalogram (EEG) [3, 4], audio-based features [5], facial expressions [6, 7], to name a few. In this paper, we will put forward various aspects of detecting emotions by means of several sensory data captured in form of biological features.

According to "The Expression of the Emotions in Man and Animals" by Darwin [8], emotions have an evolutionary history that could be traced across various species and cultures. Hence, we can infer that emotions like anger, fear, surprise, disgust, happiness, and sadness which are universal to all humans. Measuring emotions is a real-world challenge because human emotions depend on several internal and external factors of the human body. Emotions have a great impact on our daily activities. HCI plays a significant role in day-to-day life, and thus, the need to design accurate human emotion recognition systems has increased over a period of time.

Although there have been some promising results in determining emotions, we still do not have satisfactory results in this emerging field. Surely, the tools that measure emotions have been greatly revolutionized by the advancement in the field of cognitive computing, neuroscience, and development of sensors that capture human behavioral and psychological features. Today's instruments range from simple pen and paper rating scales to complex high-tech equipment that measure brain waves or eye movements [9].

We classify the survey carried out in the field of emotion recognition on the basis of behavioral, physiological, psychological, and fusion of features considered to recognize emotions [10–12].

2   Classification of Emotion Recognition

In this section, we describe the classification of features for emotion detection on the basis of various properties of human anatomy [9] which include behavioral features, physiological features, and psychological features.

2.1   Behavioral Features

In this section, we have classified the emotions using speech and facial expressions. We have also discussed few of the models found in the literature.

2.1.1   Speech and Voice

Speech exhibits the most natural way of communication and a simple way of expressing the emotional state of a person. Emotion detection from speech has become crucial in HCI domain because of its inherent capability to reveal emotion, intention, and motivation [3]. In [5], authors have analyzed different prosodic and spectral features for audio-based emotion recognition considering six universal emotions, viz. happy, angry, sad, disgust, fear, and surprise. Several researchers have attempted to use prosodic features such as pitch, energy, and duration [1, 13–15]. Speech emotion recognition systems that perform well on acted speech datasets do not perform very well in natural speech scenarios [1]. An excellent review of speech emotion recognition is given in [16].

Models for emotion recognition from speech and voice

Support Vector Machine

In [10], authors showed that best results are obtained by the 12-dimensional feature vector of MFCC when trained on EmoSTAR database [10] using SVM classifier. Mel-frequency cepstral coefficients are the most prominent and distinctive features for representing audio data for any applications including emotion recognition problems [12]. MFCC features have been effective due to their ability to capture the working mechanism of the human hearing system.

Data-Driven Fuzzy Inference System

Using speech for emotion detection can create a system where the sample contains information of multiple states of emotions. This linguistic vagueness in the definition of emotion categories implies that emotions can overlap each other in their acoustic information spaces [11]. This problem can be overcome using fuzzy inference system (FIS) as depicted in Fig. 1. The work focuses on a data-driven approach since no expert is available to determine the number of rules in the system. In this paper, 21 acoustic correlates including those providing prosodic information from the speech signal are used. Input features included in this methodology are a subset of features used in [17].

Using Hidden Markov Model (HMM)

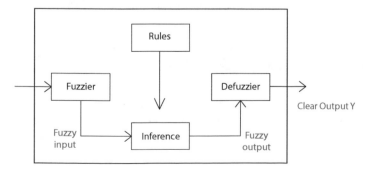

Fig. 1 Fuzzy inference system [11]

Building an HMM-based recognition framework not only demands the features to be used, but it also must fit the structure of HMM. In [18], authors analyzed various models with instant values and contours of pitch and energy as measures. They observed that these features were able to carry significant information about emotional state of the speaker. Moreover, these pitch and energy features were adaptive and invariant to channel distortion, speaker, gender, and language. An accuracy above 80% in the speaker-dependent recognition with seven emotional styles was achieved in [18]. Thus, it can be stated that speech-based features play a significant role in determining an emotion.

2.1.2 Facial Expressions

Facial expressions are known to be more effective for expressing the emotions. The movement of facial feature points is caused by a set of systems as in the facial score [6]. To determine the facial score, geometric proportions of facial features such as the eyes, nose, lips and the distance between them are used and then the score is calculated. Different databases are available according to different features taken from faces.

MPEG4 stream can also be used to take facial parameters, and HMM is used to recognize the emotion [7]. In [7], it is suggested that the proposed method will be able to scale up with future developments in technology. The method extracts the facial animation parameters which psychologists have already mentioned in their researches. The facial features are converted to MPEG4-compliant facial animation parameters (FAP). The HMM for these emotions is temporal in nature, and each emotion has four states as shown in Fig. 2. An HMM is trained with the extracted FAPs for every defined facial expression, namely sad, anger, fear, joy, disgust, and surprise.

Active appearance model (AAM) [6, 19] is another technique for capturing facial feature points. AAM detects deformable objects in images by building a statistical model with the help of shapes and appearances from training images with manually added feature points. Image parameters are found which adapt to this model. The error

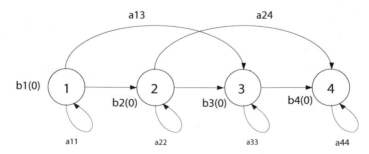

Fig. 2 Topology of Hidden Markov models [7]

is computed as the difference of actual facial points and estimated feature points for each emotion, and so a minimum error will show the emotion. This approach could not discriminate between emotions like disgust and happy, and fear and sadness.

2.2 Physiological Features

This section provides an overview of data obtained from various signals including electroencephalogram (EEG), electrodermal activity, electromyograohy (EMG). We also show extraction process of respiration and heart rate data using wireless signals.

2.2.1 Electroencephalogram

An electroencephalogram (EEG) is used to detect electrical activity in the brain using small, flat metal disks (electrodes) attached to the scalp. The brain cells communicate via electrical impulses and are active all the time, even when the human object is asleep. So monitoring the electrical impulse can result in monitoring of emotions using involuntary body response. The emotions can be categorized on the basis of arousal and valence scales as shown in Fig. 3. A study in [20] demonstrates that EEG signals fused with facial expressions convey complementary information for continuous emotion detection.

2.2.2 Electrodermal Activity

Electrodermal activity (EDA) monitors the activity of automatic nervous system and is measured from the surface of the body. It is also known as galvanic skin response [21]. The EDA characterizes changes in the skin's electrical as a result of the activity of sweat glands and is physically interpreted as conductance. These sweat glands receive input from the sympathetic nervous system and helps a good indicator of arousal level due to external sensory and cognitive stimuli.

2.2.3 Electromyography

Electromyography (EMG) is an electrodiagnostic technique which records the electrical activity produced by skeletal muscles. Using a Bayesian network, users' emotions based on EMG and skin conductance signals are determined [22].

2.2.4   Respiration and Heart Rate

Instead of using different physiological sensors, use of wireless signals to measure breathing and heart rate for detecting emotions may be explored. EQ-Radio [23] is a system which extracts the individual heartbeats from the wireless signal. Furthermore, its accuracy is also found to be competent with on-body ECG monitors. The system consists of three major components as depicted in Fig. 4. The frequency modulated continuous wave (FMCW) radio for transmitting RF signal and receiving back those signals, and ignoring reactions from other objects; the signal is sent to beat extraction algorithm to return a series of signal segments that correspond to the individual heartbeats, and these elements are sent to emotion classification system.

A high accuracy is possible to achieve using physiological features as involuntary features are measured in this domain. But it is seen that by making the test subjects to wear all these sensors and makes them conscious about the act of measuring. Thus, making them aware of the situation a neural state is not possible to achieve, which leads to an undesirable situation. It must also be taken into account that measuring is an expensive process consuming a lot of resources and a device which is feasible with the set of sensors to detect human emotion is not easily available.

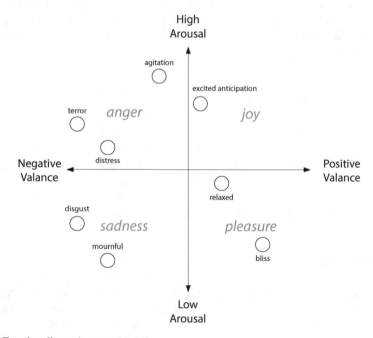

Fig. 3 Emotion dimensions models [4]

Fig. 4 Radio frequency
system [23]

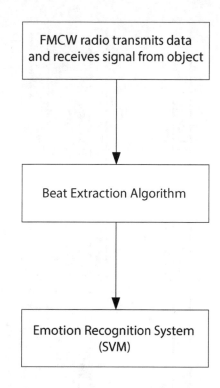

2.3 Psychological Features

Physiological features involve verbal self-reporting and nonverbal self-reporting
which is discussed in this section.

2.3.1 Verbal Self-Reporting

The affective state of mind can be determined by using a questionnaire with predefined, open-ended questions, analyzing interviews, and maintaining an emotion
diary. Some of the examples of this technique are the Academic Emotions Questionnaire (AEQ), The Positive and Negative Affect Schedule (PANAS), etc. Although
this method is effective, the language and cultural barrier make it difficult [9].

2.3.2 Nonverbal Self-Reporting

This includes tools which are language-independent and unobtrusive and can be used
in various cultures. They overcome the barriers of verbal self-reporting. However,
the range of emotions assessed by them is limited. International Affective Picture
System [9] is one such tool.

Table 1 Recent work on emotion recognition systems

References	Sensory input	Dataset	Emotions	Classifier	Accuracy
[24]	Blood volume pulse (BVP), Electromyography (EMG), Skin conductance (SC), Skin temperature (SKT) and Respiration (RESP)	Not available	Amusement, contentment, disgust, fear, neutrality and sadness	Support vector machine	92%
				Fisher linear discriminant	90%
[21]	EMG and EDA	ProComp Infiniti data capture library	Not available	Bayesian network on basis of EMG, GSR, and game status (skip-bo game)	Unknown
[4]	ECG, EMG, EDA, RESP	MIT database	Joy, anger, sad, pleasure	Linear discriminant analysis	70%
[25]	ECG, EMG	Not available	Anxiety, boredom, engagement, frustration, anger	kNN	75.6%
				Regression tree	83.5%
				Bayesian networks	74.03%
[10]	Speech signal	EmoSTAR [26]	Happy, angry, sad and neutral	SVM	96.9%
				KNN	92.3%
[23]	RF signals (EQ-Radio)	Not available	Not available	SVM	87% (Person dependent) 72% (person independent)
[27]	Embedded camera of smartphone	Japanese female facial expressions database	Anger, happy, sad, disgust, afraid, surprised, neutral	GMM based classifier	99.8%
		Canade-Kohn database [28]			
[29]	Speech signal	IVR-SERES calls CALL CENTER calls	Anger, happy, sad, neutral	SVM	85.2% (IVR-SERES)
		EMODB		ANN	82.1% (CALL CENTER)
				kNN	89.1% (EMO-DB)

2.4 Fusion of Multiple Features

As shown in Table 1, there are several studies in which multiple sensory inputs are fused to design fusion models for various applications. Thus, the new array of coefficients obtained from fusion of multiple sensory inputs can prove to be useful in recognizing emotions with a better scope. The number of coefficients can be down-sampled by using machine learning algorithms for optimization. Authors in [30] have used a hybrid deep model for audio–visual recognition applying deep networks on audio and video independently and then combining them to form a hybrid network which is then classified using SVM.

In another study, [31] have combined audio, video, and lexical modalities. The temporal estimation model mentioned employs estimated biases to control the simulation of emotions at the beginning of the conversation which is correlated with time.

3 Conclusion

In this paper, an elaborate review on emotion detection from sensory and bio-signals is given. The survey focused on classifying the features used for emotion detection on the basis of human anatomical parameters such as behavioral, physiological, and psychological aspects. We also provided a summary of features used for various emotion detection tasks with different classifiers and datasets.

Acknowledgements This survey is carried out as a part of Idea Lab project funded by the Institute of Technology, Nirma University.

References

1. Alonso JB, Cabrera J, Medina M, Travieso CM (2015) New approach in quantification of emotional intensity from the speech signal: emotional temperature. Expert Syst Appl 42:9554–9564
2. Agrafioti F, Hatzinakos D, Anderson AK (2012) ECG pattern analysis for emotion detection. IEEE Trans Affect Comput 3(1):102–115. https://doi.org/10.1109/T-AFFC.2011.28
3. Hartmann K, Siegert I, Philippou-Hubner D, Wendemuth A (2013) Emotion detection in HCI: from speech features to emotion space. In: 12th IFAC symposium on analysis, design, and evaluation of human-machine systems, pp 288–295
4. Kim J, André E (2008) Emotion recognition based on physiological changes in music listening. IEEE Trans Pattern Anal Mach Intell 30(12):2067–2083
5. Ooi CS, Seng KP, Ang LM, Chew LW (2014) A new approach of audio emotion recognition. Expert Syst Appl 41(13):5858–5869. https://doi.org/10.1016/j.eswa.2014.03.026, http://www.sciencedirect.com/science/article/pii/S0957417414001638
6. Pantic M, Rothkrantz LJM (2000) Automatic analysis of facial expressions: the state of the art. IEEE Trans Pattern Anal Mach Intell 22(12):1424–1445

7. Pardàs M, Bonafonte A, Landabaso JL (2002) Emotion recognition based on MPEG-4 facial animation parameters. In: 2002 IEEE International conference on acoustics, speech, and signal processing (ICASSP), vol 4. IEEE, New York, p 3624
8. Darwin C (1872) The expression of the emotions in man and animals. John Murray, London
9. Feidakis M, Daradoumis T, Caballé S (2011) Emotion measurement in intelligent tutoring systems: what, when and how to measure. In: 2011 Third international conference on intelligent networking and collaborative systems (INCoS). IEEE, New York, pp 807–812
10. Korkmaz OE, Atasoy A (2015) Emotion recognition from speech signal using mel-frequency cepstral coefficients. In: 2015 9th International conference on electrical and electronics engineering (ELECO). IEEE, New York, pp 1254–1257
11. Lee CM, Narayanan S (2003) Emotion recognition using a data-driven fuzzy inference system. In: Eighth European conference on speech communication and technology
12. Sato N, Obuchi Y (2007) Emotion recognition using mel-frequency cepstral coefficients. Inf Media Technol 2(3):835–848
13. ten Bosch L (2003) Emotions, speech and the ASR framework. Speech Commun 40(1):213–225. https://doi.org/10.1016/S0167-6393(02)00083-3, http://www.sciencedirect.com/science/article/pii/S0167639302000833
14. Burkhardt F, Polzehl T, Stegmann J, Metze F, Huber R (2009) Detecting real life anger. In: 2009 IEEE International conference on acoustics, speech, and signal processing (ICASSP), vol 4. IEEE, New York, pp 4761–4764
15. Koolagudi SG, Reddy R, Yadav J, Rao KS (2011) IITKGP-SEHSC: Hindi speech corpus for emotion analysis. In: 2011 International conference on devices and communications (ICDeCom), pp 1–5. https://doi.org/10.1109/ICDECOM.2011.5738540
16. Ayadi ME, Kamel MS, Karray F (2011) Survey on speech emotion recognition: features, classification schemes, and databases. Pattern Recogn 44:572–587. https://doi.org/10.1016/j.patcog.2010.09.020
17. Cowie R, Douglas-Cowie E, Tsapatsoulis N, Votsis G, Kollias S, Fellenz W, Taylor JG (2001) Emotion recognition in human-computer interaction. IEEE Signal Process Mag 18(1):32–80
18. Nwe TL, Foo SW, De Silva LC (2003) Speech emotion recognition using hidden Markov models. Speech Commun 41(4):603–623
19. Nishiyama M, Kawashima H, Hirayama T, Matsuyama T (2005) Facial expression representation based on timing structures in faces. In: International workshop on analysis and modeling of faces and gestures. Springer, Berlin, pp 140–154
20. Soleymani M, Asghari-Esfeden S, Fu Y, Pantic M (2016) Analysis of EEG signals and facial expressions for continuous emotion detection. IEEE Trans Affect Comput 7(1):17–28. https://doi.org/10.1109/TAFFC.2015.2436926
21. Nakasone A, Prendinger H, Ishizuka M (2005) Emotion recognition from electromyography and skin conductance. In: Proceedings of the 5th international workshop on biosignal interpretation, pp 219–222
22. Xu Y, Hübener I, Seipp AK, Ohly S, David K (2017) From the lab to the real-world: an investigation on the influence of human movement on emotion recognition using physiological signals. In: 2017 IEEE International conference on pervasive computing and communications workshops (PerCom Workshops). IEEE, New York, pp 345–350
23. Zhao M, Adib F, Katabi D (2016) Emotion recognition using wireless signals. In: Proceedings of the 22nd annual international conference on mobile computing and networking. ACM, New York, pp 95–108
24. Maaoui C, Pruski A (2010) Emotion recognition through physiological signals for human-machine communication. In: Cutting edge robotics 2010. InTech
25. Rani P, Liu C, Sarkar N, Vanman E (2006) An empirical study of machine learning techniques for affect recognition in human-robot interaction. Pattern Anal Appl 9(1):58–69
26. Parlak C, Diri B, Gürgen F (2014) A cross-corpus experiment in speech emotion recognition. In: SLAM@ INTERSPEECH, pp 58–61
27. Hossain MS, Muhammad G (2017) An emotion recognition system for mobile applications. IEEE Access 5:2281–2287

28. Kahou SE, Bouthillier X, Lamblin P, Gulcehre C, Michalski V, Konda K, Jean S, Froumenty P, Dauphin Y, Boulanger-Lewandowski N et al (2016) Emonets: multimodal deep learning approaches for emotion recognition in video. J. Multimodal User Interfaces 10(2):99–111
29. Chakraborty R, Pandharipande M, Kopparapu SK (2016) Spontaneous speech emotion recognition using prior knowledge. In: 2016 23rd International conference on pattern recognition (ICPR). IEEE, New York, pp 2866–2871
30. Zhang S, Zhang S, Huang T, Gao W, Tian Q (2017) Learning affective features with a hybrid deep model for audio-visual emotion recognition. IEEE Trans Circ Syst Video Technol
31. Savran A, Cao H, Nenkova A, Verma R (2015) Temporal Bayesian fusion for affect sensing: combining video, audio, and lexical modalities. IEEE Trans Cybern 45(9):1927–1941

Frequent Itemsets in Data Streams Using Dynamically Generated Minimum Support

Shankar B. Naik and Jyoti D. Pawar

Abstract We have presented an approach that generates frequent itemsets from data stream. The itemsets are compressed and then stored in the memory. The decision to whether or not compress an itemset is based on the utility of the itemset. In this chapter, the utility of an itemset is defined in terms of the amount of memory saved by its compression. Beside this, we have presented an approach to dynamically generate the value of minimum support threshold based on the data in the data streams. It avoids having a fixed minimum support threshold throughout the data stream. Since the value is generated from the latest elements in the data stream, it suits to be an appropriate measure to separate the frequent itemsets from the non-frequent ones.

Keywords Datamining · Data streams · Itemsets · Minimum support · Sliding window

1 Introduction

The aim of this study is to generate frequent patterns from transactional data streams. The elements of a transactional data stream are itemsets. The frequent patterns are in terms of frequent itemsets. For example, a data stream of tweets posted by users where each tweet is an element of the data stream and is an itemset of words. The frequent patterns are the sets of words that occur in most of the tweets. The significance of a frequent pattern, in this case, is that it contains words related to a very significant event that day.

The recent advancement in hardware and software has resulted in the generation of enormous amounts of data leading to the concept of data streams [5]. A data

S. B. Naik (✉)
Sant Sohirobanath Ambiye Government College, Pernem, Goa, India
e-mail: xekhar@rediffmail.com

J. D. Pawar
DCST, Goa University, Taleigão, India
e-mail: jdp@gmail.com

© Springer Nature Singapore Pte Ltd. 2019   357
A. J. Kulkarni et al. (eds.), *Proceedings of the 2nd International Conference on Data Engineering and Communication Technology*, Advances in Intelligent Systems and Computing 828, https://doi.org/10.1007/978-981-13-1610-4_36

stream is a collection of huge and mostly unbound set of data elements arriving at high rate. It is not possible to store all the elements of data stream at once in computer memory. The elements of data stream are processed as they arrive, and the results of this processing are stored as summary in the memory. The elements discarded after processing are not available again. There are three approaches used to process a data stream, viz landmark model, damped window model, and sliding window model [2–4]. The approach in this paper uses the sliding window model. In sliding window model, the latest w elements of the data stream, where w is the size of the sliding window, are processed.

The size of sliding window depends upon the memory available in the system. A larger-sized sliding window allows processing of more data stream elements at a time resulting in more accurate results. In this paper, we propose a method to store elements of the sliding window in a compressed form by using the utility of the pattern. The utility of a pattern is a profit value associated with it. Both the elements of the sliding window and the patterns generated are itemsets. The utility of an itemset in this paper reflects the memory savings offered by an itemsets when replaced by its compressed version.

Users are often interested in frequent patterns. The frequency of occurrence of frequent patterns is above a threshold called as minimum support s_0 [1, 2]. The value s_0 separates the frequent patterns from the others. Usually, the value of s_0 is given by the user which can be quite accepted in static database environment where multiple scans of the database are possible. In case of a streaming environment, where the characteristics of data are more uncertain, the user may not be able to specify the value of s_0. In this paper, we also present a method to dynamically generate value of s_0 from the information about itemsets in the sliding window.

The remainder of the paper is organized as follows. Section 3 describes the related work. The problem definition is given in Sect. 3. The proposed approach is presented in Sect. 4. Section 5 describes the experiments, while Sect. 6 concludes the paper.

2 Related Work

High utility itemset mining is proposed in [8, 9]. The approach in [8] uses information in the form of codes from previous batch to compresses elements. Information about the frequency of the elements in the previous batch is used to generate these codes which are stored per every batch. The approach in [9] compresses elements by dividing the data stream into same-sized batches. It uses information about the frequent patterns in each batch and uses these patterns to compress the elements of the batches. The algorithms batch-wise analyze the elements of the data and then use this information to compress them.

The framework presented in this paper incrementally compresses the elements of the data stream. As the elements of the data stream arrive, they are compressed immediately and stored in the sliding window, unlike storing all the elements of the sliding window and then compressing them at once. The elements are compressed based on their utility. The utility in this case is the amount of memory saved by replacing the element in its compressed form. This approach is suitable in a data stream environment as it allows more data stream elements to be stored in the memory for analysis.

3 Problem Definition

3.1 Preliminaries

Let $DS = (T_1, T_2, \ldots)$ be a data stream where T_i is an itemset. S is a sliding window of size w which slides across the data stream DS. Associated with each itemset X is a support measure which is the number of elements containing itemset X in the sliding window S and is denoted as $supp(X)$.

The essence of the approach is the replacement of the itemsets in the sliding window with a compressed version of it. This can be achieved in two ways—first, by codings the itemsets and then replacing them in the sliding window by their respective codes, and second, by storing pointers to the itemsets stored in the summary data structure. An intermediate summary data structure is used to store the results of intermediate processing of data stream, in this case the itemsets with their support for a sliding window.

It may happen that replacing all the itemsets in the sliding window is costly. This is because the size of code and pointer is more than the size of itemset itself. Hence, we modified the approach to replace the itemset with high utility as described in Sect. 3.2.

3.2 Utility of an Itemset

Utility of an itemset is a profit value associated with it. The utility of the itemset is calculated considering the memory saved by compressing the itemset and the support of itemset in the current sliding window.

The utility $utility(X)$ of itemset X is calculated in as follows. If v_i is the size of an item in X, then size of X is calculated as

$$v(X) = \sum v_i, \forall v_i \varepsilon X \tag{1}$$

If γ is the memory required by X in summary data structure, then $utility(X)$ is given by

$$utility(X) = support(X) * (v(X) - \gamma) - v(X) \tag{2}$$

where $v(X)$ is the size of X and is memory size of X. The term $u(X) - \gamma$ is the memory saved by compressing X once. The term $support(X) * (v(X) - \gamma)$ is the memory saved in compressing $Xsupport(X)$ number of times in the sliding window. $utility(X)$ is memory saving offered by X if compressed. If $utility(X) > 0$, then X should be compressed and not otherwise.

4 Frequent Itemset Mining Using Itemset Utility

4.1 The Intermediate Summary Data Structure

As the data stream elements are processed, the results generated are stored in an intermediate summary data structure. We propose an intermediate summary data structure which is a table *Isets* having fields—*ItemsetId*, *ItemSet*, *Support* and *Utility* (Fig. 1).

4.2 The Approach

In this paper, the approach generates closed itemsets using the algorithms presented in [6, 7]. The algorithm presented in [7] uses a sliding window model to maintain closed itemsets in the summary data structure, which is a list of itemsets with their support.

Upon the arrival of a data stream element at the sliding window, the algorithm inserts new closed itemset in the intermediate summary data if it does not exist and increases the support of the existing itemsets by one, otherwise. $utility(X)$ is calculated, while itemset X is inserted into the *Isets*. If the itemset X already exists, then $utility(X)$ is updated.

If $utility(X) > 0$, then the itemset in the sliding window is replaced by its *ItemsetId* in *Isets* table while it enters the sliding window.

Data Stream			ItemList			
tid	Itemset		Id	ItemSet	Support	Utility
1	{a,b}					
2	{a,d,e,f}					
3	{a}					
4	{b,c}					
.	.					
.	.					

Fig. 1 The framework—data stream and summary data structure

4.3 Frequent Pattern Generation Using Dynamically Generated Minimum Support S_0

A pattern is frequent if its frequency of occurrence is more than the minimum support threshold s_0. Usually, the value of s_0 is specified by the user and remains constant for the entire analysis. In a streaming environment, where the uncertainties in data are high, it is difficult for a user to specify such a value of s_0 which is appropriate to highlight the frequent patterns for a particular sliding window. This is due to the fact that the user has no idea about the data in the sliding window. The algorithm may not be able to catch frequent patterns if s_0 is high. Similarly, insignificant patterns may be presented as frequent if s_0 is low.

We present a method to generate s_0 from the support of the itemsets in the sliding window.

4.3.1 Defining Value of s_0

The value of s_0 is defined as

$$s_0 = \frac{\sum supp(X)}{N} + \delta, \forall X \, \varepsilon Isets \tag{3}$$

where N is the number of itemsets in *ItemList* and δ is the standard deviation of the support of all itemsets in *ItemList* at that moment.

Deciding which pattern is interesting is a matter of discretion of the user. However, the user can specify his/her own minimum support threshold based on dynamically generated value of s_0.

4.4 Frequent Pattern Generation

The proposed algorithm traverses through the itemsets in *Isets* table and displays the itemsets having support greater than the minimum support s_0.

5 Experiments

The experimental study was performed on a computer system with 2.26 GHz Intel Core i3 processor, 3 GB memory and Windows 7 OS and implemented in C++. The compiler used is GNU GCC compiler.

Table 1 Dataset parameters

Parameter	Value
Number of transactions	210K
Average items per transaction	10
Number of items	250

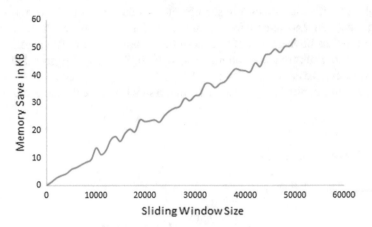

Fig. 2 Memory saved against sliding window size

5.1 Frequent Itemset Mining Using Itemset Utility

The dataset used for these experiments is generated using IBM synthetic data generator [1] and is a synthetic dataset. The parameters of the dataset are mentioned in Table 1.

5.1.1 Varying Sliding Window Size

The sliding window size was changed from 1 to 50,000 with intervals of $1K$. Figure 2 depicts that the amount of memory saved in storing the elements of sliding window is increases with the sliding window size increases.

5.1.2 Fixed Sliding Window Size

The sliding window was slided over the data stream by one element. The amount of memory saved was observed for each sliding window for 100 transitions. Figure 3 depicts that the amount of memory saved for sliding windows in the beginning is less and increases to become almost stable later. This can be explained as the algorithm

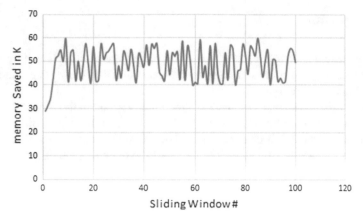

Fig. 3 Memory saved against windows sliding over data stream

learns about the itemset utility at the beginning. Thus, the number of itemsets replaced (compressed) is less.

The amount of memory saved by our approach depends upon the kind of data in the sliding window. The results of the proposed algorithm need not be promising always. It shall work best if the number and size of frequent itemsets are large.

5.2 Frequent Pattern Generation Using Dynamically Generated S_0

The dataset used in this experiment is a real dataset available on www.kaggle.com. The dataset is a set elements where each element is a set words and is ball-by-ball description (commentary) of 577 matches in IPL. These elements were processed to have only significant information alike matchid, team details, bowler, batsman, runs scored. There were 136,598 elements in the data stream. Figure 4 depicts a glimpse of the dataset.

It was observed that 44,665 itemsets were generated having their support from 1 to 65. The value assigned by the approach to s_0 is 6.52. Out of 44,665 patterns, 4565 patterns were frequent for $s_0 = 6.52$.

Figure 5 depicts the glimpse of the experimental results. The pattern at 44,665 shows that the probability of *SK Raina* scoring 1 run is 64 of 113 balls by *Harbhajan Singh*.

6 Conclusion

We have presented an approach to use utility of an itemset to replace it by its compressed version. This allows more elements to be stored in the sliding window and in the computer memory.

We have also presented a method to dynamically generate the value of minimum support threshold based on the support of the itemsets in the intermediate summary data structure. This approach does not need to access the data in the sliding window. This allows the user to gain an insight into the kind of data in the data stream which in turn helps the user to specify a minimum support value which is appropriate to data in the data stream at that point.

1	match_id	inning	batting_team	bowling_team	over	ball	batsman	non_striker	bowler	is_super_over
2	1	1	Kolkata Knight Ric	Royal Challenge	1	1	SC Ganguly	BB McCullum	P Kumar	0
3	1	1	Kolkata Knight Ric	Royal Challenge	1	2	BB McCullum	SC Ganguly	P Kumar	0
4	1	1	Kolkata Knight Ric	Royal Challenge	1	3	BB McCullum	SC Ganguly	P Kumar	0
5	1	1	Kolkata Knight Ric	Royal Challenge	1	4	BB McCullum	SC Ganguly	P Kumar	0
6	1	1	Kolkata Knight Ric	Royal Challenge	1	5	BB McCullum	SC Ganguly	P Kumar	0
7	1	1	Kolkata Knight Ric	Royal Challenge	1	6	BB McCullum	SC Ganguly	P Kumar	0
8	1	1	Kolkata Knight Ric	Royal Challenge	1	7	BB McCullum	SC Ganguly	P Kumar	0
9	1	1	Kolkata Knight Ric	Royal Challenge	2	1	BB McCullum	SC Ganguly	Z Khan	0
10	1	1	Kolkata Knight Ric	Royal Challenge	2	2	BB McCullum	SC Ganguly	Z Khan	0
11	1	1	Kolkata Knight Ric	Royal Challenge	2	3	BB McCullum	SC Ganguly	Z Khan	0
12	1	1	Kolkata Knight Ric	Royal Challenge	2	4	BB McCullum	SC Ganguly	Z Khan	0
13	1	1	Kolkata Knight Ric	Royal Challenge	2	5	BB McCullum	SC Ganguly	Z Khan	0
14	1	1	Kolkata Knight Ric	Royal Challenge	2	6	BB McCullum	SC Ganguly	Z Khan	0
15	1	1	Kolkata Knight Ric	Royal Challenge	3	1	SC Ganguly	BB McCullum	P Kumar	0
16	1	1	Kolkata Knight Ric	Royal Challenge	3	2	SC Ganguly	BB McCullum	P Kumar	0
17	1	1	Kolkata Knight Ric	Royal Challenge	3	3	SC Ganguly	BB McCullum	P Kumar	0

Fig. 4 Dataset—IPL Cricket matches

1	support	bowler	batsman	total_runs
44655	41	A Mishra	V Kohli	1
44656	42	R Ashwin	RG Sharma	1
44657	42	RA Jadeja	RG Sharma	1
44658	43	P Kumar	CH Gayle	0
44659	44	PP Chawla	RG Sharma	1
44660	44	PP Ojha	MS Dhoni	1
44661	46	A Mishra	RG Sharma	1
44662	47	RA Jadeja	V Kohli	1
44663	49	R Ashwin	V Kohli	1
44664	63	Harbhajan Singh	SK Raina	1

Fig. 5 Snapshot of frequent patterns

References

1. Agrawal R, Imieliński T, Swami A (1993) Mining association rules between sets of items in large databases. In: ACM SIGMOD record, vol 22. ACM, New York, pp 207–216
2. Babcock B, Babu S, Datar M, Motwani R, Widom J (2002) Models and issues in data stream systems. In: Proceedings of the twenty-first ACM SIGMOD-SIGACT-SIGART symposium on Principles of database systems. ACM, New York, pp 1–16
3. Han J, Cheng H, Xin D, Yan X (2007) Frequent pattern mining: current status and future directions. Data Min Knowl Disc 15(1):55–86
4. Han J, Pei J, Yin Y, Mao R (2004) Mining frequent patterns without candidate generation: a frequent-pattern tree approach. Data Min Knowl Disc 8(1):53–87
5. Naik SB, Pawar JD (2012) Finding frequent item sets from data streams with supports estimated using trends. J Inf Oper Manage 3(1):153
6. Naik SB, Pawar JD (2013) An efficient incremental algorithm to mine closed frequent itemsets over data streams. In: Proceedings of the 19th International Conference on Management of Data, COMAD'13, Mumbai, India, India. Computer Society of India, pp 117–120
7. Naik SB, Pawar JD (2015) A quick algorithm for incremental mining closed frequent itemsets over data streams. In: Proceedings of the Second ACM IKDD Conference on Data Sciences, CoDS '15, New York, NY, USA. ACM, New York, pp 126–127
8. Van Leeuwen M, Siebes A (2008) Streamkrimp: detecting change in data streams. In: Joint European conference on machine learning and knowledge discovery in databases. Springer, Berlin, pp 672–687
9. Yang X, Ghoting A, Ruan Y, Parthasarathy S (2012) A framework for summarizing and analyzing twitter feeds. In: Proceedings of the 18th ACM SIGKDD international conference on Knowledge discovery and data mining. ACM, New York, pp 370–378

A Novel Approach for Tumor Segmentation for Lung Cancer Using Multi-objective Genetic Algorithm and Connected Component Analysis

Ananya Choudhury, Rajamenakshi R. Subramanian and Gaur Sunder

Abstract Treatment in oncology relies heavily on imaging. Automatic identification of cancerous tumors is a challenging and time-taking task. This paper presents a multi-objective genetic algorithm for calculating optimal threshold value for binarization of CT lung images and segmentation of the tumor region using connected component analysis. The threshold determination method is tested on two public datasets: Data Science Bowl: Stage 1 and NSCLC lung cancer dataset. The results obtained are compared with other genetic algorithm approach and traditional Otsu's method. The proposed method gives better threshold value compared to Otsu's method and outperforms other genetic algorithm approach both in terms of threshold value and computation time. Connected component analysis is done on the labeled connected component in the binarized image. Based on predetermined geometric measurements, the components are marked as tumors and non-tumors. This method when applied to 240 cancerous images, correctly identified tumors for 205 images while showed no result where there are no tumors in the image.

Keywords Thresholding · Genetic algorithm · DICOM

1 Introduction

Lung cancer is one among the four most common types of cancer, the other being the breast, colon, and prostate cancer. In 2012 alone, more than 1.6 million people died globally due to lung cancer. By 2035, it is expected that this figure will grow up

A. Choudhury (✉) · R. R. Subramanian · G. Sunder
High Performance Computing—Medical and Bioinformatics Application Group,
Centre for Development of Advanced Computing, Pune, India
e-mail: ananyac@cdac.in; ananya.aus@gmail.com

R. R. Subramanian
e-mail: menakshi@cdac.in

G. Sunder
e-mail: gaurs@cdac.in

© Springer Nature Singapore Pte Ltd. 2019   367
A. J. Kulkarni et al. (eds.), *Proceedings of the 2nd International Conference on Data Engineering and Communication Technology*, Advances in Intelligent Systems and Computing 828, https://doi.org/10.1007/978-981-13-1610-4_37

to 3 million worldwide [1]. In India too, the estimated incidence of lung cancer was 70,275 in 2012 and the overall estimated mortality was around 63,759 [2]. According to another report published by ICMR, there was an estimated 1.14 lakh (83,000 males and 31,000 females) lung cancer records in 2016 which is likely to increase to 1.4 lakhs in 2020 [3].

According to the data provided in [4], WHO ICD-O classifies lung cancer as epithelial tumors, mesenchymal tumors, lymphohistiocytic tumors, tumors of ectopic regions, and metastatic tumors. Out of these, a rather common classification of lung tumor is non-small cell lung cancer (NSCLC), small cell lung cancer (SCLC), and lung carcinoid tumor (LCT). Three main types of NSCLC are adenocarcinoma, squamous cell carcinoma, and large cell carcinoma [5–7].

For proper treatment, an accurate and early diagnosis of lung tumors is essential in order to prevent the mortality rate. Radiology imaging plays a big role in this. The traditional method of identifying tumor region within the diagnostic image is by manual intervention, where radiologist studies the image and manually marks the delineation. But this is time consuming as well as different person may perceive the region differently. It is observed that the region marked as tumor is often different when different radiologists are involved. The purpose of our research was to devise a fully automated algorithm that is robust in segmentation of various types of lung tumors.

Depending upon the amount of human interaction involved, segmentation process can be classified as manual, semi-automated, and fully automated. Figure 1 shows a generalized classification of image segmentation techniques as obtained from the literature. Among these, thresholding methods are the simplest and computationally least expensive. However, the limitation with thresholding is that it fails to deal with attenuation variations and can be applied to only isolated lesions of tumor within the lung cavity.

The paper as such proposes a genetic algorithm approach for finding the threshold value by considering two properties of the image, viz., inter-class variance and total entropy. The paper is organized into five sections. Section 2 discusses related work available in the literature. Section 3 elaborates the proposed methodology for thresholding using genetic algorithm and segmentation using connected component analysis. Section 4 demonstrates the experimental results obtained by applying the methodology to two publicly available datasets. Section 5 draws conclusion of the methodology and a brief summary of the paper.

2 Related Work

The thresholding methods can be classified as global thresholding and local thresholding based on whether the entire image is divided into foreground and background pixels or it is done locally dividing the image into a number of sub-images [8]. Traditional Otsu's method is one of the most widely used global thresholding method [9].

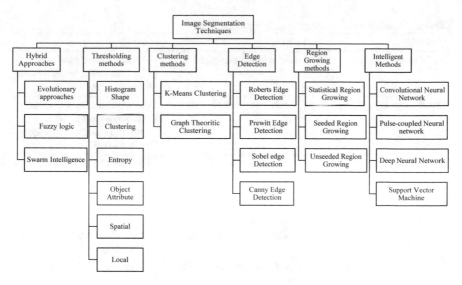

Fig. 1 Classification of image segmentation techniques

Sezgin and Sankur classified thresholding methods based on the information that is being exploited [10].

Working with medical images, the task of thresholding becomes complex. In [8], two local thresholding techniques: Niblack's algorithm [11] and Sauvola's [12] technique are explored for medical images and performance measurement in terms of peak signal-to-noise ratio (PSNR) and Jaccard Similarity Coefficient is tabulated. Jibi John and MG Mini proposed a multi-level thresholding algorithm for detecting pulmonary nodules in lung images [13]. Their method computes threshold in three levels each for segmenting thorax region, lung region, and pulmonary nodules. This increases the computation time. Also at each step, the thresholds have to be calculated efficiently in order to maintain a good true positive rate.

Manikandan et al. [14] proposed a multi-level thresholding for segmentation of medical brain images using genetic algorithm. A relatively similar approach is taken by Omar et al. [15] to compute threshold value for natural images. Table 1 lists some of the thresholding techniques along with the pros and cons of each.

Medical images, unlike natural images, have marginal differentiation between objects. Also, the contrast of the image is lower which makes a clear-cut thresholding difficult. In case of medical images, a better result and guaranteed convergence are obtained if we could take a multi-objective optimization approach where entropy and variance are both considered for determining the threshold value.

Table 1 Comparison of various thresholding techniques

Methods	Pros	Cons
Traditional OTSU [9]	Simple implementation	Do not produce good result when the image histogram has more than two peaks
Kapur's entropy [16]	Non-bimodal thresholding	Different images having same histogram gives same threshold value which may not be true in reality
Niblack's techniques [11]	Good result where the intensity variation is different along the pixel location	Threshold value depends upon the choice ok the bias value
Sauvola's technique [12]	Good result where there is foreground background variation across the image	Threshold value depends upon the value of bias and the maximum standard deviation of the image
Multi-thresholding using GA [15]	Optimal selection of threshold value	Chances of local convergence

3 Proposed Methodology

The proposed method computes a dynamic threshold by using a multi-objective genetic algorithm for binarizing the grayscale DICOM CT images. The threshold value is so obtained as to classify the pixels into two classes, one comprising the soft tissue within the lung lobes and the outer air regions and the other containing the thorax wall, bones, and other organs.

The DICOM CT image is preprocessed to remove noise by using median filter. Each chromosome is represented as a binary string of length K, where $K = \log_2(n)$ [n is the number of bits representing one pixel]. The chromosomes are initialized with random binary string with each chromosome representing a particular grayscale intensity. For reproduction, we select a portion of the fittest individuals of the current generation and pass them directly to the next generation without alteration. The rest of the individuals are taken for crossover and mutation. Two chromosomes out of k chromosomes are selected randomly for single-point crossover. The newly obtained chromosomes are then mutated by randomly selecting a bit and flipping it from 0 to 1 and vice versa.

3.1 Multi-objective Fitness Function

We propose a fitness function considering the image entropy and variance of classes, where the objective of the genetic algorithm is to minimize the function. Equations (1) and (2) depict the fitness function used for evaluating the individuals.

$$\text{Fitness}(x) = \min f(x) \tag{1}$$

$$f(x) = f_v(x)/f_h(x) \tag{2}$$

where

$f_h(x)$ ←fitness function based on entropy of the image.

$f_v(x)$ ←fitness function based on variance of intensity of the image regions

Entropy of an n-state system in information theory is a measure of randomness in the system. If a grayscale image is modeled as such a system, the states would be defined by the n-gray levels ($n = 2^8 = 256$ states for an 8-bit grayscale image). Mathematically, entropy H can be described by Eq. (3) [16].

$$H = -\sum_{i=0}^{n-1} p_i log_2 p_i \tag{3}$$

where p_i is the probability of occurrence of ith gray level. This information is used by Kapur et al. to find a threshold of a grayscale image [17]. The method assumes that the threshold value t should maximize the total entropy of the image. Total entropy is given by Eq. (4).

$$f_h(t) = H_f + H_b \tag{4}$$

$$H_f = -\sum_{i=0}^{t-1} \frac{p_i}{\omega_0} log_2 \frac{p_i}{w_0} \tag{5}$$

$$H_b = -\sum_{i=t}^{n-1} \frac{p_i}{\omega_1} log_2 \frac{p_i}{\omega_1} \tag{6}$$

$$\omega_0 = \sum_{i=0}^{t-1} p_i \tag{7}$$

$$\omega_1 = \sum_{t}^{n-1} p_i \tag{8}$$

H_f and H_b are the entropy of the foreground and background, respectively, for threshold value t.

The first objective of the genetic algorithm is to maximize the entropy for the set of chromosomes in a particular generation and across generations. The second objective of the genetic algorithm is to minimize the class variance of the image. This is derived from the traditional OTSU's method [9]. OTSU's method minimizes the intra-class variance and maximizes the inter-class variance. The fitness function is then given by Eq. (9),

$$f_v(t) = \sigma_b^2(t)/\sigma_a^2(t) \tag{9}$$

where $\sigma_a^2(t)$ and $\sigma_b^2(t)$ are the intra-class and inter-class variance, respectively. The variances are described by the following Eqs. (10) and (11).

$$\sigma_a^2(t) = \omega_0(t)\sigma_0^2(t) + \omega_1(t)\sigma_1^2(t) \tag{10}$$

$$\sigma_b^2(t) = \omega_0(t)\omega_1(t)[\mu_0(t) - \mu_1(t)]^2 \tag{11}$$

where ω and μ are the class probabilities and class mean, respectively.

$$\mu_0(t) = \sum_{i=0}^{t-1} i\frac{p_i}{\omega_0} \tag{12}$$

$$\mu_1(t) = \sum_{i=0}^{n-1} i\frac{p_i}{\omega_1} \tag{13}$$

The genetic algorithm converges to a global minimum of the fitness function and the best individual at the end of nth iteration is stored as the thresholding value. This thresholding value is used to binarize the image. The binary image is then subjected to morphological operations of erosion and dilation.

3.2 Connected Component Analysis

The pixels in the binary image are grouped into different connected components based on the pixel connectivity between them. The first level connected components are analyzed to segment out the lung lobes. Area and boundary information of the labeled components are used to make the decision. The second level of connected components is labeled, and geometric features are extracted from each component. The geometric features taken are area, bounding box, eccentricity, equivalent diameter, major axis length, minor axis length and perimeter.

The information extracted from the components is analyzed for marking them as tumor and non-tumor. In [18], an overall assessment of lung nodules is provided. Based on size, a lung nodule of size more than 5 mm has high probability of being a malignant cancer. Based on their study and it is evident that a component having overall size around 5 mm and more, which is eccentric is likely to be cancer. These assessments are used to mark the component as cancerous and discard the rest of the components.

4 Experimental Results

The algorithm has been tested on the public datasets NSCLC [19–21] and Data Science Bowl: Stage 1 [22] with 20 randomly generated chromosomes and a generation size of 50. The crossover rate and mutation rate are set at 0.80 and 0.10, respectively.

The algorithm is applied to various types of lung images those with good contrast as well as low contrast. A good threshold value is obtained for each of the images which clearly segregates the foreground and background pixels. Table 2 compares the results obtained with the proposed method for four test images taken from both the datasets and compares the result with multi-thresholding image segmentation using genetic algorithm [15] and traditional OTSU's method. It is observed that the proposed method outperforms the other genetic algorithm approach for binarization in terms of execution time, threshold value and gives better threshold value compared to traditional OTSU's method. Figure 2 shows the binarization result of four test images using each of the above-mentioned techniques. The segmentation result obtained is evaluated against the ground truth available with the datasets and result

Table 2 Comparison of image thresholding techniques

Image	Image specifications	Methods	Normalized threshold	Execution time	PSNR
Image 1	Adenocarcinoma [−1024, 1507]	Traditional OTSU	0.0087	0.009723	−7.9642
		Proposed method	0.0080	11.376948	−7.9639
		GA with class variance	0.0512	40.962069	−7.9719
Image 2	Large Cell Carcinoma [−1024, 1402]	Traditional OTSU	0.4904	0.012376	−10.6196
		Proposed method	0.4922	11.093747	−10.6197
		GA with class variance	0.2226	41.323635	−10.6284
Image 3	Squamous cell [−1024, 3071]	Traditional OTSU	0.5083	0.012832	−5.4371
		Proposed method	0.5067	10.769675	−5.4370
		GA with class variance	0.0163	39.253091	−5.4357
Image 4	Squamous cell [−1024, 3522]	Traditional OTSU	0.0080	0.012198	−4.2518
		Proposed method	0.0087	10.847697	−4.2519
		GA with class variance	0.8567	38.948293	−4.2599

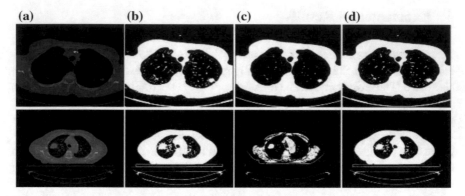

Fig. 2 CT image binarization by different techniques. **a** Original grayscale image. **b** Proposed method. **c** Multi-thresholding image segmentation using genetic algorithm. **d** Traditional OTSU's method

Table 3 Result for segmentation

Image type	No. of images	Test positive	Test negative	False positive rate	False negative rate	True positive
Cancer	240	205	35	0.08	0.03	0.91
Normal	1080	20	1060			

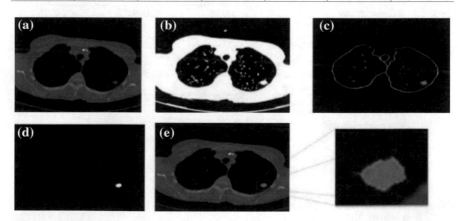

Fig. 3 Segmentation of a tumor from a typical low-contrast CT Lung image. **a** Original CT Lung image. **b** Binarized lung image. **c** Segmented lung lobe. **d** Segmented tumor. **e** Marked ROI

is tabulated in Table 3. It is observed that the overall success rate of segmentation is significantly good for all isolated lung tumors (refer to Fig. 3).

5   Conclusion and Future Work

In this paper, a genetic algorithm-based image thresholding method is proposed where both entropy and class variance are taken as optimizing parameters. The proposed methodology is tested on two publicly available datasets on more than 1000 images. The experimental results show significantly good threshold value which classifies the foreground and background pixels while preserving the region of interest. The binarized image is then subjected to connected component analysis which segments out the tumor region based on some predetermined parameters and properties. Furthermore, connected component analysis can be replaced with more sophisticated methods such as principal component analysis or edge-based methods to test for a more accurate result.

Acknowledgements The work described in this paper is carried out under the research project "BIONIC: Big Imaging Data Approach for Oncology in a Netherlands India Collaboration," funded by Ministry of Electronics and Information Technology, Government of India.

References

1. Didkowska J, Wojciechowska U, Manczuk M, Lobaszewski J (2016) Lung cancer epidemiology: contemporary and future challenges worldwide. Ann Transactional Med 1–2
2. Noronha, V, Pinninti R, Patil VM, Joshi A, Prabhash K (2016) Lung cancer in the indian subcontinent. South Asian J Cancer 1–2
3. Over 17 lakh new cancer cases in India by 2020: ICMR, 19 May 2016. [Online]. Available: http://icmr.nic.in/icmrsql/archive/2016/7.pdf
4. The 2015 world health organization classification of lung tumors: impact of genetic, clinical and radiologic advances since the 2004 Classification. J Thoracic Oncol 1243, 10(9) 2015
5. American cancer society. Lung cancer (Non-small cell), American cancer society, 2016. [Online]. Available: https://www.cancer.org/content/dam/CRC/PDF/Public/8708.00.pdf
6. American Cancer Society. Lung Cancer (Non-Small Cell) American cancer society, 2016. [Online]. Available: https://www.cancer.org/content/dam/CRC/PDF/Public/8722.00.pdf
7. American cancer society. Lung cancer (Non-small cell). American cancer society, 2016. [Online]. Available: https://www.cancer.org/content/dam/CRC/PDF/Public/8703.00.pdf
8. Senthilkumaran N, Vaithegi S (2016) Image segmentation by using thresholding techniques for medical images. Comput Sci Eng: An Int J (CSEIJ) 6(1):1–4
9. Otsu N (1979) A threshold selection method from gray-level histogram. IEEE Trans Syst Man, and Cyber 9(1)
10. Sezgin M, Sankur B (2004) Survey over image thresholding techniques and quantitative performance evaluation. J Electron Imaging 13(1):146–165
11. Niblack W (1986) An introduction to digital image processing. Prentice Hall pp 115–116
12. Sauvola J, Peitikainen M (2000) Adaptive document image binarization. Pattern Recognitiion 33(2):225–236
13. John J, Mini MG (2015) Multilevel thresholding based segmentation and feature extraction for pulmonary nodule detection. In: International conference on emerging trends in engineering, science and technolgoy (ICETEST)
14. Manikandan S, Ramar K, Iruthayaraja MW, Srinivasagan KG (2014) Multilevel thresholding for segmentation of medical brain images using real coded genetic algorithm. Measurement 47:558–568

15. Banimelhem O, Yahya YA (2012) Multi thresholding image segmentation using genetic algorithm. Int J Adv Innovative Res 1(2)
16. Rajnikanth V, Aashiha JP, Atchaya A (2014) Gray-level histogram based multilevel threshold selection with bat algorithm. Int J Comput Appl 93
17. Raju PDR, Neelima G (2012) Image Segmentation by using histogram thresholding. IJCSET 2(1):776–779
18. Sheta A, Braik MS, Aljahdali S (2012) Genetic algorithms: a tool for image segmentation. In: IEEE
19. Aerts H, Velazquez E, Leijenaar R, Parmar C, Grossman P, Cavalho S, Lambin P (2014) Decoding tumour phenotype by non-invasive imaging using a quantitative radiomics approach. Nature Communications. Nature Publishing Group, no. June 3 2014
20. Aerts HJ, Rios Velazquez E, Leijenaar RT, Parmar C, Grossmann P, Carvalho S, Lambin P (2015) Data From NSCLC-radiomics-genomics. The Cancer Imaging Archive
21. Clark K, Vendt B, Smith K, Freymann J, Kirby J, Koppel P, Moore S, Philips S, Maffitt P, Tarbox L, Prior F (2013) The cancer archive (TCIA): maintaining and operating a public information repository. J Digit Imaging 26(6):1045–1057
22. Data science bowl 2017, Data science bowl, 2017. [Online]. Available: https://www.kaggle.com/c/data-science-bowl-2017/data

Design and Implementation of Data Background Search Model to Support Child Protection Practices in India

Shubham Kumar, Aseer Ahmad Ansari, Baidehi Ghosh,
William Rivera Hernadez and Chittaranjan Pradhan

Abstract This paper deals with the design and development procedures involved in child protection practices in India. To design the logical structure of the database which is the caricature of this verification system, entity relationship diagrams have been used and implemented. The development procedure uses open-source database connectivity with the sole motive of creating a verification system which is open, robust, secure, and encrypted and contains integrity as well as referential constraints. The data background search (DBS) database has been designed and developed with the primary motive of ensuring that a convict/accused who committed a serious crime against children is kept under surveillance. The convict might try to relocate. The database created will work irrespective of the state or city the convict lives in. In the combined recruitment drive of staffs in public schools, this DBS can be implemented as a verification system to inspect the candidates. This will be beneficial for the safety of the students as well as keeping a track on the convicts.

Keywords Child abuse · Child protection · Data background search component
JDBC · Verification system · Girl child

S. Kumar (✉) · A. A. Ansari · B. Ghosh · C. Pradhan
KIIT University, Bhubaneswar, India
e-mail: shubham.sbkr@gmail.com

A. A. Ansari
e-mail: a3ahmad.kiit@gmail.com

B. Ghosh
e-mail: baidehighosh@gmail.com

C. Pradhan
e-mail: chittaranjanfcs@kiit.ac.in

W. R. Hernadez
Laurea University of Applied Sciences, Vantaa, Finland
e-mail: william.rivera.hernandez@student.laurea.fi

© Springer Nature Singapore Pte Ltd. 2019
A. J. Kulkarni et al. (eds.), *Proceedings of the 2nd International Conference on Data Engineering and Communication Technology*, Advances in Intelligent Systems and Computing 828, https://doi.org/10.1007/978-981-13-1610-4_38

1   Introduction

The aim of the this research work can be defined as working on developing a verification system to check for serial offenders of crimes against children with special
emphasis on education sector and for selecting professionals who are free from any
strain stemming from either conviction or accusation [1]. The paper deals with design,
development, and the process of implementation of the verification system named
data background search (DBS) [1]. When a parent, guardian, or caregiver, by means
of action or failing to act, inflict injury or subject a child to emotional trauma, severe
negligence, or death, then it is termed as child abuse [2]. Child abuse can be of various forms. For example, child maltreatment which includes neglect, physical assault,
sexual abuse, exploitation, and emotional trauma. There is a potent need for a central
database which will have the details of convicts of child abuse, both physical and
sexual. It could help the government keep a check on regular offenders who try to slip
into jobs of teaching/daycare to commit these harmful crimes. This DBS database
would ensure that a convict/accused of a serious crime against children is observed
irrespective of the state or city he lives in or, in many cases, tries to relocate. It would
be also helpful in inspecting and getting a hold of serial offenders who under the
present rules easily relocate and hide from the people in charge. During combined
recruitment drives of teachers in public schools, a verification system is needed to
check the authenticity of the applicants. The DBS database will act as a boon in such
a situation. Despite India's recognition of the United Nation Convention of Child
Rights, in order to ensure that all children's needs are met and that human rights are
protected, there is still an existing gap for the implementation of the law at different
levels, which prevent many children from being able to enjoy their well-being in
remote areas of India. Despite the presence of prevention of Atrocities act, 1989, for
schedule caste and schedule tribe, as act of Indian parliament to prevent atrocities
to schedule caste and tribe, the situation on the ground is very diversify with pots of
excellence in areas of utter lack of application. In India, still offenders can relocate to
different places and join as educational institute without any hindrance. This creates
the malice of repeated incidents of crimes against children (sexual abuse, child punishment, etc.) by the same person without the system knowing that they are offenders
and prosecution.

2   Work Done

A team of students from KIIT University IEEE SIGHT group along with a group
of Finnish researchers from Laurea University attended the mega learning fair from
January 10, 2017, to January 14, 2017, held at a village in Odisha, India. There,
we assessed the experiences of children of age group 8–14 years in their school.
We implemented ethnographic and participatory methods, which used as tools Lego
building, drawing and writing sessions. This assessment ensured the independent

thinking of children/participants, in order to avoid research biases and subjective interpretations. The children were divided into separate groups based on age as well as gender. After collecting the results, we found that physical and verbal punishments are very common. It was evident that to stop child abuse, there should be a verification system to check for serial offenders of crimes against children. Special importance has to be given on the education sector, for selecting professional who are not associated with any strain, stemming from either conviction or allegation.

3 Statistics and Evaluation

Based on the ethnographic and participatory methods, we tried to break the result into statistics and tried evaluating its impact on the local population [1, 3].

In Figs. 1 and 2, we see that many of the students did fall prey to being victims of abuse, which came both physically and verbally. Sixty-one percent of such including both boys and girls did report the issue either to friends or trustworthy teachers, and 39% were either introvert or too victimized. Seventy-three percent of them did claim to be a victim of either abuse due to lack of attention in classroom during studying or due to the initiatives which in their cases were wrong answers, and 27% of them even faced abuse. The complications did not seem to end here. When, it came to sanity and security, girls were profoundly the one to be afflicted. The defiled and disheveled toilets stimulated a lot. 82% of all girls apart from just finding them inconvenient also felt fearsome, such was the stew that 12% girl just waited till recess to visit their toilet homes. The precarious environment even led 26% of girls to skip school during their periods. On a combined level, 27% of students gave evidence of lack of safety in the infrastructure of school such as missing boundaries over roofs and broken walls and such. Despite of many differences and atrocities, trust factor did not seem to tumble and 89% said that they shared their sufferings and spoke about physical punishment with their friends. Undeterred by the lack of positivity, around 93% at school still could say they trusted their friends or a particular friend and a small proportion of 7% trusted a particular teacher. Fun and amusement did not seem to omit itself despite of non-promising conditions. Playing on the ground or playing with their peers was the aspect liked by almost 91% of the participants, and 9% enjoyed the geek rooms [1, 3].

4 Methodology

In order to develop a verification system, three solutions to this problem statement were conceptualized: The final approach was creation of a data background search (DBS) model. This methodology focuses on developing a central database containing the details of convicts of child abuse, physical as well as sexual. It could help the government keep an eye on regular offenders of the above-mentioned crime.

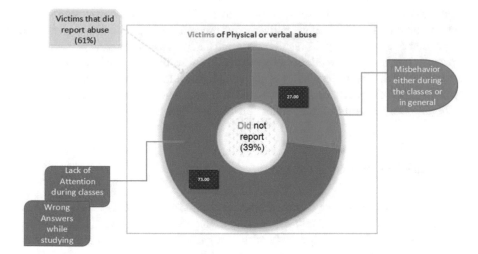

Fig. 1 Analysis of victims of physical or verbal abuse

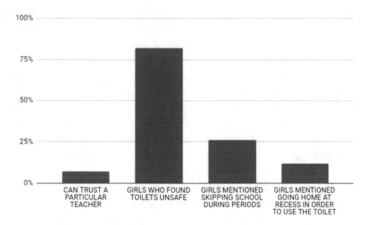

Fig. 2 Analysis of girl children who faced various problems at school

This DBS database would warrant that a convict/accused of a serious crime against children is monitored irrespective of the city or state he lives in or, in many cases, tries to move to another place. It would be beneficial for catching and inspecting those offenders, who, under the existing rules, easily shift from their native place and hide from the officials. This system has the potential of being a benison for the recruitment drive of schools/colleges/day care as it can identify the sporadic convict who tries to be a part of these institutions to carry out their heinous crime.

5 Design

The design process involves defining the tuples and attributes to be used in the database and their appropriate relations with each other. To facilitate with this process, we used a data model. A database model [4] is used to show how database is represented logically. The model contains the relationships between records and key constraints that determine how data can be stored and retrieved. The rules, regulations, and abstractions of a large data model are adopted by designers to create database models. Data models are mostly represented by an assisting database diagram. In the very initial of the database design, we need to identify the requirements of the prospective database users which will depend upon a textual explanation and detail of the database model. After selecting a particular data model and administering to it the rules, is the model translated to a conceptual schema. The type of data diagram used over here is the entity relationship (ER) model [5]. The ER model is mainly used for representation of conceptual design. To create a database, the logical relationship of entities (or objects) is diagrammatically represented by this method.

A data model [6] is a fundamental entity to introduce abstraction in a DBMS. Various objects and their relationship with entities and ethics over which they operate comprise of the data structure. It can be said that the blueprint of the database is the data model. In our data model as shown in Fig. 3, "offender" is an entity. The offender will have features like name, date of birth, crimes, time served, convictions (number of convictions and under which section of the Indian Penal Code), permanent address, birthplace, and appearance. The appearance is a composite attribute consisting of defining features such as birthmark and height. These will be the identifying characteristics of the offender.

Fig. 3 Entity relationship model of the verification system

6 Development

To establish a database connection, a server and client software need to communicate with each other. The platform, i.e., the machine, does not matter for the connection to take place. Communication between the client and the server where commands need to be sent requires a database connection. In our research work, we used Java Database Connectivity which is an application programming interface (API) of java. Queries such as update, insertion, deletion are provided by the API which is very much oriented toward relational database [7]. A popular IDE of java, NetBeans, is a nice interface to connect to DB. A wide range of databases such as Oracle, MySQL, DB4 can be used to connect to JDBC, but for our proposal, we used the in-built database called the Apache Derby. It can either start or stop from within NetBeans and runs on a virtual network.

A popular relational database management system MySQL is commonly used for stand-alone and web-based applications due to its speed, reliability, and flexibility. The data contained in the database is accessed and processed by the structured query language employed by MySQL. JDBC classes and interfaces submit SQL statements and results called by the Java Application. Connection object such as PreparedStatement, CallableStatement is used as per need by the queries to create statements [8]. NetBeans IDE is well comprised of support for MySQL such as the equipped SQL editor which is a common way of interacting with database. A new database instance can be created using the SQL editor after you are connected to the MySQL server. For a database connectivity using Java and JDBC, forName() method invoked by class takes driver path. A software component called as a "Driver" enables a Java application to interact with a database. Driver class for the MySQL database is "com.mysql.jdbc.Driver". After that getConnection() method of Driver-Manager class takes path of database, username, password of DBMS, this does the registration pat. All this is done in a user created method inside the main class. Classes of SQL such as Connection, DriverManager, PreparedStatement, ResultSet need to be imported. Then, data manipulation commands such as INSERT, UPDATE, DELETE, SELECT can be used by using reference identifier of Connection class. After the connection sets up, IP addresses of hosts are used to create a network that enables centralized updating. On a large scale as is aimed by our work, a server will be set that will be centralized. The result is shown in Figs. 4 and 5.

7 Related Work

An attached office of Ministry of Home Affairs, Government of India the National Crime Records Bureau (NCRB) is capable of a mandate to empower Indian Police with Information Technology solutions and criminal intelligence to enable them to enforce in India started in 1971. NCRB also compiles and publishes National Crime Statistics, i.e., Crime in India [9], Accidental Deaths and Suicides [10], and Prison

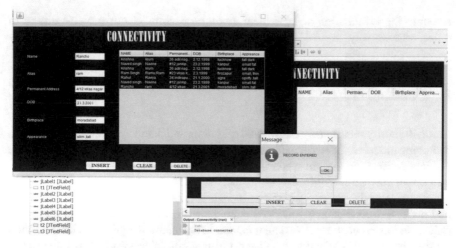

Fig. 4 Prototype testing (screenshot of interface)

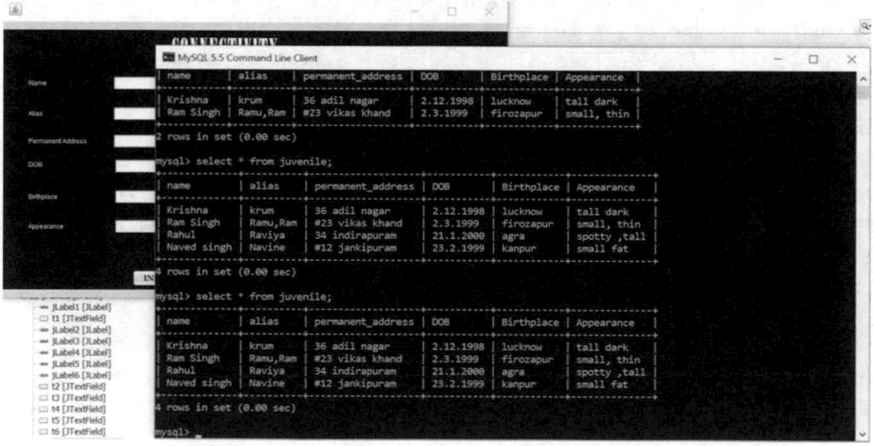

Fig. 5 Prototype testing (screenshot of interface)

Statistics [11]. These publications serve as principal reference points by policy makers, police, criminologists, researchers, and media both in India and abroad. The Central Finger Print Bureau (CFPB) is also embedded in NCRB and is a national repository of all fingerprints in the country and has more than one million ten digit fingerprints database of criminals (both convicted and arrested), provides for search facility on FACTS (Fingerprint Analysis and Criminal Tracing System) contrast to which Unique Identification Authority of India database will assist our juvenile database [11–13]. Our proposed database of convicts of child crimes (both physical and sexual) is in itself a separate unit aimed at improving the education sector profoundly. Regardless of the same approach police stations, districts, and state acquire to equip the information and circulate among the globule, our projects differs in objective adversely.

8 Conclusion

Using the entity relationship data model diagrams and database connectivity tools, a prototype model of the verification system was developed, implemented, and tested.

This verification system when implemented has the capability to become utilitarian tool for recruitment drive of teachers and keeping a track of repeated child offenders. It can be updated and maintained under the ambit of a government or judicial body. Requests can be generated from a registered educational institution or a police station to check for a particular person in the database which can be facilitated by the government/judicial body. Overall, this verification system has three major functionalities. Firstly, this database can be used as a verification system for recruitment of teachers, counselors, and non-teaching staffs of government/public schools. Secondly, it can be used to find out accused persons by a particular department of police without taking the burden of contacting different departments. A particular police station can generate a query to government/judicial body which is in charge of the DBS and can get all the required information. At last, it can serve as an evidentiary proof which can be by the different judicial bodies to track and punish serial offenders. However, this system alone is not going to solve the issue. We recognize that the involvement of good governance, civil society, social activists, groups, etc., is important to stop child abuse in schools and society. So, this DBS model is part of a systematic and holistic approach toward eradication of child abuse in schools and to a certain extent, in the society.

Ethical Statement The authors declare that we do not have any conflict of interest. This article does not contain any studies with human participants performed by any of the authors. This article does not contain any studies with animals performed by any of the authors.

References

1. Kumar S, Rivera Hernandez W, Fuschi DL (2017) DBS (Data background search) Model to support child protection practices in India. In: 26th EDEN Annual Conference. Jonkoping University, Sweden, pp 13–16 (June 2017)
2. The issue of child abuse, childhelp, https://childhelp.org/child-abuse
3. Rivera W (2017) Child friendly school experience by elementary school students aged 11–13 Years, within and Institution in Odisha, India, COHEHRE Staff Conference, Portugal
4. Aljarallah M (2014) Comparative study of database modeling approaches. Algoma University
5. Chen P (2002) Entity-relationship modeling: historical events, future trends, and lessons learned. Software pioneers. Springer, pp 296–310
6. Morris S, Coronel C (2016) Database systems: design, implementation and management. Cengage, 12th edn

7. Jatiwal M, Arora C, Arora C (2016) A study on advance java with database management. J Comput Sci Eng 1:1–9
8. Saikia A, Joy S, Dolma D, Mary R (2015) Comparative performance analysis of MySQL and SQL server relational database management systems in windows environment. Int J Adv Res Comput Commun Eng 4(3):160–164
9. National crime records bureau, crime in India, ministry of home affairs, New Delhi
10. National crime records bureau, accidental deaths & suicides in India (ADSI), Ministry of Home Affairs, New Delhi
11. National crime records bureau, Prison Statistics India (PSI), Ministry of Home Affairs, New Delhi
12. Central Finger Print Bureau, Ministry of Home Affairs, New Delhi
13. Modernisation of State Police Forces Scheme, Ministry of Home Affairs, New Delhi

The Fog Computing Paradigm: A Rising Need of IoT World

Shivani Desai and Ankit Thakkar

Abstract An increasing demand of connected devices in the field of IoT put forth the requirement of timely availability of data for the delay sensitive applications. To mitigate such requirements, fog computing plays a key role in IoT world. Fog computing supports the applications and services that do not fit with the cloud paradigm. It acts as a bridge between the underlying networks and the cloud. In this paper, we put forward the various benefits and the driving forces behind the concept of fog computing and analyze the real-world applications that include smart grid, smart traffic lights in vehicular networks, and a smarter building control.

Keywords Fog computing · Internet of things (IoT) · Cloud computing
Analytics · Edge computing

1 Introduction

CISCO has introduced a new paradigm of fog computing in the recent past. This paradigm connects billions of devices within the framework of the Internet of things (IoT). These gadgets can be equipped with the capability to perform data preprocessing work in the smarter way. Due to this reason, fog computing is also termed as edge processing. Many researchers are working in this emerging area because of the availability of the open-source environments. For example, one such open-source environment named Cisco IOx framework by CISCO that can be used with various network devices such as routers, switches, and IP cameras [1].

S. Desai (✉)
Department of Computer Engineering, Institute of Technology,
Nirma University, Ahmedabad 382481, Gujarat, India
e-mail: shivani.desai@nirmauni.ac.in

A. Thakkar (✉)
Department of Information Technology, Institute of Technology,
Nirma University, Ahmedabad 382481, Gujarat, India
e-mail: ankit.thakkar@nirmauni.ac.in

© Springer Nature Singapore Pte Ltd. 2019   387
A. J. Kulkarni et al. (eds.), *Proceedings of the 2nd International Conference on Data
Engineering and Communication Technology*, Advances in Intelligent Systems
and Computing 828, https://doi.org/10.1007/978-981-13-1610-4_39

A large amount of interconnected IoT devices produces voluminous data that requires a large number of devices with a huge amount of storage and processing capabilities. This requirement can be easily dealt by using fog computing devices in IoT networks. Both fog and cloud computing provide similar kind of services such as effective data management by providing efficient storage as well as application services as per the user needs, though fog can be recognized from the cloud by its dense geographical distribution, closeness to end-clients, and support for versatility [2]. Fog computing supports three levels of hierarchy: smart nodes, fog nodes, and cloud. This can be evident in Fig. 1 [3]. The sensor collects the required data and does preprocess before forwarding data to the cloud through the attached fog device. This paper focuses on applications of fog computing along with its advantages and dissimilarities between fog computing and cloud computing. The rest of the paper is organized as follows: Introduction, issues, and challenges related to IoT and cloud computing are presented in Sects. 2 and 3, respectively; advantages of fog computing over cloud computing along with dissimilarities between the two are explained in Sect. 4; applications of fog computing are discussed in Sect. 5, and concluding remarks are given in Sect. 6.

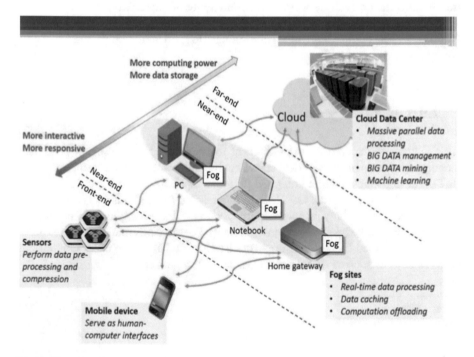

Fig. 1 Three level hierarchy of fog computing [3]

2 Internet of Things

With the utilization of Internet of things (IoT), regular objects can be associated with each other to form a network consisting of radio-frequency identification (RFID) labels, sensors, actuators, cell phones, etc. These devices can associate and collaborate with the neighbors to achieve common goals [4]. IoT provides omnipresence of the Internet by providing interaction between every object via embedded systems. This results in a distributed network of devices which is capable of communicating with humans as well as devices. IoT is opening staggering doors for countless applications that guarantee to enhance the nature of our lives [5]. In a number of applications, IoT is using 6LoWPAN [6] standard that allows a huge number of smart objects to be deployed using the IPv6 addressing scheme which supports huge address space. In IoT, non-smart items known as "things" in IoT wording, turn into the imparting hubs. These non-associated items can turn out to be a piece of IoT, with an information conveying gadget, as Bluetooth, standardized identification, or an RFID tag, etc. [7].

Gartner, Inc. figures that 20.8 billion associated things will be used worldwide by 2020 [8]. It was also forecasted that 5.5 million new things will get associated every day by 2016 [8]. The speed at which associated gadgets are expanding will add tons of information. It would be difficult to process such a huge amount of information at IoT end, where gadgets are lightweight with low memory and low computational capability. Additionally, this tremendous measure of information should likewise be used in the way it merits. These issues can be settled by coordinating IoT with distributed computing that makes another worldview, named Cloud of Things (CoT) [9, 10].

3 Cloud Computing

Cloud computing is a paradigm responsible for providing highly scalable virtual servers, virtual networks, storage and computing power. It is a user-centric paradigm which is one of the on-going trends in the computing industries. This relaxes the user from the woes of resource management and maintenance. It provides the pay-as-you-use service, under which the user pays only for the services requested. Cloud computing is an extended form of parallel computing, distributed computing, and the grid computing [11–14]. With the onset of the cloud, it is possible for the users to utilize services like storage, management, and sharing of huge quantities of digital media [15]. In the domain of processing content specifically in distributed environments, the cloud comes in handy. It provides ubiquitous access to various contents across a network without the overhead of large storage requirements or more powerful computing devices. Figure 2 shows the integration of IoT and cloud computing

Fig. 2 Integration of IoT
and cloud computing [7]

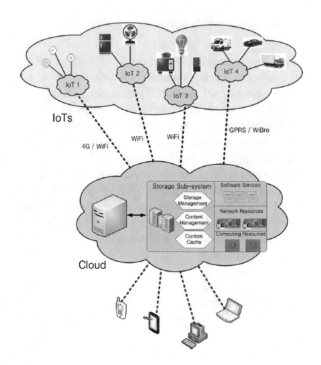

[7]. CoT has created a larger and more extensive pool of services with an added advantage of efficient accessibility. This makes an economic model based on these services more viable and profitable. The analysis of the data generated by the IoT nodes becomes more efficient many-a-fold with the use of the CoT.

Cloud computing has the following limitations and to overcome these limitations there is a requirement of a new computing model:

- IoT generates variety, volume, and velocity of data which is not handled by today's cloud models.
- IoT gadgets create a large volume of data that needs to be analyzed within a stipulated period of time. With the cloud computing model, it is not possible to analyze/process data very fast as data needs to transit between the end node and cloud which is located at far from source node where data is generated.
- Applications demand a variety of requirements viz. bounded latency, security, a high bandwidth that should be satisfied with variety, volume, and velocity of IoT data. This put forth the design of a new computing model to mitigate varied needs of the applications.
- Cloud understands IP networks only. Hence, there is a need to translate other protocol to IP before presenting it to cloud [16].

4 Fog Computing

Fog computing is termed as edge computing owing to the latter satisfying and providing various services in the domains of computation, storage, and networking at the edge of the network instead of the entire process on the cloud [2]. The cloud happens to be more centralized when compared to the fog which provides computational power and storage capabilities at various locations. The fog computing paradigm is capable of providing high QoS to mobile nodes (moving vehicles) by deploying various access points along the highway tracks. Fog acts a very valuable addition to applications with low response time requirements like gaming, video streaming, and providing assistance in ubiquitous computing [7].

Fog computing does following functionality which satisfies today's requirements with the use of IoT:

- Analysis of the time sensitive data at the network edge where IoT data is created rather sending a huge amount of data to the cloud.
- Fog nodes effectively communicate with peers, share data to make decisions using locally available information.
- As shown in Fig. 1, fog computing has three-level chain of importance which reduces the amount of data transmission by not sending all the data over a cloud, and rather aggregating them at certain access points [3].

The following are characteristics of fog computing:

- Fog computing systems can be used for the applications with low latency requirements.
- It provides privacy by protecting sensitive IoT data by analyzing it locally at fog node itself.
- It offloads gigabytes of system activity from the core network.
- Instead of batch scheduling, the fog paradigm tends to focus on the real-time interactions, and thus, it is possible to perform real-time data analytics using fog.

The difference between fog computing and cloud computing is listed in Table 1 based on the different requirements of the underlying application.

5 Applications of Fog Computing with Internet of Things

We exquisite on the part of fog figuring in the accompanying propelling situations. Few of them are as follows:

- **Smart Grid**: In smart grid, energy load balancing algorithm run at the edge of the network by smart meters and/or microgrids. These devices are intelligent enough to switch to alternative energy sources based on the energy demand, availability, and cost [1]. Fog collectors issue necessary control signals after processing data generated by the grid sensors. It also segregates the data to be consumed locally

Table 1 Difference between cloud and fog

Requirement	Cloud computing	Fog computing
TLatency	High	Low
Delay jitter	High	Very low
Location of server nodes	Within internet	At the edge of the local network
Location awareness	No	Yes
Attack on data encounter	High probability	Very less probability
Geographical distribution	Centralized	Distributed
Support of mobility	Limited	Supported
Real-time interactions	Supported	Supported

and communicates the remaining data to the higher levels for visualization and real-time reports [2, 17].

- **Smart Traffic Lights and Connected Vehicles**: In smart traffic light system, the light of the emergency vehicles can be detected by sensors and communicated to traffic control system. This helps vehicles to pass through the cross-road signals with a minimum delay [1].
- **Wireless Sensor and Actuator Networks (WSAN)**: In WSANs, sensors are used to sense the physical phenomenon of interest and/or to track the elements, while the actuator is used to take physical actions. In WSANs, actuators can be configured as fog devices to take required physical action by analyzing data provided by the sensors [1].
- **Decentralized Smart Building Control**: Wireless sensor nodes can be deployed at various floors of the building that measure temperature, humidity, and other required parameters to manage required air quality within the building. To design a smart building, fog devices can be put on each floor that manages the required environment within each floor by optimally utilizing resources such as water, energy [1].

6 Concluding Remarks

Fog computing is the emerging era that protects sensitive data by not sending them to the cloud. This helps to improve the response time of the delay sensitive applications and provides better avoidance and/or management of the disaster applications. This paper summarized the difference between fog computing and cloud computing. The paper discussed few of the applications where the use of fog computing gives better results compared to the cloud.

References

1. Stojmenovic I, Wen S (2014) The fog computing paradigm: scenarios and security issues. In: Federated Conference on Computer science and information systems (FedCSIS), pp 1–8. IEEE
2. Bonomi F, Milito R, Zhu J, Addepalli S (2012) Fog computing and its role in the internet of things. In: Proceedings of the first edition of the MCC workshop on Mobile cloud computing, pp 13 16. ACM
3. Zao JK, Gan TT, You CK, Chung CE, Wang YT, Méndez SJR, Mullen T, Yu C, Kothe C, Hsiao CT, et al (2014) Pervasive brain monitoring and data sharing based on multi-tier distributed computing and linked data technology. Frontiers in human neuroscience, vol 8
4. Atzori L, Iera A, Morabito G (2010) The internet of things: a survey. Comput Networks 54(15):2787–2805
5. Xia F, Yang LT, Wang L, Vinel A (2012) Internet of things. Int J Commun Syst 25(9):1101
6. Mulligan G (2007) The 6LoWPAN architecture. In: Proceedings of the 4th workshop on Embedded networked sensors, pp 78–82. ACM
7. Aazam M, Huh EN (204) Fog computing and smart gateway based communication for cloud of things. In: Future internet of things and cloud (FiCloud), 2014 international conference on, pp 464–470. IEEE (2014)
8. Gartner Says 8.4 Billion connected "Things" will be in use in 2017, up 31 percent from 2016. https://www.gartner.com/newsroom/id/3598917/. 15 Sep 2017
9. Aazam M, Hung PP, Huh EN (2014) Smart gateway based communication for cloud of things. In: 2014 IEEE ninth international conference on intelligent sensors, sensor networks and information processing (ISSNIP), pp 1–6. IEEE
10. Aazam M, Khan I, Alsaffar AA, Huh EN (2014) Cloud of things: integrating internet of things and cloud computing and the issues involved. In: Applied sciences and technology (IBCAST), 2014 11th International Bhurban Conference on, pp 414–419. IEEE (2014)
11. Jadeja Y, Modi K (2012) Cloud computing-concepts, architecture and challenges. In: 2012 International Conference on Computing, Electronics and Electrical Technologies (ICCEET), pp 877–880. IEEE
12. Ma W, Zhang J (2012) The survey and research on application of cloud computing. In: 2012 7th International Conference on Computer Science & Education (ICCSE), pp 203–206. IEEE
13. Zhang S, Zhang S, Chen X, Huo X (2010) Cloud computing research and development trend. In: Future Networks, 2010. ICFN'10. Second international conference on, pp 93–97. Ieee
14. Zhou M, Zhang R, Zeng D, Qian W (2010) Services in the cloud computing era: a survey. In: 2010 4th International Universal communication symposium (IUCS), pp 40–46. IEEE
15. Aazam M, Huh EN (2014) Inter-cloud architecture and media cloud storage design considerations. In: The proceedings of 7th IEEE CLOUD, Anchorage, Alaska, USA, vol 27
16. Computing F (2016) the Internet of things: extend the cloud to where the things are
17. Wei C, Fadlullah ZM, Kato N, Stojmenovic I (2014) On optimally reducing y power loss in micro-grids with power storage devices. IEEE J Sel Areas Commun 32(7):1361–1370

EEG Signal Analysis for Mild Alzheimer's Disease Diagnosis by Means of Spectral- and Complexity-Based Features and Machine Learning Techniques

Nilesh Kulkarni

Abstract Alzheimer's disease (AD) is one of the common and fastest growing neurological diseases in the modern society. Biomarker techniques for diagnosis of Alzheimer's disease and its progression in early stage are key issues for development. Electroencephalogram is one of the powerful techniques which can be used for screening of Alzheimer's disease and dementia in early stage. The objective of this paper is to analyze the EEG signal by means of spectral and complexity features to serve EEG as a biomarker for Alzheimer's diagnosis. The research is carried on experimental database obtained from hospital. EEG relative power, spectral entropy, spectral flux, and spectral centroid are analyzed, compared, and classified for separating the data between two groups by means of support vector machine (SVM) classifier and K-nearest neighbor (KNN) classifier. The obtained results indicate severity observed in AD patients reflected in EEG signals which can be treated as benchmark for Alzheimer's diagnosis.

Keywords Alzheimer's disease · Electroencephalogram · Classifier · Machine learning

1 Introduction

Alzheimer's disease (AD) is basically characterized by impaired state of memory leading toward severe Alzheimer's in which medications are not helpful for saving the life of an individual. Statistical studies have reported that around 6–25% of mild cognitive impairment (MCI) patients are transformed toward Alzheimer's and 0.2–4% from normal to mild Alzheimer's every year [1, 2]. Due to this, early diagnosis of Alzheimer's disease and its progression is key challenges in Alzheimer's diagnosis. It is expected that strength of individuals with Alzheimer's disease is about to increase in the future. To search a computationally efficient technique for early detection of

N. Kulkarni (✉)
STES's Sou. Venutai Chavan Polytechnic, Pune, India
e-mail: nileshkulkarni992@gmail.com

© Springer Nature Singapore Pte Ltd. 2019   395
A. J. Kulkarni et al. (eds.), *Proceedings of the 2nd International Conference on Data Engineering and Communication Technology*, Advances in Intelligent Systems and Computing 828, https://doi.org/10.1007/978-981-13-1610-4_40

patients who are in progress toward Alzheimer's disease but do not show any clinical symptoms of Alzheimer's is an important as well as new challenge. Although neuroimaging techniques are helpful in screening of Alzheimer's disease, they are much expensive and time consuming. EEG on other side is cheap and a promising tool for detection of several neurological disorders such as Alzheimer's, epilepsy, brain strokes [3, 4]. Review of previous research highlights that developing efficient system for automated Alzheimer's disease and dementia diagnosis using various biomarkers such as electroencephalography (EEG), magnetic resonance imaging (MRI), positron emission tomography (PET) has become difficult as well as challenging task and it has increased interests of researchers, clinicians, and scientists worldwide in this field. Earlier research findings indicate that feature extraction techniques and classification accuracy are main issues in Alzheimer's diagnosis. This paper presents EEG signal analysis in time and frequency domain for detection of Alzheimer's disease in early stage using spectral- and complexity-based features by use of suitable machine learning algorithms. Abnormalities in EEG signals of Alzheimer's disease patients not only reflect the anatomical deficits but also reflect the functional deficits of the cerebral cortex damaged by the disease. Nonlinear dynamic analysis (NDA) of EEG reveals out the loss of complexity of EEG signals and reduction in functional connections in Alzheimer's patients [3–5]. Thus, NDA of EEG signals provides important information about the progress of the disease as compared to other conventional techniques [3, 5].

The paper is organized as follows: Materials and Methods are described in Sect. 2. Database details and performance analysis of EEG data are explored in Sect. 3. Machine Learning and Classification Techniques are discussed in Sect. 4, and Conclusion is summarized in Sect. 5.

2 Materials and Methods

2.1 Participants Information

EEG data in present research work was taken from Smt. Kashibai Navale Medical College and General Hospital comprising of both mild Alzheimer's disease and age-matched healthy patients termed as normal subjects under the supervision of experienced neurologists. Alzheimer's patients were diagnosed by experienced neurologists according to NINCDS-ADRDA criteria, based on Indian version of Mini-Mental State Examination (MMSE) and Clinical Dementia Rating (CDR). Multichannel EEG signal data was collected from 100 patients classified into 2 groups. Fifty subjects in group 1 consisted of: 40 males and 10 females (mean age: 65 years) indicating functional as well as behavioral decline. Similarly, 50 subjects in group 2 consisted of: 30 male patients and 20 female patients (mean age: 62.5 years) giving no indication of functional as well as behavioral decline. Patients belonging to the abnormal group also underwent single-photon emission computed tomography

(SPECT) scanning which was followed for 14–15 months since only functional tests such as MMSE and CDR tests are not enough to set Alzheimer's diagnosis. An additional criterion used was the presence of functional and cognitive decline for last 10 months based on interview with knowledgeable informants. The patients were also tested for different medical disorders such as diabetes, kidney disease, thyroid tests, and vitamin B12 deficiency since it can also result in cognitive decline. EEG data recordings and the study details were approved by ethical committee of the hospital and participants.

2.2 EEG Data Acquisition and Recordings

EEG recording was collected according to international 10–20 electrode placement system as recommended by American EEG society. Recorders and Medicare Systems (RMS) EEG machine was used for recording. Twenty-four-channel EEG signals were acquired with participants awake, relaxed, and their eyes closed for 15–20 min. EEG signals were recorded using RMS, India EEG machine with 12-bit resolutions and 200 Hz sampling rate. Impedance of the EEG machine was maintained below 10 Mohms. EEG signals were filtered using third-order Butterworth band-pass filter between 0.5 and 30 Hz. After successful recording, EEG data has been successfully inspected by clinical technician. EEG recordings are susceptible to certain artifacts such as electronic smog, head movement, and muscular activity. For each subject, one EEG segment of 20 s (termed as "epochs") was extracted for analysis. These epochs are further used in the study for analysis.

3 Analysis of EEG Data

EEG signal is a nonstationary signal, and it is quite difficult to analyze it clinically. Brain rhythms exist in time as well as frequency domain. In present research, EEG signal is explored in time, frequency, and time–frequency domain [5, 6]. MATLAB (2013b version) software is used for implementing the algorithms proposed in present research. In this section, spectral- and complexity-based features are explored in detail. These features provide better performance results for distinguishing the subjects between two groups.

3.1 Feature Extraction and Proposed Features

The block diagram of the system methodology is shown in the following Fig. 1.

Alzheimer's disease diagnosis system is processed as (i) preprocessing of raw EEG signal, (ii) feature extraction of EEG data, and (iii) classification. Initially, raw

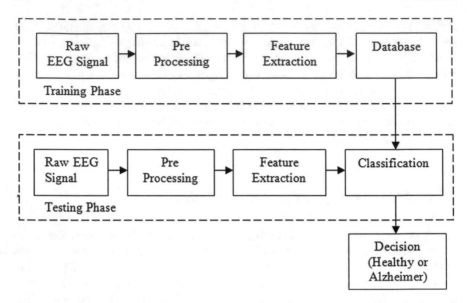

Fig. 1 Proposed system methodology used in present research work

EEG signal is preprocessed to eliminate an artifact which gets added during signal acquisition. In preprocessing, artifacts of the signal are removed. In preprocessing stage, the EEG signal is filtered using third-order Butterworth band-pass filter between 0.5 and 30 Hz. In feature extraction stage, spectral and complexity features have been computed and it is given as an input to classifier.

3.2 Spectral-Based Features

Spectrum of EEG signal is affected by neurodegenerative diseases such as mild cognitive impairment and Alzheimer's. Recent research works have reported that Alzheimer's and mild cognitive impairment (MCI) cause EEG signal to slow down. Slowing effect in EEG signal of AD patients is observed by computing power spectral density (PSD) in different EEG frequency bands. EEG signal frequency bands are helpful in obtaining important information of the patients. From the results obtained in the study, it is seen that relative power (RP) is high in low-frequency bands (delta and theta bands), i.e., frequency range between 0.5 and 8 Hz. This effect of increase in power of low-frequency bands and decrease of power in high-frequency bands of EEG signal is observed in AD patients [5–8]. This irregularity of EEG signal is also quantified by using various standard measures; one of those is Lempel–Ziv complexity discussed in [6, 7, 9]. Spectrum of EEG is useful in understanding brain activity. For extracting spectral features, i.e., relative power, EEG signal is separated into four frequency bands, namely: 0.5–4 Hz (delta), 4–8 Hz (theta), 8–12 Hz (alpha), and 30–100 Hz (gamma) using *wavedec* function. In present study, decomposition of EEG signal into five frequency sub-bands was done by use of "Daubechies" wavelet. Daubechies wavelet (db2, level 5) was used since (i) it has wide smoothing characteristics, (ii) easy understanding of the nature of the signal, and (iii) changes in the EEG signals are easily observed [10]. EEG signals are decomposed using "db2" wavelet at level of 5. In present work, relative power of EEG signal of four electrodes, namely frontal (F3 and F4), central, parietal, and temporal, is computed for all five EEG sub-bands.

The power spectral density (PSD) function helps in assessment of spectral characteristics of EEG activity of each epoch. Each sub-band power of EEG signal is calculated as Fourier transform of its autocorrelation function [11]. The normalized PSD in frequency range of 0.1–40 Hz is given as

$$RP = \sum_{flow}^{fhigh} PSD_n(f) \tag{1}$$

3.3 Complexity-Based Features

As benchmark, several features have been reported in the literature for analyzing the EEG signals in Alzheimer's diagnosis. For each EEG electrode, namely central (C3), frontal (F3 and F4), parietal (P4), different complexity features such as spectral centroid, spectral entropy, and spectral flux are computed. In previous studies, Staudinger et al. [12] used some of these features for Alzheimer's diagnosis for severe Alzheimer's disease patients using event-related potentials (ERP) EEG signals. But, in this study some of those features are computed and tested for non-ERP EEG recordings to increase the performance of the system for getting better classification as well as diagnostic accuracy. These features depict the nonlinear changes introduced in the brain activity in case of Alzheimer's disease patients. Let us discuss these features in more detail.

Spectral Entropy (SE): Spectral entropy indicates the amount of irregularity and disorder in spectrum of EEG. Higher complexity is achieved if higher amount of spectral entropy is observed [12, 13]. It is computed in the following manner:

Spectral entropy is computed by using formula

$$\text{Spectral Entropy} = \sum_{f=0.5}^{30} S(f) * \ln \frac{1}{S(f)} \tag{2}$$

where $S(f)$ is the power spectral density (PSD) of the signal $x(t)$ in given frequency band between 0.5 and 30 Hz.

Spectral Centroid (SC): Spectral centroid measures shape and position of the spectrum of EEG signals [12, 13]. Spectral centroid is calculated as

$$\text{Spectral Centroid} = \frac{\sum_{k=0}^{N-1} X(f_k) f_k}{\sum_{k=0}^{N-1} X(f_k)} \tag{3}$$

where $X(f_k)$ is spectral magnitude of kth sample and f_k is frequency corresponding to each magnitude element.

Spectral Flux (SF): Spectral flux counts the change in spectral information between two successive frames [13]. It is computed in the following manner

$$\text{Spectral Flux} = \left(\sum_{k=2}^{k} |X(f_k) - X(f_{k-1})| \right) \tag{4}$$

4   Machine Learning and Classification Techniques

Machine learning is technique of programming to optimize a performance criterion based on past experience. The performance of the system is analyzed by means of various machine learning algorithms. For testing the performance of system, different classifiers are available in machine learning and pattern recognition field. Waikato environment for knowledge analysis (WEKA) is also one of the techniques for classifying the data using JAVA platform. But, in present research work, MATLAB pattern recognition toolbox is used for classification. During classification process, 70% of the data was used for the purpose of training the data using tenfold cross-validation technique and 30% data was left for testing to check the system performance. Confusion matrix is created from these machine learning algorithms which can be helpful in calculating different parameters such as accuracy, sensitivity, and specificity [14]. In present research work, supervised learning approach is used, which requires a large database. Figure 2 shows the supervised recognition flow.

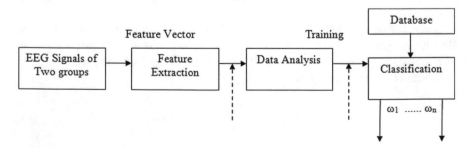

Fig. 2 Statistical recognition flow for supervised classification

4.1 Support Vector Machine (SVM) Classifier

This classifier relies on the principle of statistical learning. Classification and regression of data are typically carried out by SVM classifier. It is robust, efficient, and very effective in case of classification of high dimensionality data. In this classifier, an optimal hyperplane linearly separates data points belonging to two or more classes in case of higher dimensionality data. Researchers and scientists used SVM classifier for classification of data since it is simple and efficient [15]. Leave-one-subject-out (LOSO) cross-validation method is used for classification performance calculation. LOSO cross-validation technique is helpful since it avoids overfitting problem during separation of data and ensures the ease of the classifier to unseen data.

4.2 K-Nearest Neighbor (KNN) Classifier

Like SVM classifier, KNN classifier is also simple, efficient, and robust. It searches neighborhood of K in the training data, and a class is assigned that appears in the neighborhood of K. The value of K can be changed for each testing phase to find the match class between training and testing data, which is also used to obtain better classification rates. Basically, "1" is taken as the default value of K. But, the K value can be varied from 1 to 10 for each testing phase. "Euclidean" and "nearest" are default neighborhood setting values. For searching the object similarity in the neighborhood of K, Euclidean distance is normally used [15].

4.3 Performance Analysis of the EEG Data

The spectral and complexity algorithms are analyzed and evaluated successfully on EEG data. Results in terms of classification rates as well as diagnostic accuracy are compared with each spectral and complexity features using SVM and KNN classifiers. The obtained results are shown in Table 1.

Table 1 Classification accuracy obtained for different features

Features	Accuracy (SVM)	Accuracy (KNN)
Spectral	88	82
Spectral entropy	79	78.33
Spectral centroid	74	73
Spectral flux	76	75
Combination of all features	96	94

5 Conclusion

In this paper, spectral- and complexity-based features are explored briefly for Alzheimer's disease diagnosis. The objective of current study was to provide better results in terms of classification rates and/or diagnostic accuracy. Although the proposed features in this paper are not novel, some improvements in results in terms of classification rates are observed. Each individual feature was computed and classified by use of both the classifiers discussed. It is also to note that EEG data analysis is mainly done for four brain regions such as frontal, parietal, central, and temporal as these are the regions where changes in the brain tend to occur at initial stage in case of Alzheimer's. EEG spectrum provides us useful information clearly stating the changes occurring in the Alzheimer's patients. Spectral features such as relative power (RP) are useful in discriminating the two subjects for classification in terms of accuracy obtained. Relative power is computed in each sub-band of EEG signal to highlight the difference between spectrum of EEG signal of normal and Alzheimer's affected patients. From this EEG spectrum, we also computed different nonlinear features which also provided satisfactory results in terms of diagnostic accuracy. Appearance of plagues and neurofibrillary tangles in the cortex and decrease in volume of hippocampus slows down the EEG of Alzheimer's disease patients. In present study, SPECT scans of Alzheimer's patients were also studied with expert clinicians since only functional tests such as MMSE and CDR are not enough to validate the results for diagnosis obtained in present research. It is also to highlight that each band of EEG signal equally contributes for analysis. Support vector machine and K-nearest neighbor classifier also helped to obtain satisfactory diagnostic results on the features selected for diagnosis purpose.

It is to conclude that when we combine all features together, classification rate as well as diagnostic accuracy obtained is more and it provides comparatively better results in terms of accuracy. Future work in this study involves making analysis of each frequency bands in depth to observe whether they carry any other significant information for better diagnosis by means of various signal processing algorithms and features extraction techniques. Our future work also includes implementing present algorithms on hardware devices such as DSP processors (application-specific integrated circuits), field-programmable gate array (FPGA) devices to make a stand-alone device for diagnosis which might be useful for doctors for correctly diagnosing the patients in early stage [16]. This is a difficult task but possible to implement, and it may come out with some new technology in the market since there is no such device available in the market for Alzheimer's diagnosis. This work explores new tool for Alzheimer's disease diagnosis, and further research using some more of these features can report remarkable achievements in this field.

Ethical Statement The database in present research work was collected from Smt. Kashibai Navale Medical College and General Hospital, Pune, under the supervision of Dr. Nilima Bhalerao, Neurosurgeon. The EEG data set and its details were approved by ethical committee of the hospital and participants.

References

1. Mattson M (2004) Pathways towards and away from Alzheimer's disease. Nature 430:631–639
2. Meek PD, McKeithan K, Shumock GT (1998) Economics considerations of Alzheimer's disease. Pharmacotherapy 18:68–73
3. Wan J, Zhang Z, Rao BD, Fang S, Yan J, Saykin AJ, Shen L (2014) Identifying the neuroanatomical basis of cognitive impairment in Alzheimer's disease by correlation and nonlinearity-aware sparse bayesian learning. IEEE Trans Med Imaging 33(7):1475–1487
4. Kulkarni Kulkarni, Rathod PP, Nanavare VV (2017) The Role of Neuroimaging and Electroencephalogram in diagnosis of Alzheimer disease. Int J Comput Appl (IJCA) 3(4): 40–46
5. Jeong J (2004) EEG dynamics in patients with Alzheimer's disease. Clin Neurophysiol 15(7):1490–1505
6. Dauwels J, Srinivasan K, Ramasubba Reddy M, Musha T, Vialatte F-B, Latchoumane C, Jeong J, Cichocki A (2011) Slowing and loss of complexity in Alzheimer's EEG: two sides of the same coin? Int J Alzheimers Dis 2011:539621
7. Cassani Raymundo, Falk Tiago H, Fraga Francisco J, Kanda PAM, Anghinah R (2014) The effects of automated artifact removal algorithms on electroencephalography-based Alzheimer's disease diagnosis. Frontiers in Aging Neuroscience 6:1–13
8. Van der Hiele K, Vein AA, Reijntjes RH, Westendorp RG, Bollen EL, van Buchem MA, van Dijk JG, Middelkoop HA (2007) EEG correlates in the spectrum of cognitive decline. Clin Neurophysiol 118(9):1931–1939
9. Czigler B, Csikos D, Hidasi Z, Anna Gaal Z, Csibri E, Kiss E, Salacz P, Molnar M (2008) Quantitative EEG in early Alzheimer's disease patients—power spectrum and complexity features. Int J Psychophysiol 68(1):75–80
10. Daubechies I (1992) Ten lectures on wavelets. Society for Industrial and Applied Mathematics, Philadelphia, PA
11. Kang Yue, Escudero Javier, Shin Dae (2015) Principal dynamic mode analysis of EEG data for assisting the diagnosis of Alzheimer's disease. IEEE J of Trans Eng Health Med 3:1–10
12. Staudinger T, Polikar R (2011) Analysis of complexity based eeg features for diagnosis of alzheimer disease. In: Proceedings International Conference of the IEEE-EMBC. Boston, USA pp 2033–2036
13. Giannakopoulos T, Pikrakis A (2014) Introduction to audio analysis: a MATLAB approach. Elsevier
14. Rueda Andrea, Gonzalez Fabio A (2014) Extracting salient brain patterns for imaging based classification of neurodegenerative diseases. IEEE Trans Med Imaging 33(6):1262–1274
15. Suresh M, Ravikumar M (2013) Dimensionality reduction and classification of color features data using SVM and KNN. Int J Image Process Visual Commun 1(4):2319–1724
16. Kulkarni N (2017) Int J Inf Tecnol. https://doi.org/10.1007/s41870-017-0057-0

An Efficient Machine Learning Technique for Protein Classification Using Probabilistic Approach

Babasaheb S. Satpute and Raghav Yadav

Abstract Classifying an unknown protein into a known protein family is a challenging task and one of the demanding problems is bioinformatics and computational biology. As proteins are the main target molecules while designing the drug for any disease, it is important to study and classify unknown proteins. Machine learning algorithms like support vector machines, decision tree classifier, naïve Bayes classifier, and artificial neural networks have been effectively used for such kind of problems. In this paper, our aim is to classify an unknown protein sequence into known protein family using machine learning algorithms and to compare their performance. Here, the protein feature used for classification purpose is the probability of occurrence of a particular amino acid in the protein sequence. There are mainly 20 amino acids which form a protein. The idea here is proteins having nearly similar probability of occurrence of amino acids belong to the same family.

Keywords Classification of proteins · SCOP · UniProt · Bioinformatics
Machine learning · Probabilistic feature of proteins

1 Introduction

Machine learning is a subfield of artificial intelligence which is centered on the impression that machines or computer systems can learn from data or they can be trained using data and can be used to solve newer problems of same kind. The genesis of this field is in pattern recognition and computational learning theory in artificial intelligence. It gives machines the ability to learn and enhance or optimize the performance from past experience. Here, we focus on developing algorithms and computer programs which can access data and use it to autolearn. The main objective here is

B. S. Satpute (✉) · R. Yadav
Department of Computer Science & IT, SIET, SHUATS, Allahabad 211007, India
e-mail: satputebs@gmail.com

R. Yadav
e-mail: raghav.yadav@shiats.edu.in

© Springer Nature Singapore Pte Ltd. 2019
A. J. Kulkarni et al. (eds.), *Proceedings of the 2nd International Conference on Data Engineering and Communication Technology*, Advances in Intelligent Systems and Computing 828, https://doi.org/10.1007/978-981-13-1610-4_41

to make computers learn without human interference or support and adjust actions consequently. Since the biological data is growing with exponential rate, it becomes very difficult and costly to use traditional laboratory methods to analyze it. Machine learning algorithms [1–4] like naïve Bayes, support vector machines, decision tree classifier, artificial neural networks can apply composite mathematical calculations to large volume of data recurrently and more quickly. Hence, computational biologists are moving toward machine learning for complex sequence analysis or classification problems. Machine learning algorithms are mainly of two types, i.e., supervised and unsupervised algorithms. In supervised algorithms, machines are trained with data whose classes are known producing an inferred function which can be used to make predictions about output. On the contrary in unsupervised learning algorithms, dataset used to train is neither labeled nor classified. Here, our objective is to classify unknown proteins into currently known families based on the probabilities of occurrences of amino acid residues using various machine learning algorithms.

Protein is an important biomolecule which is formed from amino acids. There are mainly 20 amino acids which constitute proteins. Also, proteins are the main target while designing drug for any disease. Amino acids are main organic molecules which form proteins. They contain alpha (central) carbon connected to amino group, a carboxyl group, hydrogen atom, and as side chain. Several amino acids are interconnected together by peptide bond inside a protein in this manner making a long chain. Peptide bonds are formed by biochemical reaction in which a water molecule is extracted and the amino group of one amino acid is joined to the carboxyl group of adjacent amino acid. Thus, an in lines sequence of amino acids is formed which is also called as the primary structure of proteins Fig. 1.

In this paper, our claim is proteins with higher sequence similarity, i.e., proteins having same amino acids residues belong to the same family. For our work, we downloaded the sequences from Web site http://www.uniprot.org [5]. For family information of proteins, we used SCOP [6] classification of proteins. For all proteins, we calculated the odds of occurrence of each amino acid and made the database of those protein AA probabilities. In total, we made a database of 520 protein sequences belonging to 32 protein families. Thus, each tuple in the database belongs to one protein and have 20 columns each corresponding to different amino acid. Then, we divided the dataset into training and testing set and after designing the classifiers we tested their performance.

2 Literature Survey

This section gives brief review of the work done in the past on classification of proteins.

Datta and Talukdar in their paper [7] used the protein features extracted by using physicochemical properties and compositions of amino acids. They used artificial neural networks and nearest neighbor classifier for classification purpose. The efficiency achieved was 77.18%.

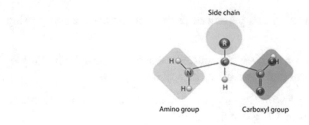

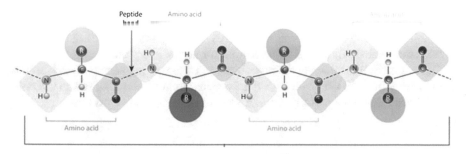

Fig. 1 Protein primary structure (chain of amino acids) [16]

Bandopadhyay in his paper [8] stressed the position of occurrence of amino acids in each protein belonging to a specific superfamily. He used nearest neighbor algorithm for the purpose of classification and the accuracy obtained was 81.3%.

Chan and team in their work developed an algorithm which is called as unaligned protein sequence classifier (UPSEC) [9]. This is a probabilistic approach to find the patterns in the sequences to classify proteins.

An algorithm AMOGA [10] was proposed by Vipsita S. to enhance the performance of radial basis network used for protein classification purpose. The authors used principal component analysis for selecting features.

Angadi U.B and Venkatesulu used unsupervised learning technique to classify proteins using structural classification of proteins. They used BLAST to create database of p values [11]. ART2 algorithm which is unsupervised was used for classification purpose, and the performance was compared with other algorithms like SVM, random forest, and spectral clustering.

3 Materials and Methodology

We downloaded the protein sequences from UniProt protein database. For each sequence, we computed the probability or chance of occurrence of each amino acid.

The probability of an amino acid in a particular protein can be computed by using the formula.

Probability of amino acid(X) = number of occurrences of amino acid(X) in the protein sequence/Total number of amino acids in the protein sequence.

That is, for a given protein S with length L, the probability of occurrence of amino acid x which occurs N times in S is given by Eq. (1).

$$P(x) = \frac{N}{L} \tag{1}$$

There are 20 amino acids in total; thus for each protein sequence, we had 20 probability values which we call as features of that particular protein. Thus, we created the database in the following format for nearly 520 proteins.

Table 1 is the illustration of the probability or chance of happening of each amino acid in every protein. We divided the database in training and testing sets in the ratio of 70:30 and tested the performances of various classification algorithms.

4   Classification Algorithms

Following classification algorithms were applied to the above-created dataset, and the accuracy was calculated.

4.1   Artificial Neural Networks (ANN)

Artificial neural networks [12] are the networks of the computing elements called neurons. The functioning of ANN Fig. 2 is similar to the human nervous system. The ANN learns by calculating the variance between the output it got and the expected output. The error is sent back into the network and weights on the connection are modified for improving results.

Feed-forward Back Propagation Algorithm [12]
There are three layers of neurons in the feed-forward back propagation network (BPN), namely an input layer, hidden layers, and output layer as shown in Fig. 2.

During training of the BPN, after the inputs are provided to the input layer, the network computes the variance between the output at the output layer and the actual expected output. The variance between the two is called as error. That error is propagated back and the weights of the network are adjusted in order to get the expected output.

Table 1 Sample probabilities of amino acids

Protein sequence →	Probabilities of amino acids																			
	A	C	D	E	F	G	H	I	K	L	M	N	P	Q	R	S	T	V	W	Y
1	0.1	0.1	0	0.2	0.01	0.09	0	0.1	0	0	0	0.1	0.02	0.08	0	0.1	0	0	0.1	0
2	0.05	0.1	0.1	0	0.05	0.1	0.2	0	0	0.1	0	0	0.1	0.04	0	0.06	0	0	0	0

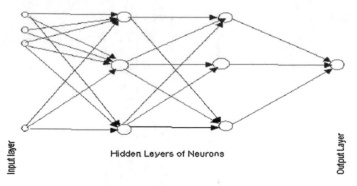

Input layer

Hidden Layers of Neurons

Output Layer

Fig. 2 Architecture of feed-forward network [12]

4.2 Naïve Bayes Classifier Algorithm

Naïve Bayes classifier classifies [13, 14] the dataset based on the features which are independent of each other. Equation (2) is the equation of the naïve Bayes classifier.

$$P(\text{Class A}|\text{Feature1, Feature2}) = \frac{P(\text{Feature1}|\text{Class A}).P(\text{Feature2}|\text{Class A}).P(\text{Class A})}{P(\text{Feature1}).P(\text{Feature2})} \quad (2)$$

Equation (2) finds the probability of data item being classified in Class A given Feature1 and Feature2.

The equation states that the probability of Class A with Feature1 and Feature2 is the ratio where the numerator is probability of Feature1 given Class A multiplied by probability of Feature2 given Class A multiplied by probability of Class A, and the denominator is probability of Feature1 multiplied by probability of Feature2.

4.3 Support Vector Machines [13–15]

The classification problem can be limited to thought of the two-class problem without loss of generality. In such problems, the objective is to isolate the two classes by using a function that is made from existing samples. The aim is to create a classifier that will work fine on unknown samples; i.e., it can be generalized.

The SVM classier is a twofold classier which increases the margin. The partition hyperplane is parallel to the margin planes and is equidistant from the planes. Each margin plane passes through points of the training set that belong to a specific class and is neighboring to the margin plane of the other class. The distance between the margin planes is called the margin. Several pairs of margin planes are likely with dissimilar margins. The algorithm discovers the extreme margin splitting hyperplane. The points from each class that decide the margin planes are called the support vectors (SVs).

4.4 Decision Tree Classifier Algorithm

A decision tree classifier [13, 14] is articulated as a recursive partition of the dataset. The decision tree contains nodes which form a binary tree with one node called as a root. In a decision tree, each internal node splits the dataset into two or more sets based on a certain discrete function of the input attributes. Each leaf is given to one class on behalf of the most suitable target value.

5 Proposed Algorithm

i.   Begin
ii. Download protein sequences from protein database.
iii. Find the chance or probability of each amino acid occurrence in a particular protein.
iv. Repeat step (iii) for every amino acid and each protein.
v.   Prepare the dataset as shown in Table 1.
vi. Divide the dataset into training and testing set in the ratio of 70:30.
vii. Repeat steps (viii) to (ix) for each classifier
viii. Train the classifier with training set
ix. Test the performance of the classifier with test set.
x.   Compare the performances of all
xi. End.

6 Results and Discussion

6.1 Dataset Used

We downloaded 520 proteins corresponding to different families.

6.2 Efficiency

The efficiency is calculated as follows

Efficiency = (no of correct family prediction of the test data/total no of proteins in the test dataset).

Table 2 Efficiency comparison of classifiers

Performance parameter ↓	Classifiers			
	ANN	Naïve Bayes	SVM	Decision tree
Efficiency (%)	63	59	68	84

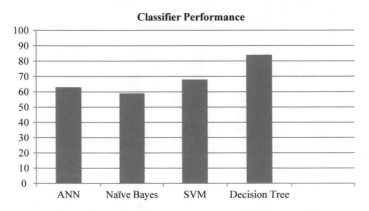

Fig. 3 Classifier performance comparison

6.3 Classifier Comparison

We tested the performances of all the classifiers using the test dataset containing 150 proteins. Table 2 compares the efficiency of all the classifiers.

6.4 Graphical Representation of Performance of Classifiers

Figure 3 shows the performance of all four classifiers used. It can be seen from the graph that decision tree performed better than others, whereas the performance of other three is nearly same.

7 Conclusion

Machine learning algorithms are becoming popular to solve biological problems and are being widely used for solving numerous biological problems especially classification problems which otherwise take lot of time and money to solve by using traditional laboratory methods. Here, we presented a probability-based method to extract the features of proteins. In this method, we computed the chance or odds of existence of each amino acid in a particular protein. Our premise here is that protein sequences with nearly similar probability of amino acids belong to same family. And the results obtained support our premise to larger extent. The performance of decision tree classifier was more than 80% whereas other algorithms gave an efficiency between 60 and 70%. The method can be effectively used to identify the family of the unknown protein sequence.

References

1. Wang D, Huang GB (2005) Protein sequence classification using extreme learning machine. In: Proceedings of international joint conference on neural networks (IJCNN,2005), Montreal, Canada
2. Duda RO, Hart PE, Stork DG (2001) Pattern classification, 2nd ed. Wiley Inter science Publication
3. Bernardes JS, Pedreira CE (2013) A review of protein function prediction under machine learning perspective. Recent Patents on Biotechnol. 7:122–141
4. Saha S, Chaki R (2013) A brief review of data mining application involving protein sequence classification. Int. J. Database Manage. Syst. 4:469–477
5. http://www.uniprot.org/
6. http://scop.mrc-lmb.cam.ac.uk/scop/
7. Datta A, Talukdar V, Konar A, Jain LC (2009) A neural network based approach for protein structural class prediction. J. Intell. Fuzzy Syst. 20:61–71
8. Bandyopadhyay S (2005) An efficient technique for superfamily classification of amino acid sequences: Feature extraction, fuzzy clustering and prototype selection. ELSEVIER J Fuzzy-Sets Syst. 152:5–16
9. Ma PCH, Chan KCC (2008) UPSEC: An algorithm for classifying unaligned protein sequences into functional families. J Comput Biol 15:431–443. https://doi.org/10.1089/cmb.2007.0113
10. Vipsita S, Shee BK, Rath SK (2010) An efficient technique for protein classification using feature extraction by artificial neural networks. In: IEEE India conference: green energy, computing and communication, INDICON 2010
11. Angadi UB, Venkatesulu M Structural SCOP superfamily level classification using unsupervised machine learning. IEEE/ACM Trans Comput Biol Bioinformatics 9:601–608, https://doi.org/10.1109/tcbb.2011.114
12. Bishop CM (1995) Neural networks for pattern recognition. Oxford
13. Machine TM (2017) Mitchell learning. McGraw Hill Education
14. Christopher Bishop, Pattern recognition and machine learning. Springer; 1st ed. 2006. Corr. 2nd printing 2011 edition (15 February 2010)
15. Zhao XM, Huang DS, Cheung YM, Wang HQ, Huang X (2004) A novel hybrid GA/SVM system for protein sequences classification, vol 3177, pp 11–16
16. https://www.nature.com/scitable/topicpage/protein-structure-14122136

Allocation of Cloud Resources in a Dynamic Way Using an SLA-Driven Approach

S. Anithakumari and K. Chandrasekaran

Abstract Cloud computing provides a wide access to complex applications running on virtualized hardware with its support for elastic resources that are available in an on-demand manner. In cloud environment, multiple users can request resources simultaneously and so it has to be made available to them in an efficient manner. For the efficient utilization, these computing resources can be dynamically configured according to varying workload. Here in this paper, we proposed an efficient resource management system to allocate elastic resources dynamically according to dynamic workload.

Keywords Cloud computing · Service level agreement (SLA) · Resource allocation · Local manager · Universal manager · Global utility value

1 Introduction

Cloud computing has evolved as the latest computing paradigm in which computing resources can be delivered to users as services through the network (Internet) and can be acquired by the users on demand. Using cloud computing, users can dynamically use computing resources from cloud providers and they have to pay only for what they have used [1]. Usage of these cloud resources can reduce the problem of over-investment and the maintenance cost in large IT industries. Cloud technology can help industries to manage their own pool of computing resources in an efficient manner.

Computing resources are often provisioned using the virtualization technology where computing power, storage and network are encapsulated into a virtual machine (VM). Each data centre is equipped with a large set of virtual machines(VMs) built

S. Anithakumari (✉) · K. Chandrasekaran
NITK Surathkal, Karnataka, India
e-mail: lekshmi03@gmail.com

K. Chandrasekaran
e-mail: kchnitk@gmail.com

© Springer Nature Singapore Pte Ltd. 2019
A. J. Kulkarni et al. (eds.), *Proceedings of the 2nd International Conference on Data Engineering and Communication Technology*, Advances in Intelligent Systems and Computing 828, https://doi.org/10.1007/978-981-13-1610-4_42

on physical machines and has the flexibility to configure the VMs according to the diverse requirements and user applications. Multiple applications can be processed separately on dedicated VMs to reduce any conflicts that might happen when an application shares the resources with another application running on the same physical machine. Virtual machines can be migrated from one application environment to another according to the changes in the users' demand. So, we need an efficient system that allows the remote resources to be joined and used as if they were normal resources in a data centre.

Virtualization technology permits the provisioning of physical resources to applications, as the application receives the maximum capacity allotted to it, without considering the workloads generated by other applications. Virtualization helps to implement elasticity of resources by providing the flexibility of expanding or condensing the quantum of computing resources [2]. Here, we define a resource allocation decision to find out the quantum of resource capacity each VM is getting from the corresponding physical machine. This decision algorithm is explained in Sect. 4.1.

2 Related Work

The system model for on-demand provisioning of computing resources to address varying workload has been discussed in many literature. VioCluster [3] and dynamic virtual clustering [4] are such architectures proposed to borrow available resources from nearby sites. Vazquez et al. [5] proposed an architecture to extend grid infrastructure into cloud by using the GridWay meta-scheduler and different resource adapters. Murphy et al. [6] developed a dynamic provisioning system on a shared physical resource pool with Condor job scheduler. Assuncao et al. [7] developed a system which allows a user to take virtual machines from both local resource pool and cloud data centre for processing an application. Silva et al. [8] addressed the problem of finding the optimal number of virtual machines that should be provisioned to maximize the speedup under a given budget. Their proposed heuristic tries to fully utilize CPU time of virtual machines and avoid loss in the one-hour charging scheme used in Amazon EC2. The cost-based scheduling and provisioning [9, 10] policy has been discussed here.

3 Cloud Provider's Data Centre

Cloud computing environment is generally viewed as a multi-layer arrangement containing different layers such as cloud provider layer, cloud user layer and end-user layer (as shown in Fig. 1). Cloud provider layer describes the infrastructure arrangement and server organization at the cloud provider's data centre. End-user layer includes the end users, and the cloud user layer describes the interface between

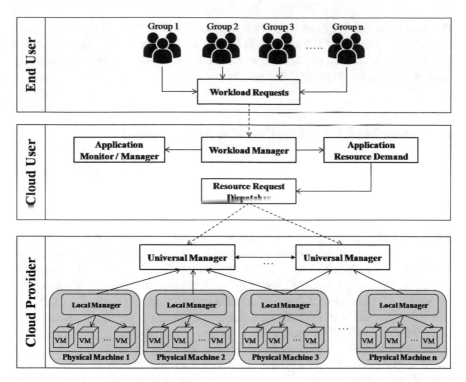

Fig. 1 Overall architecture of a data centre

cloud provider layer and end-user layer. For addressing the resource allocation problem in a dynamic way, we make use of the infrastructure organization at the cloud provider layer. The cloud provider layer contains cloud providers who provide multiple computing resources as services through a shared data centre. These computing resources are able to provide performance isolation and efficient resource sharing through virtualization technology, as per the basic feature of cloud computing. The creation of multiple virtual images from the available physical resources and the maintenance of these virtual images are done by an intermediate virtualization layer. This layer generates multiple virtual machines, on top of the physical infrastructure layer, which are isolated from one another and are capable of serving individual applications. So, these applications are also isolated like running on a dedicated machine and it uses a fraction of the entire resource capacity.

In order to proceed with the mathematical calculations, we assume a standard structure for cloud provider's data centre such as: The data centre contains a total of P physical servers, and the maximum possible VMs that can be created by all servers is taken as V. The set of applications processed by ith server is taken as A_i, and the number of VMs created on ith server is taken as n_i. That is, $\sum_{i=1}^{P} n_i = V$.

The system model for dynamic resource allocation and VM management within a single server machine is discussed in Sect. 4, and the model for global resource management, by considering all the servers in the data centre, is explored in our next paper with complete experimental analysis.

4 System Model for Adaptive VM Management

This section discusses the dynamic VM management in a single server machine by considering the server machines in a cloud provider's data centre where each VM is viewed as a single computing machine and makes use of an admission control policy [11, 12] for allotting or dropping incoming requests. As per the admission control policy, the VMs may drop some of the coming requests, because of the limitations in capacity and excess count in incoming requests. This control policy helps to address the remaining requests without affecting the assured QoS guarantees [13, 14].

The system model assumed for adaptive VM management in a single server machine is shown in Fig. 2. The major component in this model is the *allocation management* module which is to take care of all incoming workload requests and to service these requests by considering system characteristics, application's properties, VM availability and SLA contracts for maximizing the revenue of the service provider. The *allocation management* module is configured with quantitative measures of the application and SLA metrics. These measures are updated according to application change or SLA change.

The *requested workload* module is for monitoring and reading the workload requirement of each application. It contains provision to keep track of all processing applications, and accordingly it predicts the workload requirement of the current scenario. This input is forwarded to the *allocation management*. The *allocation management* decides on the allocation decision and in consultation with the *middleware control*, initiates the virtual resource mappings and generates VMs in the *virtualization layer*. The admission control policies are also taken care before the allocation of the VM images. The *allocation management* decision is based on an optimized performance model which considers SLA parameters and workload conditions of the processing applications.

The adaptive resource allocation is implemented by making frequent resource allocation decisions. The interval between two adjacent decision computing is viewed as *decision interval*, and this can be taken as a fixed value or variable according to the characteristics of the system. By choosing a smaller value for decision interval, we can make a more accurate resource allocation in a single server system. The major component in the system model is *allocation management* module because this is the module responsible for making resource allocation decision. The resource allocation decision is determined based on an optimization model by considering performance and efficiency values.

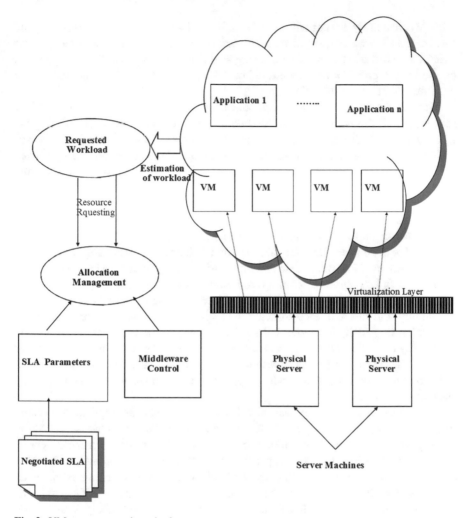

Fig. 2 VM management in a single server

4.1 Resource Allocation Decision

In our analytical system model, the resource allocation is controlled by SLA parameters and some system parameters. The estimated measures calculated from SLA parameters and applications' workload also play a role in this decision-making. The service level efficiency in cloud computing is very much dependent on SLA parameters, and so the SLA performance has related to the VM's ability to service applications by satisfying the application's response time specified in the agreement. The SLA parameters we mainly focusing are throughput(T^{TH}), response time threshold (R^{TH}), probability values(P()), and the system parameters are the total number of

VMs(V) created by the virtualization layer, the utilization value(u) of each VM and the service time average(S) of the application on a physical server. Among these values, the utilization value we are seeing as the maximum possible value provided by the service provider and the throughput value is the throughput maximum limit or throughput threshold(T^{TH}).

The *allocation management* component gets an estimate a_i (arrival rate of requests) from the *requested workload* component for each application during the next controller interval. If there occur some deviations in arrival rate from the estimated value, then a load optimizer unit is initiated and the deviations are optimized with the optimizer unit. Here, a_i is the rate of arrival requests over the considered controller interval. From these arrived requests, some may be rejected because of resource limitations and so the actual arrival rate becomes lesser than a_i. Among the processed set of requests, some may violate the agreed response time values and so they are not taken for the calculation of actual throughput, T_i.

In case of fixed controller intervals, the events which are considerably smaller than the controller interval could mislead the allocation manager such as bulk quantum of requests (with less duration), coming from some applications can stop the allocation of resources to some other class of applications because of deficiency in resource availability and this will lead to heavy penalties to the provider. For minimizing this undesired effect, the *requested workload component* provides the estimated probability (P_i) of a class of requests having higher arrival rate for the next controller interval. The parameter P_i is to represent the certainty level to assure maximum profit to the provider for VM_i, and it can be bypassed, to consider workload changes, by assigning a value 1.

To proceed with the analytical model, we assume that the application coming from a user is a unique entity and is submitted to particular VM. That is application coming from customer i is submitted to VM_i which is serviced in a mean service time S_i, and the utilization value upper limit of VM_i is u_i. In this model, each VM is eligible for a guaranteed fraction of available physical server and so we estimate the average service time by taking f_i, the fraction of service time given to VM_i, as S_i/f_i. Correspondingly, resource allocation decision is the decision-making of allocation fractions f_i ($i = 1, 2 \ldots V$) given to each VM_i. So, f_i is viewed as the important decision variable in our resource allocation problem and analytical system model.

5 Experimental Analysis

We have conducted the experimental studies in an environment simulated by arena. The quantitative results are taken from two different VMs which are running on top of the same physical infrastructure. The experimentation of the proposed resource allocation model is done with a conservative admission control policy using tokens. That is, a fixed number of tokens are ready for a fixed slot and the transactions need to acquire these tokens for getting into the system.

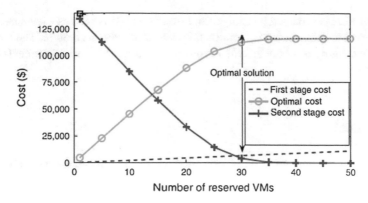

Fig. 3 VM management

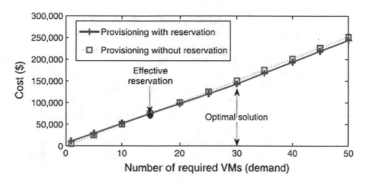

Fig. 4 VM management with reservation

The optimal allocation of created virtual machines according to varying workload is shown in Fig. 3, and the comparison between provisioning of resources within a single physical machine and a global system is illustrated in Fig. 4. The complete experimentation results are included in our next paper because of space restrictions.

6 Conclusion

We introduced a resource allocation system that is aware of the varying resource requirement in each application and adapt the system in a dynamic way. The proposed system is ready to self-organize resource provisioning according to the changing demand. We take into account both high-level performance goals of the hosted applications and objectives related to the placement of virtual machines on physical machines. An efficient way to express and quantify application satisfaction with regard to SLA is provided by using utility functions and is seen as a balanced trade-off between multiple objectives which might conflict with each other. The complete experimentation results are included in our second paper with all experimental inves-tigations.

References

1. Ali S, Jing S-Y, Kun S (2013) Profit-aware dvfs enabled resource management of iaas cloud. Int J Comput Sci Issues (IJCSI) 10:237
2. Padala P, Shin KG, Zhu X, Uysal M, Wang Z, Singhal S, Merchant A, Salem K (2007) Adaptive control of virtualized resources in utility computing environments. In: ACM SIGOPS operating systems review, vol 41, no 3. ACM, 2007, pp 289–302
3. Ruth P, McGachey P, Xu D (2005) Viocluster: virtualization for dynamic computational domains. In: IEEE international cluster computing, IEEE pp 1–10
4. Emeneker W, Stanzione D (2007) Dynamic virtual clustering. In: IEEE international conference on cluster computing (2007). IEEE pp 84–90
5. Blanco CV, Huedo E, Montero RS, Llorente IM (2009) Dynamic provision of computing resources from grid infrastructures and cloud providers. In: Grid and pervasive computing conference, (2009) GPC'09. Workshops at the. IEEE pp 113–120
6. Murphy MA, Kagey B, Fenn M, Goasguen S (2009) Dynamic provisioning of virtual organiza-tion clusters. In: Proceedings of the 2009 9th IEEE/ACM international symposium on cluster computing and the grid. IEEE computer society, pp 364–371
7. De Assunção MD, Di Costanzo A, Buyya R (2009) Evaluating the cost-benefit of using cloud computing to extend the capacity of clusters. In: Proceedings of the 18th ACM international symposium on high performance distributed computing. ACM, pp 141–150
8. Silva JN, Veiga L, Ferreira P (2008) Heuristic for resources allocation on utility computing infrastructures. In: Proceedings of the 6th international workshop on middleware for grid com-puting. ACM, p 9
9. Zhang L, Ardagna D (2004) Sla based profit optimization in web systems. In: Proceedings of the 13th international world wide web conference on alternate track papers & posters. ACM, pp 462–463
10. Salehi MA, Buyya R (2010) Adapting market-oriented scheduling policies for cloud computing. In: International conference on algorithms and architectures for parallel processing. Springer, pp 351–362
11. Perros HG, Elsayed KM (1996) Call admission control schemes: a review. IEEE Commun Mag 34(11):82–91
12. Kleinrock L (1975) Queuing systems. Wiley
13. Menasce DA, Almeida VA, Dowdy LW, Dowdy L (2004) Performance by design: computer capacity planning by example. Prentice Hall Professional
14. Papoulis A, Pillai SU (2002) Probability, random variables, and stochastic processes. Tata McGraw-Hill Education

Safe Path Identification in Landmine Area

Samarth Kapila and Parag Narkhede

Abstract Identifying the location of landmines is an important aspect considered in military and humanitarian operations in minefield regions. Advancements in technology allow to replace the sniffing animals with the autonomous machines/robots. In this paper, an autonomous robot equipped with a metal detector sensor is developed for the identification of the buried landmine. To successfully cross the minefields, a path planning algorithm is also proposed here. This algorithm requires mapping of complete navigation plane into the grid. During the motion at each location, the relative distances between the robot's location and destination along x and y directions are computed, and the algorithm decides the next movement. To check the feasibility of the proposed algorithm, it is ported on microcontroller platform and tested with the developed autonomous robot. The simultaneous checking for the presence of landmine and path planning provides the advantage of identifying the locations of landmines and a safe path to cross the minefields.

Keywords Autonomous robot · Landmine detection · Metal detector · Path planning

1 Introduction

Landmine is one of the most dangerous and widely used weapons in warfare. Landmines consist of highly explosive chemicals which explode either by contact or non-contact forces. As per the surveys, more than 100 million landmines are buried throughout the world. It is also reported that burying of landmines over 200 km of India–Pakistan border is planned by the Indian government [1].

Landmines are also laid down in the civilian and agricultural areas near international borders to increase the security of the country. The exploded landmines

S. Kapila · P. Narkhede (✉)
Symbiosis Institute of Technology, Symbiosis International (Deemed University), Pune 412115, India
e-mail: parag.narkhede@sitpune.edu.in

© Springer Nature Singapore Pte Ltd. 2019   423
A. J. Kulkarni et al. (eds.), *Proceedings of the 2nd International Conference on Data Engineering and Communication Technology*, Advances in Intelligent Systems and Computing 828, https://doi.org/10.1007/978-981-13-1610-4_43

have significant humanitarian and agricultural impacts. Hence, detection of land-mine plays a vital role in the humanitarian and military operations. Use of animals like sniffing dogs and rats is the traditional way of landmine detection [2]. How-ever, with the advancement in technology, automatic sensor-based systems are being employed in the landmine search operations. These sensor-based systems consider different characteristics of the landmine for detecting their presence.

Minefield generally consists of a number of rows of buried landmines and can be considered as a matrix of buried landmines. The successful and safe crossing of the minefields requires the knowledge of landmine locations. Knowing the locations of landmines, a safe area in the minefield can be identified and a path for crossing the minefield can be planned successfully.

Autonomous robots equipped with landmine detection mechanism are becoming important as the danger and manual detection cost is getting significantly reduced [3, 4]. Many researchers are putting their efforts to solve the problem of development of an autonomous robot for identification of buried landmine [4, 5]. Robots provide the advantage of efficient landmine search due to their lightweight. They generally pro-vide low pressure in buried landmines and does not allow landmine to get triggered. Landmine detecting COMET-II and COMET-III robots with six degrees of freedom are developed by Kenzo Nonami [6].

This paper considers the problem of identification of a safe path in the mine-fields. The considered problem is majorly divided into two parts: (1) identification of landmine location and (2) planning of the safe path in the minefield. The land-mine detection is addressed by developing the metal detector-based system, whereas a safe path identification (SPI) algorithm is proposed to solve the problem of path planning. The paper discusses the development of the prototype of the autonomous robot equipped with metal detector sensor and SPI algorithm.

The complete paper is organized as follows: Sect. 2 provides the different landmine detection technologies along with path planning techniques. Section 3 provides the hardware and software methodologies followed in the paper. Section 4 provides the results, and Sect. 5 concludes the paper.

2 Literature Overview

2.1 Landmine Detection

A wide research is currently going on in the detection of landmines. Researchers are experimenting on a number of different techniques to identify the presence of landmines.

Metal detector (MD) sensor is employed by many researchers in landmine detec-tion application [2, 4, 7]. MD sensor consists of a coil which is responsible for the generation of electromagnetic field in the nearby area. The presence of metallic object affects this induced electromagnetic field creating a signature for landmine detection [8]. Nowadays, metallic mines are getting outdated and plastic mines are being developed. Even the plastic mines contain some metallic part. Hence, identifying the metal-based detection can serve for landmine detection.

Ground-penetrating radar, thermal imaging, acoustic and millimetre waves are also used in the landmine detecting systems [9–11]. However, their performance is highly dependent on the environmental conditions. Also, the systems equipped with these detection technologies are larger in size and may not provide a low-cost solution [12]. MD sensor is cheaper and small in size and weight and hence can be the reason for being popular for landmine detection. Abeyanake et al. developed a Kalman filter-based landmine detection system employed with an array of metal detector sensors [7]. They have observed that system with multiple sensors provides better results as compared to the system with single sensor.

2.2 Path Planning

Path planning is a technique that decides a possible way from source to destination. It can basically be of two types: global planning and local planning [13]. Global planner considers the previously available knowledge of the environment and finds the optimal trajectory. However, in practice, they are computationally expensive. Local planners plan only a few steps in the near future based on current sensory data. Such planners require less computational power and are good choices for the robot with the limiting sensing abilities. However, due to limited planning, the robot may get stuck at some location and not reach the target. Thus, the problem of global convergence to the destination needs to be addressed while considering a local planner system.

Lots of approaches for robot path planning have been proposed and implemented in the literature [13–15]. Grid-based navigation [13] and interval-based navigation [15] are two of the widely used path planning techniques. An important distinction between them is in how they represent the real world.

Gonzalez et al. discussed the various motion planning techniques employed for autonomous robots [16]. They have provided the comparative analysis for the graph-based planners, sampling-based planner, interpolating curve planners and numerical optimization planners. Interpolation-based search algorithms are the most widely used techniques due to their enhanced ability to generate required coordinate points with the help of GPS data. When it comes to real-time implementation, graph-based techniques are preferred due to their fast search operations.

Grid-based navigation is extensively studied and explained by Balch [13]. In grid-based navigation, a complete navigation plane is divided into a grid, generally referred as occupancy grid, and each block of grid is logically numbered in the form of (x, y) coordinates, where x and y represent the coordinates along x and y directions, respectively. When a robot moves, its location in terms of (x, y) coordinates is updated. When the robot is present at a location (x, y), then in a grid, its position $[x, y]$ is marked as full. If a robot moves in forward direction, then the new location is indicated by $(x, y+1)$ and in a grid $[x, y+1]$ is marked as full and $[x, y]$ is marked as empty. The grid's resolution refers to how large an area in the real world is represented by one cell in the grid.

The performance of the grid-based navigation planning depends on the size of each block in grid. Larger the size of the block less is the resolution, which may result in less optimal path. So, the selection of resolution is an important aspect in grid-based navigation. The block size should be decided based on the accuracy of the position sensors and robots movement speed. High-resolution grid may represent accurate positions of the robot and obstacles, helping the optimized trajectory planning.

3   Design Methodology

This section discusses the hardware and software designs of landmine detection robot along with the design of safe path identification algorithm.

3.1   Landmine Detection Robot

The basic block-level description of the landmine detecting autonomous robot can be depicted as in Fig. 1. It consists of metal detector sensor and GPS unit as input devices, a processing unit and motors as output devices. Sunroms 1139 metal detector sensor is used to identify the metallic objects/landmine buried underground. The sensor provides the digital output when metallic object comes under its proximity. This sensor requires 5 V, 50 mA power supply for the operation. The digital signal obtained from metal detector is provided to the digital pin of Arduino for alarming. A GPS module is used to locate the position of the robot in the grid. Arduino Uno board is used for carrying out necessary computations and taking the decision for safe motion. Uno board is equipped with ATmega 328 microcontroller which is a low-power, 8-bit microcontroller and is widely used in the robotic applications. The motion of the robot is controlled by DC motors connected to it.

Fig. 1 Block diagram of landmine detecting robot

3.2 Safe Path Identification Algorithm

Algorithm plays an important role in robotic development. SPI algorithm focuses on the planning of safe path in the minefield. The stated algorithm is developed with consideration of few assumptions listed below:

– The size of robot is equal to the size of each block in grid.
– The minefield contains only landmines, and there are no other obstacles present.
– If landmine explodes, then the size of affected area is same as size of a block of a grid.
– The robot can move only along any one axis at a time.
– In one step, robot's location is only incremented by 1.

With these assumptions, the safe path identification algorithm can be stated as:

Step 1 Provide the destination coordinates to the robot (d_x, d_y).
Step 2 Measure the coordinates of robot using GPS, and locate them into considered grid let (s_x, s_y).
Step 3 Find the relative distance between source and destination along x and y directions as

$$\Delta X = d_x - s_x; \quad \Delta Y = d_y - s_y \qquad (1)$$

Step 4 Find the fractional ratios along x and y directions as

$$\Delta x = \Delta X / HCF(\Delta X, \Delta Y); \quad \Delta y = \Delta Y / HCF(\Delta X, \Delta Y) \qquad (2)$$

where HCF stands for highest common factor.
Step 5 If $\Delta X > \Delta Y$, then move the robot along x-direction; otherwise, move the robot along y direction and update the robot location in (s_x, s_y).
Step 6 Identify the presence of landmine using metal detector.
If present, note the location as danger and revert the robot to recent safe location.
Else note the location as safe, and repeat from step 2 until robot reaches destination
Step 7 The safe path can be given by combining all the points indicated by safe.

The SPI algorithm determines the safe path in the minefield which can be used by humans, militants and vehicles to safely cross the minefields. The computation of fractional ratios provides the advantage of identifying the possible short path between source and destination.

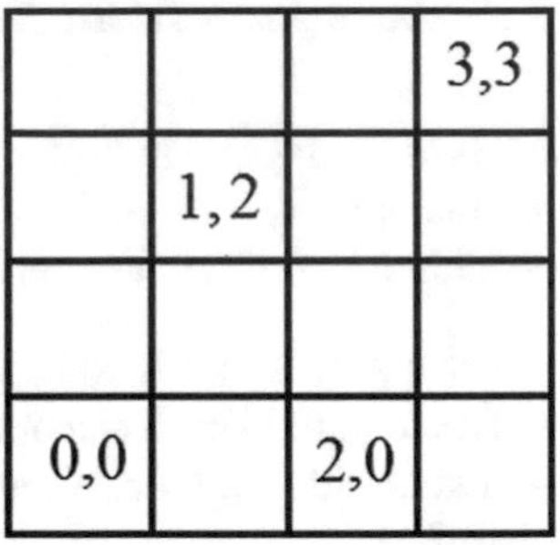

Fig. 2 Coordinates allocation in grid

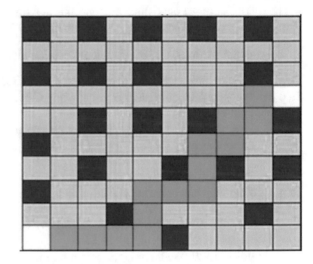

Fig. 3 Simulation results for SPI algorithm

4 Results and Discussion

The development of landmine detecting autonomous robot is carried out in two phases. In the first phase, SPI algorithm is developed in MATLAB and offline simulations are carried out to check its feasibility. In the second phase, the SPI algorithm is ported to the real hardware platform and tested with autonomous robot.

For the offline simulations, a grid of 10×10 size is taken into consideration. The locations in the grid are logically numbered as coordinates in two-dimensional matrix. A sample 4×4 grid is shown in Fig. 2.

The offline simulation result of the SPI algorithm is shown in Fig. 3. In the grid, landmines are assumed to be present at some locations; locations are shown in blue, whereas the white blocks represent the source and destination. Yellow blocks represent the safe blocks in grid. In Fig. 3, (0, 0) and (9, 6) are considered as starting point and destination points, respectively. It can observed that distance along x-direction (ΔY) is more than y (ΔY), and hence, the robot's motion is started along x and robot reaches to (0, 1). At this point, it checks for the presence of landmine, also calculates, and then decides the next movement and the process continues. When

Fig. 4 Developed
autonomous robot

robot reaches (3, 0), ΔX is equal to ΔY; i.e. distances along x and y are same and
hence change the direction to y and move to point (3, 1). But at (3, 1) landmine is
detected and robot returns back to recent safe location, i.e. (3, 0) and continuous
motion along x. Blocks in green colour represent the safe and short path identified
by the proposed SPI algorithm. Landmines identified by robot while moving from
source to destination are denoted by red blocks in grid. During the simulations, it is
assumed that width of path and the width of robot is same; also, there does not exist
any type of obstacles in the path.

The hardware setup of the SPI algorithm-enabled autonomous robot is shown in
Fig. 4. During hardware testing, the grid is prepared on a floor and small metal parts
are used in the place of landmines. From the results, it can be stated that the proposed
algorithm is suitable to find the safe path while travelling from the source location
to destination which can be used to cross the region where landmines are buried.

5 Conclusion

The development of the metal detector-based landmine detecting autonomous robot
is discussed in this paper. The path planning algorithm featured with safe path identification is also discussed in this paper here. The proposed SPI algorithm is ported
on hardware prototype developed with an autonomous robot. The tests are carried
out with offline simulations and using developed autonomous robot. The considered
metal detector sensor was able to successfully identify the metallic objects; however,
it is unable to discriminate between the landmine and metal objects. The proposed
algorithm also provides the shorter path between source and destination. The future
work will focus on removing the assumptions considered while developing the SPI
algorithm to get more robust results.

References

1. Monitor. India mine action report, Accessed March 2017, Available online at http://www.the-monitor.org/en-gb/reports/2017/india/mine-action.aspx
2. Bruschini C, Gros B (1997) A survey of current sensor technology research for the detection of landmines. In: Proceedings international workshop on sustainable humanitarian demining
3. Acar E et al (2001) Path planning for robotic demining and development of a test platform. In: International conference on field and service robotics
4. Robledo L, Carrasco M, Mery D (2009) A survey of land mine detection technology. Int. J. Remote Sens. 30(9):2399–2410
5. Cassinis R et al (1999) Strategies for navigation of robot swarms to be used in landmines detection. In: Advanced Mobile Robots, 1999. (Eurobot'99) 1999 Third European Workshop on. IEEE
6. Nonami K et al (2002) Development of mine detection robot Comet-II and Comet-III. In: The proceedings of the international conference on motion and vibration control 6.1. 2002. The Japan Society of Mechanical Engineers
7. Abeynayake C, Chant I (2001) A Kalman filter-based approach to detect landmines from metal detector data. In: Geoscience and remote sensing symposium, 2001. IGARSS'01. IEEE 2001 International. IEEE
8. Keeley R (2003) Understanding landmines and mine action. Mines Action Canada
9. Schavemaker J, Cremer F, Schutte K Den Breejen E (2000) Infrared processing and sensor fusion for anti-personnel land-mine detection. In: Proceedings Of IEEE student branch eindhoven: symposium imaging pp 61–71
10. Milisavljevic N, Bloch I (2003) Sensor fusion in anti-personnel mine detection using a two-level belief function model. In: IEEE transactions on systems, man, and cybernetics. Part C (Applications and reviews) vol 33, pp 269–283
11. Xiang N, Sabatier JM (2000) Land mine detection measurements using acoustic-to-seismic coupling. Signal 5:10
12. MacDonald J et al (2003) Alternatives for landmine detection. RAND CORP, Santa Monica CA
13. Balch T (1996) Grid-based navigation for mobile robots. The Rob. Practitioner 2(1):6–11
14. Seward D, Pace C, Agate R (2007) Safe and effective navigation of autonomous robots in hazardous environments. Autonomous Robots 22(3):223–242
15. Kieffer M et al (2000) Robust autonomous robot localization using interval analysis. Reliable Comput. 6(3):337–362
16. Gonzlez D et al (2016) A review of motion planning techniques for automated vehicles. IEEE Trans. Intell. Trans. Sys. 17(4):1135–1145

Cryptography Algorithm Based on Cohort Intelligence

Dipti Kapoor Sarmah and Ishaan R. Kale

Abstract Information security is very important in the current era as we share most of the information through digital media/Internet. Cryptography is a technique which converts the secret information into some other form which is referred to as ciphertext. In order to have a strong ciphertext, one should have a strong cryptography algorithm as well as the secured key. The performance of any cryptography algorithm can be measured through the secret text input file size (number of bytes), performance time to encrypt input file and how fast it can be retrieved the secret text through cryptanalysis attack. In this paper, a socio-inspired optimization algorithm referred as cohort intelligence (CI) is applied to the secret text to retrieve the optimized ciphertext. The efficiency of the algorithm is also analyzed in this paper with respect to the secret text capacity and time.

Keywords Cryptography · Cohort intelligence · Secret text capacity
Estimated time

1 Introduction

Nowadays, Internet dependency is widely increased for most of the applications such as email communication, ecommerce, social media. Due to this, all the respective tasks have become easier for human. At the same time, the security issues with respect to these applications have extensively increased. In order to protect the information during network communication, one of the important sciences is referred to as cryptography. Cryptography has been used to hide the secret text or to encrypt the secret text from readable to unintelligible form. In terms of cryptography, the secret text is named as plaintext and the unintelligible form is called as ciphertext.

D. K. Sarmah (✉) · I. R. Kale (✉)
Symbiosis Institute of Technology, Symbiosis International University, Pune 411042, India
e-mail: dipti.sarmah@sitpune.edu.in

I. R. Kale
e-mail: ishaan.kale@sitpune.edu.in

© Springer Nature Singapore Pte Ltd. 2019
A. J. Kulkarni et al. (eds.), *Proceedings of the 2nd International Conference on Data Engineering and Communication Technology*, Advances in Intelligent Systems and Computing 828, https://doi.org/10.1007/978-981-13-1610-4_44

To ensure the security of secret messages, the encryption process needs to be more complex so that the cryptanalysts are unable to decrypt the plaintext.

Cryptography techniques are formally divided into two different categories: (1) private key cryptography and (2) public-key cryptography. Various traditional cryptographic techniques were introduced such as Vernam cipher method, stream cipher cryptosystem, fast and secure stream cipher, RC4 stream cipher. However, these techniques are not very secure. There are some other known private and public-key cryptography algorithms such as data encryption standard (DES), triple DES, advanced encryption standard (AES), Diffie–Hellman key exchange algorithm, and RSA. Though the security level of this algorithm is high, however, the estimated time for converting plaintext to secret text is quite large due to its high complexity. In order to diminish these limitations, many researchers have introduced a nature and socio-based optimization techniques such as genetic algorithm (GA) [1–4], particle swarm optimization (PSO) [5, 6], ant colony optimization (ACO) [7] in various cipher methods [8, 9].

Formerly, GA was implemented by Coppersmith [1] using end-to-end security mechanism incorporated with threshold cryptography and Diffie–Hellman key exchange method to avoid the mobile network nodes. GA provides robust security to validate the nodes entering into the network. GA was also provided to break the mono-alphabetic substitution cipher [3] and to make the network more secure so that the cryptanalyst does not alter the original information/facts. Cryptanalyst can easily decode the information as they know well which ciphertext will be suited to decrypt the information. Therefore, Paul et al. [2] utilized GA incorporated with crossover and mutation which assist to make new ciphertext. Similar GA approach was implemented by Sindhuja and Devi [4] using symmetric key encryption technique for encryption and decryption. Sreelaja and Paib (2009) presented a distributed and decentralized swarm intelligence and ACO [10] approach for encryption. The data encryption standard (DES) was used in PSO to identify the plaintext from ciphertext [6]. This approach was based on two outputs such as particle best and error best, whereas the optimal solution is obtained by identifying the least error. Similar to these techniques, an emerging socio-inspired cohort intelligence (CI) algorithm incorporated with cipher method using the practice of mathematics and computer science is applied to encrypt the secret message. The CI algorithm was proposed by Kulkarni et al. [11] and successfully validated in health care and logistics [12], discrete and mixed variable from truss structure and design engineering domain [13], steganography [14], traveling salesman problem (TSP) [15], and various work domains. Though CI has been applied and validated in different engineering domain [16], however, it is untouched to information security domain yet. Thus, an effort has been put to apply and validate CI in this area which motivated us to develop a new cryptography algorithm.

There are totally four sections proposed in this paper. The organization of this paper is as follows: Proposed work is given in Sect. 2, Results and analysis are done in Sect. 3, conclusion and future scope are mentioned in Sect. 4, and references are written at the end.

2 Proposed Work

In this paper, a novel cryptography algorithm is proposed which is based on a socio-inspired optimization algorithm, i.e., cohort intelligence (CI). The main idea of this algorithm is based on a cohort. Cohort refers to the number of candidates competing with each other to improvise their own behavior. Each candidate is having their own abilities or potentials. During competition, each candidate tries to adapt the few qualities of other candidates or itself to make themselves as a better candidate which enables the cohort to improve its overall behavior. Though CI algorithm is validated to solve different test problems and real-world datasets, however, this algorithm is not applied yet in the information security domain especially in cryptography. Cryptography is one of the important sciences in information security which converts the secret text referred to as plaintext to the transformed text referred to as ciphertext. This is useful to make the secret text secure from different cryptanalysis attacks. There are many cryptographic algorithms available such as advanced encryption standard (AES), data encryption standard (DES), triple DES, Blowfish. AES is found more secure than the other mentioned methods; however, this algorithm is more complex and is dependent upon the number of rounds. Also, the computation time is very large to convert the plaintext to ciphertext. Due to this motivation, CI has been considered for this proposed work. The objective function of the proposed work is selected as the time function as:

$$f(t) = t \tag{1}$$

The flowchart of proposed work is presented in Fig. 1, and it is explained in the following steps.

Step 1: Accept the secret text. For example:

Secret text: "Cryptography"

Step 2: Each character is converted into binary digits such as if the first character of the secret text, i.e., "C" is considered, the binary digits are: 01000011
Step 3: Since there are 8 bits in every character, the following pair of bits are united to make a decimal number. For example:

   (i)   First and third pair bits: (00);
  (ii) Second and fourth pair bits: (10);
 (iii) Fifth and seventh pair bits: (01);
 (iv) Sixth and eighth pair bits: (01);

Step 4: The respective pair of bits are converted into decimal numbers. Thus, totally four decimal numbers are generated. As referred to step 3, the decimal numbers are:

0 2 1 1.

Step 5: In this step, the CI algorithm is applied which considers the cohort as a group of four numbers of candidates. These numbers of candidates are represented

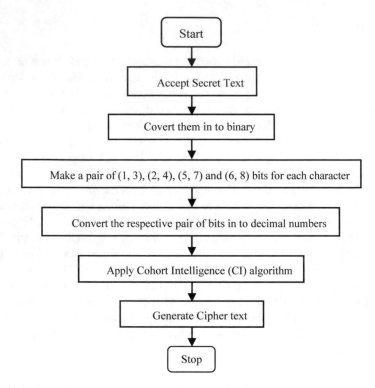

Fig. 1 Ciphertext generation using cohort intelligence algorithm

as an identity matrix and its different forms. As discussed in Step 4, the total evaluated decimal numbers are 4 for each character; thus, the dimension for each candidate is also considered as 4. The number of matrices by using 4×4 dimension by considering each pair of bits is 4! i.e.64. In order for experimental purpose, we have considered totally four candidates in this paper which are referred to as $B_1(i, j)$, $B_2(i, j)$, $B_3(i, j)$, $B_4(i, j)$ where i denotes the number of rows from 0 to 3 and j denotes the number of columns from 0 to 3. The quality of each candidate is considered as the row number where the matrix value is 1, and it is replaced with the column number to identify the cipher decimal value. For example: if $B(0, 2) = 1$, then 2 is replaced for the decimal number 0. Thus, each decimal number is having its corresponding value when it passes through different matrices. This helps to generate four different ciphertexts. In order to find the optimized ciphertext, the time function $f(t)$ is evaluated every time. One should always have less evaluation time to process this algorithm and to convert the plaintext into ciphertext. Since we have applied the same process for four different candidates, there will be four different evaluation time. To get the optimized one, the following points are considered:

a. Let us consider the computation time for the secret text to be converted it to ciphertext is $t1$, $t2$, $t3$ and $t4$.

b. Calculate the probability for computation time for each candidate; i.e.,

$$p_1 = \frac{t1}{t1 + t2 + t3 + t4}, \, p_2 = \frac{t2}{t1 + t2 + t3 + t4},$$
$$p_3 = \frac{t3}{t1 + t2 + t3 + t4}, \text{ and } p_4 = \frac{t4}{t1 + t2 + t3 + t4} \tag{2}$$

c. Calculate the cumulative probability.
d. Apply Roulette wheel approach to evaluate the candidate to be followed by other candidate.
e. To adapt the quality of the candidates, a random number from 1 to 4 is generated with respect to every candidate. This number designates the row of the following candidate to be replaced with the same row of the candidate being followed. This enables to generate a new cohort with 4 new candidates and each candidate may have different qualities.
f. The same process from step a. to step e. will be repeated till 100 times unless the saturation condition is reached. The total number of 20 runs is considered for experimental purpose.
g. The saturation condition exists where no further improvement in qualities of the candidate is identified.

Results and discussions are done in the next section. The size of the secret text and time is considered in the result section.

3 Results Analysis

Results are described in this section. The proposed work was coded in MATLAB (R2011b), and the simulations were run on Windows platform using an Intel(R) Core(TM)2Duo, 2.93 GHz processor speed, and 4 GB RAM. Time analysis with respect to the size of secret text is completed. Totally four cases are considered to capture the minimum time/optimize time taken by the candidate. Also, the total number of function evaluations and the total number of iterations of saturation condition are evaluated for a single run. Totally 20 runs are considered in the proposed work as described in the previous section. Based on the size of secret text, we have considered four cases, i.e., Case 1 for 48 bytes, Case 2 for 264 bytes, Case 3 for 480 bytes, and Case 4 for 3280 bytes. Table 1 depicted the analysis of the obtained solution with respect to evaluated parameters. We have considered five parameters for analyzing our solution. These parameters are standard deviation, minimum time, maximum time, average time, and standard deviation of function evaluations. We could see in Table 1 that the values are increased as we enlarge the secret text size for the different cases. The values against the parameters are considered with respect to 20 runs. These parameters are minimum time, iteration number on which the candidates are converged, and the number of function evaluations. As we could see from Table 1

Table 1 Result analysis of obtained solution

Parameters ↓ Cases →	Case 1	Case 2	Case 3	Case 4
Standard deviation	0.000227	0.00017	0.006735	0.000603
Min time	0.007598	0.035767	1.040855	0.092433
Max time	0.008545	0.036417	1.065062	0.094605
Average	0.008258	0.03603	1.049098	0.093233
Standard deviation function evaluations	94.72753	107.145	131.3829	124.5473

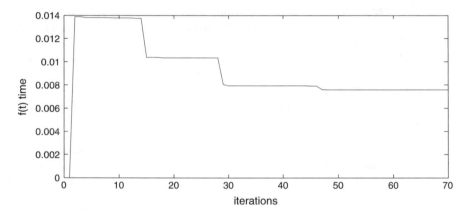

Fig. 2 Convergence plot for optimum solution 48 byte

that the standard deviation of time taken to convert 48 bytes of string (Case 1), i.e., plaintext to cipher text, using CI is quite less and the standard deviation of function evaluation is also 94.72753 which indicates that the proposed method is pretty fast. If we increase the size of secret text from 48 to 264 bytes, Case 2 is considered. We could observe that the standard deviation of time taken from Case 1 to Case 2 is decreased from 0.000227 to 0.00017. The calculated standard deviation for Case 3 is 0.000603, and the standard deviation of function evaluations is increased from Case 2 to 124.5473. Again, the secret text size is increased to 3280 bytes to see the change in the previous cases, and it is observed that the standard deviation of time taken and standard deviation of function evaluations are increased than the previous cases; however, this increment is not very substantial. Convergence plots are also considered and shown for each case as shown in the plot section.

Plots are also captured against each case as shown from Figs. 2, 3, 4, and 5. As shown in each plot, two dimensions are considered. X-axis describes the total number of iterations, and Y-axis describes the time taken. In every plot, the converged value of iteration and its corresponding captured time could be seen. In Figs. 2, 3, 4, and

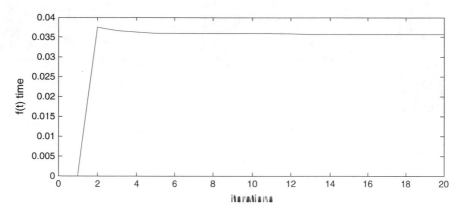

Fig. 3 Convergence plot for optimum solution 264 byte

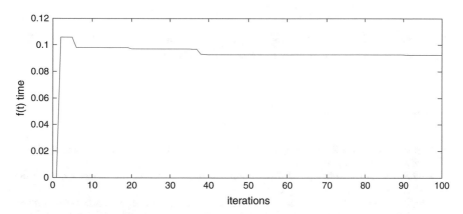

Fig. 4 Convergence plot for optimum solution 480 byte

5, one could observe that as we increase the size of the secret text, the encryption time also gets increased.

4 Conclusion and Future Scope

Due to its results as shown in the previous section, the computation time of the proposed work is very less. Even if we increase the size of the secret text to 3280 bytes, the evaluated parameters are found better than the other similar type of algorithms. Also, this method is found very secure because cryptanalyst needs to identify the optimized matrix as well as the optimized secret bits which makes the entire algorithm more secure. However, the combination of this algorithm and the other existing

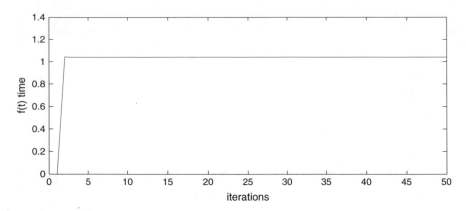

Fig. 5 Convergence plot for optimum solution 3280 byte

algorithms can also be tried to make the algorithm more secure and complex. This opens a new door for researchers to work in this direction.

References

1. Coppersmith D (1994) The data encryption standard (DES) and its strength against attacks. IBM J Res Dev 38(3):243–250
2. Paul S, Dutt I, Choudhri SN (2013) Design and implementation of network security using genetic algorithm. Int J Res Eng Technol 2(2):172–177
3. Omran SS, Al-Khalid AS, Al-Saady DM (2010) Using genetic algorithm to break a mono-alphabetic substitution cipher. In: 2010 IEEE conference open systems (ICOS), pp 63–67
4. Sindhuja K, Devi PS (2014) A symmetric key encryption technique using genetic algorithm. Int J Comput Sci Inf Technol 5(1):414–416
5. Abdul Halim MF, Bara'a, AA, Hameed SM (2008) May. a binary particle swarm optimization for attacking knapsacks cipher algorithm. In: International Conference Computer and communication engineering, 2008. ICCCE 2008, pp 77–81
6. Pandey S, Mishra M (2012) Particle swarm optimization in cryptanalysis of DES. Int J Adv Res Comput Eng Technol (IJARCET) 1(4):379
7. Khan S, Shahzad W, Khan FA (2010) Cryptanalysis of four-rounded DES using ant colony optimization. In: 2010 International conference information science and applications (ICISA), pp 1–7
8. Biham E, Seberry J (2005) Py (Roo): a fast and secure stream cipher using rolling arrays. IACR Cryptology ePrint Archive, pp 155
9. Kim H, Han J, Cho S (2007) An efficient implementation of RC4 cipher for encrypting multimedia files on mobile devices. In: Proceedings of the 2007 ACM symposium on applied computing pp 1171–1175
10. Sreelaja NK, Pai GV (2012) Stream cipher for binary image encryption using ant colony optimization based key generation. Appl Soft Comput 12(9):2879–2895
11. Kulkarni AJ, Durugkar IP, Kumar M (2013) Cohort intelligence: a self-supervised learning behavior. In: 2013 IEEE international conference, systems, man, and cybernetics (SMC), pp 1396–1400

12. Kulkarni AJ, Baki MF, Chaouch BA (2016) Application of the cohort-intelligence optimization method to three selected combinatorial optimization problems. Eur J Oper Res 250(2):427–447
13. Kale IR, Kulkarni AJ (2017) Cohort intelligence algorithm for discrete and mixed variable engineering problems. Int J Parall Emergent Distributed Syst 1–36
14. Sarmah DK, Kulkarni AJ (2017) Image steganography capacity improvement using cohort intelligence and modified multi-random start local search methods. Arabian J Sci Eng 1–24
15. Kulkarni AJ, Krishnasamy G, Abraham A (2017) Cohort intelligence: a socio-inspired optimization method. Springer, Heidelberg, Germany
16. Dhavle SV, Kulkarni AJ, Shastri A, Kale IR (2016) Design and economic optimization of shell-and-tube heat exchanger using cohort intelligence algorithm. Neural Comput Appl 1–15
17. Sreelaja NK, Vijayalakshmi Pai GA (2011) Swarm intelligence based key generation for stream cipher. Security Commun Networks 4(2):181–194

Dual-Port Multiband MSA for Airborne Vehicular Applications

Sanjeev Kumar, Debashis Adhikari and Tanay Nagar

Abstract This paper presents the design of a dual-port microstrip patch antenna for telemetry (2.2 Ghz) and GPS (1.5 Ghz) applications. The proposed antenna is suggested to be used in a low-altitude space rocket or satellite. Any data collected from the rocket would be sent to the base telemetry station through the 2.2 Ghz frequency band. The GPS band would be used to monitor the location of the spacecraft and to collect the information of its trajectory. CST Microwave Design Studio software has been used to simulate the design of the said antenna. The analysis of return loss, VSWR, gain, and radiation pattern was carried out. The proposed antenna shows return loss of -19.23 dB at 2.2 GHz and -20.31 dB at 1.5 GHz (both are orthogonally polarized to each other) which implies good results. The impedance matching is good at the desired frequencies with VSWR <2, respectively. The overall simulation results show that the antenna worked well at the desired two frequencies, hence making the antenna suitable for use. This antenna is implemented on FR4 epoxy dielectric substrate with relative permittivity $\epsilon_r = 4.3$ and thickness of the substrate $(h) = 1.6$ mm.

Keywords Dual-port microstrip patch antenna · Orthogonal polarization Telemetry · GPS

S. Kumar (✉)
Symbiosis Institute of Technology, Symbiosis International University, Pune, India
e-mail: Sanjeevkumar@sitpune.edu.in

D. Adhikari
MIT Academy of Engineering, Pune, India

T. Nagar
Software Division, L & T Infotech, Navi Mumbai, India

© Springer Nature Singapore Pte Ltd. 2019   441
A. J. Kulkarni et al. (eds.), *Proceedings of the 2nd International Conference on Data Engineering and Communication Technology*, Advances in Intelligent Systems and Computing 828, https://doi.org/10.1007/978-981-13-1610-4_45

1 Introduction

The need to save space on the chipboard of an airborne aircraft has always been paramount. The airborne applications of wireless communication require an antenna to be operated with more than one frequency, and this necessitates a dual-band operation. The design of such antennas for communication application is a tough challenge by itself.

Microstrip patch antennas, with their obvious advantage of low profile, low weight, conformability, and ease of integration, become a suitable choice for airborne applications. In this paper, a dual-band rectangular microstrip patch antenna for telemetry and GPS application is designed and simulated using CST Microwave Studio. The proposed patch antenna resonates at 2.2 GHz (telemetry) and 1.5 GHz (GPS) frequency.

The method employed to feed the proposed antenna is the discrete port excitation technique. This has an advantage that the feed can be placed at any desired position inside the feed line in order to obtain a suitable impedance matching. In the proposed design, two feed points are chosen for the antenna to operate at the two designated frequencies. To minimize interference and obtain good isolation between the two ports, one feed point is centrally fed and the other feed point is edge corner-fed. The centrally fed port attains linear polarization, and the edge corner-fed port attains circular polarization which is orthogonal to one another other resulting in good port isolation.

Polarization diversity has been employed to overcome the limitations of space and has been obtained by colocating the orthogonal polarization on the same patch. Polarization diversity requires less space compared to physically separate antennas [1, 2]. The task at hand is to obtain sufficient isolation between the ports [3] and, at the same time, striving to maintain good matching of impedance and polarization sense [4].

The use of dual-polarized antenna [5–8] for several applications has been extensive. This paper proposes an antenna able to excite two orthogonal polarizations simultaneously. The ports are excited through separate feeds with the radiating structure being the same, and the design does not require any extra dimensions. Compared to [9], a good isolation between the ports is reported.

2 Design Methodology

The characteristics of the microstrip patch antennas are defined mainly by their geometries and the material properties from which they are made of. The designed rectangular microstrip patch antenna has dimensions of 62.42×58.38 mm. The FR4 epoxy substrate has been chosen with a dielectric constant $(_r)$ of 4.3, dielectric loss tangent of $=0.002$, and a thickness of 1.6 mm. The chosen value of $_r$ gives better efficiency. Besides, it is required that the substrate material be flexible enough for it to wrap around the curved surface of a missile or that of a spacecraft. A much higher value of $_r$ can significantly reduce the antenna's radiation efficiency and also its bandwidth.

The design presented in the paper consists of an active radiating patch on one side of a dielectric substrate and the ground plane on the other. For multiband purpose, two feed points are chosen with orthogonal polarization.

The first step was designing the microstrip patch antenna operating at 2.2 GHz. Once it was achieved, different configurations of the second feed were tried out. After a few trials, Port2 resonated at 1.5 GHz when it was a diagonal corner-fed.

In the above design, two feed points are chosen for the antenna to operate at two frequencies [10]. To achieve minimum interference between the two ports and to attain good port isolation, one feed point is centrally fed and the other feed point is edge-fed [11, 12]. As a result of this, the centrally fed port attains linear polarization and the corner-fed port attains circular polarization. Both these types of polarization are orthogonal to each other, thus attaining a good port isolation.

3 Feeding Techniques

The various feeding techniques can be classified into two main categories, namely contacting and non-contacting. In the former method, power to the radiating patch is fed directly using a connecting line. A microstrip line can be chosen as an example. In the latter non-contacting method, we resort to coupling of the electromagnetic field between the microstrip line and the radiating patch. The techniques being practically employed are the microstrip line and coaxial probe for the direct contact method. For non-contacting techniques, aperture coupling and proximity coupling are employed.

The designed microstrip patch antenna incorporates a microstrip feed line to feed the RF waves to the radiating element. Microstrip feed line is easy to fabricate, and it is also easy to match the impedance of the feed line with the patch by adjusting the inset position. At times when the substrate thickness is large, surface waves and spurious feed radiation increase which results in the loss of bandwidth. But for the above design, microstrip feed line has been found suitable. Here, the conducting microstrip is directly connected to the patch antenna at its edge. The conducting strip has been designed thinner, compared to the patch. The advantage that incurs is that the feed can be etched on the same substrate in conformation to the planar structure.

4 Physical Parameters of Antenna

The different parameters of this antenna have been calculated by the transmission line method [13–18], as reflected in Table 1.

Step 1: Width of the Patch

The width is determined by

Table 1 Parameter list

Serial number	Parameter	Value(mm)
1	Length of patch (L_p)	40.71
2	Width of patch	31.14
3	Length of substrate	62.42
4	Width of substrate	58.38
5	Length of inset feed_1	9.9
6	Length of feed line_1	26.83
7	Length of inset feed_2	6.07
8	Length of feed line_2	22.54
9	Thickness of substrate	1.6
10	Thickness of patch	0.1
11	Thickness of ground	0.1

$$W = \frac{c}{2 f_o \sqrt{\left(\frac{\epsilon_r + 1}{2}\right)}} \tag{1}$$

where c is velocity of light, f_0, the resonant frequency, and ϵ_r, the relative dielectric constant.

Step 2: Knowing patch width, effective permittivity is determined as:

$$\epsilon_f = \frac{\epsilon_r + 1}{2} + \frac{\epsilon_r - 1}{2}\left[1 + 12\frac{h}{W}\right]^{-1/2} \tag{2}$$

where h is height of the dielectric substrate.

Effective permittivity must be taken into consideration the presence of fringing fields. These are a part of the electric field lines that are partly in the substrate dielectric and partly in the air and are the main contributors to radiation from the patch.

Step 3: Length extension of patch

Due to fringing effects, the electrical length patch of the microstrip antenna looks greater than its physical dimensions. This phenomenon will be represented by ΔL, which is a function of the effective dielectric constant ϵ_f and the width-to-height ratio (W/h) found as:

$$\Delta L = h(0.412)\frac{(\epsilon_r + 0.3)\left(\frac{w}{h} + 0.264\right)}{\left(\epsilon_f - 0.258\right)(h + 0.8)} \tag{3}$$

Step 4: Effective length of the patch

As length of the patch has been extended by ΔL, the effective length is found as:

$$L_f = \frac{c}{2 f_o \sqrt{\epsilon_f}} - 2\Delta L \tag{4}$$

Step 5: Calculation of total characteristic impedance of the microstrip transmission line

$$Z_l = \sqrt{Z_0 Z_{in}} \tag{5}$$

where antenna is matched to $Z_0 = 50\Omega$

Z_{in} is the input impedance of the patch antenna.

Step 6: Calculation of inset feed length of the microstrip patch antenna

$$R = \frac{L}{\pi} \cos^{-1} \sqrt{\frac{Z_0}{Z_l}} \tag{6}$$

where R is the length of the inset feed.

Step 7: Calculation of length of the microstrip transmission line

$$L_t = \frac{c}{4 f_0 \sqrt{\epsilon_r}} \tag{7}$$

where L_t is the length of the transmission line Fig. 1.

Fig. 1 Model of the proposed antenna

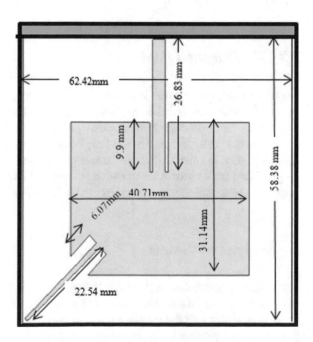

5 Design Analysis

5.1 For Inset Feed

The current is low ($\approx$zero) at the extremes of the half-wave patch and increases progressing toward the center. The impedance $Z = V/I$ could be reduced for better impedance matching if the patch was fed closer to the center. Hence, the feed is inserted at a distance R from the end. The current has a sinusoidal distribution moving a distance R from the end, and therefore, it increases by a factor $\cos(\pi R/L)$. The input impedance of the feed can then be evaluated as

$$Z_{in}(R) = \cos(\pi R/L) \cdot Z_{in}(0) \tag{8}$$

where $Z_{in}(0)$ is input impedance if patch was fed at the boundary.

Generally, the distance R is taken to be $R = L/4$. Plugging the value of R, Eq. 8 becomes

$$Z_{in}(R) = \left(\frac{1}{\sqrt{2}}\right)^2 Z_{in}(0) \tag{9}$$

The insertion of the feed by an amount of 1/8 of the wavelength would decrease input impedance by 50%.

5.2 For Diagonal Feed

When the rectangular patch is fed along the diagonal, modes TM10 and TM01 get energized, which are equal in magnitude and orthogonal in phase. These two modes add together and produce circular polarization along the diagonal of the designed patch antenna. Thus, the ratio of w/L may be adjusted to detune each mode slightly so that each mode is equal at a single resonant frequency. This enhances the bandwidth by providing better broadside radiation pattern at the resonating frequency and also improves the antenna gain, also reducing the VSWR and return loss.

6 Analysis of Results

The reflection coefficient at Port1 (**S11**), Fig. 2, was achieved as **−19.233 dB at 2248 MHz** and **−20.314 dB at 1564 Mhz** on Port2 (**S22**), Fig. 3.

This indicates that the maximum energy transfer from the source to the antenna ports has taken place at the two resonant frequencies, respectively. It can also be seen that Port1 has a bandwidth of 70 MHz and Port2 has a bandwidth of 20 MHz. The

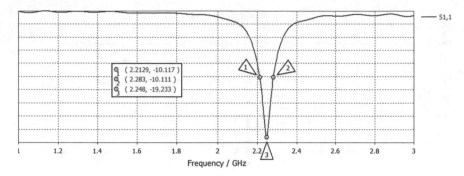

Fig. 2 S11 at Port1

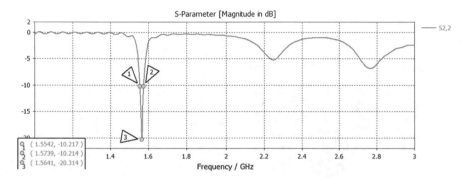

Fig. 3 S22 at Port2

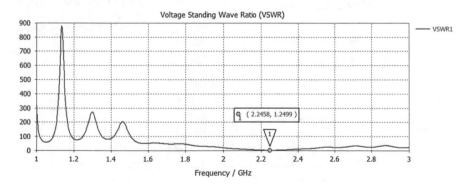

Fig. 4 VSWR at Port1

voltage standing-wave ratio (**VSWR**), Fig. 4, was achieved as **1.24 for 2.245 Ghz at Port1**(VSWR1) and **1.24 for 1.564 Ghz at Port2** (VSWR2), Fig. 5. These values indicate that the impedance values of the two ports of the antenna are well matched with the transmission line.

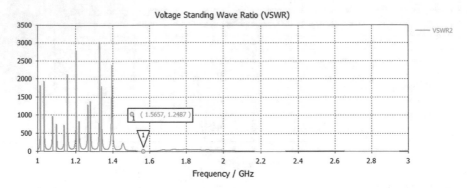

Fig. 5 VSWR at Port2

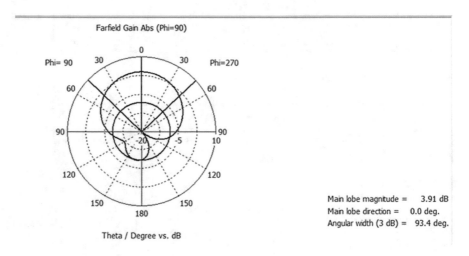

Main lobe magnitude =   3.91 dB
Main lobe direction =   0.0 deg.
Angular width (3 dB) = 93.4 deg.

Fig. 6 Radiation pattern Port1

Table 2 Comparison of results

Simulated result				Manufactured result			
Return loss		VSWR		Return loss		VSWR	
Port1	Port2	Port1	Port2	Port1	Port2	Port1	Port2
−19.233 dB	−20.314 dB	1.24	1.24	−20.14 dB	−13.631 dB	1.3	1.24

The radiation pattern, Fig. 6, attained shows a gain of 3.91 dB at Port1 and 3.59 dB at Port2, Fig. 7. The above designed antenna was fabricated in the local market, Fig. 8. On subjecting the fabricated dual-port microstrip patch antenna to tests on the VNA, the reflection coefficient was achieved as **−13.631 dB at 1.58 GHz,** Fig. 9, as compared to **−20.31 dB at 1.58 GHz** achieved in the CST software simulation, as reflected in Table 2.

This reflects a drop in the performance of Port2. This is attributed to the dimensions of the fabricated antenna not matching with the designed dimensions. The thickness

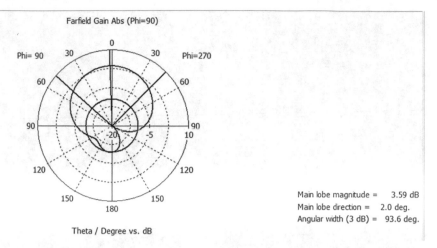

Fig. 7 Radiation pattern Port2

Fig. 8 Fabricated patch
antenna

of the substrate was observed to be varying by 1.2 mm. Hence, a drop in the reflection coefficient value at Port2 due to dimension mismatch. The voltage standing-wave ratio was achieved as **1.3 at 2.2 GHz,** Fig. 10, as compared to **1.24 at 2.2 Ghz** achieved in the software results. This shows the accuracy and the correctness of the design, and it also confirmed the simulated software results.

7 Conclusion

The design of a dual-port microstrip patch antenna for wireless applications has been proposed. It is observed that the proposed antenna effectively resonates at

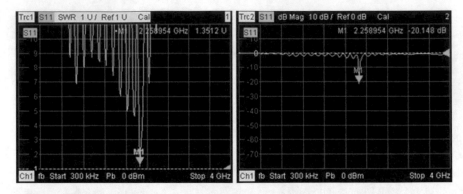

Fig. 9 Measured S11 of patch antenna

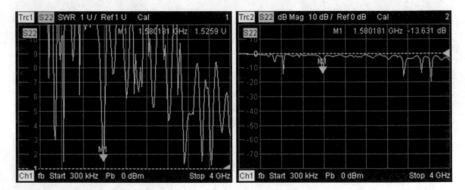

Fig. 10 Measured VSWR of patch antenna

two frequencies 2.2 Ghz (telemetry) and 1.5 Ghz (GPS). The location of the patch and the feed lines have been optimized in such a way that the antenna can operate in two frequencies at same time. Values of all the parameters, namely the return loss, radiation pattern, VSWR, and gain obtained, are considered to be good and acceptable values. The designed antenna parameters can be utilized for a conformal configuration that would make it suitable for onboard telemetry applications.

References

1. Collins BS (2000) Polarization diversity antennas for compact base stations. In: Microwave. J 43(1):76–88
2. Qin P-Y, Guo YJ, Liang C-H (2010) Effect of antenna polarization diversity on MIMO system capacity. IEEE Antennas Wireless Propagation. Lett 9:1092–1095
3. Dietrich CB, Dietze K, Nealy JR, Stutzman WL (2001) Spatial, polarization, and pattern diversity for wireless handheld terminals. IEEE Trans Antennas Propag 49(9):1271–1281

4. Piazza D, Mookiah P, Michele D, Dandekar KR (2009) Pattern, and polarization reconfigurable circular patch for MIMO systems. In: Proceedings european conference on antennas and propagation, pp 1047–1051
5. Lee C-H, Chen S-Y, Hsu P (2009) Isosceles triangular slot antenna for broadband dual polarization applications. In: IEEE Trans. Antennas Propagation 57(10):3347–3351
6. Li Y, Zang Z. Chen W, Iskander MF (2010) A dual-polarization slot antenna using a compact CPW feeding structure. In: IEEE antennas wireless propagation. Lett. vol 9, pp191–194
7. Deng C, Li P, Cao W (2012) A high-isolation dual-polarization patch antenna with omnidirectional radiation patterns. IEEE Antennas Wireless Propagation Lett 11:1273–1276
8. Wu G-L, Mu W, Zhao G, Jiao Y-C (2008) A novel design of dual circularly polarized antenna fed by L-strip. In: Progress in electromagnetics. Res, vol 79, pp 39–46
9. Narbudowicz A, Bao X, Ammann M (2013) Dual circularly-polarized patch antenna using even and odd feed-line modes. IEEE Trans Antennas Propagation 61(9):4828–4831
10. Seol K, Jung J, Choi J (2006) Multi-band monopole antenna with inverted U-shaped para sitic plane. IET Electron Lett 42(15):844–845
11. Wan YT, Yu D, Zhang FS, Zhang F (2013) Miniature multi-band monopole antenna using spiral ring resonators for radiation pattern characteristics improvement. IET Electron Lett 49(6):382–384
12. Liu WC, Wu CM, Dai Y (2011) Design of triple-frequency microstrip-fed monopole antenna using defected ground structure. IEEE Trans Antennas Propagation 59(7):2457–2463
13. Kumar G, Ray KP (2003) Broadband microstrip antennas. Aretch House
14. Narang T, Jain S (2013) Microstrip patch antenna—a historical perspective of the development. In: Conference on advances in communication and control systems
15. Adegoke OM, and Eltoum IS (2014) Analysis and design of rectangular microstrip patch antenna AT 2.4 GHz WLAN applications. In: International journal of engineering research & technology (IJERT), vol 3, Issue 8
16. James JR, Hall PS (1989) Handbook of microstrip antennas. In: I.E.E. Electromagnetic Waves Series 28- Peter Pereginus LTD
17. Balanis CA Antenna theory-analysis and design. In: Peters J (ed) John Wiley and Sons, pp 728–730
18. Mailloux RJ, McIlvenna JF, Kemweis NP (1981) Microstrip array technology. In: IEEE Transactions on Antennas and Propagation, 29(1)

Predictive Controller Design to Control Angle and Position of DIPOAC

Abhaya Pal Singh, Siddharth Joshi and Pallavi Srivastava

Abstract The paper uses model predictive control (MPC) strategy to design the controller for double-inverted pendulum on a cart (DIPOAC) system having three degrees of freedom (DOF) and an input. The main aim is to control the angles and position of the considered system in best possible minimum time. The simulation results are also shown and compared with PID controller strategy. These types of systems are normally applicable in transporting explosive materials or devices.

Keywords Control systems · Double-inverted pendulum on a cart · Model predictive control · Under-actuated systems · PID controller

1 Introduction

Under-actuated system is an important topic for research since many decades. It has many applications. The under-actuated systems are less bulky, less weight, and cost-effective, but system complexity is drastically increased because in under-actuated systems there are actuators less than degrees of freedom which make the system complex and hard to control. Under-actuated systems have a long list of applications including autobots, marine-robots, locomotives, robotics, and overhead crane.

In recent years, researchers have suggested many techniques for the control of under-actuated system. References [1–6] shows the example of control of under-actuated system. Paper [7] describes the dynamics of a class of under-actuated systems. PID controllers have been used in [8, 9] to control. Paper [10] shows how to control overhead crane using generalized predictive control (GPC). Similarly, there are many more examples which show under-actuated system is the benchmark problem on which the research is carried from a long time [11].

A. P. Singh (✉) · S. Joshi
Symbiosis Institute of Technology, Symbiosis International University, Pune, India
e-mail: abhaya.aps@gmail.com

P. Srivastava
Department of Mechatronics, Symbiosis Skills and Open University, Pune, India

© Springer Nature Singapore Pte Ltd. 2019   453
A. J. Kulkarni et al. (eds.), *Proceedings of the 2nd International Conference on Data Engineering and Communication Technology*, Advances in Intelligent Systems and Computing 828, https://doi.org/10.1007/978-981-13-1610-4_46

This paper has considered a DIPOAC system. It is a system with three DOF and one input and is highly nonlinear. There are many examples where control of pendulum on a cart (POAC), and DIPOAC are reported in [10, 12]. Paper [1–3] shows the control of DIPOAC with different strategies, and [13–15] shows the control of POAC system. The DIPOAC comprises of a double-inverted pendulum attached on a cart and stable vertically upward. This paper uses model predictive control (MPC) strategy for controlling the considered system, and the results are compared with PID controller.

MPC is a technique that predicts the future output of a plant using the plant model. It uses iteration to predict next stage of the output. These are done at subsequent control intervals. It was primarily implemented on refineries and power plants [16]. But over the years, the applications under MPC have increased and the area of applications has widened. There are many areas where MPC has been applied successfully. Survey of industrial MPC control is shown in [17]. Stability and robustness of an MPC are very important factors [18, 19]. Smoczek et al. [20] shows the trajectory tracking using MPC. References [21, 22] are some books which describe the use and application of MPC.

The outline of the paper follows here—Sect. 2 gives the modeling of DIPOAC and its linearization. Section 3 will discuss the application of MPC on the considered system and its control followed by comparative analysis with PID controller. Simulation results are obtained in Sect. 4, and in Sect. 5 conclusion is drawn.

2 Modeling

2.1 Modeling of POAC

Figure 1 shows the picture of DIPOAC system. Assumptions are—massless rod, force applied on mass M, and the pendulum's mass as m_1 and m_2 respectively. The external force $u(t)$ is applied in x-direction. Here, $\theta_1(t)$ and $\theta_2(t)$ are the pendulum angles, $x(t)$ represents position.

Modeling of DIPOAC uses Euler–Lagrangian model. Lagrangian is represented by.

$$\mathcal{L} = T - V \tag{1}$$

here,

V   potential energy (PE),
T   kinetic energy (KE).

After calculating KE and PE, and after some mathematical manipulations, the Lagrangian will be,

Fig. 1 Double-inverted
pendulum on a cart
(DIPOAC)

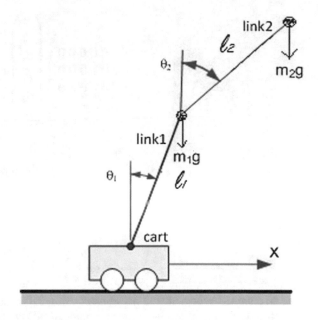

$$\mathcal{L} = \frac{1}{2}[M + m_1 + m_2]\dot{x}^2 + \frac{1}{2}[m_1 + m_2]l_1^2\dot{\theta}_1^2 + \frac{1}{2}m_2l_2^2\dot{\theta}_2^2$$
$$+ [m_1 + m_2]\dot{x}\dot{\theta}_1l_1\cos\theta_1 + m_2\dot{x}\dot{\theta}_2l_2\cos\theta_2$$
$$+ l_1l_2\dot{\theta}_1\dot{\theta}_2\cos[\theta_1 - \theta_2] - [m_1gl_1\cos\theta_1 + m_2gl_1\cos\theta_1 + m_2gl_2\cos\theta_2] \quad (2)$$

Putting the value of Eq. (2) and the parameter values, $M = 0.5$ kg, $m_1 = 0.25$ kg, $m_2 = 0.25$ kg, $l_1 = 1$ m, $l_2 = 1$ m, $g = 10(\frac{m}{s^2})$ in Euler–Lagrange formulation and casting them to nonlinear state-space and further linearizing.

We get,

$$\frac{d}{dt}(\delta\underline{z}) = \begin{pmatrix} 0 & 1 & 0 & 0 & 0 & 0 \\ 0 & 0 & -0.073 & 0 & 0.120 & 0 \\ 0 & 0 & 0 & 1 & 0 & 0 \\ 0 & 0 & -1.420 & 0 & -2.87 & 0 \\ 0 & 0 & 0 & 0 & 0 & 1 \\ 0 & 0 & 5.75 & 0 & 1.382 & 0 \end{pmatrix} \delta\underline{z} + \begin{pmatrix} 0 \\ 0.102 \\ 0 \\ -0.015 \\ 0 \\ -0.015 \end{pmatrix} \quad (3)$$

And output state-space matrix as,

$$y = \begin{pmatrix} 1 & 0 & 0 & 0 & 0 & 0 \\ 0 & 0 & 1 & 0 & 0 & 0 \\ 0 & 0 & 0 & 0 & 1 & 0 \end{pmatrix} \begin{pmatrix} x \\ \dot{x} \\ \theta_1 \\ \dot{\theta}_1 \\ \theta_2 \\ \dot{\theta}_2 \end{pmatrix} \tag{4}$$

3   MPC Design and Comparison with IOPID

Figure 2 shows the block diagram of traditional PID controller for DIPOAC and is expressed as

$$k_{pid} = k_p + \frac{k_i}{s} + k_d s \tag{5}$$

The obtained values of PID parameters using PID tuner as follows:

Tables 1, 2, and 3 show the optimum values of PID controller response for position and both the angles. Figure 3 shows the basic MPC structure overview and is designed in model predictive control toolbox [23]. From the structure overview of MPC toolbox for DIPOAC, there are three outputs and one input. Outputs are in the form of cart position, pendulum angle θ_1, and angle θ_2.

MPC has three parameters which are to be manipulated to get an optimum value of output, i.e., control interval, control horizon, and prediction horizon. This paper is taking control interval as 0.1 s, prediction horizon to be 50, and control horizon

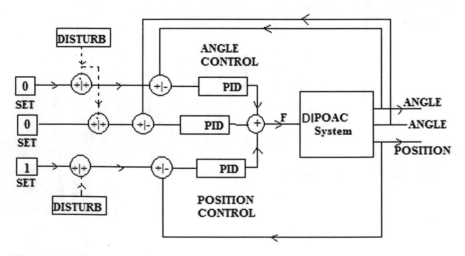

Fig. 2 Block diagram of DIPOAC system

Table 1 To control pendulum's angle θ_1

Controller	k_p	k_i	k_d
PID	−1700	−210	−560

Table 2 To control pendulum's angle θ_2

Controller	k_p	k_i	k_d
PID	−1700	−910	−5610

Table 3 To control cart's position

Controller	k_p	k_d	k_i
PID	0.001	100	0.000016

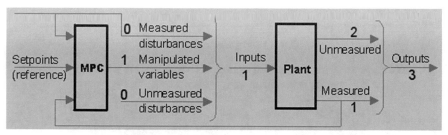

Fig. 3 MPC structure overview

to be 3 to obtain better output. The state-space equations are used further to design MPC.

4 Simulations and Results

The simulation results corresponding to Eqs. 3 and 4 are obtained using MATLAB to stable the angle and the position of the DIPOAC system at desired position and angles in minimum time with little overshoot.

We can conclude from simulation result shown in Fig. 4 that the cart position is settled at 2 m and both the angles are settled at 0°.

Figure 5 shows the output response of a PID controller for DIPOAC system; from this, we can conclude that the cart position is nearly settled at 2 m with some steady-state error, and angles are settled at 0° with oscillations (Table 4).

As we compare the performance of PID and MPC of DIPOAC system, from Figs. 4 and 5, we see that the performance of MPC is much better than that of PID. There is a steady-state error in case of a PID controller, whereas in MPC no such error exists. The oscillations are more in PID as compared to MPC. PID controller takes almost 60 s to stabilize the system, whereas MPC does the work in around 15 s. From simulation results, we observe that MPC performs good if we compare it to classical PID. From the above table, as compared to PID, MPC performs well when it comes to oscillations, settling time, and overshoot.

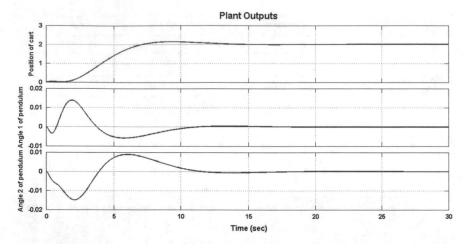

Fig. 4 MPC response with at cart position at 2 m and both the angle θ_1 and the angle θ_2 at 0°

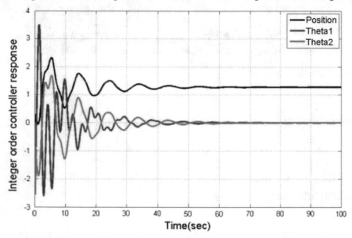

Fig. 5 PID response with at cart position at 2 m and both the angle θ_1 and the angle θ_2 at 0°

Table 4 Comparison table

Controllers specifications	PID	MPC
Oscillations (Pendulum's angle)	High	Compensated
Settling time (cart's position)	60 s	15 s
Settling time (θ_1)	60 s	12 s
Settling time (θ_2)	60 s	12 s
Overshoot	High	Very much Improved

5 Conclusion

This paper shows that MPC controller can also be used for the stabilization of DIPOAC system. If we compare with traditional PID controllers, the MPC has better response. MPC for under-actuated system settles fast with little oscillations, whereas the settling time for PID controllers of DIPOAC is large and is with oscillations and steady-state error.

References

1. Singh AP et al (2016) On modeling and analysis of launch vehicle system. Lecture Notes in Electrical Engineering. Springer, vol 396, pp 273–279
2. Furuta K et al (1978) Computer control of a double inverted pendulum. Comput Electr Eng 5(1):67–84
3. Cheng F et al (1996) Fuzzy control of a double-inverted pendulum. Fuzzy Sets Syst 79(3):315–321
4. Singh AP et al (2013) Fractional equations of motion via fractional modeling of underactuated robotic systems. IEEE ICARET
5. Singh AP et al (2012) PI α D β controller design for underactuated mechanical systems. In: 2012 12th International Conference on control automation robotics & vision (ICARCV)
6. Joshi S, Singh AP (2016) Comparison of the performance of controllers for under-actuated systems. In: 2016 IEEE 1st international conference on power electronics, intelligent control and energy systems (ICPEICES), Delhi, pp 1–5
7. Reyhanoglu M et al (1999) Dynamics and control of a class of underactuated mechanical systems. Autom Control IEEE Trans 44(9):1663–1671
8. Salih AL et al (2010) Modelling and PID controller design for a quadrotor unmanned air vehicle. In: IEEE international conference on Automation quality and testing robotics (AQTR), vol 1, pp 1–5. IEEE, 2010
9. Andary S et al (2012) A dual model-free control of underactuated mechanical systems, application to the inertia wheel inverted pendulum. In: American Control Conference (ACC), 2012, pp 1029–1034. IEEE
10. Singh AP et al (2015) Fractional order controller design for inverted pendulum on a cart system (POAC). In: Wseas transactions on systems and control, vol 10, pp 172–178
11. Shim DH et al (2003) Decentralized nonlinear model predictive control of multiple flying robots. In: Decision and control, 2003. Proceedings. 42nd IEEE conference on, vol 4, pp 3621–3626. IEEE
12. Castillo CL et al (2007) Unmanned helicopter waypoint trajectory tracking using model predictive control1. In: mediterranean conference on control and automation, 2007. MED'07, pp 1–8
13. Srivastava T et al (2015) Modeling the under-actuated mechanical system with fractional order derivative. Progr Fractional Differentiation Appl Int J 1(1):57–64
14. Abhaya PS et al (2017) On design and analysis of fractional order controller for 2-d gantry crane system. Published in Progress in Fractional Differentiation and Applications An International, vol 3, No 2, pp 155–162, 2017
15. Graichen K et al. (2007) Swing-up of the double pendulum on a cart by feedforward and feedback control with experimental validation. Automatica 43(1):63–71
16. Sandoz DJ et al (2000) Algorithms for industrial model predictive control. Comput Control Eng J 11(3):125–134
17. Qin S Joe, Badgwell Thomas A (2003) A survey of industrial model predictive control technology. Control Eng Practice 11(7):733–764

18. Kothare Mayuresh V et al (1996) Robust constrained model predictive control using linear matrix inequalities. Automatica 32(10):1361–1379
19. Fahimi F (2007) Non-linear model predictive formation control for groups of autonomous surface vessels. Int J Control 80(8):1248–1259
20. Smoczek J et al (2015) Generalized predictive control of an overhead crane. In: 2015 20th international conference on Methods and models in automation and robotics (MMAR), pp 614–619
21. Morari Manfred et al (1993) Model predictive control. Prentice Hall, Englewood Cliffs, NJ
22. Wang L (2009) *Model* predictive control system design and implementation using MATLAB®. Springer Science & Business Media
23. Bemporad A et al (2010) Model predictive control toolbox 3 User's Guide. The mathworks

Novel Machine Health Monitoring System

Mahesh S. Shewale, Sharad S. Mulik, Suhas P. Deshmukh,
Abhishek D. Patange, Hrishikesh B. Zambare and Advait P. Sundare

Abstract Machines have been an inevitable part of our life in today's era. It has become paramount to look after various machines and keep them in safe as well as efficient condition. To check the health of such machines, various devices have been developed which measures vibrations, temperature, noise level, and power consumption. Any defect in the machine is indicated by unusual behavior in the above parameters. FFT analyzers are used to measure vibrations in the machine. However, the cost of FFT analyzers is very high and it may not be affordable to small-scale industries. Also, they rarely have a provision to measure the speed, temperature, and power consumed by the machine. The present research work is an effort to provide a low-cost solution to the existing health monitoring systems along with facility to measure various parameters like vibrations, noise, temperature, rotational speed, and power consumption. A low-cost controller Arduino Mega 2560 is used along with various sensors and integrated to MATLAB GUI to store and display the acquired data. The developed device was compared with existing systems and found to have a good agreement. This device can be used as a monitoring tool in small-scale industries where high-cost FFT analyzers become obsolete due to costing issue.

Keywords FFT · Machine health · Arduino · Vibration · Time domain
Frequency domain

M. S. Shewale · H. B. Zambare
Department of Mechanical Engineering, IUPUI, Indianapolis, Indianapolis, IN, USA

S. S. Mulik · A. D. Patange (✉)
Department of Mechanical Engineering, Trinity Academy of Engineering, Pune, India
e-mail: abhipatange93@gmail.com

S. P. Deshmukh
Department of Mechanical Engineering, Government College of Engineering, Chandrapur, India

A. P. Sundare
Department of Mechanical Engineering, Sinhgad Academy of Engineering, Pune, India

© Springer Nature Singapore Pte Ltd. 2019
A. J. Kulkarni et al. (eds.), *Proceedings of the 2nd International Conference on Data
Engineering and Communication Technology*, Advances in Intelligent Systems
and Computing 828, https://doi.org/10.1007/978-981-13-1610-4_47

1 Introduction

Inertial microelectromechanical systems (MEMS) sensors assume a huge part in the huge extension of today's personal electronic gadgets. Their little size, low power, simplicity of combination, abnormal state of usefulness, and wonderful execution energize, and empower development in contraptions, for example, cell phones, gaming controllers, motion trackers, and advanced picture outlines. Likewise, inertial MEMS sensors have generously enhanced unwavering quality and lessened expense in car wellbeing frameworks, permitting them to be conveyed in many cars [1]. The object of frequency analysis is to break down a complex signal into its components at various frequencies, and in order to do this, the practical engineer needs to understand the frequency analysis parameters and how to interpret the results of spectrum measurements [2].

The continuous advancement in practical integration and performance has conjointly helped MEMS accelerometers notice their approach into various industrial systems. Some of these applications offer lower-cost alternatives to current product and services, whereas others are segregating inertial transducer action in a new and unique way. New adaptations in vibration sensing area are concluding that fast distribution and affordable price of possession are the reasons to judge the integrated MEMS devices. Vibration monitoring is coming up as associate application that has each variety of users. Conventional instruments that observe machine health for maintenance and safety usually use piezoelectric technology. High-speed automation instrumentation setup monitors vibration to trigger feedback management of lubrication, speed, or belt tension or to switch off instrument for fast attention from the operating staff [3–5].

MEMS accelerometers provide quick, efficient integration, and cost-effective solution to a rising cluster of latest users. Additionally, their advanced practical integration permits devices like the ADIS16229 digital MEMS vibration sensing element with embedded RF transmitter and receiver to supply an entire resolution inclusive of communications and signal processing [6, 7]. This kind of device will wake itself up repeatedly, record time domain vibration information data, perform fast Fourier transform (FFT) on the information recorded, apply user-configurable spectral analysis on the FFT result, provide easy pass/fail results over economical wireless transmission, offer access to that information and results and then return to hibernate or sleep mode [8]. To enhance manual measurements, embedded MEMS-based sensors give a more cost-effective method for instrumentality that needs real-time vibration information.

2 Time Domain and Frequency Domain Analysis

Time domain and frequency domain are two ways that of observing at same dynamic system. They're interchangeable, i.e., no data are lost in conversion from one

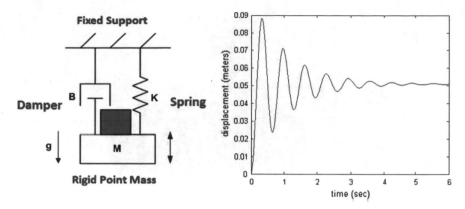

Fig. 1 Typical spring-mass-damper system and time domain response [10]

domain to other. They are unit complementary points of view that result in an entire, clear understanding of the behavior of a dynamic engineering system. Roughly speaking, in the time domain, we tend to find how long one thing takes, whereas in the frequency domain, we tend to find how fast or slow the response of system is. The time domain data are a record of the output of a dynamic system, as shown by some observed parameter, as function of time. This can be the standard way of observant the output of a dynamic system [9].

Example of time response is the displacement of the mass of the spring-mass-damper system versus time in response to the quick placement of additional mass on the hookedup mass as shown in Fig. 1. The output response is the step response of the system because sudden impact force. Usually after we examine the enactment of a dynamic system, we tend to use because the input to the system as step input [10].

3 Mechatronic Interface and Control System Design

The sensing element choice and signal-processing design depend on the application's aims. Schematic diagram of the sensors connected to the microcontroller is shown in Fig. 2.

An external power supply is provided to the Arduino. The sensors used in machine health monitoring system are K-type thermocouple, temperature sensor MAX6675, accelerometer ADXL335, current sensor AC712, noise sensor MAX4466, and a voltage divider circuit. Figure 3 denotes a sequence of operation with ADXL335 accelerometer along with Arduino mega 2560 microcontroller that uses an analog tri-axial vibration sensing element with FFT analysis and storage to observe the spectral content of apparatus vibration.

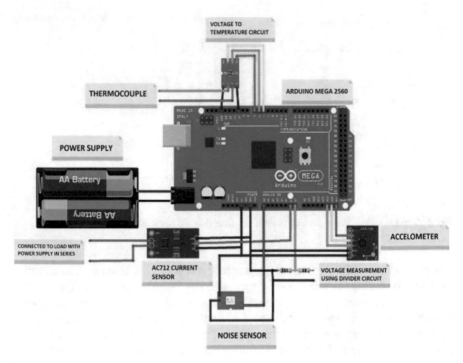

Fig. 2 Integration of sensors with controller

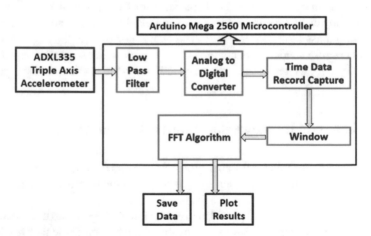

Fig. 3 Control system for vibration analysis

3.1 Core Sensor ADXL335 Accelerometer

The core sensing elements for either approach are often a MEMS accelerometer. The foremost vital attributes for choosing a core sensing element are the quantity of axes, package/assembly necessities, electrical interface (analog/digital), frequency response (bandwidth), measuring range, noise, and linearity [11]. The ADXL335 accelerometer used in this research work implements silicon on insulator (SOI) MEMS technique and takes benefit of automatically coupled, however, electrically remote differential sensing cells. Movement of the detector frame changes the differential capacitance. On-chip electronic circuit determines the change in capacitance and transforms it into an output voltage [12].

3.2 Analog Low-Pass Filter

The analog filter limits the signal content to at least one Nyquist zone that represents one half the sample rates within the example system. Even once the filter cut-off frequency is at intervals the Nyquist zone; it is not possible to possess infinite rejection of higher-frequency elements, which may still fold into the pass band [15].

3.3 Windowing

Time-coherent sampling is generally not sensible in vibration sensing applications, as nonzero sample values at the beginning and finish of the time record lead to massive spectral discharge, which may degrade the FFT resolution. Applying a window operate before scheming the FFT will facilitate manage the spectral leak [13].

3.4 Fast Fourier Transform (FFT)

FFT is an economical algorithmic program for analyzing distinct time information. The method transforms a time record into a distinct spectral record, where every sample represents a distinct frequency section of the Nyquist zone [14]. The main concept of this algorithm is that DFT of a sequence of N points can be expressed in terms of two DFT of length $N/2$. Therefore, if N is power of 2, it is easy to apply recursively this algorithm until we obtain DFT of single point [15]. The algorithm was formulated using the MATLAB analytical software. The flowchart for the algorithm is shown in Fig. 4. GUI is made using MATLAB which facilitates the connection to ARDUINO Mega 2560 and also gives flexibility to set the time for which the data have to be acquired. We can even select the axis through which we desire the time-acceleration data. Besides, we can save the results in.mat file format for later use. A suitable frequency range can be provided for maximum and minimum extremities.

Fig. 4 Flowchart for the
FFT algorithm

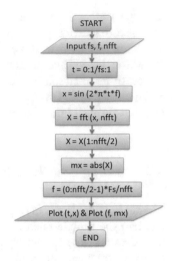

4 Results and Discussion

The above system was experimentally tested on cam follower apparatus having an involute cam profile due to which follower produced vibratory motion of a single frequency at a constant speed. At a speed of 240 rpm of cam, the analytical frequency can be determined by the formula $\frac{2\pi n}{60} = 25.13 \frac{rad}{s} = 4$ Hz. According to the results obtained from the graph in Fig. 5b, it can be clearly seen that we obtain a peak at 3.967 Hz which shows good agreement with analytical frequency. These results were compared with the results from ADASH make A4400-VA4 FFT Analyzer, and from Fig. 5c, it can be clearly seen that we get a very good agreement in amplitude as well as forced frequency for the system.

Comparison between developed device and commercially available analyzer is shown with the help of graph given in Fig. 6. The time response of a system does not give much useful information. Since the dynamic characteristics of individual components of the system are usually known, we relate the distinct frequency components (of the frequency response) to specific components. The energy concentrates near peak frequency, i.e., it is related with the rotational speed of variable motor which is connected to cam jump apparatus. Developed system reflected good accuracy with commercially available Adash 4400-VA4 analyzer up to 99.175%.

5 Conclusion

A novel health monitoring system was developed by using low-cost controller like Arduino Mega 2560. Initially, the sensors were identified to measure various parameters related to machine health. Further, these sensors were integrated with micro-

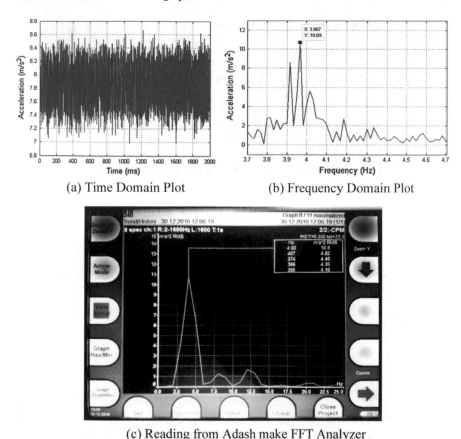

(a) Time Domain Plot (b) Frequency Domain Plot

(c) Reading from Adash make FFT Analyzer

Fig. 5 Time and frequency domain plots at 240 rpm

controller and a control system was designed to access the real-time input to be acquired from various sensors. For vibration analysis purpose, logic was developed to convert time domain data into frequency domain. The GUI was designed that could handle, process, display, and store the real-time input data from sensors. After due experiments on cam jump apparatus, this novel device reflected good accuracy with the standard FFT analyzer up to 99.175%. This device can provide more number of input channels for more numbers of sensors, and accuracy can be further increased by enhancing the resolution of input ports and controller.

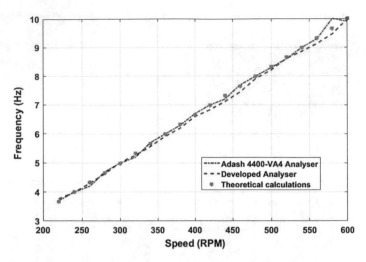

Fig. 6 Comparison of peak frequency determined by developed and commercial analyzer

References

1. Looney M (2014) An introduction to MEMS vibration monitoring. Analog Dialogue 48(06)
2. Chu F, Peng Z, Feng Z, Li Z (2009) Modern signal processing methods in machinery fault diagnosis. Science Press, Beijing in Chinese
3. Bruel & Kjaer (1982) Measuring vibration-an elementary introduction. K Larson and Son Publication; Denmark, Revised edition
4. Fan L, Qi G, He W (2016) Accurate estimation method of sinusoidal frequency based on FFT. In: Proceedings of the 35th Chinese control conference, 27–29 July, 2016, Chengdu, China
5. Weddell AS, Merrett GV, Barrow S, Al-Hashimi BM (2012) Vibration-powered sensing system for engine condition monitoring. Electronics and Computer Science, University of Southampton, Southampton, SO17 1BJ, UK
6. Maxwell JC (1892) A treatise on electricity and magnetism, 3rd ed., vol 2. Clarendon, Oxford, pp 68–73
7. Jacobs IS, Bean CP (1963) Fine particles, thin films and exchange anisotropy. In: Rado GT, Suhl H (eds) Magnetism, vol III. Academic, New York, pp 271–350
8. Feng Z, Liang M, Chu F (2013) Recent advances in time–frequency analysis methods for machinery fault diagnosis: a review with application examples. Mech Syst Sig Process 38:165–205
9. Application Note 1405-1, Introduction to time, frequency and modal domains, fundamentals of signal analysis series. Agilent Technologies, 24 May 2002
10. Taylor AP (2015) Coming to grips with the frequency domain. XPLANATION: FPGA101, Xcell Journal, Second Quarter
11. Tuck K (2008) Frequency analysis in the industrial market using accelerometer sensors. Application note by Freescale Semiconductors
12. Datasheet Rev. B, ADXL335 Accelerometer, Analog devices, D07808-0-1/10(B)
13. Sobota J, Pisl R, Balda P, Schlegel Ms (2013) 'Raspberry Pi and Arduino boards in control education. International Federation of Automatic Control
14. Fast Fourier Transform and MATLAB implementation by Wanjun Huang for Dr. Duncan L. MacFarlane
15. Math Works (2015) Fast Fourier Transform (FFT), http://se.mathworks.com/help/matlab/math/ fast-fourier-transform-fft.html. Accessed 22 May 2015

Finite Element Mesh Smoothing Using Cohort Intelligence

Mandar S. Sapre, Anand J. Kulkarni and Sumit S. Shinde

Abstract Several approaches have been developed for mesh quality improvement by considering only node movement, without disturbing the element connectivity. Laplacian smoothing is well known and easiest one to implement among all, but this approach might lead to inverted elements in concave regions. In this work, optimization-based smoothing method is proposed for the improvement of hexahedral mesh without any inverted elements. Optimization is carried out using condition number-based objective function which defines qualities of individual elements in the hexahedral mesh. Numerical optimization method used is the cohort intelligence (CI) algorithm which is socio-inspired algorithm developed by Dr. Anand Kulkarni et al. (Cohort intelligence: a self-supervised learning behavior, pp. 1396–1400, 2013 [10]). The approach is demonstrated with three examples.

Keywords Hexahedral mesh · Condition number
Optimization-based smoothing · Cohort intelligence

1 Introduction

The finite element analyses (FEAs) are required to verify the suitability of an engineering design before actual manufacturing. There are two big challenges, one is to build a sufficiently accurate model and other is to carry out its analysis, in the available time. Meshing is the essential component of any analysis or simulation and obtaining useful meshes is an important issue. All commercial FEA softwares

M. S. Sapre (✉) · A. J. Kulkarni · S. S. Shinde
Symbiosis Institute of Technology,
Symbiosis International University, Pune 412 115, MH, India
e-mail: mandar.sapre@sitpune.edu.in

A. J. Kulkarni
e-mail: anand.kulkarni@sitpune.edu.in

S. S. Shinde
e-mail: sumit.shinde@sitpune.edu.in

© Springer Nature Singapore Pte Ltd. 2019
A. J. Kulkarni et al. (eds.), *Proceedings of the 2nd International Conference on Data Engineering and Communication Technology*, Advances in Intelligent Systems and Computing 828, https://doi.org/10.1007/978-981-13-1610-4_48

are based on interpolation methods which produce approximate results and the error needs to be minimum. There are various techniques of mesh generation which are either automatic or manual [1]. Automatic mesh generators often generate meshes that are not well-shaped, and hence, a mesh smoothing technique is essential for mesh quality improvement.

Smoothing or r refinement method is a technique of correcting the poorly shaped elements. No one smoothing technique works all of the time [2]. Mesh smoothing provides a high quality of mesh and increases reliability of the solution. The most popular technique is Laplacian smoothing which uses internal nodes of the element and its connected nodes for relocation of nodes at desired place via centroid approach [3].

Knupp [4] has described the mesh quality metrics in the algebraic form which can be used as the objective functions for optimization-based smoothing. Element quality is a function of invertibility, size, and shape. Invertibility gives the measure of the positive local volume of given mesh. The quality of a mesh is defined in terms of these mesh metrics [4]. Knupp [5] proved that mesh element shape can be defined by using condition number parameter and can be improved by aspect ratio improvement and element skew reduction [5]. The mesh metrics generally used are inverse mean-ratio metric [6], aspect ratio metric [7], condition number [8], and distortion metric [9] etc.

In the present work, we discuss the application of cohort intelligence (CI) algorithm for developing a new efficient mesh smoothing technique. Objective function for individual qualities of hexahedral elements in the mesh based on the condition number is used as a quality measure. CI is based on natural tendency of the individuals or candidates, to improve its behaviour by observing and implementing behaviour of peers in the group known as cohort. This helps in the growth of overall cohort behaviour as every candidate absorbs certain qualities from one another [10].

2   Literature Review

Finite element analysis is a very powerful tool and a necessity of mechanical engineering applications like structural analyses and fluid dynamics but its usefulness is controlled by mesh quality. Meshing is a critical step in FEA and is generally carried out in two stages—point placement and mesh improvement. It is an iterative process. Kovalev [11] has shown that auto-generation of good-quality quadrilateral/hexahedral meshes is difficult [11]. Various methods have been developed for improvement of triangular, quadrilateral, tetrahedral, or hexahedral meshes. Laplace smoothing is suitable only for 2D meshes but does not guarantee good element quality 3D meshes. It produces slivered tetrahedron for 3D elements [12].

Mesh smoothing algorithms also apply optimization techniques for movement of nodes [13]. Knupp [4] explained how the quality of mesh is an important aspect in mesh smoothing algorithms and can be improved by numerical optimization of the mesh metric [4]. The optimization methods are better than Laplacian smoothing methods as they remove inverted elements on mesh improvement but are more expensive and time-consuming than Laplacian smoothing [14]. Hence, combinations of both result in effective tools for mesh improvement [8].

Both classical and heuristic techniques have been used in the literature for optimization-based smoothing. Freitag and Ollivier [15] proposed a technique which used minimum angle metric and iterative steepest descent for optimization [15]. Other classical optimization algorithms used are Newton's method, Hessian method [7], steepest gradient method [16], and linear programming [17]. Holder et al. (1998) have employed genetic algorithm for mesh optimization using distortion mesh criteria [18]. Acikgoz et al. [19] used simulated annealing technique for optimization [19]. Dittmer et al. [20] used optimization to control mesh creation parameters. [20]. Sastry and Shontz [7, 17] discussed the performance of nonlinear optimization methods for mesh quality improvement and showed that optimization solver behaviour can be improved by varying mesh size and accuracy level [7, 17].

Knupp [5] discussed condition number as a shape quality measure for hexahedral elements designed using a set of Jacobian matrices related to the particular element [5]. Yilmaz and Kuzuoglu [8] used particle swarm optimization technique for hexahedral mesh smoothing using condition number. The shape quality of all hexahedral elements is taken into account to design objective function which is modified in order to get the minimum value of objective function as zero [8].

The proposed work aims to apply condition number optimization for solving mesh smoothing using CI algorithm. The CI algorithm was proposed by Kulkarni et al. [10] and is inspired from the self-supervised learning behaviour of the candidates in a cohort [10]. All of them have the same target, and to achieve it, every candidate tries to study the behaviour of peers and improve by following a certain candidate's behaviour. Several parameters like sampling interval and reduction factor govern CI. The algorithm has been successfully applied in the areas of healthcare and logistics [21], knapsack problem [22], and mechanical design problems [23], and the present work investigates the application of CI for mesh smoothing. Throughout the journey, it will help to explore and validate various characteristics of CI [24].

3 Mesh Smoothing Algorithm Using Cohort Intelligence (CI)

Consider an objective function for hexahedral mesh smoothing (in the minimization sense) as described in Eq. 3.1. Here, k denotes the condition number of individual element and $T_{n,i}$ is the transformation matrix defined for the ith node of the nth hexahedral element. N denotes total number of elements in the problem. The attributes of the candidates are considered as the three coordinates of the eight vertices of the cube, i.e. $x, y, z = (x_1, y_1, z_1, \ldots, x_k, y_k, z_k, \ldots, x_8, y_8, z_8)$. Similarly $f(x^c y^c, z^c)$ denotes the behaviour of individual candidate $c(c = 1$ to $C)$.

$$\text{Minimize } F = f(x, y, z) = \frac{1}{8N} \sum_n \sum_i \left(k(T_{n,i})/3\right)^2 - 1 \qquad (3.1)$$

3.1 Procedure

The geometry of cube is generated using ANSYS and chosen for the preliminary analysis of the new algorithm proposed in this paper. The cube is discretized into 27 elements, 64 elements, and 729 elements, respectively. 27 elements and 64 elements have been chosen to investigate the applicability of the algorithm while 729 elements have been chosen for the comparison with previously published results. Creation of geometry file and auto-generated mesh using ANSYS APDL is the preliminary step. The analysis steps are as follows:

Step 1: (Conversion of data from ANSYS into a text file): Nodal connection data is obtained using **nlist** command and element data is obtained using **elist** command from hexahedral mesh.

Step 2: Extract node and element data: The data in this file is in complex form consisting of spaces, and alphanumeric characters. It contains data about element connectivity, real constants, and material assignment. Geometric coordinates of the nodes and elements need to be extracted from this huge piece of information. This data is initially stored in an excel file which is then extracted using textscan command. Then, the mesh data is read line by line for the removal of entries like NaN and zero values from the file. Then, the element connectivity in all directions (i.e. x, y, z) is used to define solution space.

Step 3: Develop the boundary of the object in MATLAB by edge and surface identification: The boundary of the object is required to be defined. So from the set of nodes extracted, the nodes lying on the surfaces, edges, and corners are identified. The equations of the planes for the surfaces of the cubes are developed using the three-point form of the equation of the plane. Redundant planes are excluded using the perpendicularity condition. Edges of the cube are identified as the intersection of the planes. This helps us to define external boundary of object.

Step 4: Identify the mesh metric for optimization which is the condition number: Let (x_i, y_i, z_i) be the vertices of the adjacent corners of a vertex (x, y, z), then its condition number (Eq. 3.2) is given by

$$k(T) = \text{cond} \begin{bmatrix} x_1 - x & x_2 - x & x_3 - x \\ y_1 - y & y_2 - y & y_3 - y \\ z_1 - z & z_2 - z & z_3 - z \end{bmatrix} \qquad (3.2)$$

Condition number is calculated for each corner. The maximum value represents the condition number of the element which needs to be minimized. Store the minimum condition number candidate as global best.

Step 5: Check for bad-quality element: The element is said to have good quality if condition number lies between 1 and 3 [5]. The candidates C (number of bad elements in auto-generated mesh data) are initialized. The bounds of the variables are given by Eq. 3.3

$$x_i(\text{solution space})_{\text{lower}} < x_i < x_i(\text{solution space})_{\text{upper}}$$

$$y_i(\text{solution space})_{\text{lower}} < y_i < y_i(\text{solution space})_{\text{upper}} \qquad (3.3)$$

$$z_i(\text{solution space})_{\text{lower}} < z_i < z_i(\text{solution space})_{\text{upper}}$$

Step 6: The behaviour selection probability p_c is calculated using Eq. 3.4. The behaviour of every associated candidate $c(c = 1, \ldots, C)$ to be followed is given by $f^*(x^c y^c, z^c)$, thus

$$p_c = \frac{1/f^*(x^c, y^c, z^c)}{\sum_c 1/f^*(x^c, y^c, z^c)}, (c = 1 \ldots C) \qquad (3.4)$$

The roulette wheel approach is used by every candidate $c(c = 1 \text{ to } C)$ to generate random number to follow corresponding behaviour $f(x^{c[f]}, y^{c[f]}, z^{c[f]})$ and associated qualities $x^{c[f]}, y^{c[f]}, z^{c[f]} = \left(x^{c[f]}, y^{c[f]}, z^{c[f]} = x_1^{c[f]}, y_1^{c[f]}, z_1^{c[f]}, x_2^{c[f]}, y_2^{c[f]}, z_2^{c[f]} \ldots x_8^{c[f]}, y_8^{c[f]}, z_8^{c[f]} \right)$

Step 7: Every candidate shrinks or expands attributes depending on element edge length of candidate to be followed. The relations in Eq. 3.5 are obtained using condition number criteria. Specific nodes are not permitted to move to preserve the shape of volume mesh. Corner nodes are fixed. Surface nodes are allowed to move along surface only. Edge nodes are allowed to move along edges only.

$$x_i = x_1^{c[f]} - x_{i(\text{edge})^{c[f]}} + x_{i-1(\text{edge})^{c[f]}}$$

$$y_i = y_1^{c[f]} - y_{i(\text{edge})^{c[f]}} + y_{i-1(\text{edge})^{c[f]}} \qquad (3.5)$$

$$z_i = z_1^{c[f]} - z_{i(\text{edge})^{c[f]}} + z_{i-1(\text{edge})^{c[f]}}$$

Step 8: Decide cohort behaviour to be followed by candidates and improve the quality of mesh using CI. Each candidate $c(c = 1, \ldots, C)$ tries to enrich its behaviour by following the other candidates having better behaviour associated with modified attributes $xc^{[f]}, yc^{[f]}, zc^{[f]}$. This makes the cohort available with C updated behaviours represented as $F^C = \left(f(x^1, y^1, z^1), \ldots, f(x^c, y^c, z^c) \right)$. After few learning attempts, all candidates follow global best behaviour from available behaviours which are helpful for internal mesh smoothing of hexahedral elements.

Step 9: Recheck mesh quality: The results using surface mesh plots before and after mesh improvement are compared. The objective function for global mesh quality improvement is evaluated. The cohort behaviour could be considered to be saturated if there is no significant improvement in $f^*(x^c y^c, z^c)$ of every candidate $c(c = 1, \ldots, C)$ in the cohort, and the difference between individual behaviours is not very significant for a large number of iterative learning attempts, then the convergence is achieved.

Step 10: Import the improved mesh to ANSYS. The complete flow is summarized in flowchart 1.

4 Results and Discussions

The problem was coded in MATLAB R2016a on Windows Platform with Intel Core processor with 4 GB RAM and plot mesh function developed by [25]. Mesh improvement can be validated by checking the condition number improvement within range 1–3 for number of nodes of modified mesh data.

4.1 Stage-Wise Improvements

Mesh Improvement in stage I: In the proposed work, stage 1 is 3D mesh improvement using roulette wheel approach for deciding following behaviour of candidates in the cohort. Table 1 shows the condition number analysis for the Mesh with 27 elements. Several nodes do not lie within the range of 1–3 after improvement, so current scheme needs more refinement in further stages.

Mesh Improvement in stage II: Stage 2 is 3D or overall mesh improvement using global best behaviour to be followed by all candidates in the cohort. As shown in Table 2, the maximum number of nodes lies within 1–3 after mesh improvement. It proves that current optimization-based smoothing; that is, cohort algorithm gives promising results. The positive values show that elements are not inverted.

Table 1 Comparison based on condition number count

Condition number range (k)	Original number of nodes	Condition number optimization	
		Stage 1	Stage 2
1–1.5	35	79	216
1.5–2	82	93	0
2–3	56	42	0
3–4	16	2	0
4–5	11	0	0
5–8	12	0	0
8 and above	4	0	0

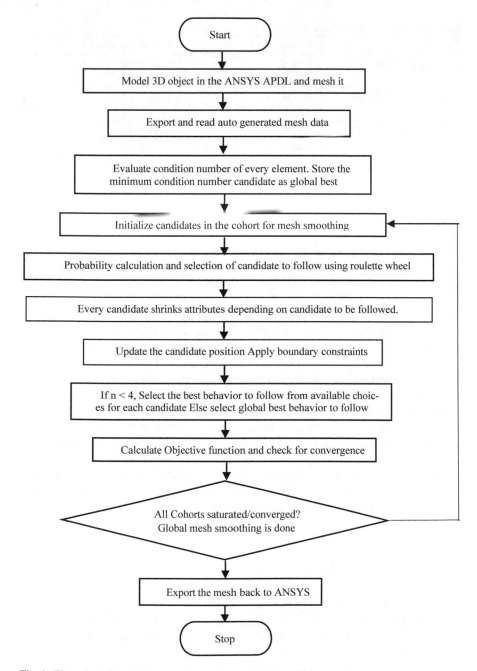

Fig. 1 Flow chart for condition number optimization using CI

Table 2 Condition number reduction

Mesh data	Results	Condition number
27 elements, 64 nodes	Before	510.1303
	After	1.0117
64 elements, 125 nodes	Before	26.0429
	After	1.2136
729 elements, 1000 nodes	Before	12.1118
	After	1.7188

4.2 Graphical and Statistical Analysis

As shown in surface mesh plot followed by convergence plot in Figs. 2 and 3, the irregular mesh having 27 and 64 elements is smoothened using CI in two stages.

The graphical plot gives an idea about condition number optimization. Y axis gives a measure of objective function value and X axis refers to the learning attempts needed. Results are plotted for a various number of element data and all shows fine refinement after several learning attempts. The plots are also obtained for 729 elements. As results stated above shows about 95% improvement in the noisy version of mesh, we can say that CI can be used for hexahedral mesh smoothing.

The mean solution, worst solution, best solution, and standard deviation (Std. Dev.), mean run time (in seconds) over the 30 runs of the algorithm for three cases of mesh improvement problems are represented in Tables 2 and 3, respectively. Condition number is a node-based quality measure for hexahedral elements while the objective function shows the overall mesh improvement. If the value of condition number lies between 1 and 8, it is considered as acceptable range for meshing, but value between 1 and 3 is the good-quality measure for the same. Since the value of the highest condition number is close to 1 and standard deviation (σ) of the objective function is closer to 0, we can say that CI can be applied for mesh smoothing if objective function and condition number are taken into account.

5 Conclusions and Future Directions

A method for mesh improvement based on condition number optimization by using CI is validated in this work. Hexahedral meshes of 3D for a cubical model of various element sizes can be smoothed with this technique. This work needs to be updated to handle complex geometries such as prismatic elements with hexahedral mesh, meshing of geometry with stress concentration regions, and enhancement in convergence time by means of adjustments in the candidate behaviour.

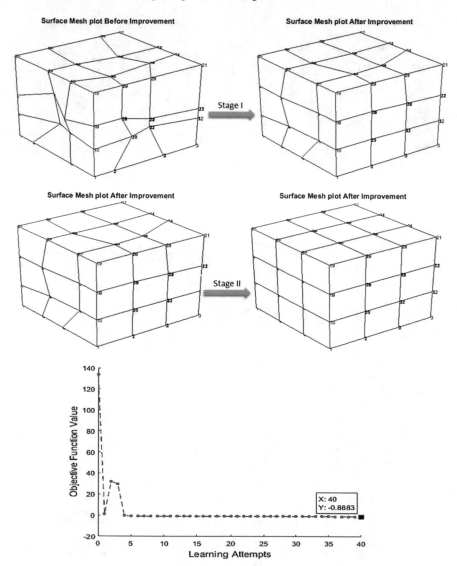

Fig. 2 Surface mesh plot and convergence plot: 27 elements, 64 nodes

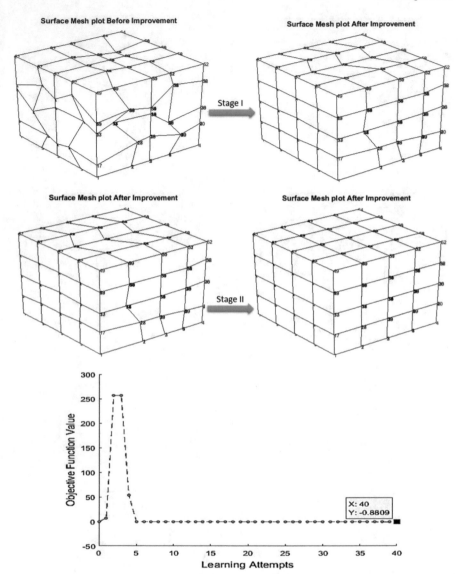

Fig. 3 Surface mesh plot and convergence plot: 64 elements, 125 nodes

Table 3 Condition number-based robustness check

Mesh data	Results	Condition number	σ
27 elements, 64 nodes	Worst	1.0117	0.001429098
	Best	1.0014	
	Mean	1.0065	
64 elements, 125 nodes	Worst	1.2136	0.006551462
	Best	1.0045	
	Mean	1.1041	
729 elements, 1000 nodes	Worst	1.7188	0.00012563
	Best	1.0216	
	Mean	1.3702	

References

1. Ho-Le K (1988) Finite element mesh generation methods: a review and classification. Comput Aided Des 20(1):27–38
2. Bern M, Eppstein D (1992) Mesh generation and optimal triangulation. Comput Euclidean Geom 1:23–90
3. Field DA (1988) Laplacian smoothing and Delaunay triangulations. Int J Numer Methods Biomed Eng 4(6):709–712
4. Knupp PM (2001) Algebraic mesh quality metrics. SIAM J Sci Comput 23(1):193–218
5. Knupp PM (2003) A method for hexahedral mesh shape optimization. Int J Numer Meth Eng 58(2):319–332
6. Munson T (2007) Mesh shape-quality optimization using the inverse mean-ratio metric. Math Program 110(3):561
7. Sastry SP, Shontz SM (2012) Performance characterization of nonlinear optimization methods for mesh quality improvement. Eng Comput 28(3):269–286
8. Yilmaz AE, Kuzuoglu M (2009) A particle swarm optimization approach for hexahedral mesh smoothing. Int J Numer Meth Fluids 60(1):55–78
9. Sarrate, J, Huerta A (2004) A new approach to minimize the distortion of quadrilateral and hexahedral meshes. In: 4th European congress on computational methods in applied sciences and engineering
10. Kulkarni AJ, Durugkar IP, Kumar M (2013) Cohort intelligence: a self-supervised learning behavior. In: 2013 IEEE international conference on systems, man, and cybernetics (SMC), pp 1396–1400
11. Kovalev K (2005) Unstructured hexahedral non-conformal mesh generation. Faculty of Engineering, Vrije Universiteit Brussel
12. Amenta N, Bern M, Eppstein D (1999) Optimal point placement for mesh smoothing. J Algorithms 30(2):302–322
13. Parthasarathy VN, Kodiyalam S (1991) A constrained optimization approach to finite element mesh smoothing. Finite Elem Anal Des 9(4):309–320
14. Canann SA, Tristano JR, Staten ML (1998) An approach to combined Laplacian and optimization-based smoothing for triangular, quadrilateral, and quad-dominant meshes. InIMR, pp 479–494
15. Freitag LA, Ollivier-Gooch C (1997) Tetrahedral mesh improvement using swapping and smoothing. Int J Numer Meth Eng 40(21):3979–4002

16. Freitag L, Jones M, Plassmann P (1999) A parallel algorithm for mesh smoothing. SIAM J Sci Comput 20(6):2023–2040
17. Sastry SP, Shontz SM (2009) A comparison of gradient-and hessian-based optimization methods for tetrahedral mesh quality improvement. In: Proceedings of 18th international meshing roundtable, pp 631–648
18. Holder M, Richardson J (1998) Genetic algorithms, another tool for quad mesh optimization. In: IMR, pp 497–504
19. Acikgoz N, Bottasso CL, Detomi D (2004) Metric based mesh optimization using simulated annealing. In: 4th European congress on computational methods in applied sciences & engineering ECCOMAS 2004
20. Dittmer JP, Jensen CG, Gottschalk M, Almy T (2006) Mesh optimization using a genetic algorithm to control mesh creation parameters. Comput-Aided Des Appl 3(6):731–740
21. Kulkarni AJ, Baki MF, Chaouch BA (2016) Application of the cohort-intelligence optimization method to three selected combinatorial optimization problems. Eur J Oper Res 250(2):427–447
22. Kulkarni AJ, Shabir H (2016) Solving 0–1 knapsack problem using cohort intelligence algorithm. Int J Mach Learn Cybernet 7(3):427–441
23. Kulkarni O, Kulkarni N, Kulkarni AJ, Kakandikar G (2016) Constrained cohort intelligence using static and dynamic penalty function approach for mechanical components design. Int J Parallel Emergent Distrib Syst, pp 1–19
24. Kulkarni AJ, Krishnasamy G, Abraham A (2017) Cohort intelligence: a socio-inspired optimization method. Springer, Heidelberg
25. Kolukula SS (2016) PlotMesh source code (version 2), Mathsworks (Source code) https://sites.google.com/site/kolukulasivasrinivas/

System Models with Threshold Cryptography for Withdrawal of Nodes Certificate in Mobile Ad Hoc Networks

Priti Swapnil Rathi, Kanika Bhalla and CH. Mallikarjuna Rao

Abstract Achieving security in mobile ad hoc network (MANET) is somewhat difficult compare to wired network because in ad hoc network malicious nodes can freely move from one location to another location and thus able to launch attacks against different nodes. In our previous work, we have explained a simple method to identify the malicious node that is collect information from, n nodes present in MANETs; but in that approach it is difficult to achieve efficient and secure key management as well as it is difficult to differentiate between valid and false accusation made by well-behaving and malicious nodes. As an extension to our previous work, this paper will help to achieve more security by using efficient key management technique. In this technique, we are dividing the functionality of the certificate authority. Threshold sharing techniques are used for distributing CA functionality. In this mechanism, every node present in the network will hold a piece of certificate authority signing key and multiple nodes in a one-hop monitoring range will jointly provide complete service. This paper also clarifies how to distinguish between legal and false accusations messages, as well as different system models will help our planned scheme for revoking the certificate of malicious node efficiently, and for avoiding false accusation attack which can occur due to inside malicious node.

Keywords MANET network · Threshold cryptography · Digital certificate

P. S. Rathi (✉) · K. Bhalla · CH. Mallikarjuna Rao
Department of Computer Science and Engineering,
GRIET College of Engineering, Hyderabad, India
e-mail: pritirathi2@gmail.com

K. Bhalla
e-mail: kanika4world@gmail.com

CH. Mallikarjuna Rao
e-mail: chmksharma@yahoo.com

© Springer Nature Singapore Pte Ltd. 2019 481
A. J. Kulkarni et al. (eds.), *Proceedings of the 2nd International Conference on Data Engineering and Communication Technology*, Advances in Intelligent Systems and Computing 828, https://doi.org/10.1007/978-981-13-1610-4_49

1 Introduction

A mobile ad hoc network (MANET) is a group of collection of more than one low-volume calculating devices such as laptops, PDA. These devices are connected to each other through wireless links and do not depend on the predefined infrastructure to keep the network connected. Mobile will act as a node in MANETs. Every node present in the MANETs can work as a sender, receiver or router who is responsible for sending and receiving of data.

There are two main types of network architectures are present, and those are wired network and wireless network. MANETs are type of wireless network. Every network, such as wired network, wireless network, local area network, will have their own features which make them unique from other types of networks. Likewise, MANETs also have their own characteristics which make it different from wired network [1, 2] and these unique characteristics of ad hoc network itself will create opportunities and challenges for achieving security in network. Following are the some of the problems which can arise due to MANETs unique features.

1.1 Absence of Centralized Entity or Server

Following are the some of the task to fulfill these tasks successfully and efficiently centralized entity or server is required that are as follows:

• To establish better trust management

Difficult to establish better trust management because in traditional network; generally, this trust management task gets done by the centralized entity. Every node which is available in MANETs needs to cooperate with the operation which gets performed inside network.

• Issuing and revocation of certificates to nodes

Difficult to perform certificate issuing and certificate revocation task because; certificates are get issued by centralized entity. Storage of certificates and retrieval of certificates in MANETs and it is difficult to perform storage and retrieval of certificates because in traditional network, the information about valid and revoked certificate get stored at some central repositories such that it will get accessed by all the nodes that are present in the network but in MANETs there is no centralized server which is present and so in the proposed system, the information about valid and revoked certificates needs to be preserved toward every node present in MANETs [0][4].

1.2 Infrastructure Support Absence

The absence of infrastructure support makes it impossible from making use of any traditional security methods such as DC for achieving security in MANETs [0]. Mostly, the DC method gets used for achieving security in wired types of network but when it comes to wireless communication network at that time these wired network security methods will not be useful for MANETs for achieving security because it will create different problems.

1.3 Dynamically Change in Network Topology

The nodes in MANETs are permitted to link and to dispensation from the network at any time. This unrestricted mobility feature of the wireless network makes always some changes in topology, with each and every single movement of nodes. This change in the network formation will affect into some route change, loss of packet from sender to receiver.

1.4 Nonexistence of Secure Boundaries

In wired network, when any outside employer wants to enter into the network for performing malicious activity, then first that node needs to go through different security mediums such as firewall, gateway before they can perform any malicious actions to the target nodes; by using these all security mediums that particular node which is adversary get found easily, and next time, it will not allow that adversary node to enter inside the network and thus target node get protected from malicious activity.

In MANETs, the adversary node does not require to go through different security mediums to access the network because once the adversary node will come within the radio range if any other node present in network, then that adversary node will be able to connect with the nodes which are present within its radio range and thus join the network immediately [3]. Once adversary node gets entered into the radio range of other node, then that respective node will be able for performing false accusation attack and try to prove genuine node as an enemy node.

These are the different challenges that can occur in ad hoc network while achieving security due to its unique characteristics. The MANETs help to set up a temporary network which will allow the user to perform instant communication due to this, and the focus on MANETs has increased lot of attention in the recent years, and due to this increased focus on MANETs, the security-related issues are considered as important and brought to front position.

Different methods such as symmetric key cryptography, digital certificates are existing for achieving security in wired network. These all methods are not able to provide security for MANETs because of the absence of central authority (CA), absence of physical connection presents between nodes, mobility. Certificate revocation theaters play a vital role for achieving sufficient amount of security in MANETs. Revocation of certificate issue in wired network is easy because once certificate of malicious node gets revoked, and then certificate authority stores this revocation information into certificate revocation list (CRLs). The CRLs are either stored on some accessible repositories or else broadcast to all the existing nodes in MANETs. Handling of certificate revocation problem in MANETs is difficult task compared to the wired network because the wired security methods are not able to provide security in MANETs. The planned system makes use of threshold-based cryptography method to revoke certificate of malicious node quickly and correctly and try to achieve security as well as protect legitimate nodes from malicious nodes in MANETs.

This paper is getting break into different section where Sect. 2 analyses and compares the proposed system with the already available techniques, next Sect. 3 gives the idea of system models, and finally, Sect. 4 describes proposed threshold-based cryptographic method with their design.

2 Related Work

This section will describe numerous approaches which are already present for MANETs and their limitations.

2.1 URSA

URSA [4] technique makes use of certified ticket-based approach. This approach, provides the tickets for the node after that the node who is consuming legal ticket those nodes are allowed to enter into MANETs whereas remaining nodes, who are not having legal tickets those nodes are not allowed to by URSA [4] for entering into network. All the tickets of nodes in URSA [4] get managed locally inside the network itself for removing adversary and malicious node form's network. The ticket of malevolent node gets cancel, when it exceeds the number of votes of its fellow citizen beyond the predefined threshold value.

2.2 Voting-Based Scheme

This scheme [5] is a transformed format of the URSA. The main modification amongst URSA and voting scheme is, it allows the nodes presents in MANETs to vote through different weight. Weight of nodes vote gets calculated from node consistency as well as nodes previous performance. To revoke the certificate of malevolent node, it estimates weights of all the polls which are received in contradiction of a particular node and after calculating sum of these reaches to a predefined threshold value, then it will remove the certificate of that particular node.

2.3 Decentralized Suicide-Based Approach

Voting-based scheme [5] is a modified version URSA. The foremost variance among URSA and decentralized suicide-based approach is that this approach permits the nodes that present in MANETs to poll with different weight value. Nodes reliability and nodes historical performance get taken into consideration for calculating weight of the nodes.

2.4 Working of Existing Methods with Proposed Methods in Comparative Analysis Form

Table 1 indicates different features and also explains the brief working of the existing methods. First column indicates the name of the existing method with its reference number in bracket. Second column indicates whether that particular excising method will make use of CA for issuing and revoking of certificate and fourth column gives information about who is going to issue certificate to nodes and revoke certificate of malicious node. Fifth column indicates the mechanism used by the existing methods to revoke the certificate of node, and last column gives information about the period required for revocation of nodes certificate by these existing methods.

Table 2 gives comparative analysis of the existing method with our new proposed methods. It gives information about advantages and limitations of the existing method whose name is indicated in first column of this Table 2 and last column of Table 2 indicated the advantages provided by planned approach. The relative study of third and fourth column indicates that the new proposed method tries to solve

Table 1 Working of CA and effect of attack on existing methods

Existing methods	CA	False accusation attack
URSA [4]	Not used	Robust for single node
Voting-based scheme [5]	Not used	–
Suicide for the common good [10] or decentralized suicide based approach	Not used	Not able to differentiate between valid and false accusation messages

Table 2 Working of existing methods for certificate revocation task

Existing methods	Mechanism to revoke certificate	Time required to revoke certificate
URSA [4]	Takes opinion from multiple neighboring nodes which are present within one-hop monitoring [12]	Less compared to [5]
Voting based scheme [5]	Allows all nodes to vote with variable weight and if sum of weight of all nodes votes against particular node exceeds a predefined threshold [13]	More compare to [4]
Suicide for the common good [10] or decentralized suicide based approach	Even if only one has made accusation against another node, then certificate of accused as well as accuser get revoked [12]	Very less compared to [4, 5]

Table 3 Comparative analysis of existing methods with new proposed method in this paper

Existing methods	Advantages of existing methods	Limitations of existing methods	Newly proposed methods
URSA [4]	Strong for false accusation attack generated by single node	The subject of spotting false accusation attack generated by more than one nodes that present in MANETs is still not resolved	Able to detect false accusation attack caused by one node as well as by more than one nodes
Voting based scheme [5]	Improves the accuracy of certificate revocation	Increases the time required to revoke certificate of malicious node	Certificate of node get revoked when it will detect the first misbehavior of that node

the limitations of these existing methods. Tables 1 and 2 show the working of the existing methods with different parameters (Table 3).

3 Planned System Models

In this unit, we will discourse about the net model, the trust model, the attack model, and the mathematical model which will be used in the proposed system.

3.1 Network Model

We consider a wireless MANETs as shown in Fig. 1 where the network is made-up from 'N' nodes and $N > 0$.

Once the network gets created with specified number of nodes 'N', then from that network our proposed system draws accusation graph $G = (V, E)$ where G represents 'Directed Graph,' V represents 'Nodes,' and 'E' represents 'Edges' as shown in Fig. 2.

In Fig. 1, nodes in the network communicate with each other if they present inside the radio or else it gets performed in multihop or ad hoc manner where more than one hop is requiring to send data from source to destination. The value of 'N' changes dynamically as new node joins or leaves the network. Each node has a unique ID which is used to identify the node. Due to the absence of structure supports the nodes that present in network and needs to get prepared with some additional net functionality those include packet forwarding, need to be fortified with a local one-hop intensive care apparatus. The one-to-one monitoring technique helps for finding neighboring nodes between its straight neighbors.

3.2 Trust Model

To achieve security in MANETs, every node which is present in the networkneeds to be genuine. This model helps to determine on which node user can keep trust and on which not. In every security design, 'trust' is very basic and important element.

Fig. 1 MANETS with six nodes

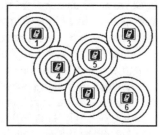

Fig. 2 Accusation graph

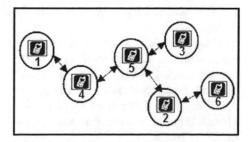

Basically, there are two main trust models are available that are trusted third-party model (TTP) [6] and the PGP 'web-of-trust' model [2].

Crepe and Davis [7] proposed a trust mechanism which will help to establish a good trust relationship among the nodes that are present in MANETs as follows: All the nodes in the MANETs maintain 'Profile Table' (PT). This table maintains different information details regarding other nodes that are present in MANETs such as nodes which they have determined as accused and their behavior index β_i. β_i is used to calculate the reputation of each node.

If A_i is large for ith node, then β_i or reputation of ith node decreases the supplementary nodes that do not believe on the accusation completed by the node whose β_i value get decreased. This model defined by [7] helps to establish good trust relationship but the β_i handles only skyjacking nodes. Opponent model is not wide and also not stated, and how the nodes are authentic, what is to be done with the malicious node as well as the process of exchanging PT each time is time-consuming.

A. **Distributed Trust Model**:

We define a scattered trust model. This model gets used to validate the nodes preset in the MANETs. It is also called as a 'trusted third-party model' where 'certificate authority' (CA) will play the role of TTP. We are using the concept of 'secret sharing,' which is based upon Shamir's 'secret sharing model' [8, 9] through threshold cryptography.

Shamir's secret sharing model helps for key management and also provides privacy among a set of nodes N. To protect message generally, we make use of encryption algorithm and encrypt that message to protect it; but what about the key protection which gets used for encrypting message? What will happen with it if it the attacker obtained it? If attacker will succeed to obtain the encryption key using which the message gets encrypted, then that attacker can easily obtain our message and misuse it. So, to avoid damage and to protect message securely we need to use efficient key management technique. Threshold scheme helps to manage encryption key so that message gets protected. In our proposed scheme, we need to protect CA signing key.

Generally, the key is kept at a central location as shown in Fig. 3, if attacker 'E' is able to access 'C' provider, then that attacker can easily obtain the CA signing key 'D' and then by making use of obtained CA signing key, the attacker is able to listen the communication which is done between source 'A' and destination 'B' and thus security get break. This will happen because the secrete key 'D' is stored at a central location and so this scheme is highly unreliable. To avoid this and to protect secrete key, threshold scheme is get used. Threshold scheme was introduced by Shamir in [1, 10].

Using this scheme, user is allowed to divide a CA signing key into 'n' different shares such that $k < n$ as shown in Fig. 4. Data 'D' is nothing but CA signing key in our proposed system. Figure 4 shows that CA signing key is get divided into 'n' parts. Figure 5 shows that it is possible to obtain CA signing key by making use of any 'k' or from more D_i ($D_1, D_2, \ldots D_{k-1}, D_k, D_{k+1}, D_{k+2}, \ldots D_n$) pieces, whereas Fig. 5 shows that it is possible to obtain CA signing key by making use of any 'k' or from more D_i ($D_1, D_2, \ldots D_{k-1}, D_k, D_{k+1}, D_{k+2}, \ldots D_n$) pieces, whereas

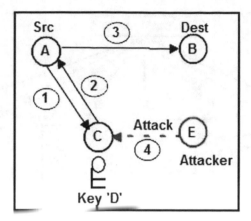

Fig. 3 Without threshold cryptography and secrete sharing (where 'A' → Source 'B' → Destination 'C' → Provider 'E' → Attacker 'M' → Message, 1 'A' makes request to 'C' for obtaining CA signing key 'D', 2 'A' makes request to 'C' for obtaining CA signing key 'D', 3 'A' makes request to 'C' for obtaining CA signing key 'D', 4 'A' makes request to 'C' for obtaining CA signing key 'D')

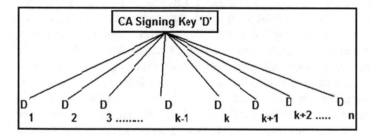

Fig. 4 Without threshold cryptography and secrete sharing

it is not possible to obtain CA signing key by making use of $k - 1$ or fewer D_i (D_1, D_2, D_3 . . .) pieces.

This type of structure is called a (k, n) threshold scheme. For efficient key management and to avoid the drawbacks occurred in Fig. 3, we make use of threshold cryptography. Consider, for example, there is one ad hoc network in which CA private key get used to sign the certificates of each and every node which are present in the network. In this situation, there are two options are available to keep CA private key secure that are.

If we are given a copy of the private key to all the nodes in MANETs so that each node is able to protect their encryption key, then in this situation the system is convenient but there are lots of chances of misuse of key.

If we make use of teamwork mechanism, then to protect CA signing key the cooperation is required among all the node's which are existing in MANETS for signing the certificate of each node. This option will keep our system safe but it is not convenient because every time the cooperation among all the network node is required to sign the certificate of newly entered node. Threshold cryptography (k, n)

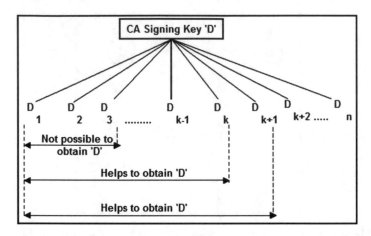

Fig. 5 Obtaining data 'D'

allows us to develop standard solution, for example, if $(k, n) = (3, n)$, then for each node a small piece D_i of CA signing key 'D' is given then to generate a temporary copy of the actual CA signing key 'D'. At least three secretes of signing key 'D' are required, and after using this temporary copy that gets destroyed and in this situation, if any malicious node want to perform any malicious activity such as obtaining CA signing key, then that malicious node requires at least two node's which also want to perform malicious activity and which will help him to obtain CA signing key.

Figure 6 shows that if attacker 'E' is able to perform attack on provider 'C', then also attacker 'E' is not able to obtain CA signing key because it will obtain only one part of signing key 'D', and due to this, attacker 'E' is not able to perform malicious activity.

3.3 Attack Model

This paper focuses on false accusation attack and to some extent with misbehaviors that can occur in network layer. The main feature of network layer is sending the data toward destination, that is, routing and packet forwarding; so our attack model tries to solve false accusation attack, packet forwarding that sends packet to well-behaving destination. In this, we are not considering other attacks that can occur at other layers.

Initially, all the 'N' nodes using which the ad hoc network get created that 'N' nodes are considered as well-behaving nodes so initially the attack can be initiated by a single node. Once the attack gets initiated, then again accusation graph $G = (V, E)$ get updated which will help to find out A_i and β_i depending on the attack that get occurred and then certificate revocation module which is discussed in our previous work [11] and get invoked to determine the value of α_i and ω_i and from all of these

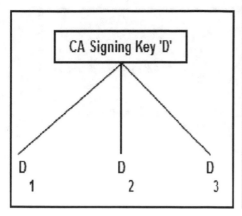 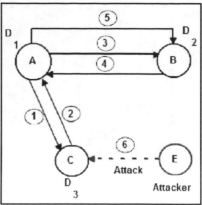

Fig. 6 With threshold cryptography and secrete sharing (Suppose $(k, n) = (3, n)$. Where 'A' → Source 'B' → Destination 'C' → Provider 'E' → Attacker 'M' → Message, 1 'A' request to 'C' for obtaining one piece of CA signing key 'D', D3, 2 'C' gives D3 piece of key 'D' to 'A' after verification and authentication, 3 'A' makes request to 'B' for obtaining other piece of CA key 'D' that is D2, 4 'B' gives D2 piece of key 'D' to 'A' after verification and authentication, 5 'A' will generate temporary key and communicate with 'B', 6 Although 'E' is able to access 'C', 'E' will not be able to perform malicious Activity)

values R_j get calculated. This calculated R_j will help to find out whether it is needed to revoke the certificate of jth node or not. Single or more than one malicious nodes can also target other well-behaving node and then false accusation attack get occurs. When attack gets occurred due to multiple misbehaving nodes, then it is called as joint accusation against well-behaving node.

In this paper, our focus is more toward the insider attacks which can occur due to malicious or selfish nodes. We consider malicious or selfish nodes which are already present in the system. We are considering the attacks caused by single node, as well as by multiple nodes which works within collaboration.

False accusation attack means the spiteful node drive to show the well behaving, nodes as attacker and owing to this the well-behaving node gets removed from MANETs Fig. 7. Demonstrate that user has designed MANETs with four nodes that are $i, j, k,$ and L where well-behaving nodes are indicated by white color and malicious nodes by red color. Initially, all the nodes are considered as a well-behaving node and attack gets initiated by single node suppose that node is 'j'. Then in this situation, if node j will try to prove any well-behaving nodes from the available well-behaving nodes, $i, k,$ and L as malicious node by generating and sending fake accusation message to other nodes except to the node to which that is trying to prove as malicious node and then we can say that the false accusation attack get occurred. Here we are considering that malevolent node j is trying to show genuine node L, as malevolent node then Fig. 7. For this will look like as shown in Fig. 8.

Figure 8 shows the network when malevolent node j will try to prove well-behaving node L, as malevolent node. Node j is generating fake accusation message

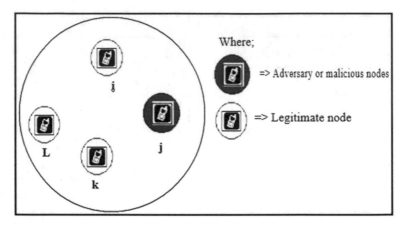

Fig. 7 Original network

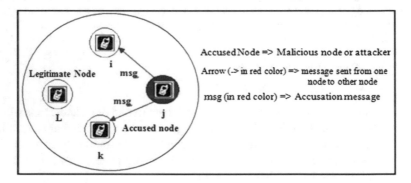

Fig. 8 False accusation attack

and then sends this fake accusation message to node *i* and node *k* to tell them that node *L* is malicious node and by doing this the malicious node *j* in this example is trying to show well-behaving node *L*, as malevolent node in such situation we can say that 'false accusation attack' get occurred in the network, and this will result in the revocation of well-behaving node *L* from the MANETs network while malevolent node leftovers in the network as shown in Fig. 9.

Figure 9 shows the effect of false accusation attack on network shown in Fig. 7. It shows that due to false accusation attack generated by malicious node j for legitimate node *L*, the well-behaving node *L* gets detached from MANETs; however, the malevolent node *j* will present in the network. The original network is shown in Fig. 7. Will result into Fig. 9. Due to false accusation attack.

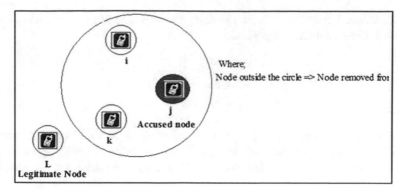

Fig. 9 Effect of false accusation attack

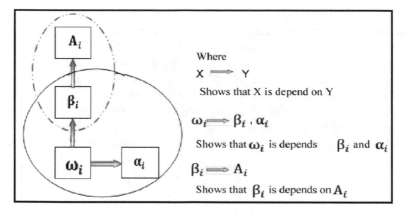

Fig. 10 Effect of false accusation attack

3.4 Planned System Mathematical Model

Figure 10 demonstrates that the mathematical model of planned system.

Parameters A_i, α_i, β_i, ω_i of ST bring variety of information. We refer to recent publication [11] for the detail discussion of the A_i, α_i, β_i, ω_i. In planned system, 'N' represents the number of nodes with which user want to design MANETs. If user has decided to design a MANETs with 'N' numbers of nodes, then extreme number of claim posts which remain allowed A_i agreed as per

$$A_i = Na - 1 \tag{1}$$

The node does not get charged for first claimed message which is generated, in the planned system. Depending on this situation, amount of claim message aimed at node gets exciting becomes $Na - 1$. Other important bit is that none of the node present in the MANETs will not produce claim note in contradiction of themselves.

Finally, reduce 1 from Na − 1, for obtaining worth aimed at total amount of claim messages aimed at node charges thus

$$\alpha_i = \text{Na} - 1$$
$$\alpha_i = \text{Na} - 1 - 1$$
$$\alpha_i = \text{Na} - 2 \tag{2}$$

Calculation of equation (1), (2) gives the value for A_i as well as on the value of α_i, as Na − 1, Na − 2 respectively. In Fig. 10, the projectile $(X \Rightarrow Y)$ determination specifies that adjustable X is dependent on Y. It also gives information about two different relationships which are present among diverse variables by manufacture use of dotted and plain ovals. Unadorned ovals from Fig. 10 demonstrate the β_i rest scheduled the A_i since β_i get used for calculating the morality of the node. The nodes' honesty is get intended by manufacturing the use total number of claim messages, which are made in contradiction of particular node ith, A_i.

If large number of nodes such as $A_i = $ Na − 1, when Na is the number of node count which are in MANETs has generated accusations message against any node n_i, at that time node n_i is in query that whether it is malevolent node or well-behaving node and due to this less weight get assigned to the claim messages that are generated by node n_i which is in question. So to assign weight to the accusation messages generated by any node n_i proposed system needs to first determine the morality of that node i that β_i which depends on A_i. Therefore, from this obtained β_i

$$\beta_i = 1 - \lambda A_i \tag{3}$$

Scattered oval in Fig. 10 protests that ω_i is rest on β_i and α_i since we know that ω_i is used to assign weight to the claims memo that made by the node i who has detected misconduct of another node j. Then to assign weight to the accusation, we need to determine the honesty of that node i firstly that means β_i which will help to find out the honesty of the node So that depending on that weight is get assigned to claim message made by that node i. So Fig. 10 shows that ω_i depends on β_i which will help to find out behavior of node i. Therefore, from this gained ω_i is:

$$\omega_i = \beta_i - \lambda \alpha_i \tag{4}$$

The value of λ in Eqs. (3) and (4) got as shadow:

$$A_i = \text{Na} - 1 \tag{1}$$
$$\alpha_i = \text{Na} - 2 \tag{2}$$
$$\beta_i = 1 - \lambda A_i \tag{3}$$
$$\omega_i = \beta_i - \lambda \alpha_i \tag{4}$$

Place value of A_i obtained in Eq. (1) into Eq. (3)

$$\beta_i = 1 - \lambda A_i \tag{3}$$

$$\beta_i = 1 - \lambda(Na - 1) \tag{5}$$

Place value of α_i obtained in Eq. (2) into Eq. (4)

$$\omega_i = \beta_i - \lambda \alpha_i \tag{4}$$

$$\omega_i = \beta_i - \lambda(Na - 2) \tag{6}$$

Place value of β_i obtained in Eq. (5) into Eq. (6)

$$
\begin{aligned}
\omega_i &= \beta_i - \lambda(Na - 2) \\
&= 1 - \lambda(Na - 1) - \lambda(Na - 2) \\
&= 1 - \lambda[(Na - 1) + (Na - 2)] \\
&= 1 - \lambda[(Na - 1 + Na - 2)] \\
&= 1 - \lambda[(Na + Na - 1 - 2)] \\
&= 1 - \lambda[(2Na - 3)]
\end{aligned}
\tag{6}
$$

$$1 = \lambda[(2Na - 3)] \quad \text{Thus} \quad \lambda = \frac{1}{2Na - 3}$$

4 Conclusion

In this paper, we have extended our previous work on certificate revocation to achieve better security and efficient key management by distributing certificate authority's functionality through threshold secrete sharing. Our work is divided into five phases that consists of creating ad hoc network, establishing trust among the nodes, distributing certificate authority's functionality, generating attack, and removing it. We used Shamir's secret sharing model with severance to reduce the effect of malicious node because it states that by using less than $(k - 1)$ pieces, it is not possible by attacker to recreate the certificate authority key, and thus, it will help to improve the integrity of the ad hoc network. The proposed scheme removes window of opportunity problem, false accusation attack, improves reliability of network, achieve efficient key management. For certificate revocation, our proposed scheme takes the reliability of each node into consideration and depending on the nodes reliability it assigns weight ω_i to each accusation message A_i made by nodes which will help to take decision about whether to revoke the certificate of node or not.

References

1. Shamir A (1979) How to share a secret. Commun ACM 22(11):612–613
2. Zimmermann P (1995) The official PGP user's guide. MIT Press, Cambridge
3. Liu W, Nishiyama H, Ansari, Nakao (2011) A study on certificate revocation in mobile ad hoc networks. IEEE
4. Luo H, Kong J, Zerfos P, Lu S, Zhang L (2004) URSA: ubiquitous and robust access control for mobile ad hoc networks. IEEE/ACM Trans Netw 12(6):1049–1063
5. Arboit G, Crepeau C, Davis CR, Maheswaran M (2008) A localized certificate revocation scheme for mobile ad hoc networks. Ad Hoc Netw 6(1):17–31
6. Perlman R (1999) An overview of PKI trust models. IEEE Netw 13(6):38–43
7. Crepe C, Davis C (2003) A certificate revocation scheme for wireless Ad hoc networks. 1st ACM workshop on security of ad hoc and sensor networks, pp 54–61
8. Stamatios V (2009) Security of information and communication networks. Wiley-IEEE Publications, USA
9. Wikipedia (2011) Shamir's secret sharing. Available at: http://en.wikipedia.org/wiki/Shamir''s_Secret_Sharing, last visited 2011
10. Luo J, Hubaux JP, Eugster PT (2005) DICTATE: Distributed CerTification Authority with probabilisTic frEshness for ad hoc networks. IEEE Trans Dependable Secure Comput 2(4):311–323
11. Rathi P, Mahalle P (2013) Proposed threshold based certificate revocation in mobile ad hoc networks. ICACNI 243(2014):377–388
12. http://essaymonster.net/technology/13772-proposed-threshold-based-certificate-revocation.html
13. Sabeena S (2014) Reduced overhead based approach for secure communication in mobile ad hoc networks. IOSIR-JCE 16(5), Ver. III:73–78

Priti Rathi has completed her BE in computer science and engineering in 2010 and M. Tech in 2014 from Pune University. She has achieved distinction throughout academic examination in BE and having 2 years of experience in teaching. Currently, she is working in 'Gokaraju Rangaraju Institute of Engineering and Technology,' in Hyderabad, as 'Assistant Professor.'

Kanika Bhalla has completed her B. Tech in computer science and engineering in 2010 and M. Tech in Software Engineering in 2014, Engineering College, Rajasthan. Currently, she is working in 'Gokaraju Rangaraju Institute of Engineering and Technology,' in Hyderabad, as 'Assistant Professor' and having 5 years of teaching experience.

Dr. Ch. Mallikarjuna Rao received the B. Tech degree in computer science at Maharashtra in 1998 and M. Tech from JNTU Anantapur, Andhra Pradesh, in 2007, and Ph.D. from JNTU Anantapur. Currently, he is working in 'Gokaraju Rangaraju Institute of Engineering and Technology,' in Hyderabad. His areas of interest are data mining, big data, and software engineering.

A QoS-Aware Hybrid TOPSIS–Plurality Method for Multi-criteria Decision Model in Mobile Cloud Service Selection

Rajeshkannan Regunathan, Aramudhan Murugaiyan and K. Lavanya

Abstract The business framework architecture and evaluation are changing the growth of mobile cloud computing (MCC). The business service users required the facilities for selecting the mobile cloud services according to their quality of service (QoS) values in the business environment. In such a business environment, QoS parameter values help to build consumer confidence and provide a reliable environment for them. So the proposed system suggests a hybrid based on TOPSIS algorithm and plurality voting method which is more efficient, trustable for selecting the best cloud services. This algorithm has three phases to identify the service in the ranking process. The first phase is used to make a group the parameters based on the user's requirements. The second phase applied the TOPSIS method on each of the parameters to get the service ranking. The final phase employed the plurality method which is counted the voting for each service to determine the best services. The hybrid algorithm takes $O(n^2)$ time in the best service selection process which is better than existing popular AHP, ANP multi-criteria decision methods in terms of time complexity.

Keywords Mobile cloud computing · Quality of service · Time complexity
Service selection

1 Introduction

A business environment has been changed in the last two decades for the globalization. There are the two major emerging technologies, i.e., mobile computing and cloud computing, which play in a business marketplace. The business organization

R. Regunathan (✉) · K. Lavanya
School of Computer Science and Engineering,
Vellore Institute of Technology, Vellore, Tamil Nadu, India
e-mail: rajeshkannan.r@vit.ac.in

A. Murugaiyan
Perunthalaivar Kamarajar Institute of Engineering and Technology, Karaikal, India

© Springer Nature Singapore Pte Ltd. 2019   499
A. J. Kulkarni et al. (eds.), *Proceedings of the 2nd International Conference on Data Engineering and Communication Technology*, Advances in Intelligent Systems and Computing 828, https://doi.org/10.1007/978-981-13-1610-4_50

needs to handle the huge secured data with mobility so that the firm takes the advantage using cloud services and mobile application. The mobile has limited power processing and memory storage capacity. Therefore, the business environment needed the cloud computing technologies to complete the business requirement and process the services. Mobile cloud services are immensely used in the business environment for various tasks such as shopping, marketing, payment.

The important challenges in service-oriented architecture are finding the best service and ranking the cloud services. The method of selecting the best service from available services in a dynamic environment is challenging task [1]. The recent research solved the service selection problem by using multi-criteria decision making (MCDM) method. For example, the popular methods AHP, ANP, and TOPSIS were applied in service selection. This paper proposed a hybrid model by using TOPSIS and the plurality method to select a better service by using QoS values of services. The plurality method used most frequently in the voting system [2]. It is used in legislative elections in the USA and India.

2 Related Work

In recent years, the mobile cloud computing (MCC) is a most popular paradigm in the business environment. The user wants to access the service from anywhere and anytime. It is very hard for users to find the best services in a dynamic environment. The service selection process is solved by MCDM method. The user can customize the QoS attribute to select a best alternative service. The most popular MCDM methods are AHP, ANP, and TOPSIS [3]. It is the most familiar technique applied in the cloud service selection. In this scenario, the service selection explored via the aggregation of finite QoS attributes. Each QoS parameter is representing the service performance aspects [4]. The main challenges of MCDM-based selection are comparable to QoS attribute among the alternatives.

Prof. Thomas Saaty developed analytic hierarchy process (AHP) method which is most popular techniques in decision-making problems. This method decomposes a problem in a structured way. This method solved a decision-making problem in three levels. The Level 0 is used to define a goal. The Level 1 defines multi-criteria of decision making, and Level 2 finds the alternative services based on Level 1 [5]. The time complexity of AHP method is $O(n^3)$ in the worst case [6]. In this scenario, the proposed hybrid model serves better for finding the best alternatives. The analytic network process (ANP) is proposed by Prof. Thomas L. Saaty. This method solved a complex problem in a network structure. The AHP and ANP have used pairwise comparison for calculating the weights of the attribute values. This method consists the three phases of selecting the services. The first phase identifies the goal of the services. The second phase is finding the criteria of the model. Finally, the third phase is rank the service based on criteria [7]. This method produces the best result in a static environment, but not in a dynamic environment. The worst case time complexity of this method is $O(n^3)$ time. This method is more suitable in nonhierarchical problem for ranking the services based on multi-criteria in the mobile cloud [8]. In this scenario, the proposed hybrid model serves better in a cloud service model.

In the year of 1981, Hwang et al. proposed the technique for order preference by similarity to ideal solution (TOPSIS) method which is one of the most popular MCDM methods which is used to find the ideal and non-ideal solution. The ideal solution is near to the best alternative and non-ideal solution far from the best solution. The ideal solution minimizes the cost criteria and maximizes the benefit whereas non-ideal solution maximizes the cost criteria and minimizes the benefit [9, 10]. The time complexity of this method is $O(n^2)$ in the worst case. The proposed hybrid model gives the best result compared to TOPSIS method. The reason is after getting the TOSIS result, apply the plurality voting method to get more reliable results.

The plurality voting method is a common method of selecting candidates for public office. It is applied most frequently in the voting system. It is used in the legislative election in the USA and India. The electoral process is the candidate who polls more votes than any other candidate who is elected. The advantages of this method are easily understood by voters. It provides a quick decision and less cost to operate another voting methods. The plurality method has been used in the proposed hybrid model in the cloud service selection. This method is applied after getting the TOPSIS result of cloud service alternative. The plurality voting system has been conducted between services. This method produces the ranking result among the services [2].

3 Methodology

3.1 Mobile Service Selection Architecture

Figure 1 mobile service architecture consists of three major parts—mobile client, mobile cloud, and cloud services. The mobile client tries to perform the tasks through the mobile application in the mobile unit. The application transfers the request to the agent who forwards the request to mobile cloud through the Internet. In the mobile cloud, the service manager identifies the tasks and determines the possible cloud services that could potentially solve the tasks. Afterward, the service manager transfers the list of cloud services to the hybrid service selection algorithm. Here the parameters are divided into groups of 2, 3, ... m parameters and then are applied on TOPSIS algorithm. Each individual ranking of groups of parameters is used to determine the final ranking using the plurality voting method to get final rankings. Based on these rankings, the service manager selects the best cloud service and patches the selected service with mobile client.

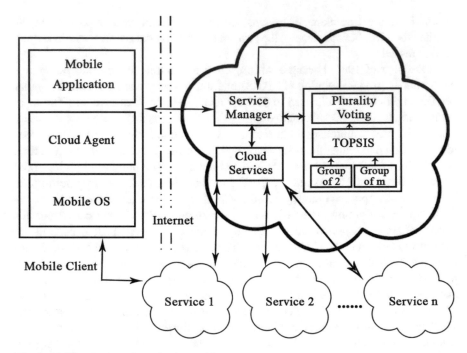

Fig. 1 Mobile cloud service selection architecture

3.2 Hybrid TOPSIS–Plurality Method for Multi-criteria Decision Model

Hybrid TOPSIS—plurality algorithm is based on the concept of improving the results of TOPSIS algorithm using the plurality voting method. The method is divided into three phases. In first phase, the service parameter data are extracted from datasets. Then, the parameters are divided into groups of 2, 3, … to m parameters. In phase 2, we will apply TOPSIS algorithm on all the different groups to generate a ranking for each group. In phase 3, we will utilize these rankings as a base to apply the plurality voting method to determine the final ranking of services.

   Hybrid TOPSIS_Plurality() Algorithm

1.   Read the no. of services—n and the no. of parameters—m from the client c.
2.   Read the values of A_{ij}, i.e., value of jth parameter for ith service.
3.   Divide the parameters into groups from 2 to m with each represented as $P_2, P_3,$ …P_m ..
4.   For j = 0 to m in P_j
     Call to TOPSIS()
     //topsis() will rank the services based on parameters that are in group j and then generate a ranking which will be stored and noted as R_j.

5. Call Plurality_Voting() with input as rankings R_j from TOPSIS() and P_j for $j = 2$ to m
6. Print the final ranking of services R_s generated from call to Plurality_Voting()

4 Experiment and Result

Test data Table 1 has been taken from QWS dataset developed by Al—Masri and Dr. Mahmoud [1, 11].

Phase 1—Grouping of parameters

a. Service—Parameters data from QWS dataset—Extract the data from the QWS dataset and convert into a tabular form with each column representing an attribute and each row representing a service [11] in Table 1.
b. Divide parameters in the groups—Parameters will divide into groups of 2, 4 up to m, where m is the total no. of parameters. Each group should have a proper distribution of minimizing and maximizing attributes (Tables 2, 3 and 4).

Minimizing attributes—response time (RT) and latency (L)
Maximizing attributes—availability (Av), throughput (T), successability (Su), and reliability (Rel)
Group of 2—response time and availability; group of 4—response time, latency, availability, and throughput; group of 6—response time, latency, availability, throughput, successability, and reliability

Phase 2—Application of TOPSIS on groups of parameters

Table 1 Sample QWS dataset

Service	RT (in ms)	Av (in %)	T (in invokes/second)	Su (in %)	Rel (in %)	L
DictionaryService	45	83	27.2	50	97.4	43
MyService	71.75	100	14.6	88	85.5	64.42
Aba	117	100	23.4	83	88	111
AlexaWebSearch	70	100	5.4	83	79.3	63
ErrorMailer	105.2	100	18.2	80	92.2	104.6

Table 2 Group of 4

Service name	RT	L	Av	T
DictionaryService	45	43	83	27.2
MyService	71.75	64.42	100	14.6
Aba	117	111	100	23.4
AlexaWebSearch	70	63	100	5.4
ErrorMailer	105.2	104.6	100	18.2

Table 3 Group of 6

Service name	RT	L	Av	T	Su	Re
Dic.Ser	45	43	83	27.2	50	97.4
MySer	71.75	64.42	100	14.6	88	85.5
Aba	117	111	100	23.4	83	88
Al.Web	70	63	100	5.4	83	79.3
Err	105.2	104.6	100	18.2	80	92.2

Table 4 Group of 2

Service name	RT	Av
DictionaryService	45	83
MyService	71.75	100
Aba	117	100
AlexaWebSearch	70	100
ErrorMailer	105.2	100

Table 5 Standardize the decision matrix—each column of the decision matrix

Criteria	Dic	Myser	Aba	Alex	Err	
RT	7	8	6.33	7	8.66	16.64
Av	7	7.33	6	7	9	16.39

Table 6 Determine ideal solution and negative ideal solution

Cri	Dic	My	Aba	Alex	Err
RT	2.94	3.84	2.40	2.94	4.5
Av	2.989	3.276	2.196	2.989	4.94

c. Prepare a service parameter table on which the TOPSIS algorithm would be applied.

d. Apply TOPSIS algorithm—After preparing service—parameter table for each group of parameters—we apply TOPSIS algorithm on each of them to get the service rankings (Tables 4, 5, 6 and 7).

Table 7 Determine relative closeness to ideal solution

Cri	Dic	My	Aba	Alex	Err
S_i^*	1.949	1.8	3.11	2.495	0
S_i'	0.958	1.798	0	0.958	3.455
$S_i^* + S_i'$	2.907	3.598	3.11	3.553	3.455
$S_i'(S_i^* + S_i')$	0.329	0.499	0	0.269	1

Table 8 Service ranking through the TOPSIS algorithm

Service	Group of 2	Group of 4	Group of 6
DictionaryService	3	4	2
MyService	2	1	3
Aba	5	3	5
AlexaWebSearch	4	5	4
ErrorMailer	1	2	1

Table 9 Voters ranking

Service	VR1	VR2	VR3	VR4	VR5	Ser. rank
DictionaryService	0	6	2	4	0	?
MyService	4	?	0	0	0	2
Aba	0	0	4	0	8	4
AlexaWebSearch	0	0	0	8	4	5
ErrorMailer	8	4	0	0	0	1

Ideal solution $= \{4.5, 4.94\}$

Negative ideal solution $= \{2.4, 2.196\}$

Ranking of services for group of 2—Errormailer > Myservice > Dictionaryservice > Alexawebsearch > Aba; similarly ranking of services for group of 4—Myservice > Errormailer > Aba > Dictionaryservice > Alexawebsearch; similarly ranking of services for group of 6—Errormailer > Dictionaryservice > Myservice > Alexawebsearch > Aba

Phase 3—Plurality voting method

A plurality voting table is prepared from Table 8 where each column represents the ranking of the group of parameters through the TOPSIS algorithm. Vote counting Table 9 is counted for each service to determine the best services. The service with the most votes as rank 1 is ranked first, and in case of a tie, the vote for the next rank is considered.

Ranking of services—Errormailer > Myservice > Dictionaryservice > Aba > Alexawebsearch

5 Performance Analysis

In HA, phase 1—Group the m parameters in k groups with each group having size of 2, 4 up to m. So time complexity will be $O(k)$.

Phase 2—Apply TOPSIS algorithm on each group of parameters. With each TOPSIS algorithm being of $O(n^2)$ [12]. Time complexity will be $O(k.n^2)$.

Table 10 Time complexity

Algorithm	AHP	ANP	HA
Time complexity	$O(n^3)$	$O(n^3)$	$O(n^2)$

Phase 3—Apply the plurality voting method to determine the final ranking [2, 13]. Time complexity will be $O(k.n^2)$. Time complexity $= O(k + k.n^2 + k.n^2) = O(k.n^2)$ (Table 10).

6 Conclusion

The proposed hybrid algorithm improved the existing TOPSIS algorithm, making the ranking determination more decisive and trustable. The plurality voting method helps in finding a reliable ranking and selection of service by finding an average ranking among the different rankings produced from different groups of parameters applied over TOPSIS algorithm. The hybrid algorithm is more efficient and easier to understand compared to other multi-criteria decision-making algorithm.

References

1. Al-Masri E, Mahmoud QH (2007) Discovering the best web service. In: Proceedings of the 16th international conference on world wide web. ACM, pp 1257–1258
2. Parhami B (1994) Voting algorithms. IEEE Trans Reliab 43(4):617–629
3. Zavadskas EK, Turskis Z, Kildienė S (2014) State of art surveys of overviews on MCDM/MADM methods. Technol Econ Dev Econ 20(1):165–179
4. Triantaphyllou E (2000) Multi-criteria decision making methods. In: Multi-criteria decision making methods: a comparative study. Springer US, pp 5–21
5. Kumar RD, Zayaraz G (2011) A qos aware quantitative web service selection model. Int J Comput Sci Eng 3(4):1534–1538
6. Garg SK, Versteeg S, Buyya R (2013) A framework for ranking of cloud computing services. Future Gener Comput Syst 29(4):1012–1023
7. Sakthivel G, Ilangkumaran M, Gaikwad A (2015) A hybrid multi-criteria decision modeling approach for the best biodiesel blend selection based on ANP-TOPSIS analysis. Ain Shams Eng J 6(1):239–256
8. Yang Y-PO, Shieh H-M, Leu J-D, Tzeng G-H (2008) A novel hybrid MCDM model combined with DEMATEL and ANP with applications. Int J Oper Res 5(3):160–168
9. Wang Y-M, Elhag TMS (2006) Fuzzy TOPSIS method based on alpha level sets with an application to bridge risk assessment. Expert Syst Appl 31(2):309–319
10. Ho W, Xu X, Dey PK (2010) Multi-criteria decision making approaches for supplier evaluation and selection: a literature review. Eur J Oper Res 202(1):16–24
11. Al-Masri E, Mahmoud QH (2007) QoS-based discovery and ranking of web services. In: IEEE 16th international conference on computer communications and networks (ICCCN), pp 529–534
12. Hamdani, Wardoyo R (2016) The complexity calculation for group decision making using TOPSIS algorithm. In: AIP conference proceedings, vol 1755, no 1, pp 070007

13. Parhami B (1991) The parallel complexity of weighted voting. In: Proceedings of Int'l symposium parallel and distributed computing and systems, pp 382–385
14. Krohling RA, Pacheco AGC (2015) A-TOPSIS–an approach based on TOPSIS for ranking evolutionary algorithms. Procedia Comput Sci 55:308–317
15. Savitha K, Chandrasekar C (2011) Trusted network selection using SAW and TOPSIS algorithms for heterogeneous wireless networks. arXiv preprint arXiv 26(8):22–29

A Rough Entropy-Based Weighted Density Outlier Detection Method for Two Universal Sets

T. Sangeetha and Amalanathan Geetha Mary

Abstract Data mining is the process of examining large databases to extract patterns, knowledge and to establish relationships to provide solution for the complex problems. Researchers have shown their interest in different areas particularly in identifying outliers. An object which significantly deviates from other objects or by their normal behaviour is termed as outliers. Real-world data have vagueness and uncertainty which can be handled by rough set theory. Research works that have been carried so far were focussed only on single universal set to detect outliers. By using intuitionistic fuzzy soft set relation, this paper proposes an idea for detecting outliers in two universal sets in which the attribute-based and object-based weighted density values are determined to detect outliers. The hiring dataset has taken for example and shown the validity of the proposed approach for outlier detection.

Keywords Data mining · Outliers · Rough set · Intuitionistic fuzzy
Two universal sets

1 Introduction

A dataset may be of mixed type which contains numerical and categorical data, and some of the objects that do not accompany with their general behaviour are identified as outliers [1]. Sometimes, outliers are discarded as noise or exceptions. Outlier detection is important in many fields such as fraudulent activities, hospital management, people's safeties, equipment damage detection in industries, artificial intelligence and in cryptographic issues.

Outliers can be easily identified in exceptional cases and novel pattern from the dataset [2]. Outliers may be detected using statistical techniques, probability model,

T. Sangeetha · A. Geetha Mary (✉)
SCOPE, Vellore Institute of Technology, Vellore, India
e-mail: geethamary@vit.ac.in

T. Sangeetha
e-mail: sangee_arasu05@yahoo.co.in

© Springer Nature Singapore Pte Ltd. 2019
A. J. Kulkarni et al. (eds.), *Proceedings of the 2nd International Conference on Data Engineering and Communication Technology*, Advances in Intelligent Systems and Computing 828, https://doi.org/10.1007/978-981-13-1610-4_51

or any distance measure with clustering technique [3]. When compared with this technique, density-based methods will identify outliers in intra-region, even though it looks normal from other approaches. Output data are the manner in which the outliers are reported. It is divided into two categories. (1) Labelling technique: For each instance, the label is assigned where the normal objects and outliers are identified easily. (2) Scoring technique: The outlier score is assigned to each pattern which depends on degree to determine outliers. The threshold value will be fixed to detect outliers in different regions.

Major domains need outlier detection in high-dimensional data. Much noise exists between two objects so that the similarity cannot be measured properly. Subsequently, the earlier outlier detection methods mainly use similarity and density measures to detect outliers, where the dimension increases [4]. The rough entropy k-means outlier detection algorithm detects outliers only for numerical data, and weighted density-based outlier detection algorithm was carried out only for categorical data. Our proposed model will work out for outlier detection in two universal sets with high level of performance.

2 Rough Set Theory

Rough set theory, a powerful mathematical tool, was introduced by Zdzislaw Pawlak during 1980s [5]. It was particularly developed by using the concepts of lower and upper approximation to deal with imprecise information in decision situation. Let us consider the knowledge base $S=(V, T)$ with each subset $Y \subseteq V$ and an equivalence relation $T \in IND(S)$. The subsets can be defined as follows:

$$\underline{T}Y = \cup\{Z \in V/T : Z \subseteq Y\} \tag{1}$$

$$\overline{T}Y = \cup\{Z \in V/T : Z \cap Y = \emptyset\} \tag{2}$$

as minimal and maximal approximation of Y, respectively. From this, boundary region of Y can be calculated as Boundary$(Y) = \overline{T}Y - \underline{T}Y$.

2.1 Approximation and Membership Relation

From the approximation space, membership relation also be derived [1]. Because both membership and sets are related to knowledge only. It can be represented as:

$$l \underline{\in}_T L \quad \text{then } l \in \underline{T}L \tag{3}$$

$$l \overline{\in}_T L \quad \text{then } l \in \overline{T}L \tag{4}$$

in which $\in_T$ reads "l surely belongs to L with respect to T" and $\overline{\in}_T$, "l possibly belongs to L with respect to T," is the lower and upper membership relation, respectively.

2.2 Two Universal Sets

Let P and Q be two non-empty finite universal sets, a binary relation (X) can be derived based on the degree of membership (μ) and non-membership (γ) levels by using intuitionistic fuzzy soft set relation $(\overline{FR})$. It can be defined as

$$\overline{FR} = \{\langle (a, b), \mu_{\overline{FR}}(a, b), \gamma_{\overline{FR}}(a, b)\rangle | (a, b) \in P \times Q\} \tag{3}$$

where $\mu_{\overline{FR}} : P \times Q \to [0, 1]$ and $\gamma_{\overline{FR}} : P \times Q \to [0, 1]$ which satisfies the condition when $0 \leq \mu_{\overline{FR}}(a, b) + \gamma_{\overline{FR}}(a, b) \leq 1$ for any $(a, b) \in P \times Q$. If $P = Q$, then the intuitionistic fuzzy relation $\overline{FR} \in IFR(P \times P)$ is known as intuitionistic fuzzy relation on P.

3 Related Work

Outliers have also been detected using rough membership function [6] to demonstrate on two publicly available datasets. From the dataset, a sample data have been clustered by using different labelling techniques and the remaining isolated data points are compared with the grouped data from which the deviated data points are termed as outliers. Manhattan distance technique was based on distance-based approach [7]. This technique outperformed the other techniques like statistical-based and distance-based approaches when the threshold value increases. The overall efficiency of the algorithm can be improved by selecting the valid outlier score. When compared with distance method, the cluster-based approach for outlier detection provides better accuracy [8]. The partitioning around medoids (PAM) are used to construct small clusters which are termed as outliers [5]. Rough sets have also been employed in neural networks to train and test the dataset using backpropagation algorithm to avoid inconsistencies within data [9].

Outlier detection method [10] is used for measuring the uncertainty and density of each object which was identified for categorical data. But clustering of data had not been done. The cluster method is enhanced by adapting preliminary centroid selection method on rough k-means (RKM) algorithm [1]. The entropy-based rough k-means (ERKM) method is developed by adapting entropy-based preliminary centroids selection on RKM and executed and also validated by cluster validity indexes [11]. In multigrain rough set, only rule extraction methods were applied to refine data from the dataset and decision rule was formed using "OR" rather than "AND" logic. The rough set framework for multi granulation was complementary in many practical applications, when two attributes from dataset had a contradiction or inconsistency relation. At that time, effective computation will be required [4]. By using intuition-istic fuzzy soft set in two universal sets, many researchers provided a solution for decision-making problems [12]. Also, precision and rough degree were used to get the optimal alternatives. They also introduced binary relation X_μ^γ with intuitionistic fuzzy relation $\overline{FR}$ between the two non-empty universe sets P and Q [7].

4 Proposed Approach

Consider the two universal sets such as P and Q which are finite and nonempty, then by using intuitionistic fuzzy soft set relation $\overline{FR}$, the binary relation between membership and non-membership levels can be derived. In general, $\mu_{\overline{FR}}$ represents the degree of membership and $\gamma_{\overline{FR}}$ represents the degree of non-membership of a and b under the relation $\overline{FR}$. The degree of membership of the object with respect to the parameter σ which should be greater than μ and the degree of non-membership of the object with respect to the parameter σ which should not be more than γ. It can be represented as $\overline{X}_\mu^\gamma$ to construct upper approximation and lower approximation $\underline{X}_\mu^\gamma$ of T. The boundary region can be determined by $\underline{X}_\mu^\gamma - \underline{X}_\mu^\gamma$.

4.1 Rough Entropy Weighted Density-Based Outlier Detection Algorithm

To compute the uncertainty of a dataset, rough entropy has been used. The weighted density value of an each object and attribute will be calculated. The objects which are having similar weighted density value will be fixed as a threshold value. The values of the object which are significantly lesser than the threshold value will be detected as outliers.

Proposition 1: Dataset Definition *A categorical dataset DS is defined by the function $DS = (W, \alpha, \beta)$ where W represents the universe, α represents the objects, and β represents the attributes in a dataset.*

Proposition 2: Indiscernibility Function *Let $DS = (W, \alpha, \beta)$ and $R \subseteq \beta$. The indis-cernibility relation R with respect to z in α or y in β is represented as*

$$\{U|IND(R)\} = \{[z]_R | z \in W\} \tag{6}$$

or

$$[z]_R = \{y \in W | (z, y) \in IND(R)\} \tag{7}$$

Proposition 3: Complement Entropy *Let* $DS = (W, \alpha, \beta)$, *and* $R \subseteq \beta$ *and* $\frac{U}{IND(R)} = \{\alpha_1, \alpha_2, \ldots \alpha_m\}$. *The complement entropy (CE) with respect to R is defined as* α

$$CE(R) = \sum_{i=1}^{m} \frac{|\alpha_i|}{|W|}\left(1 - \frac{|\alpha_i|}{|W|}\right) \tag{8}$$

Proposition 4: Weighted density value for each attribute *Let* $DS = (W, \alpha, \beta)$, *the weight of every attribute with respect to* β *is defined as*

$$W(\beta) = \frac{1 - CE(R)}{\sum_{i=1}^{m} (\beta_i)} \tag{9}$$

where $\sum_{i=1}^{m} (\beta_i)$ *represent the total number of attributes.*

Proposition 5: Weighted density value for each object *The average density of each attribute will be determined as*

$$Average\ Dens(\alpha_i) = \frac{|[\alpha_i]_\beta|}{|W|} \tag{10}$$

From that, the weighted density of each object will be determined as follows:

$$Weighted\ Density(\alpha) = \sum_{a_i \in A} (Average\ Dens(\alpha_i).W(\beta)) \tag{11}$$

Proposition 6: Fixation of threshold value *Let us consider the categorical dataset* $DS = (W, \alpha, \beta)$, *and* θ *is a fixed threshold value from the weighted density objects. If the value of Weighted Density* $(\alpha) < \theta$, *then* α *is termed to be an outlier.*

4.2   An Empirical Study on Hiring Dataset

Assume a company hire candidates based on their requirements. Ten candidates were applied for the position offered by the company. Let the candidates be represented as $S = \{S_1, S_2, S_3, S_4, S_5, S_6, S_7, S_8, S_9, S_{10}\}$ and attributes as $A = \{A_1, A_2, A_3, A_4\}$ where $A_1 = $ Degree, $A_2 = $ Experience, $A_3 = $ French, and $A_4 = $ Reference. The binary relation can be obtained by fixing the degree of membership value $\mu = 0.5$ and non-membership value $\gamma = 0.3$ which is shown in Table 2 (Table 1).

Now, the Boolean values 0s and 1s will be ordered to get categorical values. The attribute degree is ordered to UG and PG, Experience is ordered to Low and High, French is ordered to Yes and No, and Reference is ordered to Excellent and Good (Table 3).

Table 1 Hiring dataset

FR	A_1	A_2	A_3	A_4
S_1	(0.4,0.5)	(0.3,0.8)	(0.5,0.6)	(0.9,0.2)
S_2	(0.5,0.4)	(0.6,0.2)	(0.7,0.1)	(0.3,0.8)
S_3	(0.5,0.2)	(0.9,0.2)	(0.9,0.1)	(0.3,0.6)
S_4	(0.4,0.5)	(1.1,0.1)	(0.8,0.2)	(1.0,0.2)
S_5	(0.9,0.1)	(0.8,0.0)	(0.8,0.2)	(0.6,0.0)
S_6	(0.1,1.1)	(0.4,0.6)	(0.3,0.7)	(0.8,0.1)
S_7	(0.6,0.5)	(0.3,0.7)	(0.5,0.6)	(0.9,0.2)
S_8	(0.5,0.5)	(0.8,0.2)	(0.7,0.2)	(0.3,0.8)
S_9	(0.5,0.6)	(0.7,0.1)	(0.6,0.2)	(0.2,0.7)
S_{10}	(0.2,0.5)	(1.1,0.1)	(0.9,0.3)	(0.8,0.0)

Table 2 Binary relation $X_{0.5}^{0.3}$

$X_{0.5}^{0.3}$	A_1	A_2	A_3	A_4
S_1	0	0	0	1
S_2	0	1	1	0
S_3	0	1	1	0
S_4	0	1	1	1
S_5	1	1	1	1
S_6	0	0	0	1
S_7	0	0	0	1
S_8	0	1	1	0
S_9	0	1	1	0
S_{10}	0	1	1	1

Table 3 Conversion of Boolean to categorical data

Objects	Degree	Experience	French	Reference
S_1	UG	Low	No	Excellent
S_2	UG	High	Yes	Good
S_3	UG	High	Yes	Good
S_4	UG	High	Yes	Excellent
S_5	PG	High	Yes	Excellent
S_6	UG	Low	No	Excellent
S_7	UG	Low	No	Excellent
S_8	UG	High	Yes	Good
S_9	UG	High	Yes	Good
S_{10}	UG	High	Yes	Excellent

By applying the preposition's defined in rough entropy-based weighted density outlier detection algorithm, outliers can be easily detected from the dataset. The indiscernibility function for each attribute is defined as follows:

$$S/IND(\text{Degree}) = \{S_1, S_2, S_3, S_4, S_6, S_7, S_8, S_9, S_{10}\}\{S_5\}$$
$$S/IND(\text{Experience}) = \{S_1, S_6, S_7\}\{S_2, S_3, S_4, S_5, S_8, S_9, S_{10}\}$$
$$S/IND(\text{French}) = \{S_1, S_6, S_7\}\{S_2, S_3, S_4, S_5, S_8, S_9, S_{10}\}$$
$$S/IND(\text{Reference}) = \{S_2, S_3, S_8, S_9\}\{S_1, S_4, S_5, S_6, S_7, S_{10}\}$$

From this, we calculate complement entropy function for each attribute from the indiscernibility function.

$$CE(\text{Degree}) = \frac{9}{10}\left(1 - \frac{9}{10}\right) + \frac{1}{10}\left(1 - \frac{1}{10}\right) = \frac{9}{50}$$
$$CE(\text{Experience}) = \frac{3}{10}\left(1 - \frac{3}{10}\right) + \frac{7}{10}\left(1 - \frac{7}{10}\right) = \frac{21}{50}$$
$$CE(\text{French}) = \frac{21}{50}; CE(\text{Reference}) = \frac{24}{50}$$

The weight of each attribute is calculated with the derived complement entropy function by adding total number of attributes.

$$\text{Weight of Attribute(Degree)} = \frac{43}{52}; \text{Weight of Attribute(Experience)} = \frac{31}{52}$$
$$\text{Weight of Attribute(French)} = \frac{31}{52}; \text{Weight of Attribute(Reference)} = \frac{28}{52}$$

Then, the weighted density value for each object is calculated. From this, threshold value will be fixed to detect outliers.

$$W(S_1) = \frac{9}{10} \times \frac{43}{52} + \frac{3}{10} \times \frac{31}{52} + \frac{3}{10} \times \frac{31}{52} + \frac{6}{10} \times \frac{28}{52} = 1.42;$$
$$W(S_2) = 1.79; W(S_3) = 1.79; W(S_4) = 1.90; W(S_5) = 1.24;$$
$$W(S_6) = 1.42; W(S_7) = 1.42; W(S_8) = 1.72; W(S_9) = 1.79;$$
$$W(S_{10}) = 1.90.$$

If $\theta < 1.42$, then the object S_5 is an outlier.

5 Conclusion

The main idea of this paper is to propose a rough entropy-based outlier detection method for two universal sets. In two universal sets, the categorical values can be built based on membership and non-membership levels by using an intuitionistic fuzzy soft set relation. Then, the rough entropy-based weighted density outlier detection method will be applied to detect outliers. The weight of each object and every attribute will be calculated. From this, the threshold value will be fixed. An object which deviates from the fixed threshold value will be detected as an outlier. The level of performance and efficiency of this method will be high when compared to existing approaches. We have taken hiring dataset as an example to test the validity of this approach. In future, the working procedure of this proposed idea will be explored in detail. Further, the outlier detection method will be implemented in dynamic datasets and also in single granulation and multi-granulation.

References

1. Pawlak Z (1982) Rough sets. Int J Comput Inf Sci, 341–356
2. Chandola V, Banerjee A, Kumar V (2013) Outlier detection-a survey. ACM Comput Surv 41(3), Article No. 15
3. Barnett V, Lewis T (1994) Outliers in statistical data. Wiley, New York
4. Knorr E, Ng R (1998) Algorithms for mining distance-based outliers in large datasets. In: VLDB conference proceedings
5. Pawlak Z (1991) Rough sets: theoretical aspects of reasoning about data. Kluwer Academic Publishers, Dordrecht
6. Hawkins D (1980) Identifications of outliers. Chapman and Hall, London
7. Liu G (2010) Rough set theory based on two universal sets and its applications. J Sci Direct Knowl based Syst 23:110–115
8. Komorowski J, Pawlak Z, Polkowski L, Skowron A (1999) Rough sets: a tutorial, pp 3–98
9. Mitra S, Pabitra Mitra SK (2004) Data mining in soft computing framework: a survey. IEEE Trans Neural Netw 13(1)
10. Zhao X, Liang J, Cao F (2014) A simple and effective outlier detection algorithm for categorical data. Int J Mach Learn Cybern 5(3):469–477
11. Ashok P, Kadhar Nawaz GM (2016) Outlier detection method on UCI repository dataset by entropy based rough K-means. Defence Sci J 66(2):113–121
12. Nanda S, Majumdar S (1992) Fuzzy rough sets. Fuzzy Sets Syst, 157–160

Automated Traffic Light System Using Blob Detection

Rohan Prasad, T. S. Gowtham, Mayank Satnalika, A. Krishnamoorthy
and V. Vijayarajan

Abstract This paper aims at overcoming the issues posed by fixed duration traffic light systems. These traditional systems are not robust and do not make optimal use of the road user's time. We propose a simple, yet powerful solution to this issue. Through the use of image processing, the number of vehicles at any point of time is calculated. Using this information, traffic is controlled in a dynamic manner using a priority-based system. This results in a system that is efficient, easy to implement, and cost-effective.

Keywords Image processing · Object detection · Traffic lights · Raspberry Pi

1 Introduction

Traffic light systems have been used to control the flow of traffic since the 1860s. From gaslit traffic lights to manual traffic lights, we have to come to the current system of a completely automatic system of traffic signals. These signals require minimal human intervention and operate on a fixed set of predefined rules. However, these systems are not the most efficient implementation. Due to the rather rigid behavior of current systems, they are not able to handle the various cases that can arise over the

R. Prasad · T. S. Gowtham · M. Satnalika · A. Krishnamoorthy (✉) · V. Vijayarajan
Vellore Institute of Technology, Vellore 632014, Tamil Nadu, India
e-mail: krishnamoorthy.arasu@vit.ac.in

R. Prasad
e-mail: rohanprasad013@gmail.com

T. S. Gowtham
e-mail: tsgowtham97@gmail.com

M. Satnalika
e-mail: satnalikamayank12@gmail.com

V. Vijayarajan
e-mail: vijayarajan.v@vit.ac.in

© Springer Nature Singapore Pte Ltd. 2019   517
A. J. Kulkarni et al. (eds.), *Proceedings of the 2nd International Conference on Data Engineering and Communication Technology*, Advances in Intelligent Systems and Computing 828, https://doi.org/10.1007/978-981-13-1610-4_52

course of operation. Consider the scenario of a four-lane intersection. If two lanes are empty and the other two lanes are piling up with a large number of vehicles, there is no way for the system to obtain this information. The system would treat all lanes with the same priority. This results in a traffic congestion that would take longer to clear up, which in turn frustrates the drivers. The model proposed here is to use a more dynamic control of the lights based on the density of vehicles in a lane, thereby increasing the efficiency of traffic flow. This would have multiple impacts such as reducing the time wasted in commuting, reducing the emissions per vehicle, and making travel a more pleasant experience.

2   Related Work

A number of people have attempted to address this issue from various viewpoints. In 1991, Rowe [1] proposed a model which was better than any previously existing system at the time. The aim of it was to use computer and communication networks to improve traffic management. A Critical Intersection Control (CIC) method was used to control traffic lights based on the traffic demand. It modified the green-time at signals based on relative demand of each of the lane. It reduced operator time spent at traffic controls, and around 25% of traffic control systems started using CIC at that time. Later, an automated highway system to manage the problem of freeway congestion was proposed in [2]. It aimed to reduce accidents during lane switching by setting controls on vehicles. It assigned lanes to vehicles based on their destination and grouped vehicles into platoons. Each platoon was to travel at a certain speed and had to maintain a specific distance among cars thus increasing safety.

Ho in [3] discussed the use of neural networks to determine speed rules for vehicles based on the traffic density in that area. They considered an analogy with vehicular traffic and a fluid flowing through a constriction. Rather than controlling the traffic lights, they assigned a range of speed to travel at for the vehicles and this was found by training a feed-forward sigmoidal neural network which adapts according to the traffic density at that time and the lane width at that section to set a flow velocity. Beymer [4] proposed ways to improve traffic detection on highways which would work well in situations of congested traffic. It suggested using a feature-based approach and rather than tracking the entire vehicle at once: They track individual parts of vehicles as separate features and distinguish vehicles based on the uniformity of motion. This approach was tried on real-time traffic, and it demonstrated good results for situations under various lighting situations and traffic density.

Zhou [5] used a wireless sensor network to collect real-time traffic data including traffic volume, waiting time, vehicle density, and proposed a control algorithm that adjusts both the sequence and length of traffic lights. The algorithm achieved a greater throughput and lower average waiting time compared to fixed waiting time traffic lights. Chattaraj in [6] proposed a RFID-based idea to collect traffic data. Each vehicle was assigned a tag carrying a unique ID. A RFID sensor present at each intersection would detect the traffic flow based on the signals received. They take into account vehicle type, priority of the vehicle, priority of the lane in which the vehicle is traveling, and the time of the day.

3 Implementation

The system proposed here uses cameras and image processing to identify the number of vehicles in a lane. The cameras are connected to a Raspberry Pi 3B. Again, consider the case of a four-lane intersection. Four cameras are placed at each lane. These are used to identify the number of vehicles in a lane using image processing. Based on this information, the system decides which lane gets the highest priority; highest priority is going to the lane with highest number of vehicles. This lane is given the maximum time for vehicles to pass (green signal).

The model was tested on a system of four interconnected Raspberry Pi Model 3Bs. They were networked using SSH. The Pi's were running Raspbian OS and had OpenCV 3 for Python 3.5 installed. A miniature model of a traffic junction was constructed using LEDs placed on a perforated board. The data from each webcam was also uploaded to a server and could be monitored remotely from a Web browser. This model could be easily scaled up to include more junctions and be implemented on a larger scale (Fig. 1).

4 Working

The four video feeds were obtained from the four cameras and were processed frame by frame. Each frame was resized to a width of 500 to ensure uniformity. The frame was converted from the BGR color model to gray scale. The following processing techniques were used:

- A Gaussian Blur was applied to smoothen the image and reduce noise. The Gaussian function is given by

$$G(x, y) = \frac{1}{2\pi\sigma^2}e^{-\frac{x^2+y^2}{2\sigma^2}} \tag{1}$$

- To generate the background, a weighted average of the frame was taken. This allows us to generate a model of the background.
- To extract the vehicles from the background, the absolute difference between the background and the current frame was calculated. This gives a frame with only blobs present. However, these blobs were too faint to discern.
- To overcome this issue, a thresholding function was applied. If a pixel value is greater than the specified threshold, then it was assigned the color white, otherwise it was assigned the color black.

Fig. 1 Block diagram of
overall system

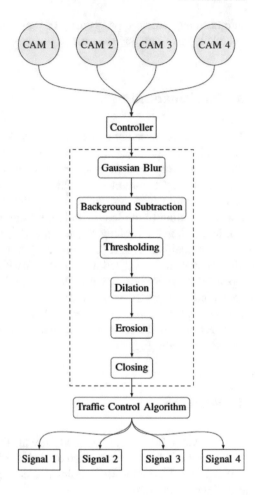

- In order to further enhance the blob and reduce the number of noise artefacts with in a blob, the frame was dilated.
- Erosion was applied to the frame. This eroded the edges between blobs. It ensured a separation between blobs and helps in distinction.
- In the end, closing was applied. This technique helped in clearing any holes within a blob. This was necessary to ensure there were no spurious blobs within a blob.

The result at the end of these processes can be seen in Fig. 2. The blobs that remained indicated the number of vehicles. To count the number of vehicles, it was required to identify the number of blobs. After rigorous testing, it was observed that blobs with an area of 200 and greater than 16,000 were found to be noise and were ignored.

A bounding rectangle was drawn around every contour whose area fell within the specified range. The moments of each contour were also found. This was used to determine whether a vehicle is on the incoming or outgoing lane. To do this, the

Fig. 2 Processed image

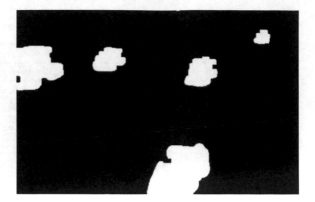

equation of the midpoint of the road was calculated using the slope intercept formula. The moments of each contour were substituted into this equation. If the value was greater than 0, then those contours belonged to the outgoing lane and, consequently, if the values were less than 0, those contours belonged to the incoming lane. In order to increase accuracy, the number of cars was averaged over 3 frames. This compensated for any spurious contours detected and improved the overall performance of the system.

Algorithm 1. Algorithm for control of traffic.

```
while True do
        get count of vehicles
        fill the queue
        sort based on number of vehicles
        for number of lanes do
                set lane with highest priority = green
                set other lanes = red
                greenTime = min(max(20, k * c[i]))
                switchLight (greenTime, index)
        end for
end while
```

To prioritize the lanes based on number of vehicles present in it, the count of vehicles in each lane was taken in a list. The list was then sorted according to the number of cars in that lane. For every swap happening during the sorting process, the indices, i.e., the lanes corresponding to that vehicle count, were also swapped accordingly (Table 1).

The indices denoted the priority of the traffic lights in a decreasing order. Using these indices, we changed the signals at the traffic light based on their priority. Every time a new cycle was initiated, the signal value for the traffic light which was at the first index (highest priority) was made green and the others were made red. The duration for a green light is given by:

$$t = \min(\max(20, k * c[i], 120)) \tag{2}$$

Table 1 Parameters for the various processing functions used

Operation	Values
Gaussian Blur	Kernel = (21, 21) pixels; iterations = 1
Weighted difference	Weight = 0.47
Thresholding	Min value = 5; output = 255
Erosion	Kernel = (3, 3) pixels; iterations = 6
Dilation	Kernel = (3, 3) pixels; iterations = 2
Closing	Kernel = (21, 21) pixels; iterations = 1

In Eq. 2, 't' is duration, 'k' is a multiplication factor, 'i' is the index of the current lane, and 'c' is the number of vehicles in lane 'i'.

After time 't', the signal for the lane having the next lower priority was made yellow and after a further 6 s, the signal for the lane at the first index was made red, and the signal for the lane with index $(i + 1)$ was made green and rest were red. This continued till all the lanes were made green once. This denoted the end of the cycle.

5 Performance

The performance of the system was nominal. There was a minor overshoot in the number of cars occasionally. This was due to the motion of the cars. When the frames were subtracted, there were areas that overlapped across frames. This affected the system by producing extra blobs which may be identified as a 'car.' This effect had been minimized by setting a lower limit on the size of a blob to be classified as a car. When it came to differentiation between incoming and outgoing vehicles, the system produced an error of 0%.

6 Conclusion

The proposed method has several practical advantages over other techniques such as Machine Learning and Haar Cascades. This method is lightweight and can be deployed instantly with an initial setup. The hardware requirements are minimal, and a Raspberry Pi would be more than enough. This model fails when the density of cars is too high, as after processing, the areas with a high density of cars get detected as a single object. But, since the model averages the number of cars detected over three frames, the error is very less.

References

1. Rowe E (1991) The Los-Angeles automated traffic surveillance and control (ATSAC) system. IEEE Trans Veh Technol 40(1):16–20
2. Hedrick J, Tomizuka M, Varaiya P (1994) Control issues in automated highway systems. IEEE Control Syst 14(6):21–32
3. Ho FS, Ioannou P (1996) Traffic flow modeling and control using artificial neural networks. IEEE Control Syst 16(5):16–26
4. Beymer D, McLauchlan P, Coifman B, Malik J (1997) A real-time computer vision system for measuring traffic parameters. In: 1997 IEEE computer society conference on computer vision and pattern recognition, 1997. Proceedings. IEEE, pp 495–501
5. Zhou B, Cao J, Zeng X, Wu H (2010) Adaptive traffic light control in wireless sensor network-based intelligent transportation system. In: 2010 IEEE 72nd vehicular technology conference fall (VTC 2010-Fall). IEEE, pp 1–5
6. Chattaraj A, Bansal S, Chandra A (2009) An intelligent traffic control system using RFID. IEEE Potentials 28(3)

Automatic Attendance System Using Face Recognition Technique

Rakshanda Agarwal, Rishabh Jain, Rajeshkannan Regunathan
and C. S. Pavan Kumar

Abstract Recognition of human face is an important domain in unique identification of humans. It is currently being widely used in many industrial applications, such as video monitoring systems, human–computer interaction, and automatic gate control systems and for securing networks. Every university uses some method of attendance to keep a record of the number of students or people who attended that particular lecture. This paper delineates a method for taking attendance of people in a classroom which integrates the face recognition technology using local binary patterns histograms (LBPH) algorithm, along with face detection by Haar feature-based cascades and distance-based clustering. The proposed system records the attendance of the people in a classroom environment autonomously and provides the user with an output as a spreadsheet describing the attendance.

Keywords Face detection · Face recognition · Clustering · Attendance

1 Introduction

Taking attendance by calling every student's name or roll number consumes around 10–15 min of time. This being a taxing job for both teachers and students, and a new methodology needs to be implemented. This saved time and can be used for other important tasks such as teaching, doubt clarification. Calling attendance normally has

R. Agarwal · R. Jain · R. Regunathan (✉) · C. S. Pavan Kumar
School of Computer Science and Engineering,
Vellore Institute of Technology, Vellore 632014, Tamil Nadu, India
e-mail: rajeshkannan.r@vit.ac.in

R. Agarwal
e-mail: rakshandapramod.agarwal2014@vit.ac.in

R. Jain
e-mail: rishabhjain.2014@vit.ac.in

C. S. Pavan Kumar
e-mail: pavan540.mic@gmail.com

© Springer Nature Singapore Pte Ltd. 2019   525
A. J. Kulkarni et al. (eds.), *Proceedings of the 2nd International Conference on Data
Engineering and Communication Technology*, Advances in Intelligent Systems
and Computing 828, https://doi.org/10.1007/978-981-13-1610-4_53

many other drawbacks also; they are marking false attendance, missing attendance. All these issues create problems for the faculty. A proper way of handling such issues is machine vision. It uses image processing, which is a way to manipulate images using mathematical functions and by higher dimensional signal processing techniques to which the input can be an image, series of images or a video while the output can be provided in the form of an image. These processes are generally digitally performed, but it can also be done via optical and analog devices [1, 2].

To take attendance through a video input, the video must first be divided into frames and faces must be extracted. Now in these extracted faces, similar faces are clustered together through basic clustering algorithm. Once clustering is successful, we have a training database which is trained, and the clusters are matched with this database. If a match is found attendance is marked and if no match is there, the new images from the input cluster are appended to the database to make out database stronger and more efficient. This paper comprises of image processing and machine learning techniques that have been used to achieve our target that is an automatic attendance system which is used to take attendance with ease and accuracy, providing us the attendance list of students. This will not only save time but will also solve the above-mentioned issues.

2 Paper Preparation

Face recognition is achieved in various steps as described in Fig. 1 (first author's (Rakshanda Agarwal) image) which include face detection and face registration, learning and training phases, clustering or classification of images and then finally accessing the database to recognize the face.

2.1 Face Detection

Detecting a face marks the onset of human face recognition. Using face detection, we can determine the coordinates and scale of face in the given input frame. Face detection can be difficult at times because face patterns have different appearances. A few factors that cause variations are expressions, skin color, or common objects such as glasses or mustache. One of the main factors is lighting changes that also can affect face detection [3].

Face detection is derived from object detection using Haar feature-based cascade classifier which was proposed by Paul Viola and Micheal Jones. This is a machine learning-based approach. To detect a face, we need a lot of positive and negative images, i.e., images with and without faces. Once we get these faces, we need to extract features from it as shown which are used to classify images. Each feature when applied to the training set a best threshold is calculated, which is then used to

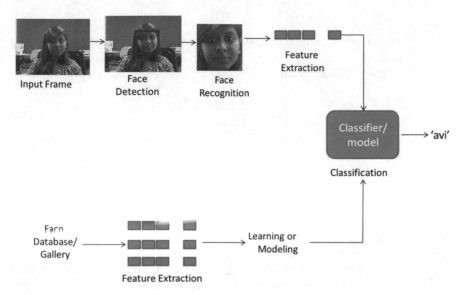

Fig. 1 Steps in face recognition (First author: Rakshanda Agarwal)

classify the face as positive or negative. This process continuous recursively until the required error rate or accuracy is achieved [4, 5].

2.2 Face Recognition

Face recognition for computer isn't as simple as it is for humans. Face recognition for computers is based on geometric features which we discussed in the face detection section above. There are various approaches to face recognition which include eigenfaces, fisherface, and local binary pattern histogram [6]. Our main solution here is obtained through local binary pattern histogram (LBPH). The main objective is to encapsulate this structure described by the local features in the image by pixel comparison to its neighboring pixels. To compute value for each pixel, compare the pixel to its eight neighbors and follow the pixels in a circular fashion, if the center pixel has a greater value in comparison with the neighbor, then give "0", else give "1". This gives us an eight digit binary number. Compute the histogram, for each combination formed. Normalize (concatenate) the histogram for every cell, this provides the feature vector for the entire face under process [7, 8].

The equation of the LBP operator is as follows:

$$\text{LBP}(x_c, y_c) = \sum_{p=0}^{p-1} 2^p s(i_p - i_c) \tag{1}$$

where

(x_c, y_c) is the central pixel, i_c is the intensity of central pixel, and i_p is the intensity of neighbor pixel.

The function $s(x)$ is defined as

$$s(x) = \begin{cases} 1 & \text{if } x \geq 0 \\ 0 & \text{else} \end{cases} \tag{2}$$

2.3 *Clustering*

The process of grouping objects into sets in such a way that in a group is called cluster, and the objects are similar or have common properties in contrast to those in other groups or clusters. One common method is the k-means clustering algorithm. This algorithm has various applications in data mining, data compression, pattern recognition, and pattern classification. In k-means clustering, k data points are classified into the groups or clusters in order to reduce the geometric mean square distance between the data point and its nearest center [9].

3 Related Work

Computer vision is a vast branch which has object detection and recognition as important aspects. Face detection and recognition is one of the foremost applications. Machine learning is also equally important for these computer vision methods. All these concepts are interrelated to each other and hence can be used in various aspects.

Support vector machine (SVM) is a machine learning technique that can also be applied to computer vision. An algorithm to decompose data is proposed that can be implemented to train SVMs on datasets of higher magnitude and guarantees optimality. Its applicability is demonstrated in systems primarily detecting faces. SVM is used since it has a well-founded mathematical point of view that follows the risk minimization principle and can handle high-dimensional input vectors, they are appropriate in computer vision [10].

Face detection has an important application in human–computer interaction, video surveillance, etc. A new algorithm is proposed to detect colored faces in different illumination conditions as well as composite backgrounds. Based on color transformations, this method distinguishes skin regions over the whole image and then produces a face based on the position of these patches of skin. The difficulty of detecting faces under low and high luminescence is overcome by applying a nonlinear transform. Thus, this detection method is better than the original one and has shown great results [11].

Another method of frontal face detection is through use of multilayered neural networks. A network is retinal connected to examine a small window of a face, and then it decides whether each window contains a part of a face or not. This process works by initiating multiple neural networks over to all portions of input images. This procedure can detect between 77.9 and 90.3% of faces with an acceptable number of false detections [12].

This method creates a model of the human face pattern by a few "face" and "non-face" clusters. A distribution-based model is built for face patterns and distance parameters are used for learning and to distinguish between "face" and "non-face" clusters. The distance matrix that is used to compute difference feature vectors and the "non-face" vectors included are both critical for the success of our system [3]. A mosaic approach for detection of human face consists of the higher two levels of the system architecture. While the lower level is an improved edge detection method, this method is efficient when the image size in unknown and can be used for black and white images without any prior info [13].

There are multiple face recognition methods: one of them is based on PCA, i.e., principal component analysis and LDA, i.e., linear discriminant analysis. In step one, a face image is projected to face subspace from original vector space through PCA, and in the second step, best linear classifier is obtained using LDA. The basic idea is to improve generalization of LDA when only a few samples per class are present. This hybrid classifier provides an useful framework for image recognition using PCA and LDA [14].

Eigenfaces are a method for recognition of faces and an approach to detect and identify human faces. This approach first tracks the human skull and then distinguishes the entire person by associating the facial features. This framework helps to detect and recognize new faces in an unsupervised manner. Also, it is efficient and relatively simple and has been observed to perform well in a sort of restricted type of environment [15].

Recognizing the frontal face with varying expressions, illumination, occlusion, and disguise is a big problem in face recognition as the results are not accurate and always different. A new method from sparse representation offers a solution to such problems. A clustering algorithm for recognition of a face is proposed and solves two main issues of face recognition: robustness to occlusion and feature extraction. The concept of sparse representation enables decision over the degree of occlusion that can be handled by the recognition algorithm and ways to maximize robustness to occlusion by selecting appropriate training images [16].

Using elastic bunch graph matching to recognize human faces from a large database wherein faces are treated as categorized graphs built using Gabor wavelet transformations. The new image distributions are mined by elastic graph matching methods and then can be matched by a similarity function. This structure is generic, flexible and is designed to recognize the members of a known group of objects. This also works on images which included mirror images and works great with faces of same pose [17].

A two way clustering is applied for data analysis of gene microarray data. Its chief purpose is to recognize the gene subgroup and model, so a stable partition emerges whenever anyone of them is utilized to classify the other. An iterative clustering method is used to perform such search. This process is used to create small groups of genes that can be used as features to cluster subsets of the samples. This is achieved through a new algorithm known as CTWC—coupled two-way clustering [18]. It is also applied to analyze a dataset comprising of feature attribute patterns of different forms of cells. This classification method also helped in classifying cancerous and non-cancerous tissues. Two-way clustering can be used for both grouping genes in functionally similar groups and in grouping tissues based on the gene feature expression [19].

4 Methodology

We recognize faces through a video frame and consequently mark the attendance and update our recognizer, i.e., database. This process has been clearly described in this section of the paper. Firstly, the input to the system is a short video of people sitting in an area such as a classroom and our initial student database. For optimal results, the input parameters are as follows: People are looking toward the camera, the faces are in an unobstructed alignment to the camera, the camera should be kept at one's shoulder height, there should be proper lighting in the area, especially over the face, and the frame rate should be high preferably in the range of 30–60 frames per second (FPS) [20].

The video input is currently in RGB format with high FPS and hence has high volume. This makes the system processing time high which needs to be reduced. Several processes are optimized to achieve this. The first process is to convert each input to grayscale reducing one of the dimensions of the data to one-third. The second step is to reduce the data by physically reducing dimension values by removing the major part of the input video feed, i.e., by extraction of faces from the video feed. This will immediately reduce the data feed size. To extract faces Haar feature-based cascade classifier is used which uses machine vision to efficiently find the faces. These faces are resized to obtain data normalization providing us with better results. Inter-cubic interpolation is used to resize the images [21] and to increase the features like in Fig. 2 (The image is taken from opencv webpage which is an open-source platform.) that are recognized by the LBPH algorithm providing better features to be found faster.

A main machine vision technique is the local binary patterns histogram (LBPH) algorithm which is used here for both the steps that are clustering and face detection. Both steps improve the accuracy of the system. Clustering of images is done by matching images with the total set, giving one cluster of images for each person. Since this is done for each video input, we do not save the recognizer but build it dynamically [2].

After obtaining cluster of images, we match each cluster with the recognition database of the group. We obtain labels for our cluster and if some label is found to be missing, the user in asked to enter the label for the cluster. This happens broadly

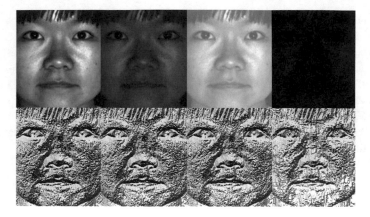

Fig. 2 Grayscale representation of LBPH features (from Opencv: free to download/open-source page mentioned in citation) [22]

for two conditions. First being when the person is scanned by the database for the first time, i.e., there is no data present about the person initially or the second case being that the features present in the new images are different in structure than what was observed before, i.e., the person could be present, but with a different facial feature or different lighting condition.

In the first case, a new label provided by the user is used to train the recognizer along with images from the cluster with various features. This increases the database length by proving a new class. While in the second case, the user provided tag is updated with these new features increasing the feature density of the recognizer leading to accurate output for a wider range of input. This provides an update to the database when the label is already present in the database, by appending this cluster to the database with same label. This will improve the efficiency of our recognizer and hence recognizing all the faces in the input video. This system has various advantages like improving the result over time while reducing the user intervention. The processes return high processor usage efficiency (PUE) when compared to other trivial methods. This ultimately leads us to achieve our goal.

4.1 Algorithm

Automatic_Attendance()

1. Create AllFaces = []
2. Load global face recognizer and Input Video
3. Open first frame in video and convert frame into grayscale
4. Detect all faces in the frame
5. For i in faces set AllFaces[i] = face

6. If next frame exists open next frame and go back to step 5
7. Create a local face recognizer and initialize TagValue = 0
8. Create a new cluster indexed by TagValue
9. Train local recognizer with top of AllFaces and TagValue
10. Remove top from AllFaces and set AllFaces[0] = NULL
11. Move to next element of AllFaces
12. Predict confidence of image by local recognizer
13. If (confidence < threshold) add image to current cluster, update local recognizer with image and remove image from AllFaces
14. If next element exists in AllFaces goto step 14
15. Increment TagValue by 1
16. If AllFaces ! = NULL goto step 11
17. Initialize index = 0
18. Open cluster with current index value
19. Predict each image in current cluster with global recognizer
20. if (image_confidence > threshold) Display an image from current cluster, ask user to enter Student ID and update global recognizer with images in current cluster and Student ID.
21. Else: Record the predicted tag as Student ID, update global recognizer with images giving confidence higher than threshold with Student ID.
22. Increment index value by 1
23. If cluster[index] exists goto step 21
24. Save global face recognizer

5 Conclusion and Future Work

Students' attendance being the foremost important task in every university is responsible for a huge amount of time consumption. Manually marking students' attendance has various drawbacks such as missing attendance, losing attendance sheet, and most importantly proxy issue. All these issues can be eradicated through our system. The only problem that our system faces is memory consumption, but since it reduces time and energy memory consumption which is not an issue. Are future endeavor includes converting this system to a software or an application so that it can be used throughout every university. We will also be working on reducing the overall time and space the system requires in execution, so that our system can be 100 percent accurate.

Declaration Images used in the paper belong to first author of the paper, and she has given her consent for use of the images. Authors take full responsibility for consequences arising from this in the case in future.

References

1. Gonzalez RS, Wintz P Digital image processing
2. Gu G, Perdisci R, Zhang J, Lee W (2008) BotMiner: clustering analysis of network traffic for protocol-and structure-independent Botnet detection. In: USENIX security symposium 2008 Jul 28, vol 5, no 2, pp 139–154
3. Sung KK, Poggio T (1998) Example-based learning for view-based human face detection. IEEE Trans Pattern Anal Mach Intell 20(1):39–51
4. © Copyright 2013, Alexander Mordvintsev & Abid K. Revision 43532856. http://opencv-pyt hon-tutroals.readthedocs.io/en/latest/py_tutorials/py_objdetect/py_face_detection/py_face_d etection.html
5. Viola P, Jones MJ (2004) Robust real-time face detection. Int J Comput Vis 57(2):137–154
6. Wagner P (2012) Face recognition with python. Tersedia dalam: www. bytefish. de (diakses pada 16 Februari 2015). 2012 Jul 18
7. ©Copyright 2011–2014, opencv dev team. http://docs.opencv.org/2.4/modules/contrib/doc/fa cerec/facerec_tutorial.html#local-binary-patterns-histograms
8. Ahonen T, Hadid A, Pietikainen M (2006) Face description with local binary patterns: application to face recognition. IEEE Trans Pattern Anal Mach Intell 28(12):2037–2041
9. Kanungo T, Mount DM, Netanyahu NS, Piatko CD, Silverman R, Wu AY (2002) An efficient k-means clustering algorithm: Analysis and implementation. IEEE Trans Pattern Anal Mach Intell 24(7):881–892
10. Osuna E, Freund R, Girosit F (1997) Training support vector machines: an application to face detection. In: 1997 IEEE computer society conference on 1997 Jun 17 computer vision and pattern recognition, 1997. Proceedings, pp 130–136. IEEE
11. Hsu RL, Abdel-Mottaleb M, Jain AK (2002) Face detection in color images. IEEE Trans Pattern Anal Mach Intell 24(5):696–706
12. Rowley HA, Baluja S, Kanade T (1998) Neural network-based face detection. IEEE Trans Pattern Anal Mach Intell 20(1):23–38
13. Yang G, Huang TS (1994) Human face detection in a complex background. Pattern Recogn 27(1):53–63
14. Zhao W, Chellappa R, Krishnaswamy A (1998) Discriminant analysis of principal components for face recognition. In: Proceedings of third IEEE international conference on 1998 Apr 14 automatic face and gesture recognition, 1998, pp 336–341. IEEE
15. Turk MA, Pentland AP (1991) Face recognition using eigenfaces. In: IEEE computer society conference on 1991 Jun 3 computer vision and pattern recognition. Proceedings CVPR'91, pp 586–591. IEEE
16. Wright J, Yang AY, Ganesh A, Sastry SS, Ma Y (2009) Robust face recognition via sparse representation. IEEE Trans Pattern Anal Mach Intell 31(2):210–227
17. Wiskott L, Krüger N, Kuiger N, Von Der Malsburg C (1997) Face recognition by elastic bunch graph matching. IEEE Trans Pattern Anal Mach Intell 19(7):775–779
18. Getz G, Levine E, Domany E (2000) Coupled two-way clustering analysis of gene microarray data. Proc Natl Acad Sci 97(22):12079–12084
19. Alon U, Barkai N, Notterman DA, Gish K, Ybarra S, Mack D, Levine AJ (1999) Broad patterns of gene expression revealed by clustering analysis of tumor and normal colon tissues probed by oligonucleotide arrays. Proc Natl Acad Sci 96(12):6745–6750
20. © 2017 Alamy Ltd. All rights reserved. http://www.alamy.com/stock-photo-school-photograp h-of-junior-girls-sitting-at-their-desks-with-open-52729837.html
21. Maeland Einar (1988) On the comparison of interpolation methods. IEEE Trans Med Imaging 7(3):213–217
22. ©Copyright 2011–2014, opencv dev team. http://docs.opencv.org/2.4/modules/contrib/doc/fa cerec/facerec_tutorial.html#id22

Cloudlet Services for Healthcare Applications in Mobile Cloud Computing

Ramasubbareddy Somula, Chunduru Anilkumar, B. Venkatesh, Aravind Karrothu, C. S. Pavan Kumar and R. Sasikala

Abstract Nowadays, the uses of mobile devices have been increasing in many aspects of our life such as playing games, sending documents, transactions, and marketing, business, and conference meetings. But this mobile device has limited the resources in terms of battery lifetime, storage, and processing capacity. The new technology known as mobile cloud computing can help users to increase utilization of mobile cloud resources. The limitation of mobile device can be overcome by MCC. The technique called as offloading can transfer the resource-intensive task to remote cloud for processing, and the result will come back to the mobile device. The mobile device connecting to remote cloud with the help of 3G (or) LTE networks causes latency-related issues bandwidth and cost. In order to overcome these problems, MCC has introduced new technology which can reduce latency problems by providing secured and efficient model based on cloudlet. In this paper, we mainly focus on healthcare applications which can be processed by new cloudlet model for reducing processing time as well as providing enough security to user's data. Initially, the user connects to the available cloudlet; if the cloudlet is not providing required resources or services, the user will redirect to remote cloud. In our model, the cloudlet is used to analyze patient medical records.

Keywords Mobile cloud computing · Cloud computing · Cloudlet model
Healthcare applications · Mobile devices

R. Somula (✉) · C. Anilkumar · B. Venkatesh · A. Karrothu · C. S. Pavan Kumar · R. Sasikala
Vellore Institute of Technology, Vellore 632014, Tamil Nadu, India
e-mail: svramasubbareddy1219@gmail.com

C. Anilkumar
e-mail: chunduru.anilkumar@vit.ac.in

B. Venkatesh
e-mail: venkatesh.cse88@gmail.com

A. Karrothu
e-mail: karrothuaravind118@gmail.com

C. S. Pavan Kumar
e-mail: pavan540.mic@gmail.com

© Springer Nature Singapore Pte Ltd. 2019   535
A. J. Kulkarni et al. (eds.), *Proceedings of the 2nd International Conference on Data Engineering and Communication Technology*, Advances in Intelligent Systems and Computing 828, https://doi.org/10.1007/978-981-13-1610-4_54

1 Introduction

Mobile devices are growing in terms of utilization in our daily life to voice conversations and video chatting with others. Especially, the smart phones became an important tool in our daily activities in e-commerce, IT industries. Even though mobile device is capable enough to handle high-end applications, it still suffers with limited resources such as short battery lifetime, storage, and processor. These changes help users to make environment where all devices share resources to run application efficiently.

The conventional computing only deals with the compute and process computation tasks. The modern technologies got birth to satisfy user requirements: big data, networking, cloud computing, fog computing, mobile cloud computing, IOT. The user will always require modern infrastructure to achieve increasing demand on both mobility and connectivity [4]. Among many technologies, mobile cloud computing became a popular model [5]. Mobile computing allows many devices interacting with other mobile devices through network technologies (Wi-Fi and 4G). The mobile devices have many advantages like portability and mobility features. The mobile computing is integrated with cloud computing technology in order to form new technology called as MCC [6]. The MCC can overcome the limitations of mobile device. In the case of implementing real MCC model, we have to take into account few challenges which cause troubles while establishing MCC environment. Mobile devices are limited by storage, battery lifetime, processing, and video streaming, augmented reality application. We should consider another important challenge in the mobility of device which is moving from one network environment to another network environment. This affects quality of performance and connectivity with remote cloud [7]. The MCC can avoid limitations of mobile device by offloading computational task into remote cloud which requires more processing power locally. As a result, the remote cloud will process it with less power consumption [8]. MCC is considered as new trend among many new technologies in coming years. Generally, the mobile devices are connecting cloud computing via various network technologies such as 3G and 4G. These technologies cause high cost, limited bandwidth, and connectivity problems as shown in Fig. 1. The important issue is nothing but security. Providing security to data from attackers over wire or wireless channel [9] is a big challenge in both cloud and mobile cloud computing. The user always expects his data to be safe and not to be affected by the attackers [10]. There are many encryption techniques to protect data from attackers [11, 12].

2 Background

The researches have been putting lot of efforts across the world for improving mobile cloud computing. The users require numerous applications in mobile device; each of these applications requires data exchange and receives as well as requires lot of pro-

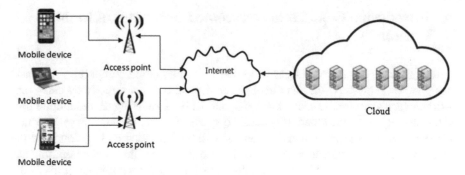

Fig. 1 Mobile computing architecture

cessing power. This paper [13] describes how the mobile computing is formed from both mobile computing and cloud computing. The author also discusses challenges, scope of MCC, and development. The sensors in network technology inspired lot of researchers in world to collect data from different useful aspects of life in clouding military, hospital, IT organization, education institutions, and crowd management [14]. The huge amount of data will be generated every day that data need to be stored efficiently all that data has to be stored in cloud server for storage and processing [15]. In paper [16], the author has analyzed main factors which cause more power consumption in mobile devices while using remote cloud. This provided an example on how to save energy between mobile device and remote cloud. They have discussed main characteristics of modern mobile device. The jobs from users can be scheduled among VMs inside cloudlet which was discussed in [17]. The key metrics are overhead of VM life cycle, scheduling of VM, job allocation to VM. The author in [15] has discussed the importance of scheduling of VMs in cloud environment in order to reduce execution time by using Cloudsim cloud environment tool. This paper proposes architecture was fine-grained cloudlet to manage all applications inside cloudlet. The cloudlet can be chosen dynamically not like previous model. The cloudlet is fixed near wireless access point. In this paper [19], the author had proposed mobile cloud computing model which is different from other previously published model in terms of scalability features. These paper experimental results have covered intended numbers of cloudlets available in covered area. The mobile device is known to acquire more power while running excessive applications. The author is motivated by the fact that optimizing power is important in MCC. In this paper [20], the author had produced mathematical model to optimize power consumption in MCC. The author in [21] had conducted experiment on mobile device by analyzing each and every component and cloudlet each component participation in total power consumption.

3   Introducing Cloudlet to Overcome Limitations of Mobile Devices

Although there are many designs proposed in mobile cloud computing, the basic design allows mobile device and network devices to connect directly by using networks which are wireless such as 3G, 4G, and LTE. Here in Fig. 2, the mobile user sends his request to the cloud. It is checked whether the sent request is valid or not. If the sent request is valid, then the request will be processed by the cloud and the result will be sent to the mobile user. As this process in mobile cloud computing has some limitations, a new technology called cloudlet model which can perform the jobs of mobile in particular range is introduced to overcome these limitations. This cloudlet model provides network remotely to the end user with less delay and higher throughput [19]. Generally, in mobile cloud computing, the users will connect to the networks like 3G, 4G, and LTE. But in cloudlet model where there are many cloudlets distributed have connection with each other through Wi-Fi. And these cloudlet levels are connected to master cloudlet. Shows in Fig. 3 the job of the master cloudlet is maintenance. This master cloudlet has the connection with cloud server.

In this proposed model, mobile user offloads tasks to the cloudlets connected to the master cloudlet through the wireless network. The master cloudlet processes the task and sends it to the end user.

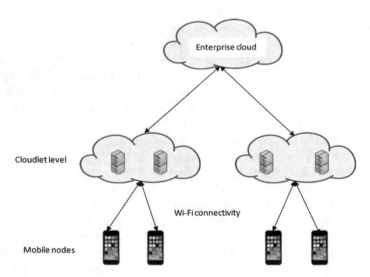

Fig. 2 Cloudlet architecture in MCC

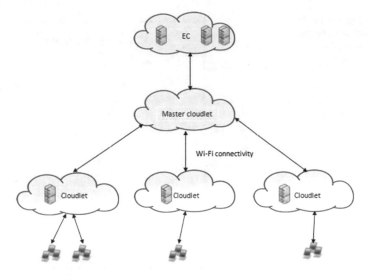

Fig. 3 Proposed cloudlet framework in MCC

4 Simulation Results

For implementing and testing our cloudlet model in mobile cloud computing, we use mobile cloud simulator (MCCSIM). In order to predict original nature of the mobile cloud environment, this tool offers to design evaluate power consuming and delay parameters. Also, it provides flexible environment to implement intended number of cloudlets dynamically through graphical user interface. We compare power consumption and delay our proposed cloudlet based on MCC model with without cloudlet in MCC. In our simulation, the no. of cloudlets is distributed evenly so that the users can access them easily. We have considered three scenarios in simulation environment which are as follows: when the mobile device directly connected to the remote cloud through 3G, (2) when the mobile device connected to remote cloudlet through master cloudlet, (3) when mobile connected to remote cloud through cloudlets which are connected to master cloudlet (Table 1; Figs. 4 and 5).

Table 1 Simulation parameters in MCCSIM

Testing time	Testing area	No. of mobile users	Mobility speed	Packet rate	Network technology	Cloudlet capacity
600 s	800 × 600 m	2000	2 m/s	0.1 Hz	3G or 4G	200

Fig. 4 Power consumption
in proposed model using
cloudlet in MCC

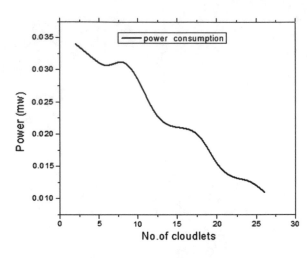

Fig. 5 Delay in proposed
model using cloudlet in
MCC

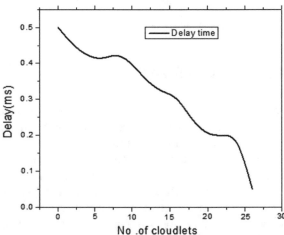

4.1 The Cloudlet Uses in e-Healthcare Systems

In e-healthcare systems, the cloud computing and cloudlet technologies are required. The cloudlet technology is available to analyze patient records and also process extract recommended features from patient database. In this system, we discussed the state of the art on healthcare systems in cloud computing the application growth on healthcare is growing day by day. Also, this application requires lot of computation and communication resources. The application requires access huge amount of data from organizations within and outside of the boundaries. The data pattern is typically dynamic. It only supports granularity interaction. In cloud environment [23], the future application will have to support heterogeneous platform inside (or) outside organization [24]. There is no wonder to say that the advantages of cloud computing can be helpful for the organization including healthcare systems. The healthcare organization depends on cloud environment for processing and storage of huge amount of data. Another important challenge is risk management. The cost of maintained health data is stored in cloud because of sensitivity of health data. The cost of health maintained data and also provide privacy and security laws is gradually growing. Taking decision for storing data about healthcare and other organizations into cloud requires lot of confidence. In this paper, the author had proposed transport load and capacity sharing. The healthcare systems in smart city meet demands by adapting cloud computing. The cloud computing is beneficial because of the distribution of cloudlet across the world. The study on grid computing shared poor accessing remotely. How it useful in healthcare applications deployment is presented. Though the paper is focused on grid computing, it also suits for cloud computing technology, multiple organizations, and application scenarios for deployment of grid computing based on different classes of organization as well as various types of applications. The requirements of healthcare system are identified and analyzed based on the result of iteration in terms of throughput. This platform identifies the computational and communicational requirements of healthcare system application. This analysis is important because the network traffic connecting healthcare system is dominated by analytical applications which require zero network latencies. The individual request is not heavy in terms of data but causes heavy traffic in feature communication. The cloudlet concept can be applied on this communication and computation concept in order to analyze the performance of cloud computing healthcare application. The author had discussed adaption of cloud solutions in healthcare systems in order to make health service provider move forward and also discussed privacy, security, risk management, and workflow challenges. There are many papers in cloud computing which focus on healthcare applications including foundation for health care, impact of cloud computing on healthcare.

5 Conclusion

Mobile cloud computing is a new technology which is used in different sectors. Various architectures have been proposed from last few years. The new concept cloudlet was introduced in MCC. The mobile users send application to the nearest cloudlet, and then, the cloudlet communicates with remote cloud. The cloudlet can deal with service request and responses between mobile users and remote cloud. The cloudlet can make process faster and also reduce energy consumption in mobile device. If there is no cloudlet concept, then the mobile user directly depends upon the remote cloud which drains mobile user battery completely. This challenge is overcome by cloudlet and also energy optimization.

The proposed model deals with many applications in our daily life including education, business, e-commerce, crowd management, and e-health care system. In this paper, we have considered healthcare system as a case study. This technology provides protection for healthcare data in order to stop attackers using sensitive data from cloud. Finally, we analyze the performance of our proposed model with non-cloudlet based on mobile cloud computing.

References

1. Kaur K, Kaur Walia N (2014) Survey on mobile cloud computing. Int J Sci Res (IJSR) 3(6):2536–2540
2. Tawalbeh LA, Ababneh F, Jararweh Y, AlDosari F (2016) Trust delegation based secure mobile cloud computing framework. Int J Inf Comput Secur 8(3)
3. Tawalbeh L, Mowafi M, Aljoby W (2013) Use of elliptic curve cryptography for multimedia encryption. IET Inf Secur 7(2):67–74
4. Goswami G (2013) Mobile computing. Int J Adv Res Comput Sci Softw Eng 846–855
5. Zimmerman JB (1999) Mobile computing: characteristics, business benefits, and the mobile framework. Univ Md. Eur Div-Bowie State 10:12
6. Tawalbeh LA, Alassaf N, Bakheder W, Tawalbeh A (2015) Resilience mobile cloud computing: features, applications and challenges. In: Proceedings of the 2015 IEEE international symposium on web of things and big data (WoTBD 2015), in conjunction with 5th IEEE eCONF. https://doi.org/10.1109/econf.2015.59. Bahrain, 18–20 Oct 2015, pp 280–284
7. Qi H, Gani A (2012) Research on mobile cloud computing: review, trend and perspectives. In: 2012 second international conference on digital information and communication technology and it's applications (DICTAP), pp 195–202
8. Benkhelifa E, Welsh T, Tawalbeh L, Jararweh Y, Basalamah A (2015) User profiling for energy optimization in mobile cloud computing. In: The 6th ambient systems and networks conference (ANT 2015). Procedia Computer Science 52 1159–1165, UK, June 2015
9. Moh'd, A, Aslam N, Marzi H, Tawalbeh LA (2010) Hardware implementations of secure hashing functions on FPGAs for WSNs. In: Proceedings of the 3rd international conference on the applications of digital information and web technologies (ICADIWT)
10. Tawalbeh LA, Mohammad A, Gutub A (2010) Efficient FPGA implementation of a programmable architecture for GF(p) elliptic curve crypto computations. AA J Sign Process Syst Sign Image Video Technol 59:233–244. https://doi.org/10.1007/s11265-009-0376-x
11. Tawalbeh LA, Jararweh Y, Moh'md A (2012) An integrated radix-4 modular divider/multiplier hardware architecture for cryptographic applications. Int Arab J Inf Technol 9(3):284–290
12. Tawalbeh LA, Tenca AF, Park S, Koc CK (2004) A dualfield modular division algorithm and architecture for application specific hardware. In: Thirty-Eighth Asilomar conference on signals, systems, and computers. IEEE Press, Pacific Grove, California, USA, 7–10 Nov 2004, pp 483–487
13. Huang D, others (2011) Mobile cloud computing. IEEE COMSOC Multimed Commun Tech Comm MMTC E-Lett 6(10):27–31
14. Tawabeh LA, Bakhader W (2016) A mobile cloud system for different useful applications. In: Proceedings of the 13th international conference on mobile web and intelligent information systems (MobiWis). 22–24 Aug 2016, Vienna, Austria
15. L o'ai AT, Bakhader W, Song H (2016) A mobile cloud computing model using the cloudlet scheme for bigdata applications. In: 2015 IEEE conference on connected health: applications, systems and engineering technologies (CHASE 2016). 27–29 June 2016. Washington DC, USA
16. Miettinen AP, Nurminen JK (2010) Energy efficiency of mobile clients in cloud computing. HotCloud 10:4–4
17. Shiraz M, Gani A (2012) Mobile cloud computing: critical analysis of application deployment in virtual machines. Int Proc Comput Sci Inf Tech 27:11
18. Bahwaireth K, Lo'ai AT, Benkhelifa E, Jararweh Y, Tawalbeh M (2016) Experimental comparison of simulation tools for efficient cloud and mobile cloud computing applications. EURASIP J Inf Secur 15. https://doi.org/10.1186/s13635-016-0039-y

19. Jararwah Y, Tawalbeh L, Ababneh F, Khreishah A, Dosari F (2014) Scalable cloudlet-based mobile computing model. In: The 11th international conference on mobile systems and pervasive computing-Mobi-SPC 2014. Elsevier, Procedia Computer Science 34, Niagara Falls, Canada, 17–20 Aug 2014, pp 434–441

20. Al-Ayyoub M, Jararweh Y, Lo'ai AT, Benkhelifa E, Basalamah A (2015) Power optimization of large scale mobile cloud computing systems. In: The proceedings of the 3rd IEEE international conference on future internet of things and cloud (Fi- Cloud). Rome, Italy, 24–28 Aug 2015

21. Tawalbeh M, Eardley A, Lo'ai AT (2016) Studying the energy consumption in mobile devices. In: The 13th international conference on mobile systems and pervasive computing (MobiSPC 2016). Procedia Computer Science (94), pp 183–189. https://doi.org/10.1016/j.procs.2016.08. 028. 15–18 Aug 2016, Montreal, Canada

22. Benkhelifa E, Welsh T, Tawalbeh L, Khreishah A, Jararweh Y, Al-Ayyoub M (2016) GA-based resource augmentation negotiation for energy-optimised mobile Ad-hoc cloud. In: The proceedings of the 4th IEEE international conference on mobile cloud computing, services, and engineering (MobileCloud). Oxford, UK, 28–31 Mar 2016, pp 110–116. https://doi.org/1 0.1109/mobilecloud.2016.25

23. Mehmood R, Faisal MA, Altowaijri S Future networked healthcare systems: a review and case study. In: Boucadair M, Jacquenet C (eds) Handbook research redesigning future internet architecture, pp. 564–590

24. Macias F, Thomas G (2011) Cloud computing advantages in the public sector: how today's government, education, and healthcare organizations are benefiting from cloud computing environments. [White Paper]

Customization of LTE Scheduling Based on Channel and Date Rate Selection and Neural Network Learning

Divya Mohan and Geetha Mary Amalanathan

Abstract The ever-increasing fame of wireless technologies and their purposeful-ness of usage in the communication scenario is fascinating the interest of customers day by day. As we all know that this is the recent technology within the mobile telecommunications is the 4G architecture. Long term evolution (LTE) is said to be the nominee of the next generation in the direction of 4G in mobile broadband technology, which offers a data rate of 100 Mbps and works with IP. It is recently an emerging technology and it offers enhanced speed, capacity and coverage for present mobility networks. LTE, which is considered to be an IP based network, is said to eradicate the issues present in the communication systems such as lack of resources and distributed services to the users in their day to day life. It increases the speed and capacity with improvements in core network deployment and different radio inter-faces all together. In order to provide a seamless and high speed communication, customization of channel is needed for each individual user. For channel allocation, there are several methods to predict available channel and optimal usage of channel to users. By the optimal selection of channel and customized scheduling, we can manage the data transmission to users. For this process, LTE system was used to schedule the channel. However, in LTE scheduling, the scheduler has to verify the channel information, user availability and hand off status in the network. The above mentioned criterions might create some limitation during scheduling of the channel. In order to overcome this limitation, we propose a novel feature extraction method and classification method for data mining process to retrieve the information about a network for LTE scheduling. In addition to this here in this paper we have analyzed the dataset and with the help of MATLAB.

Keywords LTE · Data mining · Feature extraction · Classification · Scheduling

D. Mohan (✉) · G. M. Amalanathan
School of Computer Science and Engineering, Vellore Institute of Technology, Vellore, India
e-mail: divya.mohan2016@vitstudent.ac.in

G. M. Amalanathan
e-mail: geethamary@vit.ac.in

© Springer Nature Singapore Pte Ltd. 2019
A. J. Kulkarni et al. (eds.), *Proceedings of the 2nd International Conference on Data Engineering and Communication Technology*, Advances in Intelligent Systems and Computing 828, https://doi.org/10.1007/978-981-13-1610-4_55

1 Introduction

LTE is the newly installed technology, which is standardized for communication networks by offering high peak throughputs, high data rates and scalable bandwidth. LTE imparts multimedia applications such as mobile TV, video and audio streaming, internet browsing etc. in wireless communication. The LTE system is premeditated to be a packet-based system containing less no of network elements, which in turn provides an advancement in capacity of the system and coverage in addition it also grants a flawless incorporation with other existing wireless communication networks by means of providing High performance prominently [1]. The foremost important key elements of the mobile LTE architecture is shown in Fig. 1 are

- User Equipment (UE), customer or the end user who is in need of Internet broadband Access and considered to be mobile in nature.
- A simply an obligatory module in the radio access network (RAC) of LTE is named as Evolved node B (eNB). The eNB is the base station of the LTE network which levers radio communications between numerous devices within the cell and incorporates handover decisions and radio resource management.
- Evolved Packet Core (EPC) or Core Network deals with user authentication, authorization and accounting (AAA) with addition to allocating of IP address, mobility related signalling, QoS (Quality of Service) and security.

As we all know that scheduling in LTE is the process of contributing the network capacity as well as the allocation of resources according to the UE's transmission and reception. Since LTE is an extension or evolution of 3G mobile network with the capabilities of higher capacity, lower latency and efficient radio access [2]. With these capabilities, the LTE utilizes following methodologies to achieve the high spectrum efficiency, greater user bit rates, reduced cost, operational simplicity and lower delay.

The Scheduling process in LTE till now is carried out by division of single frequency band into number of cluster of common orthogonal sub-carriers and the

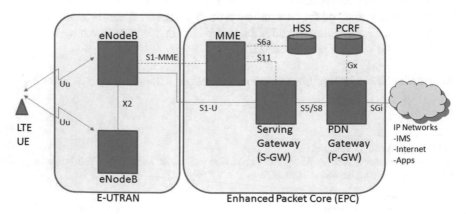

Fig. 1 LTE architecture

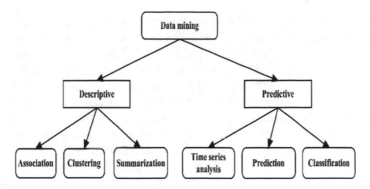

Fig. 2 Data mining task

quality enhancement in OFDMA achieves high data rate. The transfer of user's data in the sequence of frames and the optimal usage of the channels are the major objectives of OFDMA. High Peak to Average Power Ratio (PAPR) is considered to be the major limitation of OFDMA. To reduce the value of PAPR, research studies introduce the SC-FDMA in LTE for uplink data transmission. Due to the lower PAPR values, the battery power consumed by the UEs is low and hence the overall power efficiency is high. Based on variations in time and frequency domains, scheduling process is categorized into domains namely Frequency Domain Packet Scheduling (FDPS) and Time Domain Packet Scheduling (TDPS) in SC-FDMA. The MIMO technique exploits both uplink and downlink for improving the transmission reliability and data rate. The MIMO increases the spectral efficiency with low latency by simultaneous transmission of streams to many users. System capacity is increased by employing more number of antennas in the base station. Besides, it provides interference reduction capability, diversity gain, and spatial multiplexing gain.

With the variations listed above, the major objective of the information system is to gather the specific information from the large size data. The process of determining the relevant patterns from the huge dataset is known as data mining (Fig. 2).

2 Literature Survey

This section talks about the related works on the various proposal models existing traditionally. The progression of feature extraction and classification method for data mining process is to reclaim the relevant information about the network for LTE scheduling. Grondalen et al. [3] recommended the LTE scheduling downlink channel policies with time and frequency domain. They promoted several promising algorithms for scheduling with the diverse idea and allowed extensive discretion of resource allocation. They performed the experimental laboratory study of different scheduling algorithm such as saturated TCP and UDP network traffic sources. The relationship between the flat and frequency selective channels and scheduling of time and frequency domains were considered. The major issues produced in this system are low connectivity services. Chisab and Shukla [4] explored the requirements of 4G mobile networks to provide the complete solution based on Internet Protocol (IP) in a reasonable price. To achieve this, the convergence of all the wired and wireless technologies is necessary to provide the data rates in between 100 Mbps and 1 Gbps. For every block time-frequency frame, allocation of resources are done dynamically to the required users based on throughput, priority and the number of on-going transmissions. The power allocation is also minimized per unit time to minimize the energy expended for transmission. Biral et al. [5] described the revamping of GSM for Machine-to-Machine (M2M) support in terms of coverage and empty space. The available scarcity and the growing demand of new services caused the reframing of conventional services into 3G and 4G networks.

As for the preprocessing techniques, Albateineh et al. [6] highlighted the blind multi-user equalization difficulty due to the noisy multipath propagation environment. They proposed the new blind receiver plan based on variants of independent component analysis (ICA) in several filtering structures in 3G and 4G networks. Due to the licensed utilization of bands, the 3G and 4G accesses were typically high. Secci et al. [7] modelled two ICA techniques namely greedy and the game modelling. And also discussed that the greedy-utilization of right to use technologies. Bertrand et al. considered the mobility management ideology to preserve the connections between the different parts of the network. Moysen et al. [8] offered the 3G/4G network planning tool based on the Machine Learning (ML) that forecasted the QoS by using the function called Minimization of Drive Test (MDT) function. The network parameters were optimized using the GA and predicted the QoS effectively. The detection appropriate location of Radio Access Technologies (RAT) was the fundamental stage in ML-based LTE-scheduling. Tung et al. [9] Offered DXD scheme that good enough for Dynamic Scheduling with Extensible Allocation and Dispersed Offsets. This scheme revealed two algorithms such as DRX aware scheduling and DRX parameters decision. The relationship between the QoS constraints and channel condition were used to resolve the DRX period was the DRX parameters decision and the second algorithm used to decide the extended duration due to QoS would not be affected by DRX. The major issues of power consumption reduced only limited range Tiwana et al. [10].

3   Customization of LTE Scheduling Based on Channel and Data Rate Selection and Neural Network Learning

This section talks about the implementation particulars of proposed customization of neural network classification model for the recommendation of the communication by labelling the users based on the data rate usage. There are variation in data rate and speed used by the customers in real time communication. Here we are proposing

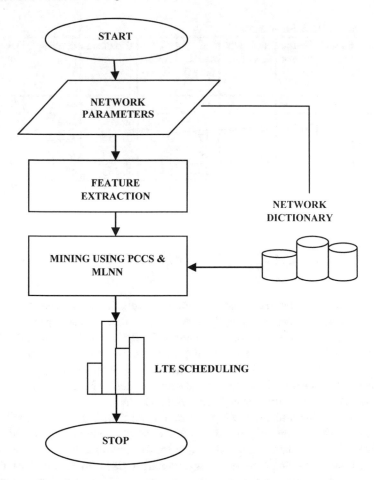

Fig. 3 Process flow diagram

a model based on which the channel is scheduled for the particular user based on the data rate usage by the particular user at that time instant. Our aim is to provide a seamless communication and suitable resource or channel to the customers, therefore by fulfilling customer satisfaction of users. Hence in our proposed method we are using the concepts of data mining into implementation.

In mining process, there are many processes that classify and retrieve the matching of data from library. These are based on match prediction of the query with trained dictionary. Here, the classification result is based on the training features of the dictionary. Since various types of retrieval present in several applications to provide matching information from the database to users (Fig. 3).

Since in the communication process, there are several numbers of users to communicate in a network. This may include signal noise and traffic on that network in channel usage. For channel allocation, there are several methods to predict available

Table 1 Pre-processed feature dataset

Customer ID	Carrier freq (GHz)	Update time of CD (ms)	HandOver delay (ms)	UE speed (km/h)	eNB power (dBm)	Distance 1	Distance 2	Distance 3
1	5	1	3	99	32	37.508	45.821	21.358
2	3	3	26	25	32	46.890	56.238	42.123
3	5	2	21	17	32	20.450	52.369	54.639
4	4	3	22	99	32	28.752	44.236	20.215
5	3	1	20	78	32	50.263	49.562	16.582
6	4	2	16	4	32	19.235	31.542	20.654
7	5	3	10	104	32	58.236	26.542	52.321
8	1	3	20	63	32	50.562	42.265	47.589
9	2	2	3	67	32	33.333	17.236	58.456
10	5	1	15	52	32	48.594	36.528	44.213

channel and to select optimal usage of channel to users. In this process, optimization and customization algorithms like, PSO, ACO, etc. were used for optimal selection of the channel to the user for communication. By the optimal selection of channel and customized scheduling, we can manage the data transmission to users. For this process, LTE system was used to schedule the channel. Since in that LTE scheduling, it has to verify the channel information and user availability in the network. This may create some limitation during scheduling of the channel.

To overcome this limitation, we perform a novel feature extraction method and classification for data mining process to retrieve the information about a network for LTE scheduling. The data mining technique for this proposed work can be done by stages the first stage is preprocessing stage, which provides dimensionality reduction in the given input feature data during the training case. The second stage is feature extraction algorithm which extracts the features of the network and its degree of angle with the distance between the features set. In this stage, we have implemented and worked through ICA method which reduces the redundant features and extracts the relevant network features and its degree of the angle with distance based estimation (Table 1).

Both the feature identification and structure of network framing increases the performance of data mining analysis. The third stage is an optimization and Customization technique for selecting the best features in the training feature set. Hence for the optimized approach of selection and classification of user attributes and labeling, we have implemented using particle search (PS) for selection and then according to that extracted features, we perform classification using multivariate learning of neural network (MLNN) to predict the category to which each individual customers belong. Also for providing best customized channel for the users by the speed of data rate used by them neural network learning is needed (Figs. 4 and 5).

Fig. 4 Reduced user
deployment based on current
data usage

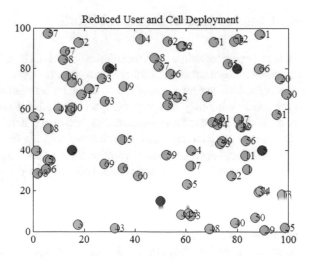

Fig. 5 Channel scheduled
for each users

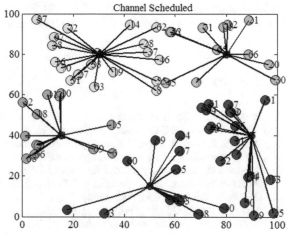

Here the multi-level classification is initially provided to categorize. After categorizing, it also predicts the label of network category from the given network information. Then this information is passed to LTE scheduling system for best communication. This type of feature extraction and classification will provide better performance rate comparing with traditional methods. Thus the performance of the system is increased by using the data mining techniques. These steps are not mandatory all the time for scheduling of channels in the LTE network. However if these steps are performed we will get a better performance system and high speed communication without delay between the users.

4 Conclusion and Extended Work

In this paper, techniques for customization of LTE scheduling by the process of dimensionality reduction, optimal feature selection, multilevel classification and data mining in LTE are proposed. Now in this LTE scheduling algorithm there has been increase in efficiency of the system; reduced packet burst loss and computational complexity. The earlier dimensionality reduction techniques do not support linear regression and linear separation and it is not suitable for smaller number of samples. The traditional optimal feature selection method has larger time for the convergence process, expensive cost function and reduced computational efficiency. Likewise, LTE in data mining techniques produces the reduced throughputs and insignificant detection results. To overcome the above mentioned drawbacks; ICA will be developed as the dimensionality reduction method during training phase. The particle search optimization algorithm has been implemented for the selection of best features. The neural network based classification method has been employed to predict the label of network category. Further the performance and the comparative analysis should be performed and discussed as an extension to this work.

References

1. Cao J, Ma M, IEEE Li H, Zhang Y, Luo Z (2014) A survey on security aspects for LTE and LTE-A networks. IEEE Commun Surv Tutorials 16
2. Ghosh A, Ratasuk R, Mondal B, Mangalvedhe N, Thomas T (2010) LTE-advanced: next-generation wireless broadband technology. IEEE Wirel Commun 10–22
3. Grondalen O, Zanella A, Mahmood K, Carpin M, Rasool J, Osterbo ON (2017) Scheduling policies in time and frequency domains for LTE downlink channel: a performance comparison. IEEE Trans Veh Technol 66:3345–3360
4. Chisab RF, Shukla C (2014) Performance evaluation of 4G-LTE-SCFDMA scheme under SUI and ITU channel models. Int J Eng Technol IJET-IJENS
5. Biral A, Centenaro M, Zanella A, Vangelista L, Zorzi M (2015) The challenges of M2M massive access in wireless cellular networks. Digit Commun Netw 1:1–19
6. Albataineh Z, Salem FM (2016) Adaptive blind CDMA receivers based on ICA filtered structures. Circ Syst Sig Process 1–29
7. Secci S, Pujolle G, Nguyen TMT, Nguyen SC (2014) Performance–cost trade-off strategic evaluation of multipath TCP communications. IEEE Trans Netw Serv Manage 11:250–263
8. Moysen J, Giupponi L, Mangues-Bafalluy J (2016) A machine learning enabled network planning tool. In: 2016 IEEE 27th annual international symposium on personal, indoor, and mobile radio communications (PIMRC), pp 1–7
9. Tung L-P, Lin Y-D, Kuo Y-H, Lai Y-C, Sivalingam KM (2014) Reducing power consumption in LTE data scheduling with the constraints of channel condition and QoS. Comput Netw 75:149–159
10. Tiwana MI, Tiwana MI (2014) A novel framework of automated RRM for LTE son using data mining: application to LTE mobility. J Netw Syst Manage 22:235–258

Edge and Texture Feature Extraction Using Canny and Haralick Textures on SPARK Cluster

D. Sudheer, R. SethuMadhavi and P. Balakrishnan

Abstract Image retrieval is the most significant technology forever. Computer vision is improving its trends and methodologies to perform like a human. System can identify any object without a human help by a simple query. To train the system, we need precise algorithms. Content-based image retrieval (CBIR) is the recent emerging trend in computer vision to retrieve relevant images from huge amount of data. This paper used a distributed and parallel processing paradigm to accelerate the retrieval process. SPARK stream processing environment is used on the top of the Hadoop Distributed File System (HDFS). To extract visual content of the images, edge and texture features are used. The distance between query image and database is measured with Mahalanobis distance metric. The performance of the retrieval system has been compared with other distributed image retrieval systems. The proposed methodology using SPARK environment outperforms the existing systems.

Keywords SPARK · HDFS · Map reduce · CBIR · Feature extraction
Texture analysis

1 Introduction

CBIR is the advanced trend in the computer vision. CBIR is not only used to retrieve the relevant images from the large-scale databases, but also to recognize objects in medical picture archiving communication systems (PACS), biomedical applications, satellite image retrieval systems, and many other applications. Currently, CBIR has achieved sophisticated accuracy using feature extraction systems based on color, shape, texture. But CBIR is lagged in efficient storage and processing paradigm due to increase in usage of smart devices. Since CBIR systems have to adapt to the commodity hardware [1]. Processing multimedia data is tedious process in terms of speed and accuracy. This paper used open-source distributed and parallel frame-

D. Sudheer (✉) · R. SethuMadhavi · P. Balakrishnan
SCOPE, VIT University, Vellore, Tamil Nadu, India
e-mail: 2498sudheer@gmail.com

© Springer Nature Singapore Pte Ltd. 2019 553
A. J. Kulkarni et al. (eds.), *Proceedings of the 2nd International Conference on Data Engineering and Communication Technology*, Advances in Intelligent Systems and Computing 828, https://doi.org/10.1007/978-981-13-1610-4_56

work Hadoop released by Apache foundation. To achieve more reliable and fastest response, SPARK is used on the top of the Hadoop. To achieve accurate retrieval result, Canny edge algorithm and Haralick textures are used in spatial domain.

This paper is organized as follows: Sect. 2 describes Big Data processing framework, Section 3 describes the related work done so far in the application, Sect. 4 describes the proposed methodology, Sect. 5 describes the experimental results and discussion, and Sect. 6 describes the conclusion of the work.

2 Big Data Processing Framework

The multimedia data is growing through social media (Flickr—75 million public images per day, Instagram—60 million images per day, Facebook—136,000 images per min, Google—57,988 query per second). Storage of these huge amounts of data very complex task using traditional systems. To overcome this issue, Google has published a white paper on Google file system. This is a distributed block-level file system. The data will be decentralized to several nodes. Each node will have numerous blocks [2]. Each block size can change by user requirement. Later, Google also published another white paper on MapReduce. MapReduce is parallel processing paradigm. Here, data locality is introduced. Instead of transforming the data through the network, code will be sent to the distributed blocks that exist in the several nodes. These two approaches were combined together and free leased as an open-source framework and named as Apache Hadoop. Hadoop is not suitable for a large amount of small files. Hadoop is designed as block-oriented file system, where default block size is 128 MB [3]. When the file size is less than block size, then it will create a separate block to that file and use the remaining space to other block. MapReduce will divide the process into two tasks as Map and Reduce, in which Map tasks are created based on the number of blocks that contain our data. So if system contains more number of blocks, then more number of Map tasks should be initiated by the Hadoop Master. This process will create a bottle neck problem to the Hadoop cluster. Hadoop introduced sequence file format to process all small files as a single large file [4]. Data sharing is not much faster in Hadoop due to replication and serialization process [5].

To simplify this problem, Hadoop introduced SPARK to processing. It is designed for fast cluster computations. SPARK will execute the scripts on the top of the Hadoop, and it extends the MapReduce mode to streaming process [6]. SPARK is also introduced by Apache foundation. SPARK was developed in 2009. The main feature of SPARK is in-memory cluster computation. SPARK not only supports MapReduce, but it also supports SQL, machine learning, etc. SPARK will create the data as Resilient Distributed Dataset. The RDDs will be divided into logical partitions which are able to process in several nodes. RDD is a read-only record; it can be created through deterministic operations. RDDs are fault-tolerant. It can reuse the intermediate results to accelerate the data sharing process.

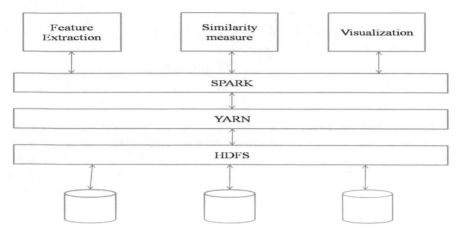

Fig. 1 Architecture of the proposed system

Image database was collected from the signal processing laboratory Web server. Ultrasound images are collected from the online Web server and stored in the HDFS [7, 8]. Feature extraction and similarity measure were done using Python and OpenCV library. The experiment was run on the SPARK integrated with Hadoop-2.6.0 single-node cluster environment. The proposed architecture is shown in below figure.

2.1 Architecture

See Fig. 1.

3 Related Work

CBIR has evolved with robust feature extraction algorithms so far. Color histogram-based features are primary and easy to represent any image. The color maps can be represented as binary, or gray scale, RGB, HSV, etc.; the extraction of color histogram is done based on the intensity levels of pixels in an image. The distribution of intensities can be represented as mathematically as follows:

$$h_c = \frac{1}{MxN} \sum_{i=0}^{M-1} \sum_{j=0}^{N-1} \delta(f_{ij} - C), \quad \forall c \in C$$

$$\text{where, } \delta(x) = \left\{ \begin{array}{l} 1, \text{if}(x = 0) \\ 0, \text{if}(x! = 0) \end{array} \right\}$$

The color histogram features were computed, and the similarity between vectors is computed using distance metrics [9]. Color features alone cannot obtain the exact required image due to high variations in the intensity levels. So texture-based features are introduced in 1973 by Haralick. The gray-level co-occurrence matrix (GLCM) will be computed from the image, and the statistical features of the GLCM will be extracted as texture of the image [10]. The important statistical features that can be extracted from GLCM were shown below:

$$\text{Energy} = \sqrt{\sum_{i=0}^{N-1} \sum_{j=0}^{N-1} M^2(i, j)}$$

In the above formula, i, j denotes the spatial position of an image. Calculating the extent of pixel pair repetition is called as energy. This energy will become large if the image pixels are same.

$$\text{Entropy} = \sum_{i=0}^{N-1} \sum_{j=0}^{N-1} M(i, j)(-\ln(M(i, j)))$$

Entropy is used to categorize the texture of an image. Randomness of the input image can be known using entropy. When the entire co-occurrence matrix values are same, then the value of entropy will also reaches utmost.

$$\text{Contrast} = \sum_{i=0}^{N-1} \sum_{j=0}^{N-1} (i - j)^2 M(i, j)$$

The intensity of a pixel and its neighbor of an image is measured using contrast. A contrast is nothing but the divergence in the color and the brightness of object which is explained by visual perception [11].

Many applications have proved experimentally that hybrid features will give more accuracy. However, the accuracy might have satisfied in traditional approach, but when dealing with real-time and large-scale datasets, it is complex to process with existing CBIR [9]. Distributed and parallelized CBIR systems have been developed and proved the response time is very less when compared to traditional systems [12].

4   Methodology

The proposed methodology used edge and texture features. The workflow of proposed methodology is shown in Fig. 2. Canny algorithm is used to extract the edge features. Canny uses the first-order derivatives of the image. To remove the noise, Canny applies the Gaussian filter first. The gradients are extracted by applying first-order derivatives [13]. Intensity thresholding is used to remove the false edges.

4.1   Canny Edge Descriptor Algorithm

Step 1   To remove the noise and to make an image smooth, Gaussian filter is to be applied.

Step 2.   Image's intensity gradients are to be found

Step 3.   Non-maximum suppression is to be applied to get exonerate from the false response of edge detection.

Step 4.   To find the potential edges, apply double threshold.

Step 5.   Those edges that are not strongly connected to the other edges are to be detected.

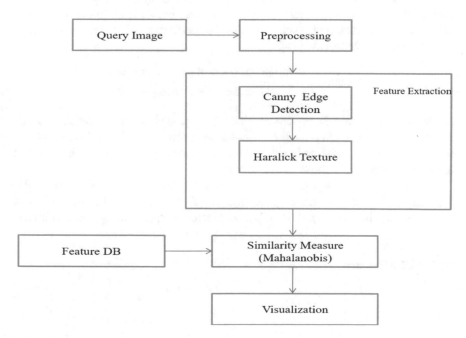

Fig. 2 Workflow of the system

4.2 Workflow

See Fig. 2.

4.3 Haralick Texture Features

To create the "texture" data, Haralick analysis is the one among the various techniques. Haralick descriptors are used to extract the texture features; those can be
evaluated from the gray-level co-occurrence matrices (GLCMs). GLCM is a 2D histogram, which apprehends the co-occurrences of the two-pixel intensities, each other
at certain offset in a region from where texture is calculated. Features like energy,
homogeneity, contrast, and dissimilarity are extracted from GLCM [10]. The GLCM
can construct in three directions (horizontal, vertical, and diagonal).

5 Results and Discussion

Figures 3 and 4 show the retrieval results of the proposed system. Figure 1 retrieved
MedPix dataset; it contains scan images of various body parts. Figure 4 contains
SPLab dataset; it contains ultrasound images. The performance of the retrieval system
is mathematically defined as [14];

$$\text{Precision} = \frac{\text{Number of relevant images retrieved}}{\text{Total Number of Retrieved Images}}$$

The proposed retrieval system obtained 75% precision for the input datasets.
Another performance measure is considered as speed. The speed of the system is
compared between single-node and multimode cluster by varying multiple dataset
sizes. The performance graph was shown below.

Figures 5 and 6 show the relationship between time and data size on single-
node and multi-node cluster, respectively. The X-axis denotes the time taken to
execute the task in milliseconds. The Y-axis denotes number of images in the dataset.

Fig. 3 GLCM directions

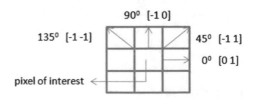

Fig. 4 Result for dataset1

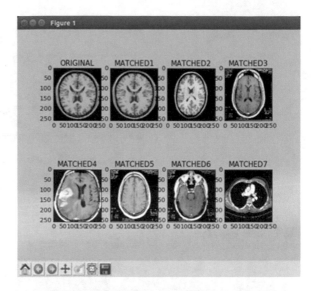

Fig. 5 Result for dataset2

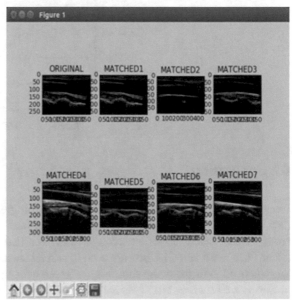

The experimental results had proved that our proposed methodology obtained better precision and faster than existing systems (Fig. 7).

Fig. 6 Time graph for
single-node cluster

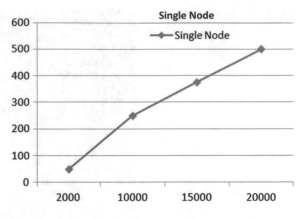

Fig. 7 Time graph for
multi-node cluster

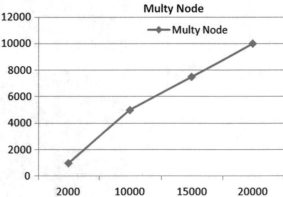

6 Conclusion

Image retrieval is the growing trend. Many technologies are emerging largely nowadays for this sake. Features like mean, standard deviation, and variance are extracted from LBPs. Canny edge detection is used for detecting edges. Gray-level co-occurrence matrices (GLCMs) are used to extract texture features. Mahalanobis distance is used to measure the similarity between query images and database. SPARK performs well for the entire system. Thus, we can able to get the related images from the database to the query image successfully.

References

1. Palle A, Kulkarni RB (2016) Classification of medical MRI brain images based on Hadoop. In: Proceedings of the second international conference on information and communication technology for competitive strategies—ICTCS '16, 1–4. https://doi.org/10.1145/2905055.290 5154

2. Alham NK, Li M, Liu Y, Hammoud S (2011) A MapReduce-based distributed SVM algorithm for automatic image annotation. Comput Math Appl 62(7):2801–2811. https://doi.org/10.101 6/j.camwa.2011.07.046
3. Sudheer D, Lakshmi AR (2015) Performance evaluation of Hadoop distributed file system. Pseudo Distrib Mode Fully Distrib Mode (9):81–86
4. Vesset D, Olofson CW, Nadkarni A, Zaidi A, Mcdonough B, Schubmehl D, ... Carnelley P (2015) IDC FutureScape: worldwide big data and analytics 2016 predictions. IDC, 1–13
5. Yamamoto M (2012) Parallel image database processing with Mapreduce and performance evaluation in pseudo distributed mode. Int J Electron Commer Stud 3(2):211–228. https://doi.org/10.7903/ijecs.1092
6. Raju USN, George S, Praneeth VS, Deo R, Jain P (2015) Content based image retrieval on Hadoop framework. IEEE Int Congr Big Data 2015:661–664. https://doi.org/10.1109/BigDat aCongress.2015.103
7. Hsieh LC, Wu GL, Hsu YM, Hsu W (2014) Online image search result grouping with MapReduce-based image clustering and graph construction for large scale photos. J Vis Commun Image Represent 25(2):384–395 https://doi.org/10.1016/j.jvcir.2013.12.010
8. Costantini L, Nicolussi R (2015) Performances evaluation of a novel Hadoop and spark based system of image retrieval for huge collections. Adv Multimedia. https://doi.org/10.1155/2015/629783
9. Gu C, Gao Y (2012) A content-based image retrieval system based on Hadoop and Lucene. In: 2012 second international conference on cloud and green computing, pp 684–687. https://doi.org/10.1109/CGC.2012.33
10. Wang X-Y, Yu Y-J, Yang H-Y (2011) An effective image retrieval scheme using color, texture and shape features. Comput Stand Interfaces 33(1):59–68. https://doi.org/10.1016/j.csi.2010.03.004, Arabi PM, Joshi G, Vamsha Deepa N (2016) Performance evaluation of GLCM and pixel intensity matrix for skin texture analysis. Perspect Sci 8:203–206. https://doi.org/10.101 6/j.pisc.2016.03.018
11. Yin D, Liu D (2013) Content-based image retrial based on Hadoop. Math Prob Eng. https://doi.org/10.1155/2013/684615
12. Chen M, Mao S, Liu Y (2014) Big data: a survey. Mobile Netw Appl 19(2):171–209. https://doi.org/10.1007/s11036-013-0489-0
13. Moise D, Shestakov D, Gudmundsson G, Amsaleg L (2013) Indexing and searching 100 M images with map-reduce. ICMR 2013:17–24. https://doi.org/10.1145/2461466.2461470
14. Sakr NA, ELdesouky AI, Arafat H (2016) An efficient fast-response content-based image retrieval framework for big data. Comput Electr Eng 54:522–538. https://doi.org/10.1016/j.co mpeleceng.2016.04.015
15. Bolón-Canedo V, Sánchez-Maroño N, Alonso-Betanzos A (2015) Recent advances and emerging challenges of feature selection in the context of big data. Knowl-Based Syst 86(May):33–45. https://doi.org/10.1016/j.knosys.2015.05.014

Improving the Response Time in Smart Grid Using Fog Computing

Shubham Kumar, Suyash Agarwal, A. Krishnamoorthy, V. Vijayarajan
and R. Kannadasan

Abstract The conventional grid system the world has been using ever since its advent in 1890s, but this aging infrastructure of demand driven control structure, has made in way for more robust, fast, inter-connected and a smarter system with a cutting-edge technology called the Smart Grid. Smart grid uses a bidirectional communication model between the consumer and the utility service. The sensing along the communication line is what makes the grid system smart. Linking the smart grid with fog and cloud not only makes the process extra firm but also more secure and reliable. This paper presents a mechanism to reduce the response time in the smart grid architecture so as the system is more responsive. We have introduced a system which reduces the response time in grid system using fog computing.

Keywords Smart grid · FOG computing · Cloud computing · Throughput
Internet of things (IoT) · FOG node · Grid

1 Introduction

Organizations nowadays are rapidly changing in all the aspects, and one of the major areas of transformation is in the field of data storage. As big organizations has huge

S. Kumar · S. Agarwal · A. Krishnamoorthy (✉) · V. Vijayarajan · R. Kannadasan
School of Computer Science and Engineering,
Vellore Institute of Technology, Vellore 632014, Tamil Nadu, India
e-mail: krishnamoorthy.arasu@vit.ac.in

S. Kumar
e-mail: shubhamkumar2014@vit.ac.in

S. Agarwal
e-mail: suyash.agarwal2014@vit.ac.in

V. Vijayarajan
e-mail: vijayarajan.v@vit.ac.in

R. Kannadasan
e-mail: kannadasan.r@vit.ac.in

© Springer Nature Singapore Pte Ltd. 2019   563
A. J. Kulkarni et al. (eds.), *Proceedings of the 2nd International Conference on Data Engineering and Communication Technology*, Advances in Intelligent Systems and Computing 828, https://doi.org/10.1007/978-981-13-1610-4_57

amount of data to handle with very limited time to organize and manage it. Conventional cloud computing architectures do not meet the requirements to provide smooth and fast access to such huge data from the cloud. Since most of the major work done by the cloud service provider so the consumer has limited control over its data, making it more vulnerable to security as your sensitive data is with the third party. Hence, the best place for moving the sensitive data from the consumer devices and Internet of things (IoT) devices depends on how securely and rapidly the data is processed. The ideal place to withhold most of the data from the IoT devices is near the devices that produce and act on that data. To bring cloud computing to next level by overcoming its disadvantage is fog computing.

We propose how the smart grid could be incorporated with the fog nodes to provide a better-managed electricity to all. Smart grid which generates electricity through the renewable and non-renewable resources delivers electrical energy to the consumers. Each customer place will be equipped with a smart meter which is liable for checking the amount of electricity being used by the particular consumer. The data produced by the smart meter is being sent to the fog server nodes which are the stationed at the edge of the network. If the fog node where the data is being received is found to be overloaded, then the data will be redirected to different fog node which is in an underloaded state, so as to distribute the load evenly between the fog nodes. The computed data is now send to the cloud server where data from all the fog nodes are stored. Finally, the smart grid receives the amount of load that needs to be distributed efficiently based on the inputs received from the cloud server. The smart grid will now process the data with respect to each consumer and accordingly will map the information to deliver the energy across everyone in an efficient manner. See Fig. 1.

Fog computing is a distributed computing architecture which acts a mediator between the cloud datacenters and IoT devices. Fog provides networking and storage facilities so that the cloud-based servers could be extended further till the IoT devices and sensors. Fog computing environment is composed of various networking devices such as bus, routers, proxy servers, switches, base station. Fog receives the real-time feeds from the IoT devices into the fog nodes. Fog nodes run IoT-aided applications for a real-time control and analytics. Fog nodes provide a transient storage for around 1–2 h and then send the periodic summaries of data to the cloud server. Fog works by developing the IoT applications at the network edge.

Smart grid's advantage over the conventional grid is that the conventional grid is a one-way communication, i.e., transmission of electrical energy from the utility to customer, whereas the smart grid allows for two-way communication between the utility to its customers. Implementing smart grid ensures that electricity can be efficiently used and hence saved for the future.

2 Related Work

Stojmenovic in [1] expresses the application of fog computing in the field of decentralized building control and software-defined network (SDN) and about how SDN

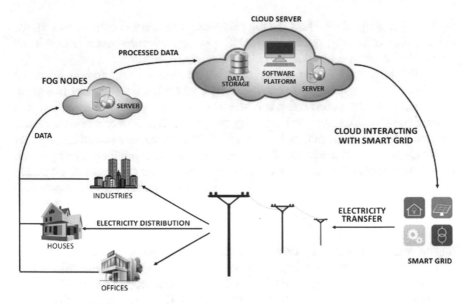

Fig. 1 Integration of smart grid with fog

is used in the different field and security and safety issue of fog computing. [2], Yang et al. present improved ways on how to renovate the existing metering infrastructure. He proposes a storage processing method for refining the current metering infrastructure, because as the number of meters keeps on increasing at one point it would be difficult for centralized information processing architecture to handle such huge explosion of data. Kannan et al. [3, 4] the latest advancements in the field of fog computing and analyses the challenges in Fog which is acting as an intermediate layer in between the IOT device architecture and the Cloud datacentres. They map the existing work in domain to the taxonomy in order to recognize the research gaps in the area of fog computing. A method in which the fog computing service providers will provide a privacy protection data compression technique for eliminating duplicate copies of repeating data to minimize the resource overindulgence (data deduplication) has been presented in [5, 6]. Samanthula et al. [7] propose an effectual and secure data sharing (SDS) structure which uses the proxy re-encryption and homomorphic encryption arrangements which stops the escape of unauthorised information and data when the system is returned is invalidated.

In [8], Huaqun et al. propose a concept of anonymous and secure aggregation scheme (ASAS) which combines the data from the terminal nodes and forwards the data to the public cloud server. The ASAS technique also protects the identity of the user by implementing a homomorphic technique in addition to saving of bandwidth between the fog nodes and the devices. Paper [9] discusses how the power line communication (PLC) works and gives an alternative of using the smart grid technology. Vangelis et al. through [10] demonstrate how to use adaptive operations platform to manage the fog computing platform and implement it in different industries. Bonomi et al. [11] give the idea of how the fog computing is better in comparison with the cloud in term of latency, application, mobility, etc. Authors also stress on how it can be used in the smart grid, smart cities, and wireless sensors.

In [12], titled "A Fog Computing Based Smart Grid Model," authors tell about the use of fog computing in the field of smart grid model. Smart grid is a green power house model which offers two-way communication such as electricity and information through single line. And fog computing is a way of storing data and information before transmitting back to cloud. In [13], Author Hathal in his paper "FOG Computing: The new Paradigm" conveys how fog computing is being useful in real-time application and how integrating it with cloud can increase efficiency and lower latency and time of respond as it will work on the network edge. With the increase of IOT, fog computing can be very handy in real-time application. In [14], Michael et al. explain the smart grid as a newer modified platform of power grids which is used in improving efficiency of power supply and power consumption. Implementing renewable resources of energy on the supply side and modifying the usage on the consumption side powers the smart grid. Authors devise three things in smart grid: (i) communications and data management necessities in smart grid, (ii) systems and techniques for context awareness, (iii) implementing a middleware platform for smart grid with context awareness in it [15].

3 System Architecture

From the figure below, we can see how fog computing interacts with the user. The cloud server is situated not known to the user, but with fog computing, it is more in periphery of the user, and hence, it is more efficient and can be dealt in time of any issues or breakdown. The advantage of fog computing over cloud is it has all the similarities of cloud such as server and database and also has very low latency; that is, the transfer and receiving of messages will be faster; it would be wirelessly accessed, and the most important part of security is also improved in fog computing.

Figure 2 shows how the consumers are connected to the fog nodes and cloud servers. It shows how the data usage of particular user is transferred between smart meter and cloud server through fog nodes. Smart meter is a device which keeps track of all the electricity usage by the client and keeps on transmitting the data wirelessly to the fog nodes at regular intervals of time. Cloud consists of data and server, and in the next level on the edge network, we have the fog nodes which are installed to make the structure efficient, secure, and faster as compared from the cloud servers.

4 Proposed Model

- Volume capacity (V_k) of a particular Fog Node, where C_{numk} is the number of consumers connected to the fog node k, TP_k is (throughput) the amount of data

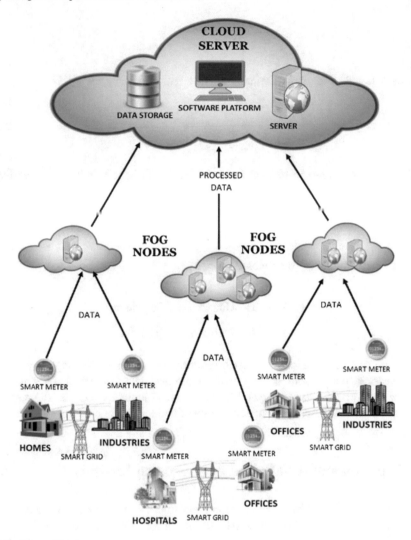

Fig. 2 Fog architecture

transferred from customers place to the fog node k successfully in a given period and $BW_{\text{fog}k}$ is the Bandwidth of fog node.

$$V_k = C_{\text{num}k} \times TP_k + BW_{\text{fog}k} \qquad (1)$$

- Total volume capacity (V) of all the fog nodes combined together is

$$V = \sum_{k=1}^{n} V_k \qquad (2)$$

- Total amount of load on a fog node in the smart grid is

$$L = \sum_{i=1}^{n} L_i \qquad (3)$$

- The maximum amount of load that a fog node can handle is its threshold limit, here Th_i is the threshold limit of fog node,

$$Th_i = \text{Load per fog node} \times V_i \qquad (4)$$

- Imbalance on load of fog nodes is given by

$$\text{If, } V_i < Th_i \ \text{Underloaded Fog Node,} \qquad (5)$$

$$V_i > Th_i \ \text{Overloaded Fog Node.} \qquad (6)$$

The imbalance in the load between different fog nodes can be managed by migrating the excess of load from the overloaded fog nodes to the underloaded fog nodes so that there are no overloaded fog nodes. Overloaded fog nodes decrease the response time and hence making the architecture slower and less efficient.

5 Experimental Results and Performance Analysis

The experimental analysis for the parameters such as throughput and load on the fog servers was obtained based on the amount of load and number of users it has, respectively.

The data in Table 1 shows the amount of load that each fog server has on them based on the number of active users.

The data set shown in Table 2 is the various throughput that is applied to the fog nodes based on the load.

6 Conclusion

In this paper, we proposed algorithm which would reduce the response time for the transfer of data from the fog node to cloud to smart grid. The smart grid produces electricity on the basis of the usage data. Here the fog nodes acts as an edge network

Table 1 Load on fog server versus no. of users

No. of users (in thousands)	Fog server 1	Fog server 2	Fog server 3	Fog server 4
100	18	26	30	30
200	20	40	55	48
300	20	40	68	60
400	18	40	70	70
500	17	45	80	80
600	18	45	75	85
700	17	50	80	95
800	15	50	80	100

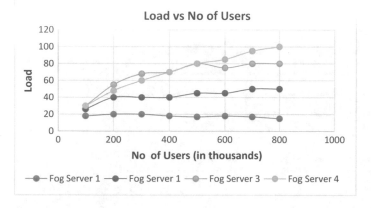

and speed up the transfer of data. We have the concept of overload and underload to detect whether the given fog node is full or vacant for devices to connect. We have used different parameters such as response time, throughput, efficiency to compare them and generate data.

Table 2 Load versus throughput

Load	Throughput 1	Throughput 2	Throughput 3
100	0.89	0.89	0.89
200	0.81	0.79	0.77
300	0.85	0.83	0.81
400	0.84	0.77	0.78
500	0.79	0.72	0.72
600	0.83	0.75	0.74
700	0.82	0.67	0.7
800	0.77	0.68	0.62

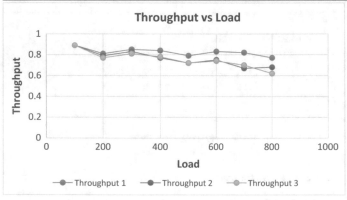

References

1. Stojmenovic I (2014) Fog computing: a cloud to the ground support for smart things and machine-to-machine networks. Telecommunication networks and applications conference (ATNAC) 2014, 26 Jan 2015 Australasian. https://doi.org/10.1109/atnac.2014.7020884
2. Yan Y, Su W University of Michigan-Dearborn (2016) A fog computing solution for advanced metering infrastructure. 978-1-5090-2157-4/16/$31.00 ©2016 IEEE
3. Kannan N, Arasu K, Jagadeesh Kannan R, Ganesan R (2017) An efficient system model for multicasting measured noise value of polluting industries. In: Modi N, Verma P, Trivedi B (eds) Proceedings of international conference on communication and networks. Advances in intelligent systems and computing, vol 508. Springer, Singapore
4. Zhanikeev M Fog cloud caching at network edge via local hardware awareness spaces 2332-5666/16 $31.00 © 2016 IEEE https://doi.org/10.1109/icdcsw.2016.41
5. Mahmud R, Buyya R (2016) Fog computing: a taxonomy, survey and future directions. arXiv: 1611.05539v3[cs.DC] 24 Nov 2016
6. Koo D, Hur J (2018) Privacy-preserving deduplication of encrypted data with dynamic ownership management in fog computing. Future Gener Comput Syst
7. Samanthula BK, Elmehdwi Y, Howser G, Madria S (2015) A secure data sharing and query processing framework via federation of cloud computing. Inf Syst
8. Yu Z, Au MH, Xu Q, Yang R, Han J (2018) Towards leakage-resilient fine-grained access control in fog computing. Future Gener Comput Syst
9. Galli S, Senior Member, IEEE, Scaglione A, Fellow, IEEE, Wang Z, Member, IEEE (2011) For the grid and through the grid: the role of power line communications in the smart grid. Proc IEEE 99(6)

10. Gazis V, Leonardi A, Mathioudakis K, Sasloglou K, Kikiras P, Sudhaakar R (2015) Components of fog computing in an industrial internet of things context. https://doi.org/10.1109/seconw.2015.7328144
11. Bonomi F, Milito R, Zhu J, Addepalli S (2004) Fog computing and its role in the internet of things
12. Wang H, Wang Z, Domingo-Ferrer J (2018) Anonymous and secure aggregation scheme in fog-based public cloud computing
13. Okay FY, Ozdemir S (2016) A fog computing based smart grid model. Netw Comput Commun (ISNCC). https://doi.org/10.1109/isncc.2016.7746062
14. Alwageed HSA (2015) FOG computing: the new paradigm. Communications on Applied Electronics (CAE)
15. Donohoe M, Jennings B, Balasubramaniam S (2015) Context-awareness and the smart grid: requirements and challenges

Machine Learning Algorithm-Based Minimisation of Network Traffic in Mobile Cloud Computing

Praveena Akki and V. Vijayarajan

Abstract Mobile cloud computing is an emerging technology where mobile device is integrated with cloud computing. It has many applications such as social network, online shopping, Flickr, Picasa. Besides these applications, it suffers from network traffic issue. The demand of the users has been increasing day by day but due to the limited density on base stations, it has become an overhead to the network service providers to provide the service. In this paper, we have applied machine learning techniques on the preprocessed data to classify client requests and generated rules to accept or to discard a client request. We aimed to minimize network traffic. We have applied J48, Naïve Bayes, Multi-Boosting AB, Simple Logistic Regression, Random Forest. It is observed that Random Forest has highest accuracy rate of 86.36% compared with other algorithms.

Keywords Network traffic · Machine learning techniques · Preprocessing

1 Introduction

Mobile handsets are the biggest inventions of the early 1990s. They have changed human life. Users have migrated to portable computers such as smart phones from desktop computers due to the advancement in the technology. These advancements have made the users to access online services from anywhere at any time. Cloud computing is an emerging technology which provides different services like applications, infrastructure, and platform to the users on demand. In the traditional mobile computing, mobile devices had limited computing and storage capabilities. Applications were built and managed by the developers. But users prefer to access the services

Praveena Akki · V. Vijayarajan (✉)
School of Computer Science and Engineering,
Vellore Institute of Technology, Vellore 632014, Tamil Nadu, India
e-mail: vijayarajan.v@vit.ac.in

Praveena Akki
e-mail: praveena.akki2016@vitstudent.ac.in

© Springer Nature Singapore Pte Ltd. 2019
A. J. Kulkarni et al. (eds.), *Proceedings of the 2nd International Conference on Data Engineering and Communication Technology*, Advances in Intelligent Systems and Computing 828, https://doi.org/10.1007/978-981-13-1610-4_58

at anytime from anywhere, store the data, share the data with their friends. To meet the requirements of the users, new technology called mobile cloud computing was introduced, cloud computing in combination with mobile devices over the network.

1.1 Features of Mobile Cloud Computing

The description of features of mobile cloud computing is shown in Fig. 1 Resource Management: Mobile clouds can enable resource provisioning and demand provisioning automatically. The resources are network resources, computing resources, mobile device resources.

Security and Privacy: This feature includes security capabilities, technologies, process. This feature protects the data, devices, and network from unauthorized access and damage.

Scalability: This feature relies on cloud, network, and mobile scalability.

Convenience: This feature is designed for end users to provide access to the cloud resources at any time and from anywhere.

Connectivity: This feature chooses different well-designed APIs, protocols, standards to provide secure and easy connection between different networks and third-party applications.

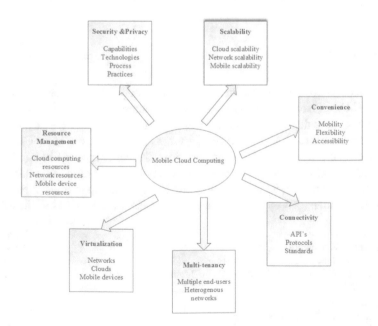

Fig. 1 Features of mobile cloud computing

Virtualization: Mainly three types of virtualization are used in mobile cloud computing namely network virtualization, cloud virtualization, mobile devices virtualization.

2 Related Work

In [2] soft air, a new SDN architecture for 5G has been proposed to overcome the limitations of software-defined networking (SDN) and network function virtualization (NFV). The software-defined network has been introduced for data center networks helps to deploy new apps and services. NFV abstracts network functionalities and implements them in software. The different aspects in wireless cellular network have been analyzed quantitively. To overcome limitations, soft air architecture was proposed. The scalability was improved by using high-performance controllers and network traffic was optimized. In [3], the resources will be dynamically allocated to the users by network virtualization. Network virtualization provides more efficient resource usage. The virtualization enables reliability and overcome the ossification of the Internet. How to assign the incoming service request to the server has become a challenge in mobile cloud computing. Migrating the service request can provide better utilization of resources; but it involves operational cost. In this paper, online migration algorithms have been used and the costs were compared without migration. In [1], various network centric parameters were analyzed such as traffic load, mobility speed, transfer type, energy consumed on migration, total execution time, end-to-end packed delay, packet delivery ratio. It was concluded that the transfer time, energy consumed on migration, and execution time increase with the traffic load.

By optimizing running states' size and applications, the network transfer time can be reduced. In [8], Rate Adaptive Topology-Aware Heuristic (RA-TAH) algorithm, an optimization algorithm, was proposed to minimize energy consumption. The network will be automatically adjusted according to the utilization. The performance of the algorithm is maximized when the utilization is high. A utilization threshold is defined as 0.7. If the load crossed threshold then another switch will be turned on, so the load will be balanced. In [12], a range-free centroid localization algorithm was proposed. Malguki algorithm was used to find location of an unknown node in the network. But its suffers from localization error. In traditional Malguki algorithm, the node is selected randomly, but in the proposed algorithm the unknown node is computed based on centroid of anchor points. The algorithm is simple, and the computed node is like the real node. This method is useful where GPS does not work to estimate nodes position. In [11], greedy algorithm was used for finding optimal service area. Greedy algorithm selects a tracking area and assigns to service area. This process will be repeated until a traffic load of service does not exceed service gateway capacity.

Limitation of greedy algorithm is due to its inefficiency since it does not consider the cost. To overcome the limitation, repeated greedy algorithm was proposed. It minimizes the number of Service Gateways (S-GW) relocations. The advantage is it can be launched periodically when there is change in traffic load. It can scale up or down the virtual service gateways. In [6], a dynamic programming is used to select the efficient communications between mobile devices and cloud servers. The large problems are divided into smaller sub-problems to solve whole problem and the best solution is selected. The cost function of each cloudlet is computed and assigns the devices to the cloudlet having minimum energy cost. In [4], the current research activities and further research challenges were presented. MANET paradigm and protocols were discussed. Mainly four adhoc networking paradigms, mesh, opportunistic, vehicular, and sensor networks, were discussed. The effect of increasing smartphones in everyday life on network traffic was discussed. In [7], a new mobile cloud computing framework, mobicloud was proposed which is used to enhance communication by addressing secure routing, risk management, trust management issues in the network. The development of new applications with enhanced processing power is the advantage of mobicloud. In [9], intelligent access schemes and heterogenous access management schemes were proposed inorder to improve quality of mobile connectivity and service availability to the users when required. In [5], an optimization approach, VM planner, was proposed to minimize the network power cost without affecting network performance. It also controls traffic flow routing by turning off as many unnecessary network elements to minimize power consumption. In [10], offloading strategies were proposed to minimize power consumption in femtocells. Femtocell is a small base station which is used to expand mobile network capacity. But the power consumption increases as the number of mobile users increases which leads to poor quality of service. The proposed offloading strategies have minimized energy consumption from 12.27–15.46%.

3 Proposed Architecture

The needs of mobile users have been growing exponentially, for example, Facebook, twitter, Picasa, eBay, amazon. Facebook, twitter have been using to share activity of users. The Picasa, Flickr are used for sharing photos. The eBay, amazon is used for online shopping. In the past era, people used to purchase items from shops in person. Due to the advancement of mobile cloud computing technology, people are preferring online transactions now. This has increased the demand for mobile data capacity. Mobile network operators carry large amount of internet traffic. It was observed that more than half of the mobile data traffic constitutes video streaming, and the remaining is images and text. Due to the limitation on the density of mobile base stations the network traffic has become a bottleneck. Many techniques were proposed to reduce the network traffic.

Prefetch is the technique where the proxy server retrieves the data from cloud server in advance based on the previous users log data. But not all the data retrieved from the server will be used by the user. Some of it will be discarded. Clients interact with server through API. The data returned by backend service should be understandable form. Mostly XML is used for its simplicity and interoperability, but it suffers with higher encoding overheads and repetitive nature of start–end tags, numerical values, binary data encoded in tags as text messages. This contributes to the increased size of messages across the network. Techniques have been proposed to eliminate the data redundancy but detecting the unnecessary tags and eliminating them increases additional overhead. Mobile clients can access cloud storage services and from everywhere. The proposed method aimed to reduce the network traffic. It predicts the client requests based on machine learning techniques. The data was collected from access log data, IRCache. Then the data was preprocessed to eliminate missing values and irrelevant information. The data was filtered by selecting frequently accessed values based on IP address and URLs. The data was analyzed by applying machine learning algorithms. The algorithm with good accuracy was selected to apply in mobile cloud computing environment. The machine learning algorithm predicts whether to accept the client request or to discard the request (Fig. 2).

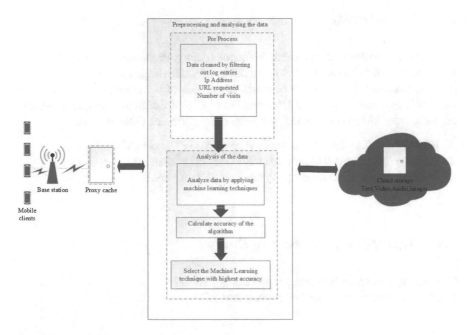

Fig. 2 Proposed architecture

3.1 Data Collection

Access log data set was taken in the proposed scheme from a proxy CC. Squid was used to collect log data based on cloud computing log data.The various fields in the client requests are

$1157689312.049500610.105.21.199TCP_H IT/2001357$

GET http://google.com/image.gif

1. Time stamp of the request
2. Time required to process the request in msec
3. IP address of the mobile client
4. Cache Hit/Miss
5. Requested data size in bytes
6. Type of method
7. Requested URL
8. Content type

3.2 Preprocessing

It is important to preprocess the data before analyzing. Preprocessing is used to clean and normalize the data to improve the quality of data. In the proposed method, mainly three attributes were taken into concern. IP address of the client, number of visits, and URL requested. Different IP addresses represent different users, and frequent requests were filtered out by analyzing the three attributes (Tables 1 and 2).

1. Remove entries with request methods except GET and POST.
2. Remove the records which have status code other than 200.
3. Remove uncatchable requests which have missing fields.

3.3 Preprocessing Access Log Dataset

Total number of records before preprocessing: 10000
Total number of records after preprocessing: 7635

Table 1 Before preprocessing

Time stamp	Elapsed time	IP address	Log tag and HTTP code	Request method	URL
1157689320	2864	10.105.21.199	TCP_MISS/304	GET	http://www.goonernews.com/
1157689359	1982	10.105.33.214	TCP_MISS/200	POST	http://shttp.msg.yahoo.com/notify/
1157689379	4	10.105.33.214	TCP_IMS_HIT/304	GET	http://a1568.g.akamai.net/7/1568/1600/20040405222807/radio.launch.yahoo.com/radio/common_radio/resources/images/t.gif
1157689388	60	10.105.21.199	TCP_HIT/200	GET	http://us.i1.yimg.com/us.yimg.com/i/us/pim/dclient/d/img/liam_ball_1.gif
1157689388	60	10.105.21.199	TCP_HIT/200	GET	http://us.i1.yimg.com/us.yimg.com/i/us/pim/dclient/d/img/liam_ball_1.gif
1157689409	15888	10.105.37.180	TCP_MISS/200	POST	http://gateway.messenger.hotmail.com/gateway/gateway.dll?
1157689437	752	10.105.37.180	TCP_REFRESH_HIT/200	GET	http://eur.i1.yimg.com/eur.yimg.com/i/fr/hp/sunsh.jpg
1157689446	429	10.105.37.184	TCP_HIT/200	GET	http://eur.i1.yimg.com/java.europe.yahoo.com/eu/hp/fu/jsbase003.js

Table 2 After preprocessing

Time stamp	Elapsed time	IP address	Log tag and HTTP code	Request method	URL
1157689359	1982	10.105.33.214	TCP_MISS/200	POST	http://shttp.msg.yahoo.com/notify/
1157689379	4	10.105.33.214	TCP_IMS_HIT/304	GET	http://a1568.g.akamai.net/7/1568/1600/ 20040405222807/radio.launch.yahoo.com/radio/ common_radio/resources/images/t.gif
1157689388	60	10.105.21.199	TCP_HIT/200	GET	http://us.i1.yimg.com/us.yimg.com/i/us/pim/ dclient/d/img/liam_ball_1.gif
1157689409	15888	10.105.37.180	TCP_MISS/200	POST	http://gateway.messenger.hotmail.com/gateway/ gateway.dll?
1157689446	429	10.105.37.184	TCP_HIT/200	GET	http://eur.i1.yimg.com/java.europe.yahoo.com/eu/ hp/fu/jsbase003.js

3.4   Algorithm

Result: Preprocessed data

Pc=client IP address;
 R=URL requests;
while $i = 0$ *to total* **do**
| count=1;
| **if** *count==limit* **then**
| | delete the record;
| | **if** *log_tag! =200* **then**
| | | delete the record;
| | | **if** *Request method! = GET* || *Request method ! = POST* **then**
| | | | delete the record;
| | | **else**
| | | | Save the record in the log file and repeat the same for all the
| | | | records
| | | **end**
| | **else**
| | | Save the record in the log file
| | **end**
| **end**
| repeat the same for all the records;
end

Algorithm 1: Algorithm for Data preprocessing

3.5   Machine Learning Algorithms

Machine learning algorithms predict outcomes without being explicitly programmed. The algorithm receives input data and use some statistical analysis to predict the output within an acceptable range. After preprocessing the data, machine learning algorithms were applied. These algorithms are mainly classified into supervised and unsupervised. In the proposed method, J48, Random Forest, Naive Bayes, Multi-Boost AB, Simple Logistic Regression were applied to analyze the data and compared them with one another to identify highest accuracy.

$$\text{Random Forest Prediction } s = \frac{1}{K} \sum_{k-1}^{k} K\text{th tree response}$$

Using decision tree, it will generate rules, but the drawback is it is greedy. This problem can be overcome by using Random Forest. The main advantage of using Random Forest is the overfitting problem which can be eliminated and it is applicable to regression. It is also capable of handling missing values and categorical values. The accuracy of Random Forest is high because the number of trees in the forest is high.

4 Simulation Results

The input access log data was collected from real cloud. The file is converted to arff or csv format. The simulation was done by WEKA tool. Different IP address represent different users. The number of cross-validation iterations is set with 10, which are the number of folds. The experiment type of performance was set as classification. The performance evaluation is based on classifier evaluation which is based on classifier evaluation metrics. The accuracy is measured as,

$$\text{Accuracy} = \frac{\text{Number of Correct Data}}{\text{Total Data}} * 100 \tag{1}$$

Precision is the models ability to discern whether the data is relevant from the returned population and recall is the models ability to select relevant documents from the population. F-measure is the harmonic mean of precision and recall. These performance metrics have been calculated based on Eqs. 1–5, the results are shown in Tables 3 and 4 and the accuracy of machine learning algorithms are shown Fig. 3.

$$\text{precision} = \frac{TP}{TP + FP} \tag{2}$$

$$\text{Recall} = \frac{TP}{TP + FN} \tag{3}$$

$$\text{Error Rate} = \frac{FP + FN}{TP + FP + FN + TN} \tag{4}$$

$$\text{F-Measure} = 2 * \frac{\text{Precision} * \text{Recall}}{\text{Precision} + \text{Recall}} \tag{5}$$

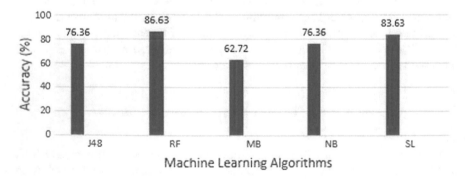

Fig. 3 Accuracy of machine learning algorithms

Table 3 Performance Metrics

Performance metrics/algorithms	J48	Random Forest	Multi-Boost AB	Naive Bayes	Simple Logistic
TP-rate	0.764	0.836	0.627	0.764	0.836
FP-rate	0.198	0.178	0.396	0.079	0.106
Precision	0.753	0.823	0.407	0.783	0.814
Recall	0.764	0.836	0.627	0.764	0.836
F-measure	0.741	0.816	0.491	0.764	0.822
ROC	0.869	0.968	0.639	0.871	0.913

Table 4 Accuracy of machine learning algorithms

Algorithm	Accuracy in %
J48	76.363
Random Forest	86.363
Multi-Boost AB	62.72
Naive Bayes	76.36
Simple Logistic	83.63

5   Conclusion

This paper presented an efficient machine learning technique for mobile cloud computing. The frequent requests from the same clients were identified and removed by applying preprocessing techniques. Then, the machine learning techniques were used to predict whether to accept the client request or to discard. Due to the limitation on the density of base stations, network traffic has become a bottleneck. By using the proposed method, the network traffic has been reduced therefore improving quality of service.

References

1. Ahmed E, Akhunzada A, Whaiduzzaman M, Gani A, Ab Hamid SH, Buyya R (2015) Network-centric performance analysis of runtime application migration in mobile cloud computing. Simul Modell Practice Theory 50:42–56
2. Akyildiz IF, Lin SC, Wang P (2015) Wireless software-defined networks (w-sdns) and network function virtualization (nfv) for 5g cellular systems: An overview and qualitative evaluation. Comput Networks 93:66–79
3. Arora D, Bienkowski M, Feldmann A, Schaffrath G, Schmid S (2011) Online strategies for intra and inter provider service migration in virtual networks. In: Proceedings of the 5th international conference on Principles, systems and applications of IP telecommunications, ACM 10 (2011)
4. Conti M, Boldrini C, Kanhere SS, Mingozzi E, Pagani E, Ruiz PM, Younis M (2015) From manet to people-centric networking: milestones and open research challenges. Comput Commun 71:1–21

5. Fang W, Liang X, Li S, Chiaraviglio L, Xiong N (2013) Vmplanner: optimizing virtual machine placement and traffic flow routing to reduce network power costs in cloud data centers. Comput Networks 57(1):179–196
6. Gai K, Qiu M, Zhao H, Tao L, Zong Z (2016) Dynamic energy-aware cloudlet-based mobile cloud computing model for green computing. J Network Comput Appl 59:46–54
7. Huang D, Zhang X, Kang M, Luo J (2010) Mobicloud: building secure cloud framework for mobile computing and communication. In: Service Oriented System Engineering (SOSE). In: 2010 Fifth IEEE international symposium on, IEEE, pp 27–34 (2010)
8. Huu TN, Ngoc NP, Thu HT, Ngoc TT, Minh DN, Tai HN, Quynh TN, Hock D, Schwartz C et al (2013) Modeling and experimenting combined smart sleep and power scaling algorithms in energy-aware data center networks. Simul Modell Practice Theory 39:20–40
9. Klein, A, Mannweiler, C, Schneider, J, Schotten, HD (2010) Access schemes for mobile cloud computing. In: 2010 Eleventh International Conference on mobile data management (MDM), IEEE, pp 387–392 (2010)
10. Mukherjee A, Gupta P, De D (2014) Mobile cloud computing based energy efficient offloading strategies for femtocell network. In: Applications and innovations in mobile computing (AIMoC), 2014, IEEE, pp 28–35 (2014)
11. Taleb T, Ksentini A (2013) Gateway relocation avoidance-aware network function placement in carrier cloud. In: Proceedings of the 16th ACM international conference on modeling, analysis & simulation of wireless and mobile systems, ACM, pp 341–346 (2013)
12. Wang Y, Jin Q, Ma J (2013) Integration of range-based and range-free localization algorithms in wireless sensor networks for mobile clouds. In: Green computing and communications (Green-Com), 2013 IEEE and internet of things (iThings/CPSCom), IEEE international conference on and IEEE cyber, physical and social computing, IEEE pp 957–961 (2013)

POUPR: Properly Utilizing User-Provided Recourses for Energy Saving in Mobile Cloud Computing

Ramasubbareddy Somula, Y Narayana, Sravani Nalluri,
Anilkumar Chunduru and K. Vinaya Sree

Abstract The mobile cloud computing (MCC) is an emerging technology, and its popularity is increasing drastically day by day. The mobile terminals (MTs) utilization can be improved by using remote cloud. In MCC technology, the MT energy consumption can be reduced, and this technology improves processing power, but still mobile terminals are suffering with dead zone issues. The previous works had not focused on this problem in mobile cloud platforms. In this paper, we address this problem by proposing energy-saving algorithm and also provide mathematical analysis. This proposed algorithm works based on user-provided resources called as properly offloading using user-provided recourses (POUPR). The mobile user shares processing power based on available devices in cluster when they are not close to MCC. The performance of the proposed algorithm POUPR is validated mathematically.

Keywords Cloud computing (CC) · Mobile cloud computing (MCC)
Networking · Energy consumption · Offloading · Algorithm · Partitioning
Self-organized criticality

R. Somula (✉) · S. Nalluri · A. Chunduru
Vellore Institute of Technology, Vellore 632014, Tamil Nadu, India
e-mail: svramasubbareddy1219@gmail.com

S. Nalluri
e-mail: sravani22me@gmail.com

A. Chunduru
e-mail: chunduru.anilkumar@vit.ac.in

Y. Narayana · K. Vinaya Sree
CSE, MIC College of Technology, Vijayawada, India
e-mail: naarayanaa808@gmail.com

© Springer Nature Singapore Pte Ltd. 2019
A. J. Kulkarni et al. (eds.), *Proceedings of the 2nd International Conference on Data Engineering and Communication Technology*, Advances in Intelligent Systems and Computing 828, https://doi.org/10.1007/978-981-13-1610-4_59

1 Introduction

The research area with the combination of cloud computing and mobile Internet is known as mobile cloud computing. According to the mobile cloud computing, the data is processed by the cloud and sent to the mobile terminals (MTs). Here, the cloud plays an important role by processing the data, and this technique is used to extend the lifetime of the battery of mobile terminals as well as develops the running speed [1]. In this technique, disconnection of the cloud leads to dead spots or coverage holes because of the wireless connection which is unstable. Due to the disconnection, the obstruction in the server from cloud is obtained [2]. To deal with the problems faced due to the disconnection, base stations which act as mediators to send service from the cloud to mobile terminals can be established. Establishment of these base stations might be an ineffective solution as it leads to high overhead in several scenarios. The other alternative for such problems which are caused due to the cloud disconnection which is cost supportive and also acts as MAC-layer used to develop the range of the base station is assignment of relay node [3]. However, engaging the relay node fills the coverage holes and develops the range with low cost. The delay Jitter and Packet toss issues arises due to establishment of really no dc. In result, the relay node losses it's capacity, this will impact on delay sensitive and real-time application [4]. Mobile cloud platform schemes ignore the problems caused due to dead spots or coverage holes [5–8]. As mentioned, mobile cloud platform schemes ignore the dead spots or coverage problems; here, mobile users ignores it's own computing capacity for future use when network is disconnected [9]. When the cloud is disconnected, then the mobile users provide their own resources, and these resources are used as a compliment of potentials to the clouds which are based on the data center. The platform with the resources is provided by the user. These platforms help to develop the quality of service (QoS) of mobile cloud computing when the cloud is disconnected. Figure 1 represents result of the problem which has been neglected. The above example describes that there are two base stations named as base station 1 and base station 2. And the three mobile terminals are named as mobile terminal X, mobile terminal Y, and mobile terminal Z. The service for the mobile terminals is provided from the cloud, and the service provided will be in the range of base stations only. If the base station 1 does not work, then the mobile terminal z is handed over to the nearest base station to the mobile terminal z that is base station 2. And the mobile terminals x and y are out of service. The solution for this problem is using relay node technique. Here, relay technique is used to assign the task of the mobile terminals x and y to the mobile terminal z, then to the base station 2, and finally to the cloud. There builds a connection between all the base stations and mobile terminals. It is represented as $X \rightarrow Y \rightarrow Z \rightarrow$ base station 2 $\rightarrow$ cloud. This technique is time-consuming due to the occurrence of multi-hop. The performance of delay-sensitive networks will be affected by multi-hop. Computing capacity of local resources like mobile terminals is ignored which leads to a lot of resource waste.

In the above-given example, the mobile terminal x is overloaded. And the computation of this mobile terminal is done separately by assigning its task to low-loaded

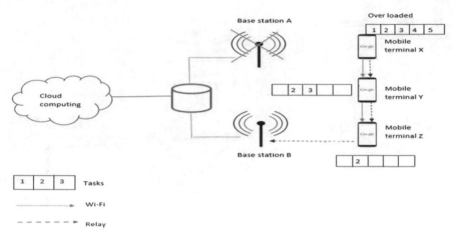

Fig. 1 User-provided resource platform

mobile terminals y and z. If the mobile terminal x has to assign its task to the nearest mobile terminal y, then the transformation of its task with one hop is sufficient. Otherwise, it has to assign its task to mobile terminal z; then, the transformation with two hops is required. The transformation of the tasks is dependent on the computing capacity of the mobile terminals comparing with the four hops (cloud $\rightarrow$ base station $\rightarrow$ MT Z $\rightarrow$ MT Y $\rightarrow$ MT X). When cloud is disconnected, the advantages that are based on the computing capacity of mobile terminals are as follows: (1) When cloud is disconnected, the service from the cloud is delayed; this could be declined by computing capacity. (2) Computing capacity of MTs saves the energy. (3) It costs less. For satisfying the required service delay constraint, an energy-saving algorithm which is known as task offloading using self-organized criticality (POUPR) is proposed to adjust the critical threshold of load.

Here, the remaining part of this paper is divided into sections where Sect. 2 is about environment given by user and method of problem formulation. In Sect. 3, we discuss the utilization of self-organized criticality for task offloading and critical threshold design. Section 4 is the conclusion of this paper.

2 Background

In this section, by using the resource platform provided by the user, system model is described and the formation of energy-saving problem is discussed.

Fig. 2 Flow of offloading at
mobile terminal

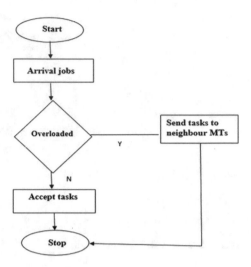

2.1 System Model

The framework of the resource platform provided by the user contains platform manager, task manager, and task sensing where platform manager is used to allocate the resource of MTs. And the work of the task manager is to run the task offloading. The information regarding resources like computing capacity, bandwidth, and storage is checked out by the task sensing. If we observe Fig. 2b, it is clear that one mobile terminal is offloading its task to the neighbor for computing even without cloud. $|R| = N$ represents the set of MTs. Each vertex $S \in R$ represents on MT, and w_R^S represents the offloading of task from MT R to MT.

2.2 Problem Formulation

For mobile terminal R, $W_R(n) \in \{0, 1, 2, 3\}$ where uniform distribution takes place. The number of arriving requests between the beginning of the nth execution and $(n + 1)$th execution is given by $W_R(n) = x$, $\{0 \leq x \leq 4\}$. The energy consumption at the nth execution when the request is executed at MT R is given by $E_R(n) = P_1 T_R(n)$. The number of arriving requests offloaded from R to S is represented by w_R^S. The total no of offload at R at the nth execution is $\sum_{S \in R} W_R^S(n)$. Otherwise, the offloading leads to the storage cost. Storage cost is neglected as energy consumed for storing is negligible than processing. Energy consumption due to offloading in idle state is represented as

$$E_x(n) \left\{ P_x \sum_{S \in R} \max w_R^S(n) \left[T_R^S(n) T_{RS}^q \right] \right\} \tag{1}$$

According to (1), the energy utilization function and response time are

$$E(n) = E_x + \sum_{S \in R} \left[W_R(n) - \sum_{S \in R} W_R^S \right] E_R(n) + (n) E_R^S(n) + \sum_{S \in R} W_R^S(n) E_{SR}(n) \quad (2)$$

And

$$T(n) = \sum_{S \in R} \left\{ \max \left[\left(W_R(n) - \sum_{S \in R} W_R^S(n) \right) \left[T_R(n) + T_R^q(n) \right] \right. \right.$$
$$\left. \left. \cdot W_R^S(n) \left(T_R^S \left(n + T_{SR}^0 + T_{SR}^q(n) \right) \right) \right] \right\} \quad (3)$$

Service delay must satisfy

$$T(n) \leq T_{\text{resp}} \quad (4)$$

where service delay is represented by T_{resp}.

Let $E(n) \lim_{T \to \infty} \frac{1}{T} \sum_n^T E(t)$ denote the median of utilizing energy

$$\min_{w(n)} E(n) \forall n = 0, 1, 2 \ldots T - 1 \quad (5)$$
$$T_{\min} \leq T(n) \leq T_{\text{resp}} \quad (6)$$
$$T_{\min \leq \theta} \quad (7)$$

Here, θ represents median of visiting demands and $T_{\min}$ represents minimum latency. Self-motivated mobile user resources are responsible for resources platform provided by self-organized criticality [10]. In favor of solving problems (5)–(7), an algorithm of offloading is presented by the self-organized criticality. This task sending innovation is trade-off between energy utilization and latency.

3 Task Offloading Using Self-organized Criticality

We go through the description of task loading using self-organized criticality (POUPR) and layout of significant threshold of POUPR.

3.1 Description of POUPR

The decision offloading procedure for any MT is whenever a new task is received by the MT, it is checked whether that task is received by MT is overloaded or underloaded. If that MT is overloaded, then some of the tasks it received are sent to the adjacent (or) nearby MTs; this process of sending that task from one mobile terminal to other is called offloading. If those neighboring mobile terminals are also overloaded, then again some of the tasks from the overloaded mobile terminals will be transferred to the neighbor mobile terminals. This process continues till the MTs are not overloaded. This cycling process of offloading is called avalanche.

Let us assume that time of the MT to offload task is much negligible than the time it takes to complete computation. Nearly, the offloading time is zero. $\left(T_{SR}^0(n) \ll T_S^R(n)\right)$. Moreover, the no of requests for Mt R at nth execution is n given by $W_{R(n)}$. The number of tasks which are offload from R to S by S_z is given by $\sum_{S \in R} W_R^S(n)$. The time taken for each MT to process its task is unit. Since $T_{SR}^0(n) \ll T_S^R(n)$, avalanche always starts at nth execution and ends at $(n+1)$th execution.

Step 1.   Start the threshold by given initial values and the threshold S_z for each MT R.

Step 2.   Tasks are changed as $W_R(t)$ at the starting of the nth execution for each MT R, and then $S_R(n) \Rightarrow S_R(n) + W_R(n)$.

Step 3.   If $S_R(n) \geq S_z$, then MT R offloads its task to surrounding MTs $R+1$ and $R-1$, and then

$$S_R(n) \Rightarrow S_R(n) - \left[S_R(n) - S_z\right]$$

$$S_{R+1}(n) \Rightarrow S_{R+1}(n) + \left[\frac{S_R(n) - S_Z}{2}\right]$$

$$S_{R-1}(n) \Rightarrow S_{R-1}(n) + \left[\frac{S_R(n) - S_Z}{2}\right]$$

Step 4.   When $S_R(n) < S_z$ for $S \in R$, avalanche completes and waits for a unit time. When nth $= (n+1)$th, then critical threshold changes from $S_R(n)$ to $S_R(n) - 1$. To continue the offloading processes, go to Step 2 or else Step 3.

In this process, critical threshold decreases; as a result, energy consumption increases and service delay decreases due to the POUPR. Here, the critical threshold is proportional to service delay and inversely proportional to energy consumption.

3.2   Critical Threshold Design

Let us assume that requests arriving to the mobile terminals are of same computational complexity and with equal processing capacity.

$$E(n) = E_x + \sum_{S \in R} W_R(n) E_R(n) + \sum_{S \in R} W_R^S(n) E_{SR}(n) D\left(W_R^S(n)\right) \qquad (8)$$

where $D\left(W_R^S(n)\right)$ represents the offloading times of requests arrived between nth and $(n + 1)$th executions. Here, a lemma related to the derivation of $E(n)$ is presented in [10].

Lemma 1 *If the distribution function of avalanche size for self-organized criticality is D (k), then that function is given by*

$$D(k) \approx K^{-T} (T \approx 3) \tag{9}$$

According to Lemma 1 and [5], theorems characterizing the performance of the POUPR can be derived. Theorems 1 and 2 prove the association of both latency and average energy utilization.

Theorem *When the minimum incoming tasks interval $\theta > T_{min}$, with the critical threshold $S_Z \leq 2\left(\frac{T_{resp}}{T_R^S} - 1\right)$, then service delay due to POUPR is lesser compared to latency $\left(i.e.; T(n) \leq T_{resp}\right)$*

Proof All incoming tasks consist of identical computational complication where every empty has same processing capacity MTs are with equal processing capacity. $T_R(n) = S_R(n) \times T_R$ and $T_{SR}(n) = S_S(n) \times T_R^S(n)$. From Lemma 1. $\lim_{T\to\infty} \frac{1}{T}\sum_{S\in R} W_S^R = E[D(k)] \approx 1$. On summing $\lim_{T\to\infty} \frac{1}{T} W_R(n) = \frac{4-2}{2} = 2$. Now, the minimum incoming task interval $\theta > T_{min}$, and POUPR is constant. The average request series length is less than $\frac{S_Z}{Z}$ [10].

$$\lim_{T\to\infty} \frac{1}{T}\left(W_R(n) - \sum_{S\in R} W_R^S(n)\right)\left(T_R(n) + T_R^Q(n)\right)$$
$$\leq (2 - 1)\left(T_R(n) + \frac{S_Z}{2}T_R(n)\right)$$
$$= \left(\frac{S_Z}{2} + 1\right)T_R(n) \tag{10}$$

Since $T_{SR}(n) \ll T_R^S(n)$,

$$\lim_{T\to\infty} \frac{1}{T}W_R^S(n)\left(T_R^S(n) + T_{SR}(n) + T_{RS}^q(n)\right)$$
$$\approx \lim_{T\to\infty} \frac{1}{T}W_R^S(n)\left(T_R^S(n) + T_{RS}^q(n)\right)$$
$$\leq \left(\frac{S_Z}{2} + 1\right)T_R^S(n) \tag{11}$$

Based on 3,

$$T(n) \leq \left(\frac{S_Z}{2} + 1\right)T_R^S(n) \tag{12}$$

Moreover, (12) must satisfy (6),

$$S_Z \leq 2\left(\frac{T_{\text{resp}}}{T_R^S} - 1\right) \tag{13}$$

Thus, $T(n) \leq T_{\text{resp}}$ when (13) is satisfied.

Theorem 2 *If we assume* $\exists S_C > 0$ *and* $\forall 0 < S_Z < S_C$, *then* $\lim_{T \to \infty} \frac{1}{T} D[W_R^S(n)] \approx |R|^{\frac{1}{S_Z}}$ *with the average arriving request interval* $\theta > T_{min}$

$$E(n) \leq 2 \times E_R(n) + E_{SR}(n) \tag{14}$$

where $|R|$ *is total no. of MTs.*

Proof When $\theta > T_{min}$, SOCTO is stable. MT R does not consume any energy in the idle state, and then the no of tasks of R will be critical value at nth execution where the no of offloading tasks is $S_R(n) - S_Z$ at $S_R(n) \geq S_Z$.

$$E(n) = \sum_{S \in R} W_R(n) E_R(n) + \sum_{S \in R} W_R^S(n) E_{SR}(n) D\left(W_R^S(n)\right) \tag{15}$$

The energy consumed in average can be written as

$$E(n) = \lim_{T \to \infty} \frac{1}{T} E(n) \tag{16}$$

Since $\lim_{T \to \infty} \frac{1}{T} D[W_R^S(n)] \approx |R|^{\frac{1}{S_Z}}$, $\lim_{T \to \infty} \frac{1}{T} W_R(n) = 2$ and $\lim_{T \to \infty} \frac{1}{T} \sum_{S \in R} W_R^S(n) \approx 1$ when $T \to \infty$ in (16), we get equality (14).

Here, (12) is the boundary for the minimum latency. S_z is proportional to the boundary where it allows requests to spend time in waiting before execution. According to (14), S_z is inversely proportional to the energy utilization which causes ability of TOSOC to minimize energy utilization, and (13) represents that time gets satisfied by the application latency. The performance of the environment (energy utilization and latency) provided by the user is given by 500 m × 500 m square area.

Let the design of the network be a collection of MTs in signal dead zone to get the performance of the TOSOC. We can get the performance of the energy utilization and latency by comparing with our designed innovation in x ($x = 4, 5, 6, 7$). Here, S_z is proportional to latency and inversely proportional to energy utilization. Our algorithm leads to minimizing energy utilization by adjusting significant threshold.

Figure 3 represents the energy consumption and completion time when S_z set to 6.

Further, to know the performance of offloading, we compare service delay and energy consumption among our platform, remote cloud platform with relay node and without mechanism which defines the tasks are executed by MT without offloading any tasks.

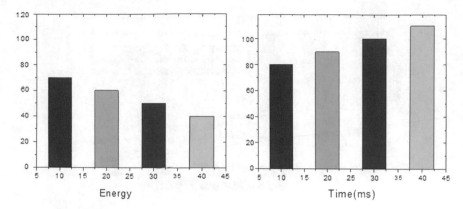

Energy   Time(ms)

Fig. 3 Comparison between energy consumption and processing time

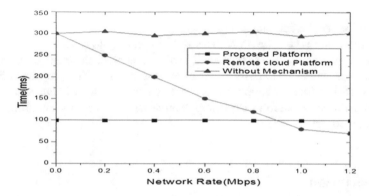

Fig. 4 Service delay comparison

Figure 4 shows the association between latency and network progression. The relationship is as follows when network rate is low, service delay is decreased by the offloading platform. This is because when network rate is low, our offloading platform helps MTs to find and process the surrounding computing resource. The remote cloud environment has the greater latency due to the occurrence of the multi-hop scheduling, delay jitter, and the excessive chances in package loss in relay nodes. Comparing to our platforms, remote cloud platforms compute much faster. But remote cloud platforms have low service than our platforms in high network rate. Service delay without mechanism will be constant 300 ms as network rate increasing. This is because when network rate is increasing, MTs do not offload their tasks instead performing themselves. The proposed platform makes trade-off between service delay and energy consumption constant latency of the designed environment with increasing network progression.

Figure 5 shows the average energy consumption. The service interruption caused to remote cloud and relay nodes due to the low network progression. This leads to the higher consumption of energy for remote cloud platforms. Mobile energy

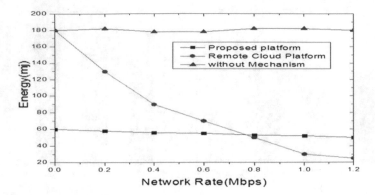

Fig. 5 Energy consumption comparison

consumption will be constant at 180mj without our mechanism, while network rate is growing. Due to the large no of tasks in MTs leading to MTs' crashing without offloading any task from the MTs, it leads to higher energy consumption. As network rate increases, the average energy consumption decreases. The proposed platform allows MTS to offload jobs to remote server when they are without relay. In this context the proposed platform acts as supplement scheme to mobile cloud computing where network rate is low.

4 Conclusion

Mobile cloud computing provides poor service and sometimes no service to the MTs when there are dead zones. We present a user-provided platform where MTs can provide their own idle computing resources when cloud is unable to provide resources or service to the MTs and increases the QOS of mobile cloud computing. The offloading technology in mobile cloud computing reduces energy consumption and response time of MTs, we presented mathematical model for it. In order to satisfy user requirements related to energy and latency, a task offloading algorithm based on self-organized criticality is proposed. There are three issues (1) service availability: We should make sure high resource accessibility as there is no surety in the user-provided resources that mobile user's local resources are always online for user-provided resources platform. (2) Mobile user incentive: The design of better incentive is necessary as there is a query that all available users are interested to share computing capacity for other users for free. (3) Business model: It is still unsure about who will organize both task offloading and the proposed model. User provided resources benefit cloud platform to process more requests coming from mobile users. But still our model requires organized business model.

References

1. Niyato, D, Wang, P, Hossain, E, Saad, W, Han, Z (2012) Game theoretic modeling of cooperation among service providers in mobile cloud computing environments. In: Wireless communications and networking conference (WCNC), 2012 IEEE, pp 3128–3133
2. Satyanarayanan M, Bahl P, Caceres R, Davies N (2009) The case for vm-based cloudlets in mobile computing. IEEE Pervasive Comput 8:4
3. Deb S, Mhatre V, Ramaiyan V (2008) WiMAX relay networks: opportunistic scheduling to exploit multiuser diversity and frequency selectivity. In: Proceedings of the 14th ACM international conference on mobile computing and networking, pp 163–174
4. Oyman O (2006) OFDM2A: a centralized resource allocation policy for cellular multi-hop networks. In: Fortieth Asilomar conference on signals, systems and computers, 2006 (ACSSC'06), pp 656–660
5. Huang D, Wang P, Niyato D (2012) A dynamic offloading algorithm for mobile computing. IEEE Trans Wirel Commun 11(6):1991–1995
6. Chang H, Kodialam M, Kompella RR, Lakshman TV, Lee M, Mukherjee S (2011) Scheduling in mapreduce-like systems for fast completion time. In: 2011 proceedings IEEE INFOCOM, pp 3074–3082
7. Maguluri ST, Srikant R, Ying L (2012) Stochastic models of load balancing and scheduling in cloud computing clusters. In: 2012 Proceedings IEEE INFOCOM, pp 702–710
8. Benslimane A, Taleb T, Sivaraj R (2011) Dynamic clustering-based adaptive mobile gateway management in integrated VANET—3G heterogeneous wireless networks. IEEE J Sel Areas Commun 29(3):559–570
9. Wang H, Wang F, Liu J, Groen J (2012) Measurement and utilization of customer-provided resources for cloud computing. In: 2012 proceedings IEEE INFOCOM, pp 442–450
10. Bak P, Tang C, Wiesenfeld K (1987) Self-organized criticality: an explanation of the 1/f noise. Phys Rev Lett 59(4):381

Experimental and Comparative Analysis of Packet Sniffing Tools

Chunduru Anilkumar, D. Paul Joseph, V. Madhu Viswanatham, Aravind Karrothu and B. Venkatesh

Abstract Intrusion detection system (IDS) is one of the applications or software that detects the malicious activities and vulnerabilities throughout the network. The purpose of IDS is to monitor the application-related activity, i.e., incoming and outgoing traffic, and to monitor the threats or attacks that are originating from the other networks; IDS detects lot of threats that are existing, but could not control or detect the new attacks and handle them. To overcome this, intrusion detection and prevention systems (IDPSs) were introduced. The main task of IDPS is to monitor, detect, and prevent the threats and attacks. Till today, there are many attacks that prevailed IDPS. This paper concentrates or peeps through the birth of IDS and IDPS, briefly describes the various attacks and threats that are behind and beyond IDPS, and tries to create a new attack by adding some rules over the network that bypasses the monitoring of IDPS. In addition to this, this paper compares all the existing IDS/IDPS techniques and different types of IDPS and analyzes all the tools existing, compared with different metrics or parameters in different environments.

Keywords Intrusion detection system (IDS) · Intrusion detection and prevention systems (IDPSs) · Intrusion prevention system (IPS) · Network behavior analysis (NBA)

C. Anilkumar (✉) · D. Paul Joseph · V. Madhu Viswanatham · A. Karrothu · B. Venkatesh
Vellore Institute of Technology, Vellore 632014, Tamil Nadu, India
e-mail: chunduru.anilkumar@vit.ac.in

D. Paul Joseph
e-mail: pauljoseph.d@vit.ac.in

V. Madhu Viswanatham
e-mail: vmadhuviswanatham@vit.ac.in

A. Karrothu
e-mail: karrothuaravind118@gmail.com

B. Venkatesh
e-mail: venkatesh.cse88@gmail.com

© Springer Nature Singapore Pte Ltd. 2019 597
A. J. Kulkarni et al. (eds.), *Proceedings of the 2nd International Conference on Data Engineering and Communication Technology*, Advances in Intelligent Systems and Computing 828, https://doi.org/10.1007/978-981-13-1610-4_60

1 Introduction

Intrusion detection can be defined as the process of observing the behavior of the computer system or networking system for possible instruction detections by detailed analysis [1]. Intrusion detection system can be automated by using software for detections. The aim of the automated software is to detect the instruction and record the events for future reference. Instruction prevention software is similar to instruction detection software; in addition, it prevents the instructions. The three main functions of the IDS's server are observing, finding, and reacting to abnormal behaviors and instructions [2]. The four main systems where IDS can be applied are as follows [3]:

- Network-based
- Wireless
- Network behavior analysis (NBA)
- Host-based

Network-Based: IDS in networking system observes the network traffic in small segment or devices to find out the instructions [4]. It also observes the working protocols in the networks for the abnormal activities.

Wireless: The wireless traffic is used for detecting the instructions by analyzing the working wireless protocols in the wireless networks [5].

Network Behavior Analysis (NBA): This system finds the attacks in traffic flows by finding out the deviation of the flow from the normal flow which can be denial of service, malware attacks, and violation of rules [6].

Host-Based: It is used for monitoring the single host for instructions by abnormal activities [7].

Attacks on IDS and IDPS: With reference to TCP/IP Protocol stack, the different attacks at different layers are given in Table 1.

2 Motivation

Since the birth of networks in computer era, different types of attacks are being performed throughout local network and global network like buffer attacks, waterhole attacks, TCP landing attacks, DoS and DDoS attacks. It is more surprising that more than networking technologies, attacks are more and highly sophisticated. So this paper is written in a way that gives a complete guide to all types of people regarding various attacks at various networking layers. Moreover, this paper concentrates on testing the various tools that capture packets, analyze them, decode them, and give the best packet analyzer and decoder tool to the people. All the tools used in these experiments are tested under various platforms of different architectures like x86, x32, and i386. With an intention of providing various sophisticated attacks, this paper has provided different attacks and different tools and gives the results from different tools.

Table 1 Attacks at different layers

Layer	Attacks
Application layer	FTP attacks
	SQL injection
	IP spoofing
	DNS attack
	Cross-site scripting
	Email spoofing
Transport layer	Session hijacking
	TCP land attack
	Port scanning
	Bind DNS
	Mail transport system
Network layer	Packet sniffing
	RIP routing attack
	Fragmentation attack
	ICMP attack
Data link layer	Cam table exhaustion attack
	ARP spoofing
	DHCP starvation attacks
	VLAN hopping
	Fake access point attack
	Hidden node attack

3 Background

3.1 Taxonomy

In this section, the birth and development of IDS is covered. Right from its approach to attack, the different stages in development and different architectures developed so far are given in Fig. 1.

3.2 Comparison of Detection Techniques

Though there are different classifications of techniques, the following are the overall detection techniques present in IDS and are explained along with their dimensions in brief.

- Anomaly-based IDS
- Specification-based IDS

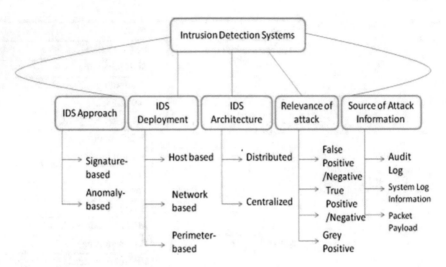

Fig. 1 Taxonomy of intrusion detection systems

- Signature-based IDS
- Dynamic-based IDS

Anomaly-Based IDS. It is different from normal methods of intrusion detection; instead of considering the previous historical test signals obtained from the unsupervised learning, this approach considers runtime characteristics [8, 9].

Specification-Based IDS. This approach finds the deviation of the system process from its normal characteristics. It does not need user's features or flow of information; instead, it detects the system based on abnormal behavior of the system. It has very less false negative detections [10].

Signature-Based IDS. Similar to anomaly intrusion detection, signature-based method also takes into account the runtime characteristics that analyze patterns of abnormal features in the system. It has very less false positive results. Formally, this detects only known attacks, but it does not analyze and find the signature [11]. Few advantages of this approach can be listed as outsider attacks, users changing profiles, permanent dictionary, and enough storage.

Dynamic-Based IDS. This approach can also be called as reputation-based detections. It has reputation manager to find out the node for selfish behavior instead of considering its security deviations [3, 12].

3.3 Pros of IDPS

See Table 2.

Table 2 Pros of IDPS

Types of IDS	WLANS	WPANS	Ad hoc networks	Mobile telephony
Signature-based	Variable CONOP, easy updates	Variable CONOP, constrained resources	High detection rate	Variable CONOP, constrained resources, easy updates
Anomaly-based	Unknown attacks	Unknown attacks	Unknown attacks	Well-defined CONOP
Specification-based	Unknown attacks	Unknown attacks	Unknown	Unknown attacks
Reputation-based	Find selfish actors	Find selfish actors	Find selfish actors	Find selfish actors
Behavior-based	Low false negatives	Minimal memory	Minimal memory	Minimal memory

Table 3 Cons of IDPS

Types of IDs	WLANS	WPANS	Ad hoc networks	Mobile telephony
Signature-based	Dictionary freshness	Dictionary size	Dictionary freshness	Dictionary size
Anomaly-based	Variable CONOP	Variable CONOP	Variable CONOP	Revenue impact
Specification-based	Lack common use case	Lack common use case	Lack common use case	Lack common use case
Reputation-based	Selfish actor	Selfish actor	Selfish actor sanctions	Revenue impact

3.4 Cons of IDPS

See Table 3.

4 Performance Measurement Analysis

Wireshark: Originally known as Ethereal, it is an open-source packet. It can be deployed in networks.

Microsoft Network Monitor: Microsoft's Network Monitor is a tool that allows capturing and protocol analysis of network traffic [12]. Network Monitor 3 is a protocol analyzer. It enables you to capture, to view, and to analyze network data. You can use it to help troubleshoot problems with applications on the network.

Interceptor: The Interceptor tool combines two processes: the recorder and the replayer. The recorder intercepts interactions destined for a Master Data Engine, logs them to an "interaction data" file, and then forwards the interactions on to the engine in a transparent manner [12]. The interactions in the file can then be replayed on a destination engine to facilitate routine maintenance or upgrade.

Fiddler: This is a tool used for modifying the HTTP traffic in case of instructions [13]. In real time, Fiddler is applied in Microsoft's WinINET HTTP(S), in which proxy is used for redirecting the traffic in run time by any browser.

5 Results and Discussion

So far in this paper, we have implemented four intrusion detection tools, and based on the different parameters like memory set, working set, response time of CPU, kernel, average CPU, and threads, their performance is measured. Based on different parameters and different platforms, these tools are deeply tested and analyzed. Under different conditions, each tool has its own strength and its disadvantages. Coming to CPU usage and kernel time response, Microsoft Network Monitor stands in first followed by Wireshark. In terms of working set, Interceptor comes in first followed by Fiddler. On overall comparison on average parameters, Microsoft Network Monitor is the best tool that captures packets because of its concurrent capturing ability (Fig. 2).

The Network Monitor tool is observed under various parameters like CPU usage, kernel time, user time, total CPU time, and memory working set, pooled quota, network transfer rate, and the resultant graph is shown in Fig. 3. It is found that CPU time, kernel time, and CPU usage are fair enough as they are in safe zone (represented in green). The usage which ranges 50–70% is shown in yellow, and above 70% usage is represented in red color, which is not found in this tool. Interceptor tool, a multi-server configuration tool is tested under same conditions like other tools, and resultant graph is shown in Fig. 4. In analysis, it is found that Interceptor tool using average CPU, page fault and memory working set and network read/write is above usage.

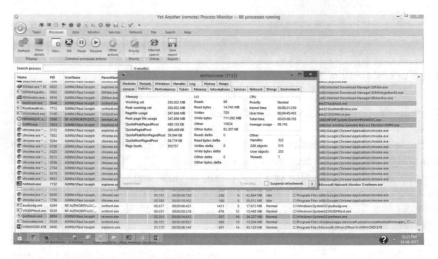

Fig. 2 Network Monitor in live analysis

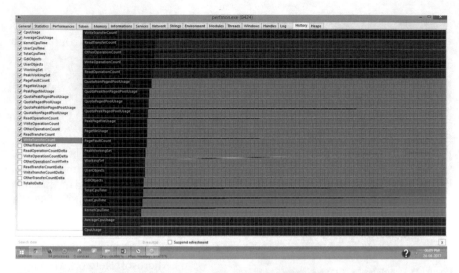

Fig. 3 Graphical analysis of Network Monitor in terms of parameters

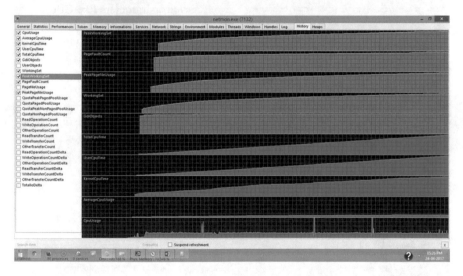

Fig. 4 Graphical analysis of Interceptor in terms of parameters

As of above described, all the tools are tested under different test conditions and different platforms. Packets are captured in live analysis as well as from benchmark datasets of US National Cyber Defense Competition MACCDC. The observed results are shown in Fig. 5. Based on different scenarios, Microsoft Network Monitor is proved to be the best tool in terms of CPU and kernel usage, and Fiddler is proved to be the best HTTP traffic decoder. Coming to terms of parallel processing, Interceptor is proved to be the best (Fig. 6).

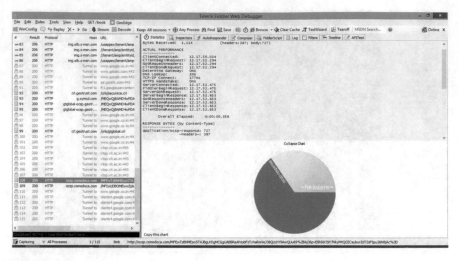

Fig. 5 Debugging of HTTP traffic by Fiddler

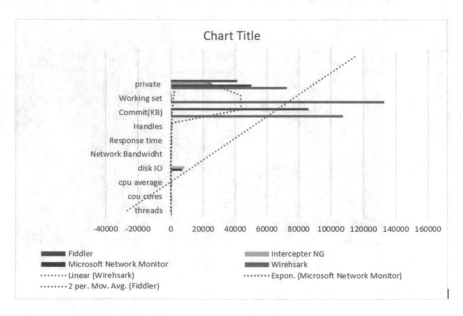

Fig. 6 Performance analysis of tools in live environment under various conditions

6 Conclusion

This paper provides a basic and in-depth comparison of various attacks at different networking layers and various IDS tools. In depth literature survey regarding various IDS and IDPS types, their classification is provided in this paper. In additional information regarding dimension, time, and detection, survey is done on IDS and IDPS. Different types of WLANs based on IDS dimensions are also identified and given in a tabular format for better understanding. The overall results declare that Microsoft Network Monitor is the best tool on an average parameter.

References

1. Pontarelli S, Bianchi G, Teofili S (2013) Traffic-aware design of a high-speed FPGA network intrusion detection system. IEEE Trans Comput 62(11):2322–2334
2. Creech G, Hu J (2014) A semantic approach to host-based intrusion detection systems using contiguous and discontiguous system call patterns. IEEE Trans Comput 63(4):807–819
3. Marchang N, Datta R, Das SK (2016) A novel approach for efficient usage of intrusion detection system in mobile ad hoc networks. IEEE Trans Veh Technol PP(99):1
4. Mitchell R (2013) Behavior-rule based intrusion detection systems for safety critical smart grid applications. IEEE Trans Smart Grid 4(3):1254–1263
5. Lin CH, Song KT (2014) Probability-based location aware design and on-demand robotic intrusion detection system. IEEE Trans Syst Man Cybern Syst 44(6):705–715
6. Kyriakopoulos KG, Aparicio-Navarro FJ, Parish DJ (2014) Manual and automatic assigned thresholds in multi-layer data fusion intrusion detection system for 802.11 attacks. IET Inf Secur 8(1):42–50
7. Faisal MA, Aung Z, Williams JR, Sanchez A (2015) Data-stream-based intrusion detection system for advanced metering infrastructure in smart grid: a feasibility study. IEEE Syst J 9(1):31–44
8. Yang Y et al (2014) Multi attribute SCADA-specific intrusion detection system for power networks. IEEE Trans Power Deliv 29(3):1092–1102
9. Spiess J, Joens YT, Dragnea R, Spencer P (2014) Using big data to improve customer experience and business performance. Bell Labs Tech J 18(4):3–17
10. Tsikoudis N, Papadogiannakis A, Markatos EP (2016) LEoNIDS : a low-latency and intrusion detection system. IEEE Trans Emerg Top Comput 4(1)
11. Zhou C et al (2015) Design and analysis of multimodel-based anomaly intrusion detection systems in industrial process automation. IEEE Trans Syst Man Cybern Syst 45(10):1345–1360
12. Liu L, Zhang W, Deng C, Yin S, Wei S (2015) BriGuard : a lightweight indoor intrusion detection system based on infrared light spot displacement. vol. 9, no. July 2014, pp. 306–314, 2015
13. Ambusaidi M, He X, Nanda P, Tan Z (2016) Building an intrusion detection system using a filter-based feature selection algorithm. IEEE Trans Comput PP(99):1
14. I. E. C. S. Networks (2017) Multidimensional intrusion detection system for 32(2):1068–1078

Emerging Trends for Effective Software Project Management Practices

J. Kamalakannan, Anurag Chakrabortty, Dhritistab Basak
and L. D. Dhinesh Babu

Abstract Developing software projects nowadays is not much of an errand. With several tools springing up, it is easy to develop software projects and if we go down with a sneak-peek into the overall project scenario, what ultimately matters is the management. Management serves as upright in whatever we do, and for software projects, it is none the less. Software projects can be developed and tested with several software and automated tools in hand, but efficient and successful management is needed, satisfying customers, stakeholders and those involved in the development and other tasks incorporated in developing software projects. This paper of ours aims to bring out newer trends, solutions, and practices in the field of management involved in software projects so as to cope up with the ever increasing diversification of projects, customer needs, demands, and preferences.

Keywords Efficient management · IT world · Management trends · Management practices · Newer management trends · Software project development · Software project management

1 Introduction

Management is the art of handling things; the more efficient, the more skillful, more are the chances of a planned, organized, and well-executed outcome. Over the last few decades, we have found that there has been a sharp growth in the software industry leading to unprecedented software projects and with the evolvement of such projects, customers too have varied demands and have adapted themselves accordingly; but surprisingly despite innovative proposals and developments, we find lack of managerial skill in the developers and the development team and environment, at large. Hence, we direct at ushering out coeval management trends and unearth

J. Kamalakannan (✉) · A. Chakrabortty · D. Basak · L. D. Dhinesh Babu
School of Information Technology and Engineering, Vellore Institute of Technology, Vellore, India
e-mail: jkamalakannan@vit.ac.in

© Springer Nature Singapore Pte Ltd. 2019 607
A. J. Kulkarni et al. (eds.), *Proceedings of the 2nd International Conference on Data Engineering and Communication Technology*, Advances in Intelligent Systems and Computing 828, https://doi.org/10.1007/978-981-13-1610-4_61

discernment of the management to indoctrinate developers which may be added feathers in successful developments.

2 Literature Survey

Biplav Srivastava, in his paper [1], states that management is of uttermost importance in case of software projects, besides development; there comes various snags when the picture of handling complex applications draws in. The author sets forth a formal model which can be handy for taking cost-effective decisions and provides automated planning and reasoning techniques. The model admires the user's commendable execution and effort that he/she puts in. The model accompanies a prototype which expresses its efficacy and practical implications.

Paweł Pierzchałka, in her paper [2], states that efficacious management is needed in software projects, and systematic and well-coordinated, well-implemented management may lead to several successful outcomes like that of "cost reduction, removing certain barriers, be it in team-building, coordination, or even communication." The author proposes "an incremental way of organizational and managerial approach" and confers about several tools and methods required in achieving the same.

Robson Marhanhao, Marcelo Marinho, and Hermano de Moura, in their paper [5], state that innovation is nowadays of great tenor to have an enduring effect on the projects and the aftermath of the same. Leaving beside orthodox methodologies and letting innovation set in, may result in sundry impediments which we can get rid of only if the management too tends to get itself revamped toward fresh approaches.

Janusz Sosnowski, Bartosz Dobrzynksi, and Pawel Janczarek, in their paper [6], state keeping aside software maintenance schemes and build-out methodologies, we need to focus on managerial facets by having perspicacity on project repositories. Data and details collected from those repositories may help in understanding and taking up ways of grasping short as well as persisting projects and dealing with them with lesser discrepancies in the "remote future".

A survey occurred among experts in IT to test whether the model proposed in the before paper is successful or not. The model was the result of an overview, led in the prior paper where the surveyees were approached to vote in favor of the undertaking related in the task procedure out of a size of 1–5 in view of their significance. The model is as follows (Fig. 1).

The same surveyees when presented this model were asked to give their ratings to the effectiveness of the above-proposed model and in view of the rate gathered, the accompanying chart was plotted (Fig. 2).

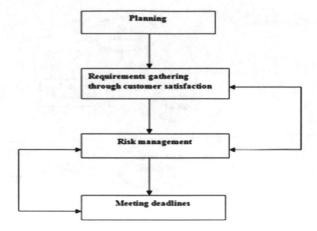

Fig. 1 Model based on the importance of the activities associated in a software project process

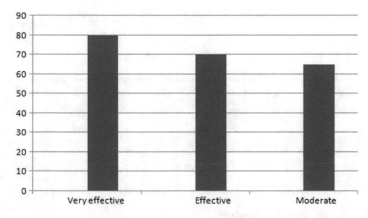

Fig. 2 Graph based on the percentages accumulated from the second survey

3 Proposed Theory

With psychology playing an important mantle in management, another riveting yet foremost feature that may come into space is "emotional intelligence." The term "emotional intelligence" refers to the ability to expose the emotions hidden in people entangled in the team, be it the developers, testers, or project managers themselves. Emotional intelligence may play a supreme role in problem solving, censorious analysis, and thinking tasks. Spreading emotional intelligence may help in restricting one's emotions while boosting up the emotion of another person, creating and maintaining a healthy and benevolent environment (Fig. 3).

   i. DevOps as the emerging model from Agile.

Fig. 3 Model for emotional intelligence

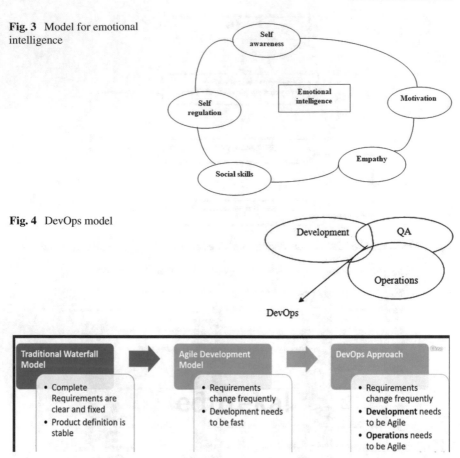

Fig. 4 DevOps model

Fig. 5 Evolutionary approach to DevOps

'Advancement with operations', habitually known as "DevOps," is a technique for correlation and connection between "IT operations groups" and "programming engineers." Currently, DevOps is associated with bringing "culture" into a work environment, and in augmentation, pulling in an adroit group. DevOps, in a conjunct effort with constant carriage, accredits Agile advancement, which is intensely centered on associations and correspondence among organizations and relational improvement (Fig. 4).

DevOps is intrinsically the aggrandizement of Agile standards to subsume frameworks, protocols, and operations as opposed to culminating its apprehensions at code scrutinizing. Aside coadjuvancy as groups of originator, analyzer, and engineer, as a constituent of an Agile group, DevOps embraces to inculcate operations also in the elucidation of cross-practical group. DevOps aspires to concentrate on either, general administration or programming, completely conveyed to the client, rather than just waging on the programming (Fig. 5 and Table 1).

Table 1 Points of difference between Agile and DevOps with corresponding description

Speed to production	Team skill sets	To Sprint or not to sprint
Nimble improvement relates generally to the way advancement is thought of any division of an organization can be a "dexterous" office through legitimate preparing and accentuation, using short dashes to finish a task. DevOps concentrates on sending programming in the most solid and most secure course, which is not really dependably the quickest, the principle of Agile	Agile development puts a large emphasis on training all team members to have a wide variety of similar and equal skills. When something goes awry, any member of the Agile team can lend assistance rather than waiting for a team lead or specialist. DevOps, on the other hand, likes to divide and conquer: spreading the skill set between the development and operation teams while maintaining consistent communication	DevOps and Agile operate on very different time tables. Agile methodology mainly focuses on "sprints," i.e., the timetable is much shorter (within a month) and numerous features are to be produced and released within that given time period. On the other hand, DevOps strives for concrete deadlines and benchmarks with major releases, rather than smaller and more frequent ones

4　Differences

See Table 1.

5　Similarities

Besides, we led another review among experts with IT foundation to choose their proper model in view of the likenesses, so they can have an outline to get which show in what circumstance. The survey table is as follows (Tables 2 and 3).

Based on survey results, the plotted graph is provided for reference (Fig. 6).

Table 2 Points of similarities between Agile and DevOps with corresponding description

IT productivity	Frequent collaboration
While DevOps may appear like the slower way to deal with conveying items, it has an indistinguishable objective as a main priority from Agile advancement: efficiency and unwavering quality. With Agile pushing DevOps to work speedier, and DevOps squeezing Agile to be exhaustive, the final product is a fruitful and well thoroughly considered item	Despite the fact that DevOps and Agile shift in their range of abilities approach, the two gatherings have a huge accentuation on correspondence and cooperation inside gatherings and inside the organization overall as well. Visit updates and benchmark evaluations are a need inside the two advancement groups. So while DevOps and Agile have some key contrasts, their coordinated effort and assertion is essential keeping in mind the end goal is to make progress

Table 3 Survey table for selection of models

Total no. of persons reviewed	20
Area/background of persons reviewed	IT
For, IT productivity, no of persons voted for Agile	14
For, IT productivity, no. of persons voted for DevOps	6
For, frequent collaboration, no. of persons voted for Agile	8
For, frequent collaboration, no. of persons voted for DevOps	12

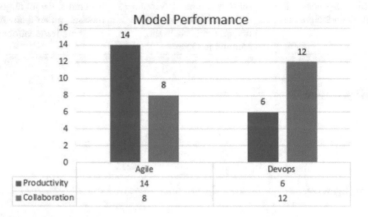

Fig. 6 Graph plot based on above survey table

Here on, we locate that Agile thinks of deftness and speed and can help or push up DevOp's exhibitions and in zones of simultaneousness and conjunction between modules in a task, DevOps guarantees for yielding enduring exertion.

ii. It is assumed that by next few decades, mobile devices will automatically enter into project-doing scenario and replace stringent time or schedules, atmosphere, and single location. Project members will have flexibility to have their work done from anywhere on any mobile device and for coordination; it won't hamper much since walkthroughs may occur from anywhere with the help of virtual tools, considering the fact of wireless Internet as well. Thus, it will help in attaining project flexibility and well-maintained team coordination, atmosphere and result in an overall flexible environment.

6　Conclusion

With the IT world growing at a quick pace, remembering developments and fast conveyance, it is essential to remember to venture administration situation which will help over long hauls as we not only need to convey tasks quickly, we likewise need quality improved, since it will have an enduring effect on the end client, whatever the result or circumstance might be. Consequently, keeping in mind the end goal to have these consolidated and have a decent proclivity with the customer, we have to experience the venture administration viewpoint and have strong administration spine which is conceivable when we pass with contemporary patterns, models, and successful practices.

References

1. Srivastava B (2004, March) A decision-support framework for component reuse and maintenance in software project management. In: Eighth European conference on software maintenance and reengineering, 2004, CSMR 2004 Proceedings, pp 125–134. IEEE, Piscataway
2. Pierzchałka P (2006) An evolutionary approach to project management process improvement for software-intensive projects. In: Software engineering techniques: design for quality, pp 127–138. Springer, Boston, MA
3. Deininger M, Schneider K (1994) Teaching software project management by simulation—experiences with a comprehensive model. Software engineering education, pp 227–242
4. Hall NG (2012) Project management: recent developments and research opportunities. J Syst Sci Syst Eng 1–15
5. Maranhão R, Marinho M, de Moura H (2015) Narrowing impact factors for innovative software project management. Procedia Comput Sci 64:957–963, ISSN 1877-0509, https://doi.org/1 0.1016/j.procs.2015.08.613. (http://www.sciencedirect.com/science/article/pii/S18770509150 27489)
6. Sosnowski J, Dobrzyński B, Janczarek P (2017) Analysing problem handling schemes in software projects. Inf Softw Technol 91:56–71, ISSN 0950-5849, https://doi.org/10.1016/j.infsof. 2017.06.006. (http://www.sciencedirect.com/science/article/pii/S0950584916304840)
7. Sundararajan S, Bhasi M, Vijayaraghavan PK (2014) Case study on risk management practice in large offshore-outsourced Agile software projects. IET Softw 8(6):245–257
8. Fernandes G, Ward S, Araújo M (2013) Identifyingusefulproject management practices: a mixed methodologyapproach. Int J Inf Syst Project Manage 1(4):5–21
9. Jones C (2004) Software project management practices: failure versus success. CrossTalk: J Defense Softw Eng 17(10):5–9
10. Rech J, Althoff KD (2004) Artificial intelligence and software engineering: status and future trends. KI 18(3):5–11

Forest Fire Prediction to Prevent Environmental Hazards Using Data Mining Approach

J. Kamalakannan, Anurag Chakrabortty, Gaurav Bothra, Prateek Pare
and C. S. Pavan Kumar

Abstract To classify forest fires dataset w.r.t burned area of the forests, we have brought data mining approach. Forest fires are a major environmental issue, causing damage in terms of economy as well as ecology, besides endangering human frailty and lives. Forest fires are usually unrestrained, and they occur over widespread forest areas taking the shape and vigor of destruction. Data mining is a way to extract the knowledge patterns and information, useful for our work; data being put and taken from the database. Classification has been applied to search and locate the particular or distinct where respective data instances have some relation in one way or the other, confined in the produced dataset.

Keywords Data mining · Data mining techniques · Environmental hazards
Environment and ecology · Forest fire prediction · Forest fire prevention

1 Introduction

Forest fires are a major environmental issue, creating damage, in terms of economy and ecology, besides endangering human frailty and lives. A forest fire is an uncontrolled fire occurring in nature over a forest or area of woodland. The old-fashioned human analysis is costly and can be easily affected by various factors (viz., analyzing of large amount of historical data is laborious) which needs a programmable solution. The satellite-based solution which helps in predicting fire is a big-budget operation;

J. Kamalakannan (✉) · A. Chakrabortty · G. Bothra · P. Pare
School of Information Technology and Engineering,
Vellore Institute of Technology, Vellore, India
e-mail: jkamalakannan@vit.ac.in

C. S. Pavan Kumar
School of Computer Science and Engineering,
Vellore Institute of Technology, Vellore 632014, Tamil Nadu, India
e-mail: pavan540.mic@gmail.com

© Springer Nature Singapore Pte Ltd. 2019   615
A. J. Kulkarni et al. (eds.), *Proceedings of the 2nd International Conference on Data
Engineering and Communication Technology*, Advances in Intelligent Systems
and Computing 828, https://doi.org/10.1007/978-981-13-1610-4_62

whereas using historical data and analyzing it is comparatively an efficient solution and does not suffer cost of maintenance.

2 Literature Survey

Miao et al. [1] described gray histogram technology. Using this technology, we can detect smoke and cloud in the satellite images. The authors proposed a sub-region detection method which used information entropy which helps in swift smoke detection. Gray histogram gives a fruitful result only when the image is covered by greater crowded area. The proposed method used the linearization technology which helps in locating the smoke area and the calculated gray histogram helps in recognizing the litter smoke easily.

Divya et al. [2] gave the solution the problem of clustering or grouping of images of forest fires already occurred. Forming suitable clusters of fire forest images and getting vital information from them was a big problem that came up. To solve this problem, the authors came up with some techniques of data mining that can go well either with wireless or with wire-free sensor networks which can detect as well as predict the forest fire much faster than the normal and traditional satellite-based approach.

Qiaoli et al. [3] designed a system that help in the stimulation dealing with forest fire spreads and visualize 3D images. Programming language used for developing this system is C# in ArcGIS platform. It provides the reflection of two-dimensional map images, and in return, it gives the real-time and dynamic images that are transmitted via GPS from fire visions on three-dimensional maps using same interfaces.

Xiao et al. [4] worked on the problem of motoring the forest fire and practical implementation of this process. The authors created a system that would help to monitor the system and work on it using the digital imaging processes and techniques. The authors found out the several characteristics, both dynamic and configuration along with color of region affected by fire with the help of algorithms on processing of digital images, and identifying corresponding fire sources as per obtained image features.

3 Methodology

We have adopted a data mining approach to predict forest fires using meteorological data. It can be used to test regression methods. Also, it could be used to test outlier detection methods, since it is not clear how many outliers are there. Yet, the number of examples of fires with a large burned area is very small.

There are 517 instances and 12+ output attributes (Tables 1 and 2).

Table 1 Dataset attributes used along with their description and corresponding ranges

S. no	Attributes	Description	Range
1	X	It is the x-axis spatial coordinate present in the Montesinho park map	1–9
2	Y	It is the y-axis spatial coordinate present in the Montesinho park map	2–9
3	Month	It defines the months of the year	January–December
4	Day	It represents the days of the week	Monday–Sunday
5	FFMC	It represents the FFMC index in the FWI system	18.7–96.20
6	DMC	It represents the DMC index in the FWI system	1.1–291.3
7	DC	It represents the DC index in the FWI system	7.9–890.6
8	ISI	It represents the ISI index in the FWI system	0.0–56.10
9	Temperature	It represents the temperature (in terms of Celsius Degrees)	2.2–33.0
10	RH	It represents the relative humidity (in %)	15.0–100
11	Wind	It represents the relative humidity (in %)	0.40–9.40
12	Rain	It is the rain occurred and measured in mm/m^2	0.0–6.4
13	Area	It represents that area of the forest which has been burned or affected by the fire and measured in terms of ha	

Table 2 Dataset particulars and details

Dataset characteristics	Multivariate
Attribute characteristics	Real
Associated tasks	Classification
Number of instances	517
Number of attributes	13
Missing values?	None
Area	Physical
Date donated	2008-02-29
Number of Web hints	527,886

Input:

Data	
Workbook	forestfires.csv
Worksheet	forestfires
Data Range	$A$1:$M$518
# Records	517

Fig. 1 Snapshot of the records used in Weka tool

Variables										
# Input Variables	10									
Input variables	Y	FFMC	DMC	DC	ISI	temp	RH	wind	rain	area
Output variable	X									

Fig. 2 Snapshot of variables used in Weka tool

3.1 Algorithms

Discriminant Analysis Algorithm
Discriminant analysis is a statistical technique which is used to build model of group based on a single variable. In this technique, we are taking one variable as a output variable and other variable take as a selected (input) variable. Based on the output variable, we are generating the report.

K-Nearest Neighbor Algorithm
K-nearest neighbor algorithm, a basic calculation which stores all the cases that are available and then classifies the sets of new cases which depend on a measurement of equality.

Naïve Bayesian Algorithm
Naïve Bayesian algorithm in mining of data is superintend learning and statistical approach for above-mentioned classification. The algorithm is successful to analyze the dataset algorithm performance testing

The following results are shown with the help of report generation out of the dataset so used using Weka tool. These are as follows (Figs. 1 and 2):

3.2 Discriminant Analysis Algorithm

See Figs. 3 and 4.

Fig. 3 Snapshot showing the working of discriminant analysis algorithm in Weka tool

Elapsed Times in Milliseconds			
Reading Data	Computatio	Writing Data	Total
1	558	39	598

Fig. 4 Snapshot giving the error report of discriminant algorithm in Weka tool

Error Report			
Class	# Cases	# Errors	% Error
1	48	31	64.58333
2	73	50	68.49315
3	55	55	100
4	91	61	67.03297
5	30	30	100
6	86	55	63.95349
7	60	48	80
8	61	11	18.03279
9	13	6	46.15385
Overall	517	347	67.11799

Fig. 5 Snapshot showing the working of K-nearest algorithm in Weka tool

Elapsed Times in Milliseconds			
Reading Data	Computatio	Writing Data	Total
37	85	5	127

Fig. 6 Snapshot giving the error report of K-nearest algorithm in Weka tool

Error Report			
Class	# Cases	# Errors	% Error
1	48	0	0
2	73	0	0
3	55	0	0
4	91	0	0
5	30	0	0
6	86	0	0
7	60	0	0
8	61	0	0
9	13	0	0
Overall	517	0	0

4 K-Nearest Algorithm

See Figs. 5 and 6.

Fig. 7 Snapshot giving the
error report of naïve
Bayesian algorithm in Weka
tool

Error Report			
Class	# Cases	# Errors	% Error
1	48	37	77.08333
2	73	39	53.42466
3	55	23	41.81818
4	91	44	48.35165
5	30	25	83.33333
6	86	16	18.60465
7	60	31	51.66667
8	61	8	13.11475
9	13	7	53.84615
Overall	517	230	44.48743

4.1 Naïve Bayesian Algorithm

See Fig. 7.

5 Implementation

5.1 Program Code

```
   library("caret")
x<- read.csv("f:/forestfires.csv")
a<- scale(x) 3 #normalisation
view(a)
train<- as.data.frame(a[1:370,])
test<- as.data.frame(a[371:517,])

model<- train(train[1:11], train$area,method ="rf",preProcess = C("center,scale")
p<- predict(model,test)
RMSE<- sqrt(mean((test$area-p)^2))
Plot (test$area,p,col="blue",xlab = Äctual", ylab ="Predicted",pch-10
```

5.2 Results

We applied the following algorithms (Figs. 8 and 9):

• Discriminant analysis algorithm.

Fig. 8 Snapshot to represent the scatter plot in *R*

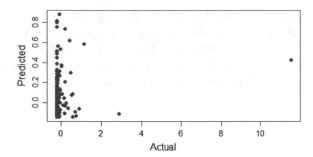

Fig. 9 Graph plot of the training dataset in *R*

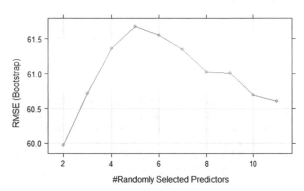

- *K*-nearest algorithm.
- Naïve Bayesian algorithm.
- Random forest algorithm.

Comparing the error rate of all the algorithms, we conclude that the *K*-nearest neighbor algorithm provides a better classification than all the other algorithms. We can say so because *K*-nearest neighbor's classification algorithm provides 0% as the overall error rate, whereas all the other algorithms provide 50% or more as the overall error rate.

Hence, with respect to forest fires dataset *K*-nearest neighbor is best-suited algorithm as it gives minimum error rate. Furthermore, we put the entire dataset in R software and adopted a second method of coding in R language. We had to normalize the dataset within the range −1 to +1 so as to obtain better and accurate results. Graphs were plotted and studied as shown in the respective figures.

6 Conclusion

Based on the above graph, using random forest algorithm after normalizing the data sheet [since using the other algorithms the graph was not clear or could not meet up the expectations we opted for random forest, which is used as a classification algorithm and can be useful in making decision trees], we can see that the graph rises up to 63.5% (approx.) and also calculating root mean square which comes to around 100% (approx.), we predict that there are chances of forest fires in our selected training and test dataset, and hence, prevention and protection are needed at the earliest as far as environmental hazards protection and at the same time environmental concern is needed.

References

1. Miao S, Hu K, Gao H, Wang X (2016) Small fire smoke region location and recognition in satellite image. In: International congress on image and signal processing, BioMedical Engineering and Informatics (CISPBMEI), pp 714–718. IEEE, Piscataway
2. Divya TL, Manjuprasad B, Vijayalakshmi MN, Dharani A (2014) An efficient and optimal clustering algorithm for real-time forest fire prediction with. In: 2014 international conference on communications and signal processing (ICCSP), pp 312–316. IEEE, Piscataway
3. Qiaoli H, Keqilao M, Qin D (2009) The system of simulation of forest fire spread and assistant decision-making based on ArcGIS. In: International conference on new trends in information and service science, 2009, NISS'09, pp. 302–305. IEEE, Piscataway
4. Xiao J, Li J, Zhang J (2009) The identification of forest fire based on digital image processing. In: 2nd international congress on image and signal processing, 2009, CISP'09, pp 1–5. IEEE, Piscataway
5. Kim HR (2009) Prediction of forest fires using data mining methods. Analysis 1:3
6. Singh R (2016) Predicting wildfire using data mining. Doctoral dissertation, Rochester Institute of Technology
7. Cortez P, Morais ADJR (2007) A data mining approach to predict forest fires using meteorological data
8. Stojanova D, Panov P, Kobler A, Džeroski S, Taškova K (2006) Learning to predict forest fires with different data mining techniques. In: Conference on data mining and data Warehouses (SiKDD 2006), Ljubljana, Slovenia, pp 255–258
9. Castelli M, Vanneschi L, Popovič A (2015) Predicting burned areas of for-est fires: an artificial intelligence approach. Fire Ecol. 11(1):106–118
10. Hamadeh N, Karouni A, Daya B (2014) Predicting forest fire hazards using data mining techniques: decision tree and neural networks. In: Advanced materials research, vol 1051, pp 466–470. Trans Tech Publications, Zürich
11. Özbayoğlu AM, Bozer R (2012) Estimation of the burned area in forest fires using computational intelligence techniques. Procedia Comput Sci 12:282–287
12. Yu YP, Omar R, Harrison RD, Sammathuria MK, Nik AR (2011) Pattern clustering of forest fires based on meteorological variables and its classification using hybrid data mining methods. J Comput Biol Bioinf Res 3(4):47–52

IoT-Based Smart Office System Architecture Using Smartphones and Smart Wears with MQTT and Razberry

R. Padma Priya, J. Marietta, D. Rekha, B. Chandra Mohan and Akshay Amolik

Abstract Internet of things due to its profound capability has been responsible for the innovative transformations in lifestyles of human. We propose a system for the smart office environment using the Internet of things. Our developed system configuration provides a comfort of handling smart office environment using Google Voice Over Protocol and Android-based application. Use of lightweight MQTT protocols, high-frequency RF trans-receivers, Raspberry 2 Z-wave-based long distance controllers makes this system feasibly suitable for wider office areas. For low power consumption, high performance and immediate time-to-time response QoS parameters, the system is equipped with PIR, gas, flame, sound, light, temperature and humidity sensors. We ensure reliable communication with a separate channel for every sensor signals, secured with AES encryption and plug-and-play provisions. Further, mail-based anomaly or emergency alert notification is also incorporated. Finally, we have performed the experiment and also tested the developed system in Android phones and smart wear watch, namely Moto360 Watch.

Keywords Internet of things · Smart office · MQTT · Raspberry · Smart city

R. Padma Priya · J. Marietta (✉) · D. Rekha · B. Chandra Mohan · A. Amolik
Vellore Institute of Technology, Vellore, India
e-mail: jmarietta1@gmail.com

R. Padma Priya
e-mail: padmapriya.r@vit.ac.in

D. Rekha
e-mail: rekha.d@vit.ac.in

B. Chandra Mohan
e-mail: Chandramohan.b@vit.ac.in

A. Amolik
e-mail: amolikakshay007@gmail.com

© Springer Nature Singapore Pte Ltd. 2019
A. J. Kulkarni et al. (eds.), *Proceedings of the 2nd International Conference on Data Engineering and Communication Technology*, Advances in Intelligent Systems and Computing 828, https://doi.org/10.1007/978-981-13-1610-4_63

1 Introduction

Internet of things has numerous applications. Many problems related to the environment and smart city could be addressed by various technologies in IoT. The concept of smart office, as an application belonging to Internet of things (IoT) domain, provides features related to the intelligent behaviour of entire work environment such as office rooms, office entrances, office parking.

A network of dedicated sensors, actuators and various specialized devices is used to control temperature, light intensity, humidity, vibration, noise level and other in-room environment parameters related to office employees. Nowadays, the need to log, an underlying business process that helps operate the daily tasks, which are regularly done in the offices, is becoming an important artefact. Smart office-based solutions have to ensure the need to communicate with the various project management systems and information systems to control the regular office work. Thus, Internet of things (IoT) not only is about improving business processes but has also the potential to profoundly impact the lives of many citizens.

This paper is divided into the following sections where Sect. 2 presents a survey of work done in this field. Section 3 gives the Proposed Work, while Sect. 4 shows Experimental Results and Analysis and Sect. 5 gives Conclusion and Future Work.

2 Literature Survey

Various literatures have been studied related to the smart office environment in the smart city. The idea of smart office environment by using the Internet of things was proposed in [1]. They have used a methodology which is based on communication between various sensors and server running remotely.

Domaszewicz et al. [2] have done the whole automation system with low size and automated hints and performed action on nearby object. Olivieri et al. [3] proposed the publish–subscribe approach to make integrated IoT-based system. They have proposed their research work with smart office as a use case. Foster et al. [4] have given intelligent methods which control the air conditioners which are provided by sensor data. Sanoob et al. [5] proposed the integrated system which uses the camera to capture the motion and sends SMS to the user's smartphone about any security theft. Karol et al. [6] have given adaptable network architecture of smart office solution based on service-oriented technology which focuses on the continuous enhancement of user experience and monitoring of evaluation of adjustment of the integrated system to user requirements which are continuously changing according to social and business parameters.

Nati et al. [7] proposed a user-centric approach for IoT in smart campus. Prasad et al. [8] proposed an innovative Hadoop-based cloud framework for observing and tracking tangible assets in smart campus. In their research paper, they experimented on the displacement or missing assets in laboratories and library by integrating cloud computing and RFID transceivers. Zanella et al. [9] presented and discussed various guidelines for best practice and technical solutions in IoT for smart cities. Their research paper provides comprehensive knowledge and survey about modern architectures, protocols and technologies used in urban IoT. Muralidhara et al. [10] proposed and developed secured and smart healthcare monitoring system.

Lucke et al. [11] developed cyber-physical system which can be effective and can be adapted by governments for their administration. This will help the government to execute public tasks effectively and efficiently. Ghazal et al. [12] proposed traffic light control system with the help of PIC microcontroller which continuously analyses traffic density using infrared sensors and provides dynamic monitoring. Hayajneh et al. [13] proposed an adaptable and resilient disaster recovery network for smart cities.

3 Proposed Work

3.1 System Architecture

The detailed design of the fully automotive system consists of IoT-based configuration and mobile application-based user interface. Scalability and flexibility have been given special importance while designing the proposed method. This allows heterogeneous devices to be connected to the system. Cloud storage provides the options to store the huge data. Wireless sensor nodes are connected through the suitable protocols. This provides wide area coverage and accuracy in sensor data.

The flow of the total system is shown in Fig. 1 where all operations are clearly specified. The following flow chart provides how exact sensor data is processed and how communication between phone and sensor happens. In the following flow chart, each operation represents a function which is performed in this system. The flow of operation is simple and easy to understand.

All sensors are controlled by the Arduino Controller, which is having RFM69HW wireless transceiver mounted over it. We can call this whole set-up as sensor node. All the sensors send data to another Arduino which has Arduino W5100 ethernet shield mounted on it. We can call this ethernet gateway.

3.2 Face Detection Algorithm

Algorithm for face detection is as follows:

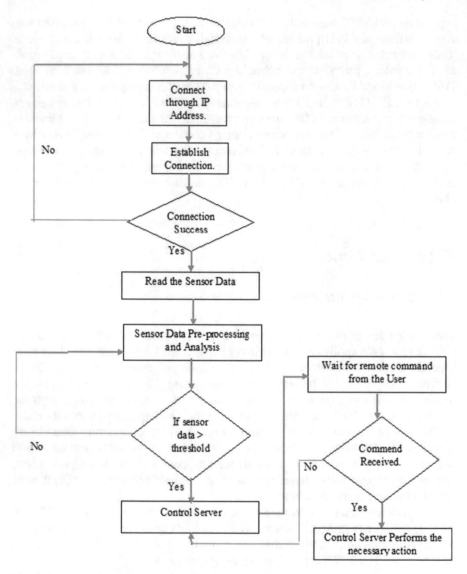

Fig. 1 Flow chart for the proposed methodology

Algorithm
1. Start.
2. Pass the input image or take a photo to detect face.
3. Click on Detect button to start detecting the human faces in the photograph.

4. Extract the face parameters from the detected face (parameters such as the distance between left eye and right eye, the distance and position of lip).
5. Compare the extracted parameters with the parameters within the database (mobile database, i.e. SQLite).
6. If the parameters are matched, then the result (i.e. the person's information is passed to the user).
7. Otherwise, the person is unknown and notified to the user.
8. Stop.

4 Experimental Results and Analysis

We were able to achieve the reliability and accuracy of the system. Due to use of MQTT protocol which is lightweight protocol, we have reduced the time of communication between sensor and mobile device.

Below scenarios are experimentally tested in our proposed model which includes fire emergency condition monitoring, theft detection, surveillance and security provision with the camera (Refer Figs. 2, 3 and Table 1 for sensor data collection, experimental setup and hardware specifications, respectively). The accuracy of results is very high, and the fully adaptable system can be easily integrated or scaled with a larger system.

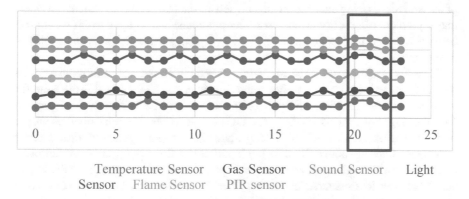

Temperature Sensor   Gas Sensor   Sound Sensor   Light Sensor   Flame Sensor   PIR sensor

X-Axis:- Sensor Values (Analogy and Digital Values).
Y-Axis:- Time in Seconds.

Fig. 2 Graph of sensor values over time

Fig. 3 Hardware setup and notification in the watch

Making an adaptable, scalable smart system with IoT is a challenging Task. The cost is also an important factor while making these systems. But a combination of cheap sensors, adaptable server with uniprocessor Raspberry Pi promotes researchers towards testing smart applications. Due to MQTT protocol and openHAB server, we have achieved good results. We have assigned each sensor data a unique key so that we can get accurate results with less amount of time. This avoids the mismatching of data and false results as observed in (Fig. 2). For each sensor, the threshold value is checked and thus it helps us to process only valid data and in reducing the time of transmission. Using encryption key, all sensor data are secured, which solves the major problem of security. Many receivers are connected to the system which can be scaled. So, the user can connect his phone, laptop, smartwatch simultaneously to the system.

The above graph shows the sensor values over time. The sensor continuously sends the valid data to the server only after some time interval. This interval is different for various sensors so by doing this space on the server is utilized properly. The graph in the following shows the time when the data is sent to server. This also ensures proper utilization of resources. The blue rectangle box shows an emergency condition where all sensor values are above the threshold, and at this time, the user is notified with notification or through mail. All these things make the whole system scalable and adaptable with less resource utilization. The server logs can be saved for getting history. Past days' logs are saved, and the older logs are deleted automatically to do proper utilization of space. PIR sensor and flame sensor send digital values. So when digital output is 1, then only sensor sends data to the server. This is shown in the emergency condition when the entire sensor sends high values. So this will also reduce the time and server space. Only valid data is processed so this reduces extra overhead on the server.

Table 1 Hardware Specifications

Device Name	Hardware Specification/Features
Raspberry pi	1GB Ram, 10/100 Ethernet, 2.4GHz 802.11n wireless LAN, Bluetooth 4.1 Classic Bluetooth Low Energy, CPU: $-4\times$ ARM Cortex-A53, 1.2GHz
Arduino UNO R3	Digital I/O Pins: -14 Flash Memory: -32 KB SRAM: -2KB Operating Voltage: -5V
W5100 Ethernet Shield For Arduino	Digital I/O Pins: -14 6 Analog Inputs, 16 MHz crystal oscillator, Rj45 connection, power jack, ICSP header
RFM69HW Transceiver	+20 dBm -100 mW power Capacity. RF is constant over voltage. FSK, MSK, OOK, GMSK, GFSK modulations. Automatic RF Sense with Ultra-Fast AFC
PIR Sensor	Sensitivity Range: - up to 20 Feet (6 Meters), 110° X 70° detection range Power Range: -5V-12V
MQ2 Gas Sensor	Detect Combustible Gas, Smoke Power Supply:- 5v Wide Detecting scope High Sensitivity and Fast Response
Sound Sensor	Easy to use Microphone Sensitivity 52–48 db Microphone Frequency 16–20 kHz

5 Comparison Results of AMQP and MQTT

MQTT is the IoT connectivity protocol. It is described as the machine-to-machine protocol. Since it supports smallest measuring and the monitoring devices. Hence termed as the lightweight protocol. The data could be transmitted in the intermittent network too. It is a publish/subscribe model. The real-world objects could be connected to the servers and consumers. The scalability issue faced by the sensors, actuators, mobiles and the tablets could be a problem, and to overcome this issue, MQTT protocol is designed. The battery power and the bandwidth are very essential for the connected devices in IoT. The developed principles and the designed MQTT protocol are helpful for the connected IoT devices. This is a protocol for collecting device data and communicating it to servers. It is TCP/IP-based messaging protocol. It is light on network bandwidth and has a low power draw.

AMQP is designed as a replacement for existing proprietary messaging middleware. The advantage of using AMQP is reliability and interoperability. The features included by AMQP are reliable queuing, topic-based publish-and-subscribe messaging, flexible routing, transactions and security. The message formats such as classic

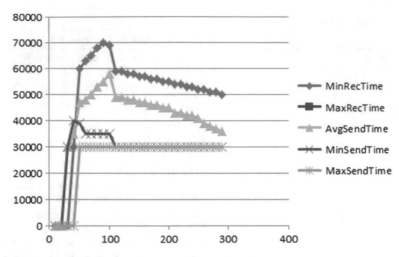

Fig. 4 Time consumed when sending message using MQTT

message queues, round-robin, store-and-forward are supported by AMQP. In order to support idempotent messages and message grouping, meta-data support used by AMQP is a queuing system designed to connect servers to each other.

CoAP runs on UDP. This can be used on extremely resource-constrained environments. It uses semantic, parallel to HTTP which is a best mechanism for local network communication. Both MQTT and CoAP are used for the device-to-service connection. CoAP is Application protocol, which requires a broker (server).

The MQTT statistic graphs provide statistics about message published from and received by the MQTT broker during a simulation run. The hardware platform for testing has two Raspberry Pi boards with the same hardware configuration to investigate rate of packet transmission and performance based on time. The measurement of performance is run by sending messages between publisher and subscriber through broker.

Graph Figs. 4 and 5 represents as simulation system model by sending messages between Raspberry Pi 1 and Raspberry Pi 2 either simulation with AMQP or MQTT. Performance is measured by sending message between publisher and subscriber through broker. Below graph represents the simulation model by sending messages through the kit for AMQP (or) MQTT. AMQP could send message up to 3000 with acknowledgement MQTT could send message up to 4000 with acknowledgement.

6 Conclusion and Future Work

The observations and experimental tests, prove that the system is designed to fulfil the smart office environment, removing all the barriers of theft and to protect the office from theft. This system can be made as hub in future work where we can use various

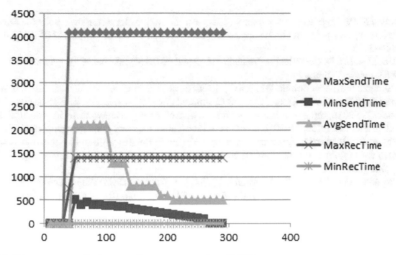

Fig. 5 Time consumed when sending message using AMQP

machine learning techniques. This will help the system to recognize speech or voice of the user and will help ON the user command. All the sensors can communicate with this hub where fuzzy logic is used to make prediction. Interfaces such as Amazon Echo (Alexa Voice Service), Google Speech API can also be integrated with Google Assistant. The voice commands to the system could be given by the device such as Amazon Echo dot. We can also add Bluetooth devices and GPS tracking to improve performance. Use of routing algorithm can help to enhance the system.

References

1. Durand M (1999) Smart office: an intelligent and interactive environment. In: Proceedings of 1st international workshop on managing interaction in smart environments, Dublin, pp 104–113, R.I, 997
2. Domaszewicz J, Lalis S, Pruszkowski A, Koutsoubelias M (2016) Soft actuation: smart home and office with human-in-the-loop. IEEE Pervasive Comput 15(1)
3. Olivieri AC, Rizzo G, Morard F (2015) A publish-subscribe approach to IoT integration: the smart office use case. In: INSPEC Accession Number-15107080, Advanced Information Networking and Applications Workshops (WAINA) IEEE Conference on Networking
4. Foster TW, Bhatt DV, Hancke GP, Silva B (2016) A web-based office climate control system using wireless sensors. IEEE Sens J 16(15):6104–6113
5. Sanoob AH, Roselin J, Latha P (2016) Smartphone enabled intelligent surveillance system. IEEE Sens J 16(5) 1361–1367
6. Furdik K, Lukac G, Sabol T, Kostelnik P (2013) The network architecture designed for an adaptable IoT-based smart office solution. Int J Comput Netw Commun Secur 1(6):216–224
7. Nati M, Gluhak A, Domaszewicz J, Lalis S, Moessner K (2014) Lessons from smartcampus: external experimenting with user-centric internet-of-things testbed. Wireless Pers Commun 93(3):709–723

8. Prasad DVV, Priya RP, Jaganathan S (2016) An innovative cloud framework for tracking and monitoring assets in smarter campus using RFID. Asian J Technol 15(11):1713–1722, ISSN: 1682-3915

9. Zanella A, Bui N, Castellani A, Vangelista L, Zorzi M (2014) Internet of things for smart cities. IEEE Internet Things J 1(1):22–32

10. Muralidhara KN, Bhoomika BK (2015) Secured smart healthcare monitoring system based on Iot. Int J Recent Innov Trends Comput Commun 3(7):4958–4961. ISSN: 2321-8169

11. Lucke JV (2016) Smart government—the potential of intelligent networking in government and public administration. In: 2016 conference for e-democracy and open government (CeDEM)

12. Ghazal B, Elkhatib K, Chahine K, Kherfan M (2016) Smart traffic light control system. In: 2016 third international conference on electrical, electronics, computer engineering and their applications (EECEA)

13. Hayajneh AM, Zaidi SAR, Mclernon DC, Ghogho M (2016) Drone empowered small cellular disaster recovery networks for resilient smart cities. In: 2016 IEEE international conference on sensing, communication and networking (SECON Workshops)

Classification of Sentiments from Movie Reviews Using KNIME

Syed Muzamil Basha, Dharmendra Singh Rajput, T. P. Thabitha, P. Srikanth and C. S. Pavan Kumar

Abstract Nowadays in entertainment, cinema industry has become one of the most popular industries, gaining the attention of public toward them by making unnecessary stunts by the production team in promoting their movie and influencing the public to watch the movie at least for one time. By deeply understanding the impact of a particular movie in advance, using reviews made after watching the movie benefits others in saving the major resources like time and money. The objective of our research is to save the time and money spent on watching the movie in theaters and motivating them to use up their valuable time with family members, especially during weekends. In this paper, we aim to demonstrate the application on sentiment classification using decision tree algorithm available in KNIME to rate the movie performance. In which, the textual data from the document are converted into strings, and these strings are preprocessed to get numerical document vectors. Later, from the document vectors the sentiment class is extracted and the predicted model is built and evaluated. In our experimental work, 93.97% of classification accuracy with 0.863 Cohen's value was achieved in classifying the sentiments from the movie reviews.

Keywords Decision tree · Classification · Sentiment · KNIME

S. M. Basha · D. S. Rajput (✉) · C. S. Pavan Kumar
Vellore Institute of Technology, Vellore 632014, Tamil Nadu, India
e-mail: dharmendrasingh@vit.ac.in

S. M. Basha
e-mail: muza.basha@gmail.com

C. S. Pavan Kumar
e-mail: pavan540.mic@gmail.com

T. P. Thabitha · P. Srikanth
MIC College of Technology, Vijayawada 521180, India

© Springer Nature Singapore Pte Ltd. 2019
A. J. Kulkarni et al. (eds.), *Proceedings of the 2nd International Conference on Data Engineering and Communication Technology*, Advances in Intelligent Systems and Computing 828, https://doi.org/10.1007/978-981-13-1610-4_64

1 Introduction

From the huge amount of text information available in the form of review sites, researchers are encouraged to do their own investigation on automatic sentiment extraction. For research purpose, sentiment analysis can be termed as "Knowing the other opinion in advance about a particular product belongs to a domain." Area of sentiment classification is broadly classified into document, sentence, topic level. In our experiment, we have focused much on document level by representing each review as a single document. As per the literature review made on multi-language sentiment classification, only 26.3% of the Internet users in 2016 are English speakers and 34% of all tweets are written in English. There is a need to focus on developing a multi-language sentiment classifier. In [1], Deriu et al. (2017) tested four languages and achieved the classification accuracy as English (63.49), Italian (67.79), German (65.09), and French (65.68). In our research, we collected a benchmarked and supervised dataset of 1400 records, containing data in the form of text as one of the attributes from UCI Machine Learning Repository. In [2], Blitzer et al. (2007) collected the reviews from Amazon for four different products, proposed an algorithm to measure the relationship between source, target domain, and obtained 85.9% of classification accuracy.

Our contribution in this paper is to understand the application workflow on sentiment classification using decision tree algorithm available in KNIME to rate the movie performance. In which, the textual data from the document are converted into strings, and these strings are preprocessed to get numerical document vectors. Later, from the document vectors the sentiment class is extracted and the predicted model is built and evaluated. In our experimental work, 93.97% of classification accuracy with 0.863 Cohen's value was achieved in classifying the sentiments from the movie reviews. This paper is organized as follows: In Sect. 2, the research is carried out in the field of sentiment classification and their achievements are discussed, In Sect. 3, the design of our experiment is discussed elaborately. In Sect. 4, the result obtained is discussed, and in conclusion finally, the summary of our contribution along with future work of the researchers is given.

2 Literature Work

In the process of applying the machine learning techniques (MLTs) to extract the sentiment from the review [3], Pang et al. (2002) examined the effectiveness of applying MLT in understanding the review made by the reviews and stated that support vector machine (SVM) performs better than maximum entropy (ME) and naïve Bayes (NB) using unigram plus bigram features with 82.9% of classification accuracy. Whereas, in [4], Go et al. (2009) also stated SVM performance is better than ME and NB and used unigram as a feature with 82.2% of classification accuracy. In [5], Pant et al. (2010) focused on domain-independent sentiment analysis, proposed an algorithm called spectral feature alignment, and achieved 87.1% of classification accuracy. Glorot et al. (2011) in [6] focused on domain adaptation, proposed a deep learning approach in extracting the sentiments using intermediate built-in representations, and stated that proposed algorithm has less transfer error compared to the other traditional deep learning algorithms and also that performance of SVM can be improved in training the SVM on target domain instead of different domain. Tang et al. (2014) in [7] proposed a learning sentiment-specific word embedding (SSWE) for sentiment analysis. In which, sentiment information is encoded into the continuous representation of words and helps in separating good from bad to either end of the spectrum. Tang et al. (2015) in [8] focused on document representation using a neural network approach for sentiment classification and achieved 2.71 mean squared error (lower is best). Tripathy et al. (2016) in [9] examined the performance of different machine learning algorithms (ME, NB, SVM) and evaluated them using statistical parameters (precision, recall, F-measure).

3 Methodology

In this section, the design process of our experiment is discussed with all the details required, making readers understand our workflow. Initially, data collection is performed using file reader node in CSV format, considering comma as a delimiter. The supervised dataset consists of reviews in the form of text of about 1400 reviews, as one of the attributes. The reviews present in the dataset are related to movies. In our experiment, we have considered each review as a document using the document creation node before starting the preprocessing stage.

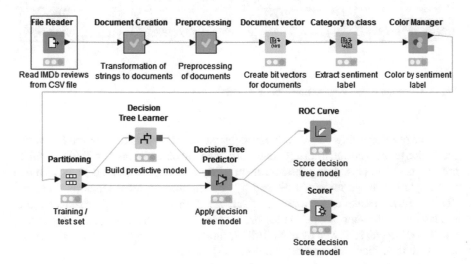

Fig. 1 Workflow of sentiment classification using KNIME

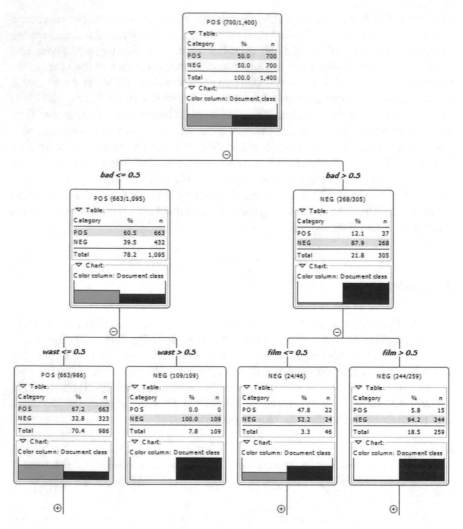

Fig. 2 Construction of decision tree using KNIME

In the preprocessing stage, all the missing values are replaced with the last value prediction technique and filtering, stemming, removal of stop words and URLs, which are not useful in classifying the sentiments. Bag of words is created to replace the missing terms in the reviews followed by construction of Term Document Matrix (TDM). Term frequency is calculated from TDM, that intern considered as feature in classifying the sentiments. In the process of extracting the sentiment labels, term frequency is used, followed by in KNIME. Using a color manager node, the sentiment labels are identified using different colors as shown in Fig. 1.

Table 1 Measure used to evaluate the application

Parameters	Formula
Accuracy (A)	$A = \frac{TP+TN}{Total}$
Misclassification rate (MCR)	$MCR = \frac{FP+FN}{Total}$
Recall (R)	$R = \frac{TP}{Actual\ True}$
Precision (P)	$P = \frac{TP}{Predicted\ True}$
Prevalence (PV)	$PV = \frac{Actual\ True}{Total}$
F-score (FS)	$FS = 2 \times \frac{R \times P}{R+P}$

Fig. 3 Confusion matrix

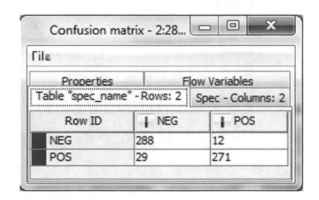

The constructed tree with three levels is shown in Fig. 2. By this, the reader can easily interpret that the classification is based on the label generated using the workflow available in KNIME.

4 Result

To evaluate the experiment carried out in this paper. We make use of confusion matrix generated using scorer node available in KNIME. To obtain evaluation parameters as shown in Table 1, we have used the values plotted in Fig. 3. Based on the literature review made, the author in [3] had achieved 82.9% of classification accuracy using unigram as a feature and stated that SVM performance is better than NB and ME algorithms [10]. In [11], Moreas et al. (2013) showed that artificial neural network (ANN) outperformed support vector machine on the benchmark dataset of movie reviews and stated ANN can be a contestant approach in the task of sentiment learning [12]. Whereas, in our application, we have achieved 93.2% of classification using term frequency as a feature.

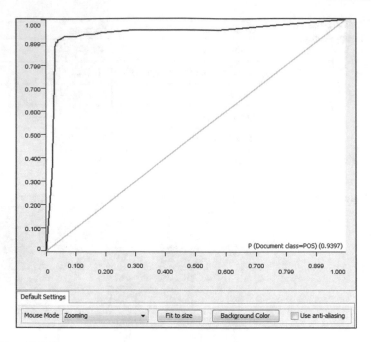

Fig. 4 Plot of ROC curve

Table 2 Evaluation of decision tree using statistical parameters

	NEG	POS
True positive	288	271
False positive	29	12
True negative	271	288
False negative	12	29
Recall	0.96	0.903
Precision	0.909	0.958
Sensitivity	0.96	0.903
Specificity	0.903	0.96
F-measure	0.934	0.93
Accuracy = 0.932	Cohen's value = 0.863	

From Fig. 4, one can easily interpret the performance of the experiment performed on assigning the sentiment labels. The closer the ROC curve is to the upper left corner of Fig. 4, the higher the overall classification accuracy of the model (Table 2).

5 Conclusion

In this paper, we stated that the performance of the decision tree algorithm can be improved by selecting the term frequency as a feature in classifying the sentiment. In the demonstrated work, the application on sentiment classification using decision tree algorithm available with KNIME is used in rating the movie performance. The sentiment class from the reviews is extracted, and the predicted model is built and evaluated. In the experimental work, we achieved 93.97% of classification accuracy with 0.863 Cohen's value in classifying the sentiments from the movie reviews. In future, we would like to compare the performance of the model built in the application with other traditional machine learning classification algorithms.

References

1. Deriu J, Lucchi A, De Luca V, Severyn A, Müller S, Cieliebak M, ... Jaggi M (2017) Leveraging large amounts of weakly supervised data for multi-language sentiment classification. In: Proceedings of the 26th international conference on world wide web, pp 1045–1052. International World Wide Web Conferences Steering Committee
2. Blitzer J, Dredze M, Pereira F (2007) Biographies, bollywood, boom-boxes and blenders: domain adaptation for sentiment classification. In: ACL, vol 7, pp 440–447
3. Pang B, Lee L, Vaithyanathan S (2002) Thumbs up?: sentiment classification using machine learning techniques. In: Proceedings of the ACL-02 conference on Empirical methods in natural language processing, vol 10, pp 79–86. Association for computational linguistics
4. Go A, Bhayani R, Huang L (2009) Twitter sentiment classification using distant supervision. CS224 N Project Report, Stanford, 1(2009), 12
5. Pan SJ, Ni X, Sun JT, Yang Q, Chen Z (2010) Cross-domain sentiment classification via spectral feature alignment. In: Proceedings of the 19th international conference on world wide web, pp 751–760. ACM, New York City
6. Glorot X, Bordes A, Bengio Y (2011) Domain adaptation for large-scale sentiment classification: a deep learning approach. In: Proceedings of the 28th international conference on machine learning (ICML-11), pp 513–520
7. Tang D, Wei F, Yang N, Zhou M, Liu T, Qin B (2014) Learning sentiment-specific word embedding for twitter sentiment classification. In: ACL, vol 1, pp 1555–1565
8. Tang D, Qin B, Liu T (2015) Document modeling with gated recurrent neural network for sentiment classification. In EMNLP, pp 1422–1432
9. Tripathy A, Agrawal A, Rath SK (2016) Classification of sentiment reviews using n-gram machine learning approach. Expert Syst Appl 57:117–126
10. Wang S, Manning CD (2012) Baselines and bigrams: simple, good sentiment and topic classification. In: Proceedings of the 50th annual meeting of the association for computational linguistics: short papers, vol 2, pp 90–94. Association for computational linguistics
11. Moraes R, Valiati JF, Neto WPG (2013) Document-level sentiment classification: an empirical comparison between SVM and ANN. Expert Syst Appl 40(2):621–633
12. Basha SM, Zhenning Y, Rajput DS, Iyengar N, Caytiles RD (2017) Weighted fuzzy rule based sentiment prediction analysis on tweets. Int J Grid Distrib Comput 10(6):41–54

Document-Level Analysis of Sentiments for Various Emotions Using Hybrid Variant of Recursive Neural Network

Akansha Shrivastava, Rajeshkannan Regunathan, Anurag Pant
and Chinta Sai Srujan

Abstract Sentiment analysis makes use of natural language processing, computational linguistics, and analysis of text in order to extract information from text. In this project, we consider machine learning algorithm for classification of longer sentences with polarity from a huge amount of data from articles, forums, consumer reviews, surveys, blogs, Twitter, Whatsapp chat, etc., and come out with suitable results sentiments (e.g., sad, happy, surprised, angry) and give intensity/degrees. Naïve Bayes and SVM are complex and inaccurate and cannot handle large amount of data. Recurrent neural network takes into consideration the sequence of words in a sentence. We use a hybrid of both LSTM and GRU to learn the long-term dependency and do sentiment analysis. Our end goal is to perform sentiment analysis on long sentences and get accurate visualization results. Sentiment analysis can be widely applied to social media and reviews for various applications, such as customer service, marketing.

Keywords Long short-term memory · Gated recurrent unit · Recurrent neural network · Sentiment analysis

1 Introduction

Machine learning is a major subtype of artificial intelligence that has evolved over the years and basically provides the functionality of training the machine without being explicitly programed. In this paper, we use machine learning techniques to perform sentiment analysis [1] which is a way to evaluate or infer whether a written or spoken language or expression is positive, negative, or neutral [2]. Based upon the past research, it is observed that the basic machine learning algorithms are efficient in varied tasks involving natural language processing such as topic categorization. These algorithms, however, cannot be used effectively for performing sentiment

A. Shrivastava · R. Regunathan (✉) · A. Pant · C. S. Srujan
School of Computer Science and Engineering, Vellore Institute of Technology, Vellore
632014, Tamil Nadu, India
e-mail: rajeshkannan.r@vit.ac.in

© Springer Nature Singapore Pte Ltd. 2019   641
A. J. Kulkarni et al. (eds.), *Proceedings of the 2nd International Conference on Data Engineering and Communication Technology*, Advances in Intelligent Systems and Computing 828, https://doi.org/10.1007/978-981-13-1610-4_65

analysis. This is because it requires opinion classification which in turn requires in-depth understanding of the various emotions and their proper analysis. We are using neural networks to perform sentiment analysis. Naïve Bayes and SVM cannot handle large amount of data unlike recurrent neural network (RNN) and are complex and inaccurate, respectively [3, 4]. RNN takes into consideration the sequence of words in a sentence. We propose a hybrid of two recurrent neural network (RNN) algorithms, namely long short-term memory (LSTM) and gated recurrent unit (GRU). The LSTM makes use of recurrently connected subnets called memory blocks which consist of memory cell, input gate, forget gate, and output gate. LSTM is different from the traditional recurrent unit that overwrites its content during each time step. LSTM, on the other hand, has the feature of deciding whether to keep the existing memory using the introduced gates. GRU is similar to LSTM but differs in that it does not have a memory cell. It consists of an update gate and a reset gate. Our end goal is to classify long sentences with polarity [5, 6] from a huge amount of data from blogs, articles, forums, consumer reviews, surveys, Twitter, etc., [7] and come out with suitable results which correspond to the sentiment. Sentiment analysis can be widely applied to reviews and social media for a variety of applications, ranging from marketing to customer service [8].

2 Literature Review

Deep learning using neural networks is better than the traditional methods like naïve Bayes, support vector machines as there is no demand for optimized hand-crafted features like lexical features as word embedding is used instead of these features. These embeddings contain context information, and the neural networks learn the features during the training phase itself [9]. Deep learning also allows for better representation leaning and adapts to task variations in different problem statements which cannot be done in traditional machine learning algorithms. Comparisons have been done on the various deep learning methods for sentiment analysis like paragraph vector, long short-term memory (LSTM), deep recursive neural network, and various datasets were used to validate the results [10].

LSTM does better analysis of emotions for long sentences compared to the tradi-tional RNN language. LSTM is used for multi-classification of different emotional attributes in the text. By training different emotion models and using them, we can find out which emotion is being reflected in the sentence. In LSTM, it learns how much past observations it should forget. This makes them efficient for long-term (i.e., longer than bi-grams) dependency retainment and noise-tolerant. There are two models which can be broadly classified as word level and character level [11]. In world level, words are looked up in a table during run time and are matched to the sentiment models. Ling et al. (2015) have rightly pointed out that this way the previ-ously unseen words are not taken into consideration, such as "Classification", even though "class" and "-ification" have been observed before. Morphologic variants of words need to be stored in the lookup table. The vectors are independent which correspond to the variations. On Twitter, people write anything casually and make up words on their own and write their own spellings; thus, this handles those types of texts. The sentiment analysis would not work in such cases, especially for scenarios where we count number of positive and negative words. A representation is stored for every character in character-based models.

A sentence might have sentiment polarity which is not only determined by its content but also because of concerned aspect. An attention-based long short-term memory network was proposed for this purpose. It focuses on the different parts of the sentence when there are different aspects present while taking input [12]. With respect to the organization or companies mentioned in tweets, positive, neutral, and negative sentiments are analyzed. This is just a three-class classification. There is another model named word2vec which is very efficient for learning word's vector representation. The vector learned by this model stores semantic information and has resulted in drastic improvements in the field of NLP.

Gated recurrent unit (GRU) is a simplified version of LSTM. GRU has sometimes performed better than LSTM. By changing initialization of LSTM, it can be converted to GRU. It predicts positive, neutral, and negative, and it comprises LSTM/GRU with 0.5 probability of drop and success if three neuron layers are completely connected. The result is fed as input in the model [13]. Our approach to sentiment analysis uses frequency/count of words in the text. The sentiment values are assigned to each word by the system. Some of these approaches do not take order of words and the complex meaning into consideration. GRU has the capability to save information for long-term dependency on sequential data. Recurrent neural network considers text of variable length. These are a bit challenging to learn, and therefore, new variants of RNN like LSTM and GRU were used [14].

3 Architectural Framework

We aim to detect more sentiments and their extent also. We will not only take the frequency of stored keywords but also the sequence and polarity of the sentence into consideration. Through character- and aspect-level model, we will give better results of sentiments, and after a deep study and comparison between LSTM and GRU, we will make a hybrid which overcomes their respective disadvantages. We will be plotting it in a bar graph and get the words in various types of word clouds. We further want to scale this project and include emojis which are very frequently being used by people in their chats. We first collect the data that we want to perform our analysis on, and then we do the pre-processing of the text by tokenization, stemming, and filtering. After the text is cleaned, we go ahead with the algorithm; the algorithm is performed in two phases. Firstly, we do the training, and then when the machine has been trained, we go ahead with the testing. In the testing phase, we obtain the results. We will be discussing in detail about the algorithm that we intend to implement (Fig. 1).

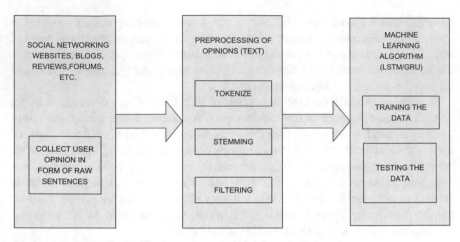

Fig. 1 Overall architecture

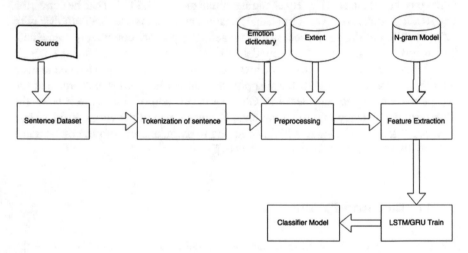

Fig. 2 System model: training

The training and testing include the same steps as discussed in the architectural framework. The training occurs with the help of emotion dictionary and then we use *n*-gram model to deal with a set of words as we want to analyze long bipolar sentences. Once we are done with the feature extraction and LSTM/GRU training, we finally get a classifier model and we mark the end of training (Fig. 2).

4 Algorithm

```
function SENTIMENT-ALALYSIS-LSTM-GRU (sentence) returns
```

```
sentiment
        static: xₜ,  input vector
                hₜ, output vector
                cₜ, cell state vector
                W, U and b parameter matrices and vector
                fₜ, iₜ and oₜ gate vectors
                        fₜ forget gate vector. Weight of
remembering old information
                        iₜ input gate vector. Weight of
 acquiring new information
                        oₜ output gate vector. Output
candidate
                σg Sigmoid function
                σc Hyperbolic tangent
                σh Hyperbolic tangent
                Zₜ  forget and input gate, Update gate
                rₜ  cell and hidden gate, Reset gate
                Hₜ candidate activation
        cell ← UPDATE CELL( cell, sentence)
        While sentences do
        Tokens ← Tokenize (sentences)
        Punctfree ← RemovePunctuations (Tokens)
        Spacefree ← RemoveWhiteSpaces (Punctfree)
        Neghandle ← HandleNegation (Spacefree)
```

$i_t \leftarrow \sigma_g\ (W_i x_t + U_i h_{t-1} + b_i)$

$f_t \leftarrow \sigma_g\ (W_f x_t + U_f h_{t-1} + b_f)$

$\vec{c}_t \leftarrow \tanh(\ W_c x_t + U_c h_{t-1} + b_c)$

$c_t \leftarrow f_t\ .\ C_{t-1} + i_t\ .\ C_t\ bar$

$o_t \leftarrow \sigma\ (\ W_o x_t + U_0 h_{t-1} + V_o C_t)$

$h_t \leftarrow o_t\ .\ \tanh\ (C_t)$

$Z_t \leftarrow \sigma\ (W_z x_t + U_z h_{t-1})$

$r_t \leftarrow \sigma(W_r x_t + U_r h_{t-1})$

$H_t \leftarrow \tanh\ (W_H x_t + U_H\ (r_t\ .\ H_{t-1}))$

$h_t \leftarrow (1-Z_t)\ h_{t-1} + Z_t H_t$

```
        return sentiment
```

5 Dataset

We just need to set the present working directory and give the command for "readlines (chat.txt)" from the text file which is chat.txt which we have linked. Therefore, there is no fixed dataset for this particular code. We can extract or export any text document, chat, tweets, blogs, etc., and perform sentiment analysis. The sample entries which we have taken in the input text file are given below for reference and in order to enhance the understanding of our obtained results.

- I am very happy today. This day is just wonderful!
- What a lovely person he is.
- I was super excited to meet him.
- I was on my way when I saw a sad kid.
- I gave him a chocolate and he got very happy. Other people look him with disgust and anger.
- While I was giving that kid a chocolate, I saw that man standing next to me and I was surprised. We introduced each other. He was glad to meet me too.
- It was fun and jolly experience to spend the day with him.

6   Results and Discussion

We then run the code and get the required plots and word clouds. The word clouds help in better visualization of the sentiments derived from the input text data and assist incorrect and efficient mood analysis. We also use the ggplot2 library to plot the word count of sentiments in the form of graphical representations that allocate better interpretability to the results obtained from our experiment (Fig. 3).

This is the end goal of our project. We get a plot of ten sentiments, and this plot can be analyzed to understand the various sentiments involved in the text. Any text file can be used [15] (Figs. 4 and 5).

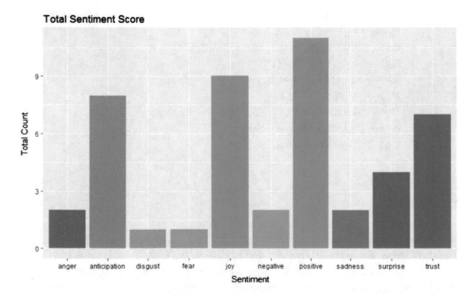

Fig. 3 Plot of bar graphs depicting different emotions and their extent

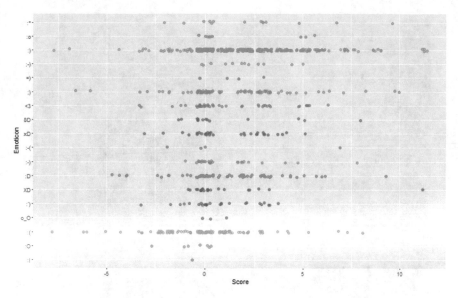

Fig. 4 Plot based on frequently occurring emoticons

7 Conclusion

This paper is aimed at obtaining a solution for performing sentiment analysis on given sentences obtained from a dataset which can be a text dataset or a dataset obtained from social networking sites like Twitter, Facebook. The proposed hybrid algorithm gives optimal results when compared to using LSTM and GRU individually to perform sentiment analysis. Hence, our proposed is a better choice to judge the mood inferred from the given text. It gives accurate results and is therefore a good choice for using this type of neural network-based machine learning for sentiment analysis. Our future work includes using some other algorithms to make the hybrid interpretation of the contributing algorithms in a different way in order to get even better results.

Fig. 5 a Word cloud based on the frequently occurring words. **b** Word cloud based on various sentiments

References

1. Basant A, Namita M, Pooja B, Garg S (2015) Sentiment analysis using common-sense and context information. Hindawi Publishing Corporation Computational Intelligence and Neuroscience
2. Doaa ME (2016) Enhancement bag-of-words model for solving the challenges of sentiment analysis. Int J Adv Comput Sci Appl 7(1)
3. Saif MM, Peter DT (2010) Emotions evoked by commonwords and phrases: using mechanical turk to create an emotion lexicon. In: Proceedings of the NAACL HLT 2010 workshop on computational approaches to analysis and generation of emotion in text, association for computational linguistics, pp 26–34, Los Angeles, California

4. Yulan H, Chenghua L, Harith A (2011) Automatically extracting polarity-bearing topics for cross-domain sentiment classification. In: Proceedings of the 49th annual meeting of the association for computational linguistics, association for computational linguistics, Portland, Oregon
5. Khairullah K, Baharum B, Aumagzeb K, Ashraf U (2014) Mining opinion components from unstructured reviews: a review. J King Saud Univ Comput Inform Sci 26(3):258–275
6. Lucie F, Eugen R, Daniel P, (2015) Analysing domain suitability of a sentiment lexicon by identifying distributionally bipolar words. In: Proceedings of the workshop on computational approaches to subjectivity, sentiment and social media analysis, EMNLP
7. Ling P, Geng C, Menghou Z, Chunya L (2014) What do seller manipulations of online product reviews mean to consumers?" HKIBS Working Paper Series 070-1314. Hong Kong Institute of Business Studies, Lingnan University, Hong Kong
8. Tawunrat C, Jeremy E (2015) Simple approaches of sentiment analysis via ensemble learning Chapter Information Science and Applications. Volume 339 of the series Lecture Notes in Electrical Engineering, DISCIPLINES Computer Science, Engineering SUBDISCIPLINES AI, Information systems and applications-computational intelligence and complexity
9. Singhal P, Bhattacharyya P (2016) Sentiment analysis and deep learning: a survey. Center for Indian Language Technology, Indian Institute of Technology, Bombay
10. Hong RJ, Fang M (2015) Sentiment analysis with deeply learned distributed representations of variable length texts. Deep learning for natural language processing technical report, Stanford University. CS224d
11. Li JD, Qian J (2016) Text sentiment analysis based on long short-term memory. In: Computer communication and the internet (ICCCI) IEEE international conference
12. Wang Y, Huang M, Zhao L, Zhu X (2016) Attention-based LSTM for aspect-level sentiment classification. Tsinghua University, Beijing, China In: Conference on empirical methods in natural language processing, pp 606–615, Austin, Texas, Nov 1–5,. c Association for Computational Linguistics
13. Arkhipenko K, Kozlovand I, Trofimovich J (2016) Comparison of neural network architectures for sentiment analysis of russian tweets. In: Computational linguistics and intellectual technologies: proceedings of the international conference, Moscow, 1–4 June (2016)
14. Biswas S, Chadda E, Ahmad F Sentiment analysis with gated recurrent units. Department of Computer Engineering. Annual Report Jamia Millia Islamia New Delhi, India
15. Ainur Y, Yisong Y, Claire C (2010) Multi-level structured models for document-level sentiment classification. In: Proceedings of the 2010 conference on empirical methods in natural language processing, pp 1046–1056. MIT, Massachusetts, Association for Computational Linguistics, USA

Optimal Rate Allocation for Multilayer Networks

Harshit Pandey, Priya Ranjan, Saumay Pushp and Malay Ranjan Tripathy

Abstract Issue of rate allocation on a variety of exotic networks is a vexing one. In this work, a multilayer network is studied for optimal rate allocation with generally speaking inter-layer links appearing as bottlenecks due to constrained capacities. Continuing the idea from our earlier research, we unfold new connections between aggression of users for a resource and clogging of a crucial link between two layers. We believe that in the situation of rampage in post-game scenario or terrorist attacks which have become all too common, and there is a new way to rethink about designing inter-layer links. They are the arteries which are going to choke first and strain the movement of crucial exits in most difficult times. Further, links with aggression and aggregate crowd behavior are illustrated.

Keywords Multilayering · Rate allocation

1 Introduction

As societies emerge and populations explode, the need for reliable, uninterrupted, and sophisticated communication grows exponentially. The Internet is an unmatched and most legitimate resource that forms the basis of all formal communication in today's era like emails, video, and audio conferencing. The Internet is combined as collection of a humongous number of nodes that choose to interact with each

H. Pandey (✉) · P. Ranjan · M. R. Tripathy
AUUP, Sector-125, Noida, Uttar Pradesh 201313, India
e-mail: harshitpandey16@gmail.com

P. Ranjan
e-mail: pranjan@amity.edu

M. R. Tripathy
e-mail: mrtripathy@amity.edu

S. Pushp
KAIST, 291 Daehak-ro, Guseong-dong, Yuseong-gu, Daejeon, South Korea
e-mail: saumay@nclab.kaist.ac.kr

© Springer Nature Singapore Pte Ltd. 2019   651
A. J. Kulkarni et al. (eds.), *Proceedings of the 2nd International Conference on Data Engineering and Communication Technology*, Advances in Intelligent Systems and Computing 828, https://doi.org/10.1007/978-981-13-1610-4_66

other depending upon the structural and strategic framework of the network [1, 2]. Graph theory, however, is found to be a very useful tool to deal and map complex networks and thus finds many applications in the analysis of complex networked systems [3]. Any real-time application like online games or video streaming poses a probable congestion in the network which leads to a packet being dropped [4] and thus the information being lost [5]. With advancements in technology, the division of a network into layers is found to drastically improve the throughput (rate of sending packets) in complex networks. Network layering provides more sagacity and robust behavior [6] to each individual layer, and thus, any layer is capable to provide reliable, efficient and smooth service to customers individually even in cases where a part of the network is not operational. Such networks are termed as multilayer networks. A multilayer network is one where the system is characterized by layers. This concept of a multilayer network can be thought of as being analogous to a city transportation system model. In the transportation model, at level 0 the subway is present for pedestrians to walk, at level 1 there exists the road for automobiles, at level 2 there is elevated path of the metro and finally at level 3 the airplanes fly. Considering this example it is seen that this model has a total of four layers that are individually capable and efficient enough to pick a man from home and drop to office. The motivation to study this system lies in the fact of changing layers while traveling, like a staircase from level 1, i.e., road to level 2, i.e., metro. Here the fascination lies in the change of layers by a man who initially was going by a car, but due to road traffic tries to take the metro. This transition from road to metro takes into account an inter-layer change. Also, if the car breaks down, the man can take a cab so this is an intra-layer transition. The nuance of transitions between inter- and intra-layer yield interesting results pertaining to the man reaching his destination depending upon the physical size of the layers. So if the transition of people from road to metro is very high and the capacity of metro is less, this yields to congestion. Hence, the underlying motive is to study the optimal rates allocated to different users depending on physical layer throughput capacities to mitigate congestion before it arises. This paper approaches to solve the concern of rate allocation to users in the network while the whole network is essentially layered. The motivation for this paper lies in the fact that due to mobile and layered nature of modern communication networks, a new kind of graph theory is needed to address modern volatile and turbulent connectivity scenarios. Networks have become dynamic giving to the axiom that the connecting edges change [7, 8] depending upon the strategic and structural formulations of users [1]. In the upcoming sections, we define rate allocation, prepare ground for multilayer networks and finally model a multilayer network with simulation results in MATLAB environment. MATLAB code can be made available for community verification and further development by interested parties. In the end, we collect the conclusions.

2 Rate Allocation in Networks

2.1 Optimal Rate Control Framework

Rate allocation is the basis on which the transmission rates, i.e., throughputs are allocated to users. The way in which allocation is done has to be compatible with some fairness criterion. Kelly [5] gives the analysis of how rate is controlled in network traffic. Rate allocation takes care of the amount of rate that is assigned to any particular user in terms of flow so as to keep the users satisfied and happy. The user who has a higher demand gets higher rate assigned, for which he has to pay more compared to the user who demands and gets lower rate which is directly in correspondence with what he pays to the network. This allocation of rates has to be handled meticulously so as to not violate any rules defined for this particular issue. If a particular user usurps all the flows, then the other users do not get any access to any data at all and this condition can be thought of as the failure of optimal rate allocation. The aim is to combine maximize the sum of all utilities [9] of all the users in the network. Mathematically, the optimization of rate allocation by a network can be denoted as.

2.2 Multilayer System (U, A, B) Problem

$$\text{maximize } (x) = \sum_{i=1}^{m} U(i)x(i) \tag{1}$$

subject to intralayer constraints $A_i x_i \leq b_i$ $i = 1 \ldots l.l$ is number of layers
interlayer constraints $Ax \leq b$, $i = 1 \ldots m.m$ is number of users
positivity $x \geq 0$

Here A is the flow routing matrix, x being the rate allocated and B represents the capacity of transmitting links. U represents the utility of the users. The utility can be thought of as the satisfaction of a user with the service [9]. There are various kinds of utilities as sigmoid or logarithmic and can be used for modeling the network. The utility considered here [10] is denoted as:

$$U(x) = -1/(ax^a) \quad \text{where } a > 0 \tag{2}$$

2.3 Multilayer Network

A network that can be split up into a number of smaller networks by layers and can form subsystems which is referred to as a multilayer network. A basic single layer network can be represented in terms of a graph $G = (V, E)$, where V is the set of all nodes and E represents the edges. If there are two nodes a and b, then the two nodes are adjacent to each other and the edge $e(a, b)$ is then called to be incident to each of the two connected nodes, where $e \in E$. This framework of a simple network can be broadened to analyzing a multilayer network, which has multiple layers in addition to having edges and vertices. Every multilayer network is a subset of a multi-aspect network. Every multi-aspect network can be mapped as a directed graph and since multilayer network belongs to the broader class of multi-aspect network; hence, every multilayer network model can be equivalently modeled as a graph [3]. From the concepts of graph theory, every directed graph can be represented in a matrix representation form and this makes it possible to map a multilayer network into the routing and adjacency matrix format [11]. A simple three-layer network without any loss of generality has been shown in Fig. 1. This figure shows three layers and has two nodes in Layer 1, 2 nodes in the Layer 2 and 3 nodes in layer 3. The nodes are connected by edges as shown. The solid black lines represent the intra-layer connections between nodes, and the dotted black lines represent the inter-layer connection between nodes. These connections are called as edges and flows essentially take place only through paths connected via edges.

Model formulation here we consider a multilayer network model as described in the basic configured figure for various layers as shown in Fig. 1. This figure is a simple representation of a three-layer network describing the inter- and intra-layer connections by means of edges connecting the vertices (solid black for intra-layer and dashed black for inter-layer). For model formulation in Fig. 1a is analyzed and studied in this section and equivalently in Fig. 1b is obtained. Figure 1b represents the detailed analysis of the simple 3 layer 7 node structure. Here three layers are seen as layer 1, layer 2, and layer 3. This model is an extension of Fig. 1a and here a number of users and their routing patterns are studied and described. The yellow

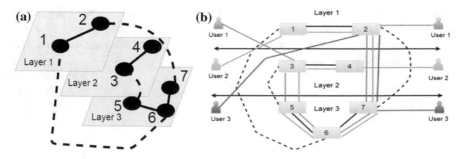

Fig. 1 **a** Flow control diagram. **b** Multilayer graph

boxes represent the nodes or vertices of the network through which the data actually travels and as seen in the model vertices 1, 2 are present in layer 1, vertices 3 and 4 are present in layer 2 and vertices 5, 6, and 7 in the layer 3. The two horizontal lines with bidirectional connectors at endpoints separate the three layers in the graph. It is seen in the graph that vertex 1 is connected to vertex 2 in layer 1, vertex 3 is connected to vertex 4 in layer 2, and similarly vertex 6 is connected to vertex 5 and 7 in layer 3. This connection of node 1 to node 2, node 3 to node 4, node 6 to node 5 and 7 represented by solid black connecting lines describe the intra-layer connections of the graph. The connections that are represented by dotted black lines, i.e., connection of node 1 in layer 1 to node 6 in layer 3, node 3 in layer 2 to node 5 in layer 3, and node 2 in layer 1 to node 7 in layer 3, all represent the inter-layer connections of the graph. These connections of the vertices are due to the structural framework of the network. The significance of these edges is the restriction of flow of data only through one of these paths which forms a route depending upon which user opts for which path. The edges of the network are present by virtue of the network structure, whereas the flows are random. This model considers three users as shown in blue, green, and red and marked as user 1, user 2, and user 3. The motive of any user in the network is data transmission and reception which has been shown in the model as user 1, user 2, and user 3 on both sides where one can be considered as sending end and the other as receiving end from the network. Our emphasis lies on the fact of studying the multilayered graph with optimal rate allocation. The information travels depending upon the routing of a user in the network. In this model, there are three flows or routes blue for user 1, green for user 2, and red for user 3. The routing matrix is given as follows:

$$
A = \begin{array}{ccc}
1 & 1 & 0 \\
1 & 1 & 0 \\
1 & 1 & 0 \\
1 & 1 & 1 \\
0 & 1 & 0 \\
0 & 1 & 0
\end{array}
$$

The A matrix is of size $6 * 3$ as there are six edges between nodes in the network and three flows of interest, namely blue, green, and red, respectively. The A matrix is the routing matrix which represents the routes taken by various users and puts a 1 in the place if a route exists [11]

$$
A_{ij} = \begin{array}{l}
1 \text{ if resource } j \text{ is on route } r \\
0 \text{ otherwise}
\end{array}
$$

3 Simulation Results

Next we outline the parameters in the table for upcoming simulation scenario set up.

The model represented in Fig. 1 was simulated and analyzed for the changes in utilities of the users. Some link capacities were defined in order to observe how allocation varies from user to user so as to handle the issue of creation of congested nodes. Level of happiness of a user by the service he receives in the network can be defined as utility. Each user has a utility of its own, and every utility has a coefficient as defined by Eq. 2. The linear increase of this coefficient is studied here for every user 1, 2, and 3 by observing its effect on the amount of rate being allocated to each user for a particular value of utility. Table 1a–c represents the variation of the value of the allocation of rate with respect to the gradual linear increase in coefficients of utility for one user at a time. Hence, Table 1a is for user 2 where the coefficient of utility for only user 2 changes, whereas for user 1 and user 3 remain constant. Similarly, Table 1b is for user 1 and Table 1c is for user 3. Figure 2a–c is plots that graphically sketch out the dependence of the rate on the utility of a user, where Y-axis denotes the value of assigned rate to the user and X-axis denotes the utility coefficient for that particular user.

Table 1 Rate allocation for user. 1. Rate allocation for user 2. Rate allocation for user 3

(a)	
Link capacities [5, 5, 3, 6, 3, 2] Utility coefficient $a = 3, c = 10$	
Coefficient of utility (b) For user 2	Rate allocated to user $2(x_2)$
1	1.6972
2	1.5864
3	1.5000
4	1.4326
5	1.3796
6	1.3372
7	1.3028
8	1.2744
9	1.2507
10	1.2306

(continued)

Table 1 (continued)

(b)

Link capacities [5,5,3,6,3,2]
Utility coefficient $b = 5$, $c = 10$

Coefficient of utility (a) For user 1	Rate allocated to user 1 (x_1)
1	1.7866
2	1.6972
3	1.6204
4	1.5552
5	1.5000
6	1.4533
7	1.4136
8	1.3796
9	1.3503
10	1.3249

(c)

Link capacities [5,5,3,6,3,2]
Utility coefficient $a = 3$, $b = 5$

Coefficient of utility (c) For user 3	Rate allocated to user 3(x_3)
1	3.0000
2	3.0000
3	3.0000
4	3.0000
5	2.9999
6	2.9998
7	2.9997
8	2.9950
9	2.9856
10	2.9612

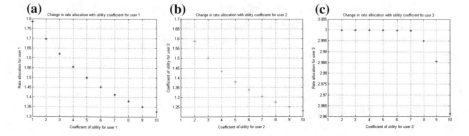

Fig. 2 **a** Rate allocation-I. **b** Rate allocation-II. **c** Rate allocation-III

Multilayer networks seem more interesting than single-layer networks as single-layer networks possess enormous redundancy, which means that even if an edge does not let the flow of data-pass through it due to link breakage, the data is still routed from another edge to reach the user. This can be analogous to the city transportation system where even if the car breaks down, the person can take another cab or bus to still reach the destination. But in case of multilayer networks, what fascinates the most is that if any link between inter-layer is not operational or fails to route data, the whole transmission into that layer from that link ends, thereby leading to loss of data. This is again analogous to the city transportation system model where people need to change layers of transport to travel. Suppose that the link shown in the model in this paper for inter-layer transition between any two layers is the staircase or escalator for people on the road to switch to the metro which is on a new layer. Here in this situation, the staircase is the bottleneck for people who are transitioning between layers and if in case the escalator stops functioning or is closed, the link between the two layers, i.e., road and metro is completely cut off so people can no more switch layers and chaos occurs which can even lead to injuries to people and in severe cases also result in casualties. If in case the staircase is not big enough to handle the amount of people, then congestion or traffic occurs. So each person must have the amount of space required for him to switch layers. This is exactly what happens in networks, if capacity is lesser than demand, congestion occurs and if the inter-layer link does not work, the whole data in the network for this layer is completely chopped off which is highly undesirable. From results of simulation, it is seen that as we go on increasing the coefficient of utility for a particular user, the rate of allocation keeps on decreasing. So if the user is aggressive (has low coefficients of utility), it has more rate but as we go on increasing its utility, the rate allocation decreases. This phenomenon can be seen in the graphs and can be thought of as when a man has no money the value of food is more for him but as he earns a lot of money, the value of food decreases where money can be thought of as utility and food as rate. The node 4 in this model is crucial as all three flows pass through it and it is an inter-layer link. Thus, rate has to be allotted in this link in such a way so that every user receives data and at the same time no user is denied his data due to congestion. Simulating the model for value of coefficients of utility of user 1, 2, 3 as 3, 5, 10 with link capacity having value 6 for node 4 gives rate as 1.6204, 1.3796, and 2.9912 which sums up to 5.9912 which is lesser than the actual capacity. Thus, this is how rate is being allocated and the inter-layer link forms a crucial part of this network.

3.1 Future Directions

In this paper, the rate allocation for a multilayer network has been developed and the graphs between the varying coefficient of utility functions of a particular user and the subsequent value of rate assigned are seen. Intuitive analogies of this multilayer network have been made in terms of a real-time physical model for the transportation system. Future work will approach to aim at modeling and analyzing multi-aspect graphs which are a superset of the multilayer graph. Mapping a network not only in the perspective of layers but also other eminent aspects like data security, speed, delay, and throughput that are the backbone of today's Internet communication, will be analyzed and focussed upon. Essentially, the future works aim to find the rate allocation to users while now the network does not only pertain to being multilayered but in a way being multi-aspect so as to be more reliable and efficient and to make the users more satisfied. Further, we would like to explore the deeper connection and impact of aggression in psycho-economic legal framework [12] and the same in sports context [13].

References

1. Ranjan P, Tripathy MR, Pushp S, Pandey H (2017) Understanding rate allocation mechanism in strategic and structural communication network via dynamic adjacency, to appear in Proceedings of PIERS
2. Wilson JD, Palowitch J, Bhamidi S, AB Nobel (2016) Community extraction in multilayer networks with heterogeneous community structure. arXiv preprint arXiv:1610.06511
3. Wehmuth K, Fleury R, Ziviani A (2016) Multi aspect graphs: algebraic representation and algorithms. Algorithms 10(1):1
4. Jacobson Van (1988) Congestion avoidance and control. ACM SIGCOMM Comput Commun Rev 18(4):314–329. ACM, New York City
5. Kelly F (2001) Mathematical modelling of the Internet. In: Mathematics unlimited 2001 and beyond, pp. 685–702. Springer, Berlin, Heidelberg
6. Liakh S (2009) Multilayer network survivability
7. Skyrms B, Pemantle R (2009) A dynamic model of social network formation. In: Adaptive networks, pp 231–251. Springer, Berlin, Heidelberg
8. Kumari S, Singh A, Ranjan P (2016) Towards a framework for rate control on dynamic communication networks. In: Proceedings of the international conference on internet of things and cloud computing, p 12. ACM, New York City
9. Abdelhadi A, Shajaiah H (2016) Optimal resource allocation for cellular networks with MATLAB instructions. arXiv preprint arXiv:1612.07862
10. Ranjan P, La RJ, Abed EH (2006) Global stability conditions for rate control with arbitrary communication delays. IEEE/ACM Trans Netw (TON) 14(1):94–107
11. Matta I (2013) Optimizing and modeling dynamics in networks. In: Haddadi H, Bonaventure O (eds) eBook on Recent Advances in Networking. ACM SIGCOMM 1
12. Kinsella NS, Tinsley P (2004) Causation and aggression. Q J Austrian Econ 7(4):97–112
13. Wanjiku MA (2016) Assessment of crowd management strategies used for football events in government-owned sports stadia in Nairobi county, Kenya. Ph.D. dissertation, School of Applied Human Sciences, Kenyatta University

Folding Automaton for Paths and Cycles

N. Subashini, K. Thiagarajan and J. Padmashree

Abstract In this paper, folding automata for paths and cycles are obtained. For given paths (P_n) and cycles (C_n), automata are studied and folding concepts are applied to paths and cycles and also obtained folding automata. Also obtained the result that for P_n, n is odd and $n > 2$ the vertex folding is possible (vertex symmetry is available), and if n is even, the edge folding is possible since there is an availability of edge symmetry. And for cycle C_n, edge folding is possible only for even number of vertices.

Keywords Automaton · Path · Cycle · Folding

1 Introduction

The class of automata contains many interesting and complicated groups. Its thorough investigation was started with automata with small number of states. An automaton plays an important role in theory of computation, compiler construction, AI, motion, and formal verification. Conjointly finite automata are utilized in text process, compilers, and hardware.

N. Subashini
Mother Teresa Women's University, Kodaikanal, Tamil Nadu, India
e-mail: subashini.sams@gmail.com

K. Thiagarajan (✉)
Pacheri Sri Nallathangal Amman College of Engineering and Technology, Dindigul,
Tamil Nadu, India
e-mail: vidhyamannan@yahoo.com

J. Padmashree (✉)
Bharathiar University, Coimbatore, Tamil Nadu, India
e-mail: padmasathiru@gmail.com

© Springer Nature Singapore Pte Ltd. 2019
A. J. Kulkarni et al. (eds.), *Proceedings of the 2nd International Conference on Data Engineering and Communication Technology*, Advances in Intelligent Systems and Computing 828, https://doi.org/10.1007/978-981-13-1610-4_67

The study of finite automata (FA) respectable, or automatic, structures began within the works by Hodgson (Théories décidables par automatize fini, Annales DE Sciences Mathématiques, 1983), [1] then carried on in Khoussainov and Nerode. A structure is named Finite Automata is delineate if its domain and put down relations square measure recognized by finite automata operative synchronously supported passing input. So, these structures have finite shows the category of automatic structures is of interest group within the field of theoretical engineering science. In this paper, we tend to correlate graph and automata for the trail, cycle, and straightforward graph in recursive means.

2 Preliminaries

2.1 Definition: Finite Automaton [2]

A finite automaton is pictured formally by the 5-tuple

$$(Q, \Sigma, \delta, q_0, F),$$

where

Q may be a finite set of states.
Σ may be a finite set of input symbols.
δ is the transition performed, that is, $\delta: Q \times \Sigma \to Q$.
q_0 is the begin state.
F may be a set of ultimate states of Q

2.2 Definition: Path [3]

A path in a graph could be a finite or infinite sequence of edges that connect a sequence of vertices.

2.3 Definition: Cycle [3]

A cycle or circular graph could be a graph that consists of vertices that are connected in a very chemical chain.

3 Folding Automata for Paths

a. Graph for path P_2:

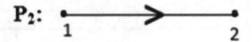

Figure 1

Automata-2 states with inputs (0, 1):

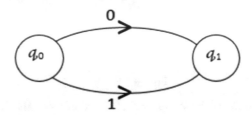

Figure 2

In this graph, vertex folding is not possible. Hence, we will get only edge folding. For vertex folding 'n' must be >2, while for edge folding 'n' must be ≥2.

b. Graph for path P_3:

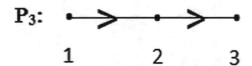

Figure 3

Automata-3 states with inputs (0, 1):

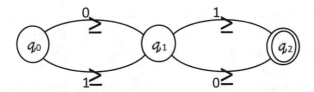

Figure 4

Transition table for automata-3 states with inputs (0, 1):

Table 1

	q_0	q_1	q_2
q_0	–	0, 1	–
q_1	–	–	1, 0
q_2	–	–	–

Folding graph of P_3:

$$(1, 2, 3)$$

Figure 5

This is point symmetry. Here, vertex folding is only possible but edge folding is not possible.

Folding automata for P_3:

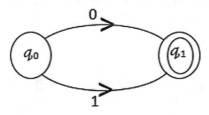

Figure 6

Transition table for automata-2 states with inputs (0, 1):

Table-2

	q_0	q_1
q_0	–	0,1
q_1	–	–

∴ In general, when $n =$ odd and >2 then vertex folding is possible since vertex symmetry is possible. Similarly, when $n =$ even and ≥ 2, then edge folding is possible since edge symmetry is available.

4   Folding Automata for Cycles

a. Graph for cycle C_3:

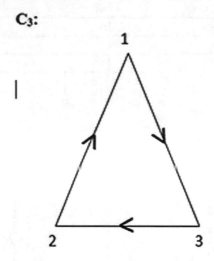

Figure 7

In this graph, folding is not possible. Hence, we observed that folding is possible for cycle C_n, $n > 3$.

Automata for C_3:

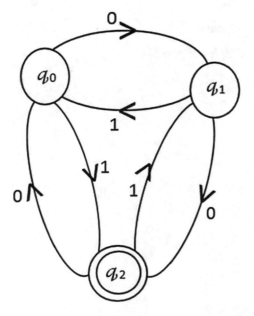

Figure 8

Transition table for automata-3 states with inputs (0, 1):

Table-3

	q_0	q_1	q_2
q_0	–	0	1
q_1	1	–	0
q_2	0	1	–

b. Graph for cycle C_4:

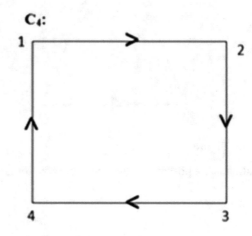

Figure 9

In this graph, folding is not possible,

Automata for C$_4$:

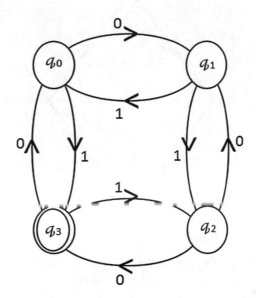

Figure 10

Transition table for automata-4 states with inputs (0, 1):

Table-4

	q_0	q_1	q_2	q_3
q_0	–	0	–	1
q_1	1	–	1	–
q_2	–	0	–	0
q_3	0	–	1	–

If cycle C$_4$ is considered in the following way,

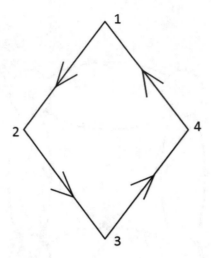

Figure 11

then we have edge folding but not vertex folding for four vertices and number of edges which is equal to $a + (n-1)\, d = 4 + (1-1)1 = 4$.

Automata for C_4:

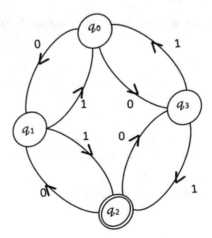

Figure 12

Transition table for automata-4 states with inputs (0, 1):

Table-5

	q_0	q_1	q_2	q_3
q_0	–	0	–	0
q_1	1	–	1	–
q_2	–	0	–	0
q_3	1	–	1	–

Folding graph for C_4:

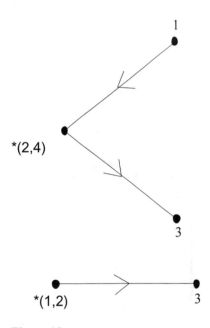

Figure 13

In Figure 13, edge folding is possible but vertex folding is not applicable.

- Indicates point 2 goes to 4
- Indicates point 1 goes to 2.

Folding Automata for C₄:

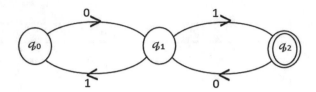

Figure 14

Transition table for automata-3 states with inputs (0, 1):

Table-6

	q_0	q_1	q_2
q_0	–	0	–
q_1	1	–	1
q_2	–	0	–

c. Graph for cycle C₆:

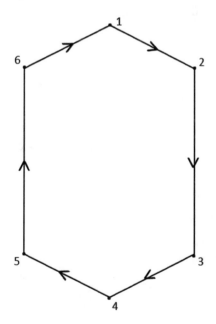

Figure 15

In this graph, edge folding is possible with six vertices. Generally, this is observed as the formula $a + (n-1)d$, with $a = 4$, $d = 2$, $n \geq 1$.

Automata for C_6:

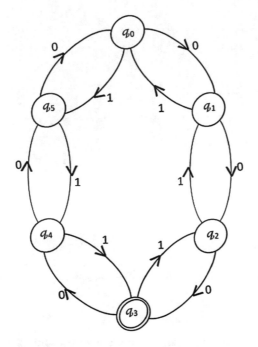

Figure 16

Transition table for automata-6 states with inputs (0, 1):

Table-7

	q_0	q_1	q_2	q_3	q_4	q_5
q_0	–	0	–	–	–	1
q_1	1	–	0	–	–	–
q_2	–	1	–	0	–	–
q_3	–	–	1	–	0	–
q_4	–	–	–	1	–	0
q_5	0	–	–	–	1	–

Folding graph for C$_6$:

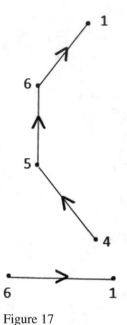

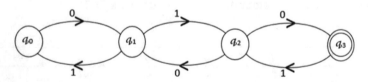

Figure 17

In this graph, edge folding is possible but not vertex folding.
Folding automata for C$_6$:

Figure 18

Transition table for automata-4 states with inputs (0, 1):

Table-8

	q_0	q_1	q_2	q_3
q_0	–	0	–	–
q_1	1	–	1	–
q_2	–	0	–	0
q_3	–	–	1	–

In general, for cycles edge folding is possible only for even number of vertices with even number of edges of the form $a + (n-1)d$ that is, starts with minimum $a = 4$ number of vertices, edges and common difference $d = 1, 2$.

Hence, this will generate an arithmetic progression 4, 6, 8, ...

In cycle C_n, edge folding is not possible for odd number of vertices 'n'.

Note: The edge folding is all applicable for connected graph with even number of vertices.

Now, we can convert this concept for connected graph into automata concepts.

5 Conclusions

We intend different applications which are available in automata theory for generalization of graphs which is connected. The relation of connected graph and automaton is observed for paths and cycles. Also observed the folding technique is applied to obtain folding automata for paths and cycles.

Future work In future, this approach may be extended for tree and connected graph also.

Acknowledgements The authors would like to thank Dr. Ponnammal Natarajan, Former Director—Research and Development, Anna University Chennai, India, for her initiative ideas and fruitful discussions with respect to the paper's contribution.

References

1. Hodgson BR (1983) Théories décidables par automate fini. Annales de Sciences Mathématiques 7(39–57):14
2. Harary F (1972) Graph theory, Third Printing. Addison Wesley Publishing Company, Boston
3. Bondy JA, Murthy USR (2015) Graph theory with applications
4. Deo N (1995) Graph theory with applications to engineering and computer science. PHI Publications, New Delhi
5. Hopcroft JE, Motwani R, Ullman JD Introduction to automata theory, languages, and computation, 3rd edn. Pearson Publication, London
6. Gallian JA (2010) A dynamic survey of graph labelling. Electronic Journal of combinatorics

Decisive Tissue Segmentation in MR Images: Classification Analysis of Alzheimer's Disease Using Patch Differential Clustering

P. Rajesh Kumar, T. Arun Prasath, M. Pallikonda Rajasekaran and G. Vishnuvarthanan

Abstract Alzheimer's disease usually occurs in the elderly, and over a course period of time, it contributes to dementia. Substantiation of cerebrovascular disease progressively develops to cognitive performance in the very earlier stage of Alzheimer's disease. Structural brain imaging plays a vital role in recognition of changes that appear in brain relevant to Alzheimer's disease and various brain diseases. Magnetic resonance imaging takes place a major task in identification and analyzing of Alzheimer's disease diagnosing process before the surgical planning. An eminent segmentation process of disease-related tissues leads the physician in making the decision on diagnosing. In this work, segmentation task has been performed with patch image differential clustering (PIDC) based on the labeling over the patches. According to the segmentation operation, the various brain matters like gray matter (GM), white matter (WM), and cerebrospinal fluid (CSF) are grouped individually. Finally, the volume estimation of above-mentioned brain matters is adopted for initializing the nearest particle interconnect (NPI) classifier for classifying the abnormality of AD.

Keywords Alzheimer's disease (AD) · Nearest particle interconnect (NPI) Patch image differential clustering (PIDC)

P. Rajesh Kumar (✉) · T. Arun Prasath · M. Pallikonda Rajasekaran · G. Vishnuvarthanan
Department of Electronics and Communication Engineering,
Kalasalingam Academy of Research and Education, Virudhunagar 626126, Tamil Nadu, India
e-mail: rkrrajesh74@gmail.com

T. Arun Prasath
e-mail: arun.aklu@gmail.com

M. Pallikonda Rajasekaran
e-mail: m.p.raja@klu.ac.in

G. Vishnuvarthanan
e-mail: gvvarthanan@gmail.com

© Springer Nature Singapore Pte Ltd. 2019   675
A. J. Kulkarni et al. (eds.), *Proceedings of the 2nd International Conference on Data Engineering and Communication Technology*, Advances in Intelligent Systems and Computing 828, https://doi.org/10.1007/978-981-13-1610-4_68

1 Introduction

Alzheimer's disease was analyzed and identified 100 years ago. But during the past 30 years, only researches have been developing in its risk factors, symptoms, causes, and treatments. Nowadays throughout the world, more than 35 million people have been affected by Alzheimer's disease with its various stages [1]. Among the total population in a country, 54% of people who have affected by AD at the stage of beginning or mild level. The number of affected ratio has been increasing double every 20 years. Alzheimer's is an ultimate prevalent type of dementia. The term dementia describes variety of diseases which begin when the nerve cells of brain stop functioning or brain tissues start to die progressively [2]. The major reason for AD is aggregation of beta amyloidal neurofibrillary tangle, the lacking of TAU protein in the brain. The most recent diagnostic benchmark of AD is depending upon the clinical dementia ratio (CDR) and Mini-Mental State Examination (MMSE) for psychometric test which are more useful for classifying various stages of AD [3]. For earlier detection and treatment of AD, there is an extensive need to develop a new concept. The related more meaningful information about the disease can be gathered from the brain only at autopsy. And, this similar view of a number of various brain imaging techniques devotes to noninvasive ways of brain atrophy visualization [4]. The segmentation process has been achieved with patch image differential clustering (PIDC) method for segmenting the different brain matters such as GM, WM, and CSF. And, the stages of AD have been classified with the nearest particle interconnect classifier [5].

2 Previous Work

Platero and Tobar [6] proposed a novel multiple-atlas-based image segmentation principle, which helps to associate a patch image differential-based labeling technique with the combination of an atlas warping by the implementation of nonrigid registrations. Valverde et al. [7] introduced an approach which has a robust segmentation of certain amount of volume measured from tissue with WM elimination as well as substantial combination of both the intensities of morphological- and probabilistic-based mapping structures. Intensity modeling schemes of the dispute and clear achievement have been done. Beheshti et al. [8] developed a novel computer-aided diagnosis (CAD) structure that helps make the feature ranking with genetic algorithm. CAD which has been developed with the help of a morphometry based on voxels for investigating the local gray matter (GM), global atrophy, and objective function has been achieved by the genetic algorithm, and the classification has been executed with the help of support vector machine (SVM). Liu et al. [9] introduced morphometric pattern analysis based on multi-templates for multi-view of the morphological operator produced from number of relevant pixels that are selected as feature demonstration for the brain MR images. An intrinsic multi-view depends on structure predilection technique with different templates for the purpose of classifying AD/MCI. Zhan et al. [10] used a network-based statistic (NBS) technique to develop the cluster structure of associated regions in AD/MCI. It analyzed abnormality of the association mechanism by the use of NBS depending upon the resting-state functional magnetic resonance imaging (fMRI). The results provide that patients were having decreased performance in connectivity model and healthy in some components.

3 Materials and Methods

In this research work, T1-weighted MR scans of brain images have been chosen because they produce gray matter (GM) and white matter (WM) with good contrast as compared with T2-weighted MRI. For our work, MRI scan images have been accessed from Open Access Series of Imaging Studies (OASIS) dataset for analyzing the brain image. This work has been focused on mild stage of AD patients, and 150 brain scan images are downloaded for our research work from OASIS database including the scans of younger healthy controls, Alzheimer's affected patients, and healthy older controls [11]. After the differential clustering segmentation process based on CDR and MMSE values, 24 scans of older subjects are categorized as Alzheimer' s affected out of 50 MR scan images.

4 Proposed Methodology

4.1 Intensity Normalization

Normalization of an image is the process of changing the range of pixel value. The intensities of all the input images were fixed in the range of [0–100], and the normalization was performed by the stated method [N]. By the help of this technique, the luminance and contrast of every tissue category are dependable across the taken training image dataset [12]. The linear intensity normalization can be written as follows:

$$I_N = (I - \text{Min})\frac{\text{newMax} - \text{newMin}}{\text{Max} - \text{Min}} + \text{newMin} \tag{1}$$

4.2 Laplacian of Windowing Filter

The Laplacian of windowing filter helps to measure 2D isotropic second-order spatial derivative of the taken input image. Mostly, the Laplacian is applied for edge preventing, and sudden intensity changes in the region of an image can be highlighted with this. Generally, when the Laplacian is applied for an image to be smoothened and it should be limited, because continues convolution process leads an image to blurring.

$$\nabla_\gamma^2 = (1 - \gamma)\nabla_5^2 + \gamma\nabla_\times^2 = (1 - \gamma) \begin{bmatrix} 0 & 1 & 0 \\ 1 & -4 & 1 \\ 0 & 1 & 0 \end{bmatrix} + \gamma \begin{bmatrix} \frac{1}{2} & 0 & \frac{1}{2} \\ 0 & -2 & 0 \\ \frac{1}{2} & 0 & \frac{1}{2} \end{bmatrix} \tag{2}$$

By using the above kernel, the value of Laplacian has been calculated with the help of standard convolution principle. Generally, the Laplacian kernel is very smaller than images and is very sensitive to noise.

4.3 Segmentation Using Patch Image Differential Clustering and Volume Calculation

Patch preselection and Patch Propagation

The sum of the squared difference (SSD) computation between every selected input image and its objective image has been chosen for ranking and selecting the first images. This measure selected for the reason of SSD is associated to make comparison between the patches depending upon the intensities. The patches of the entire subset of the images, which have been listed for the registration by the use of affine transformations, are preassigned with a regional similarity measurement which includes both the labeling and the intensities of the applicant patches into account. In this proposed research work, we have carried out contrast and luminance criteria to attain the best patch preselection [13]. Depending upon the first and second provisions of the renowned structural similarity measure (SSIM), the patch preselection process can be designed as follows:

$$ss = \frac{2\mu_i\mu_{s,j}}{\mu_i^2 \mu_{s,j}^2} \times \frac{2\sigma_i 2\sigma_{s,j}}{\sigma_i^2\sigma_{s,j}^2} \tag{3}$$

A conciliation between the performance analysis of patch image differential labeling technique which reduces the execution time and efficiency is to select the first 10 images that are most favorable to the objective image. In label propagation task, the analyzing has been performed regarding nonlocal patches-based approach. Patch propagation has been done with three different categories which are given as follows:

Label propagation based on graph weight

w_i denotes the weight of the graph. Let us assume the image analyzing domain Ω that is associated with each other voxel x of the taken input image I. And, the voxel of y with its respective weights are

$$w_{i(x,y)} = f\left(\frac{\sum_{x'\epsilon p_I(x),\, y'\epsilon p_{I_i}(y)}\left(I(x') - I_i(y')\right)^2}{2N\beta\hat{\sigma}^2}\right) \tag{4}$$

The principle label propagation has been implemented by using the graph set. $\{w_i\}_{i=1,...n}$ values indicate the local similarity between $\{I_i\}_{j=1...n}$ and I.

Label propagation based on pair similarity

Let the process pair-wise label starting with input image I and T which are represented as an anatomical image for the respective labeled image [14]. According to those above values, the patch image differential labeling technique is designed at the beginning in Eq. (4).

$$\forall x \epsilon \Omega,\, L(x) = \frac{\sum_{y\in N(x)} w(x,\,y)l(y)}{\sum_{y\in N(x)} w(x,\,y)} \tag{5}$$

L represents the each label's proportion, and $l(y)$ is the total number of labels.

Label propagation based on group similarity

This label propagation has been constructed depending upon the fuzzy labeling principle which is the similarity measure between the group labeled images of arbitrary numbers.

$$\forall x \epsilon \Omega,\, L(x) = \frac{\sum_{i=1}^{n} \sum_{y\in N(x)} w_i(x,\,y)l_i(y)}{\sum_{i=1}^{n} \sum_{y\in N(x)} w_i(x,\,y)} \tag{6}$$

The final segmentation process has been fixed as L. If any of one final label is considerable then it will not come under the pair-wise concept.

$$\forall x \epsilon \Omega,\, P_L(x) = \sum_{i=1}^{n} \sum_{y\in N(x)} w_i(x,\,y)p_{l_i}(y) \tag{7}$$

The average weight of the label is fixed as default with the help of similarity measure among the local patches.

The severity of AD can be classified by the radiologist according to changes in the volume of WM and GM. Previous works showed that the main reason of AD is reduction in GM due to brain atrophy [15].

$$\text{Volume}_{\text{WM}} = \sum_{\text{slice}=1}^{n} \sum_{i=1}^{x} \sum_{j=1}^{y} f(i,\,j) > \text{thres} \tag{8}$$

$$\text{Volume}_{\text{GM}} = \sum_{\text{slice}=1}^{n} \sum_{i=1}^{x} \sum_{j=1}^{y} f(i,\,j) == \text{thres} \tag{9}$$

5   Nearest Particle Interconnect Classifier

The various stages of AD can be categorized with the help of nearest particle interconnect classifier. The classifier has been designed with the help of the label characteristics which are related to brain atrophy changes in GM. Generally, the disease-related
decisions are taken according to the changes in brain atrophy which leads to the GM
reduction. The reduction in GM volume provides better information to radiologist
for identifying the AD. The labels and the volume of GM have been compared with
standard medical state of the art such as clinical dementia ratio (CDR) and Mini-
Mental State Examination (MMSE). From CDR value, 0.25 is normal and 0.73 is
abnormal behavior, and from MMSE value, 29.52 ± 0.5 is normal and 20.79 ± 2.58 is
classified as cognitive impaired.

6   Results and Discussion

From Fig. 1, 1.1(b) represents the enhanced output image with the process of intensity
normalization to get better enhancement for the image using Eq. (1) for the corresponding sagittal view of input image 1.1(a). Laplacian windowing filter process has
been performed by the suitable kernel mask using Eq. (2) for removing the additive
speckle noises and also performs smoothening for enhancing the pixel intensity of
the images. The labeling process has been implemented to the filtered image output
1.1(c). Three various strategies are implemented for labeling task which is based
on different labeling propagations in nonlocal patch. Structure similarity has been
calculated according to Eq. (3).

After the structure similarity calculation, the initial level of label propagation starts
with graph weight approach which is designed using the Eq. (4). Another type of
label propagation is depending upon the pair similarity measure by the use of Eq. (5).
The final label propagation is following group similarity-based approach which has
been processed with Eqs. (6) and (7). The above-mentioned various types of patch
propagation have been applied, and the output of labeled image was shown in 1.1(d).
There are three different types of labels that are designed for segmenting the various
brain matters like white matter (WM) and gray matter(GM) and for the skull region
also. Based on different intensity levels, the above brain matters are clustered, and
the higher intensity pixels are clustered as skull region which are presented in edge
of the image.

Table 1 shows the various features selection, and the classifier performance values
are measured with the help of above values. Four features were selected, and it has
been fed as input to the classifier. The classifier efficiency has been validated with
above-mentioned values, and it was compared with SVM classifier and proved that
the classifier performance is higher than it.

From Fig. 2, 2.1(b) shows the segmentation output of gray matter (GM) with skull
region, and fig. 2.1(c) shows the segmented output of white matter (WM) with skull

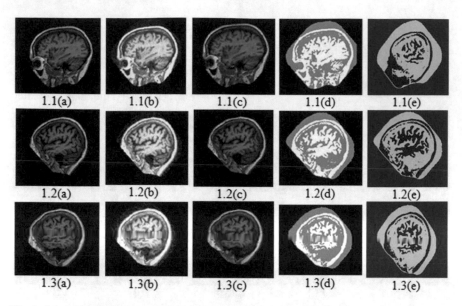

1.1(a)   1.1(b)   1.1(c)   1.1(d)   1.1(e)

1.2(a)   1.2(b)   1.2(c)   1.2(d)   1.2(e)

1.3(a)   1.3(b)   1.3(c)   1.3(d)   1.3(e)

Fig. 1 1.1–1.3(**a**) labeled images, 1.1–1.3(**b**) gray matter with skull, 1.1–1.3(**c**) white matter with skull, 1.1–1.3(**d**) segmented gray matter, 1.1–1.3(**e**) segmented white matter

Table 1 Estimated values of features

Features	Estimated values				
Cluster prominence	2.6521e+03	2.4477e+03	2.4722e+03	2.3564e+03	2.5672e+03
Correlation	0.8674	0.8624	0.8841	0.8765	0.8854
Entropy	0.5386	0.4818	0.4612	0.5762	0.4723
Energy	0.7252	0.7677	0.7711	0.7354	0.7522
Sensitivity	95.2381	95.2521	95.2318	95.1154	95.8564
Specificity	97.9167	97.5462	97.8624	97.9167	97.6524
Recall	95.2381	95.6452	95.2381	95.8564	95.2384
Accuracy	96.3964	95.5462	96.5562	96.2312	95.6423

region. Figure 2.1(d) shows the exact segmentation output of GM, and fig. 2.1(e) represents the segmented of WM. And, the volumes of WM and GM are estimated with the help of Eqs. (8) and (9). The above-mentioned processing steps are applied to the outputs 2.2(b–e) and 2.3(b–e).

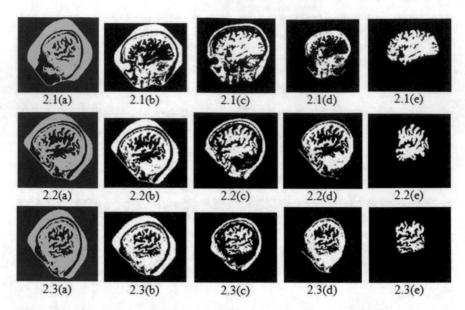

Fig. 2 2.1–2.3(**a**) Labeled images, 2.1–2.3(**b**) gray matter with skull, 2.1–2.3(**c**) white matter with skull, 2.1–2.3(**d**) segmented gray matter, 2.1–2.3(**e**) segmented white matter

7 Conclusion

The present study has been done with differential clustering using patch-based labeling principle, and this technique has been compared with PSO K-mean clustering and proved that it has provided better segmentation and accuracy, and execution time is also compared with state of the art. According to various protocols, the patch preselection task has been completed for achieving exact segmentation process of GM and WM. Volumetric information of GM and WM provides more information than previous work for staging the severity of AD, and the compression is also done with the normal subjects. So these results will be more useful to medical society for classifying the stages of AD. Nearest particle interconnect classifier provided the subject's NCI and MCI, and further work is going to deal with ensemble of classifiers for accurate staging of AD.

Acknowledgements The author would like to thank the Sir C.V. RAMAN KRISHNAN International Research Center for providing financial assistance under the University Research Fellowship. Also we thank the Department of Electronics and Communication Engineering of Kalasalingam University, Kalasalingam Academy of Research and Education, Tamil Nadu, India, for permitting to use the computational facilities available in Centre for Research in Signal Processing and VLSI Design which was set up with the support of the Department of Science and Technology (DST), New Delhi, under FIST Program in 2013 (Reference No: SR/FST/ETI-336/2013 dated November 2013).

References

1. Khanal B, Lorenzi M, Ayache N, Pennec X (2016) A biophysical model of brain deformation to simulate and analyze longitudinal MRIs of patients with Alzheimer's disease. NeuroImage 134:35–52
2. Tokuchi R, Hishikawa N, Sato K, Hatanaka N, Fukui Y, Takemotoa M, Ohta Y, Yamashita T, Abe K (2016) Age-dependent cognitive and affective differences in Alzheimer's and Parkinson's diseases in relation to MRI findings. J Neurol Sci 365:3–8
3. Mirzaei G, Adeli A, Adeli H (2016) Imaging and machine learning techniques for diagnosis of Alzheimer's disease. Rev Neurosci 27(8):857–870. https://doi.org/10.1515/revneuro-2016-0029
4. Liu M, Zhang D (2016) Relationship induced multi-template learning for diagnosis of Alzheimer's disease and mild cognitive impairment. IEEE Trans Med Imaging 35(6):1463–1474
5. Muneeswaran V, Rajasekaran MP (2016) Performance evaluation of radial basis function networks based on tree seed algorithm. In: 2016 International conference on circuit, power and computing technologies (ICCPCT). IEEE
6. Platero C, Tobar MC (2016) A fast approach for hippocampal segmentation from T1-MRI for predicting progression in Alzheimer's disease from elderly controls. J Neurosci Methods 270:61–75
7. Valverde S, Oliver A, Roura E, González-Villá S, Pareto D, Vilanova JC, Ramio-Torrent L, Rovira A, Llado X (2017) Automated tissue segmentation of MR brain images in the presence of white matter lesions. Med Image Anal 35:446–457
8. Beheshti I, Demirel H, Matsuda H (2017) Classification of Alzheimer's disease and prediction of mild cognitive impairment-to-Alzheimer's conversion from structural magnetic resource imaging using feature ranking and a genetic algorithm. Comput Biol Med 83:109–119
9. Liu M, Adeli E, Zhang D (2016) Inherent structure-based multiview learning with multitemplate feature representation for Alzheimer's disease diagnosis. IEEE Trans Biomed Eng 63:7
10. Zhan Y, Yao H, Wang P, Zhou B, Zhang Z, Guo YE, An N, Ma J, Zhang X, Liu Y (2016) Network-based statistic show aberrant functional connectivity in Alzheimer's disease. IEEE J Sel Top Signal Process 10(7):1182–1188
11. Govindaraj V, Murugan PK (2014) A complete automated algorithm for segmentation of tissues and identification of tumor region in T1, T2, and FLAIR brain images using optimization and clustering techniques. Int J Imaging Syst Technol 24(4): 313–325
12. Chatterje P, Milanfar P (2012) Patch-based near-optimal image denoising. IEEE Trans Image Process 21:4
13. Coupé P, Manjón JV, Fonov V, Pruessner J, Robles M, Collins DL (2011) Patch-based segmentation using expert priors: application to hippocampus and ventricle segmentation. NeuroImage 54:940–954
14. Rousseau F, Habas PA, Studholme C (2011) A supervised patch-based approach for human brain labeling. IEEE Trans Med Imaging 30:10
15. Rajesh Kumar P, Arun Prasath T, Pallikonda Rajasekaran M, Vishnuvarthanan G (2018) Brain subject estimation using PSO K-means clustering—an automated aid for the assessment of clinical dementia. In: Satapathy S, Joshi A (eds) Information and communication technology for intelligent systems (ICTIS 2017)—Volume 1. ICTIS 2017. Smart innovation, systems and technologies, vol 83. Springer

Verifiable Delegation for Secure Outsourcing in Cloud computing

Nalini Sri Mallela and Nagaraju Devarakonda

Abstract In distributed environment, to accomplish access privileges and maintain information as secret, the data proprietors could receive attribute-based encryption to encode the put-away information. Users with constrained-figuring power are, however, more prone to appoint the veil of the decryption task to the cloud servers to decrease the computing cost. During the consignment of the cloud servers, they can take the related encrypted text and may give it to the third-party people for any malicious activity; in addition to this, we have another problem of key generation of the simple attribute-based encryption (ABE). The existing ABE data outsourcing techniques are not capable of restricting the third-party people for accessing the encrypted text and outsourced decrypted text. These third-party issues should be addressed. We propose a new scheme with secure outsourced key generations and decryption by using "key-generation service center (KGSC)" and "decryption service center (DSC)" for secure data outsourcing system. We solve the problem in the existing system and also propose verifiable delegation for secure outsourcing. The unauthorized users are restricted to access the data onto achieving the data confidentiality. Our scheme also provides the fine-grain access control and allows the secure data outsourcing.

Keywords Secure data outsourcing · Attribute-based encryption
Outsourcing computation · Checkability · Verifiable delegation · Hybrid encryption

N. S. Mallela (✉) · N. Devarakonda
Lakireddy Bali Reddy College of Engineering, Mylavarm, India
e-mail: mallela.nalini@gmail.com

N. Devarakonda
e-mail: dnagaraj_dnr@yahoo.co.in

© Springer Nature Singapore Pte Ltd. 2019   685
A. J. Kulkarni et al. (eds.), *Proceedings of the 2nd International Conference on Data Engineering and Communication Technology*, Advances in Intelligent Systems and Computing 828, https://doi.org/10.1007/978-981-13-1610-4_69

1   Introduction

Attribute-based encryption (ABE) [1] has the capability of providing the data confidentiality and fine-grained access control with public key cryptography. In ABE the keys and corresponding ciphertext have been generated by the help of set attributes and access policies and any ciphertext anybody want to decrypt they should satisfy the attributes set and the defined access structure. In recent trend, cloud computing data sources should be managed effectively. In distributed environment, the cloud servers provide different data services like storage of the data and data outsourcing. The stored data can be accessed only by the authenticated users. Cloud servers restrict the unauthorized users to access the data. Only after proper authentication and checking only the users are allowed to access the data. For delegation computation, the cloud server should be able to provide the data required by the users according to the user request. Cloud computing maintains different types of information, maybe personal or organizations and business information in cloud with the help of the cloud servers [3]. We need secure outsourcing of the data on the comprised storage servers, instead of the secure servers. The group of the users should share the data securely. This can be achieved by the help of the ABE scheme. This makes the number of users to access the data on the storage server by following the defined secure policies. In cloud computing, data were shared based on the access policies and specially ciphertext policy (CP-ABE) [4, 5] and the verifiable delegation [6, 7] are utilized to provide the data confidentiality and the verifiability of delegation on comprised cloud servers. Two different forms of attribute-based encryptions are available. They are "key-policy attribute-based encryption" (KP-ABE) [8–10] and "ciphertext-policy attribute-based encryption" (CP-ABE). In a "KP-ABE" system, the keys are distributed by the key distribution center and not by the sender or the receiver; this scenario limits the achievability and applicability in the real time applications. In a "CP-ABE" system, the ciphertext is conjoined with the access policy and the access structure, and the set of keys are affiliated with the set of attributes used for defining the access structure. In an ABE system, for the access, policy circuits are added; it could be the better approach for policy description. Another attractive service provided by the cloud servers is delegation computing. Because of the delegation any untrusted cloud trying translate the actual cipher text to original plain text they did not get anything from that. Data outsourcing is the considerable issue [11, 12]. So, combine ABE scheme with the data outsourcing. Waters [2] combined ABE with data outsourcing with separate decryption scheme to make the decryption process easier and reduce the computing cost. After that, Parno [3] introduced a scheme ABE with verifiable outsourcing and decryption. Any user can register with his identity to store or send the data, and then the comprised servers can take the identity of user whom message they want to forge, because the identity is not containing any secret value. With this, they can easily capture the ciphertext related to that message. Only securing the identity related to that message is not sufficient. The cloud server should be capable of restricting the access of the user toward data stored in the servers.

2 Scheme Description

2.1 Procedure

Our scheme is implemented on integers. In delegation computing, the user applies the validation procedure to check the cloud servers that are giving corresponding encrypted text to decrypt immediately and efficiently. The ABE with outsourcing allows the decryption of outsourced data and generates the key for delegation. Here, we are introducing new policy controlled by the default set of attributes and connector is used for linking new policy and the user's policy. The keys required for the user's policy are generated by the key-generation service center (KGSC) [11] with the help of attribute authority. The private key of the user is masked by the key-blinding method. This blinded key will be given to the decryption service center (DSC) for partial decryption, and complete decryption will be done at the local center. The proposed scheme is able to provide the efficiency and the security, in case of attributes and the users. Our scheme ensures that the untrusted cloud was unable to learn the content from the encrypted text.

2.2 System Model

See Fig. 1 and Table 1.

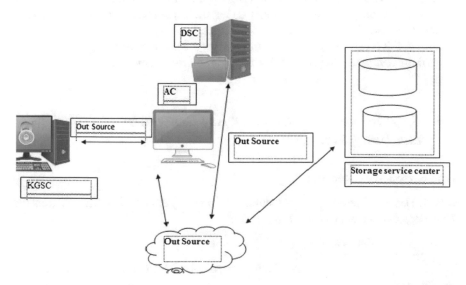

Fig. 1 Outsourced ABE model

Table 1 Notations

Role	Description
AC	Authentication center
KGSC	Key-generation service center
DSC	Decryption service center
SSC	Storage service center

- **Set up algorithm**: It generates the set of public key P_K and a master key M_K by considering the input V_s as a security parameter.
- **Keygen$_{init}$**: This is for requesting the private key of the user to generate the key for delegation computing. It results the key pair K_{KGSC}, K_{AC}. By considering V_s and access policy (attribute set) A_{key}, and the master key M_K.
- **Keygen$_{out}$** (A_{key}, K_{KGSC}): This algorithm generates key for the delegation and takes the input V_s, access structure (set of attributes) A_{key}, and K_{KGSC} key for KGSC. It results a partial transformation key TK_{KGSC}.
- **Keygen$_{in}$** (A_{key}, K_{AC}): It is a key-generation algorithm that results TK_{AC} (partial transformation key). By considering input V_s the access structure (or attribute set) A_{key}, and the key K_{AC} for authority.
- **KeyBlind** (TK): It takes the input V_s and the transformation key $TK = (TK_{KGSC}, TK_{AC})$. It results SK (private key) and a f (TK) (blinded transformation key).
- **Encrypt** (M, Text): It takes as input V_s message M and an access structure. It results the CT (ciphertext).
- **Decrypt$_{out}$** (CT, TK): It results the partially decrypted ciphertext CT part, if (Ikey, Ienc) = 1. The delegated decryption algorithm takes as input V_s ciphertext CT which was thought to be encrypted under the access structure and the blinded transformation key f(TK) for access structure.
- **Decrypt** (CTpart, SK): It results the message M. By considering input V_s, the partially decrypted ciphertext CT part, and the private key SK.

2.3 Steps Involved in Outsourced Key Generation and Decryption

Outsourced key generation

Step 1: User send request w to authentication center (AC) for keys. The AC runs the Keygen$_{init}$ and generates K_{AC}, K_{KGSC}

U   Request w   **AC**

$\longrightarrow$

AC Run keygen $_{init}$(w,M_K) to obtain K_{AC},K_{KGSC}

Step 2: Authentication center sends the user request w and K_{KGSC} to the KGSC to obtain TK_{KGSC}, then the key-generation service center sends TK_{KGSC} to the authentication center. The authentication center sends TK_{KGSC}, TK_{AC} to user

AC   w,K_{KGSC} **KGSC**

    ——————————▶ **KGSC Run keygen$_{out}$(w,K_{KGSC}) to obtain TK$_{KGSC}$**

Decryption

Step 1: Storage service center (SSC) sends ciphertext (CT) to the user (U)

   U   **CT**   **SSC**

   ◀——————————

Step 2: User request (w), ciphertext (CT),transformation key (TK) send to decryption service center (DSC). Then, DSC runs Decrypt$_{out}$ (CT, TK) to obtain CT$_{part}$

  DSC (w,CT,TK)   **U**

  ◀——————————

               Run Decrypt$_{out}$(CT,TK) to obtain CT$_{part}$

Step 3: User (U) runs Decrypt (CT$_{part,}$ SK) to obtain M

3 Hybrid VD-CPABE Scheme

3.1 System Model

See Fig. 2 and Table 2.

  Our Verifiable Delegation we use Hybrid VD-CPABE with set of algorithms

- Setup (S_p, n): Authority executes the algorithm. It results the set of parameters PK and MK which are kept secret, this algorithm takes as input a security parameter S_p, the number of attributes n.

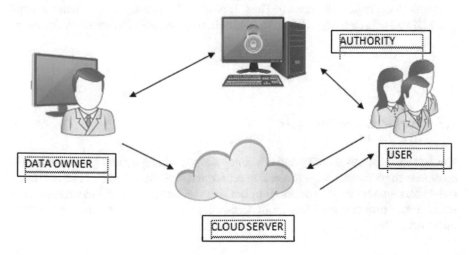

Fig. 2 System model

Table 2 Role description

Role	Description
Authority	Who checks the authentication details
Data owner	The data files are uploaded into the cloud in encrypted form
User	Who are trying to access the data from the cloud
Cloud server	Data files are stored

- Hybrid-Encrypt (PK, M, f): Data owner executes the algorithm by two parts—key encapsulation mechanism and authenticated symmetric encryption.
- The "KEM" algorithm takes parameters PK and an access structure f. It computes f' and chooses a R (unique random string). Then, it generates $K_M = (dk_m, vk_m)$, $K_R = (dk_r, vk_r)$, and the CP-ABE ciphertext (CK_M, CK_R).
- The "AE" algorithm takes as input a message M, the random string R, and the symmetric key K_M and K_R. Then, it results the ciphertext (C_M, C_R, S_{ID}, vk_m(CM‖CR), S_{ID}, vk_r (C_M‖C_R)).
- The ciphertext CT (CK_M, CK_R, C_M, C_R, S_{ID}, vk_m(C_M‖C_R), S_{ID}, vk_r (C_M‖C_R)).
- KeyGen (MK,$r \in \{0,1\}^n$): The user's private keys are generated by the authority. This algorithm accepts input MK (master key) and x (bit string). It results a private key SK and a transformation key TK.
- Transform (TK, CT): This algorithm considers the input encrypted TK (transformation key) and CT (ciphertext) with respect to f and f'. It results the fractionally decrypted ciphertext CT' = (CK'_M, C_M, C_R, S_{ID}, vk_m(C_M‖C_R)) or CT' = (CK'_R, C_M, C_R, S_{ID}, vk_r (C_M‖C_R)). This algorithm was executed by the cloud servers.
- Verify-Decrypt (SK, CT'): This considers inputs the SK (secret key) and the CT' (partially decrypted ciphertext). Then, it outputs the message M_b which satisfies that if f(x) = 1, then M_b = M, and if f(x) = 0, then M_b = R. Executed by the users

4 Results

4.1 Performance Analysis

Our scheme achieves security and efficiency in outsourcing by using verifiable delegation with outsourced key-generation center and decryption center. The unauthorized users are restricted to access the data onto achieving data confidentiality. Our scheme also provides the "fine-grain access control" and allows the "secure data outsourcing."

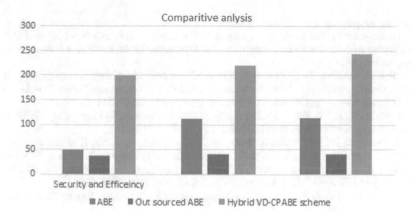

5 Conclusion

We have considered issue is amid designation the server may alter or supplant the assigned figure message and react a produced registering comes about with malignant expectation. They may likewise cheat the qualified clients by reacting them. This problem is overcome by using verifiable-delegation model. It does not allow to cheat authorized user, because the data owner maintains a secret key; without having the key, user cannot access the data. This verifiable-delegation model provides security and does not allow unauthorized user. Here, also we present verifiable delegation and encryption mechanism with our hybrid encryption, and we could delegate the verifiable decryption paradigm to the server.

6 Future Scope

The future scope of this paper is for security, KEM consolidates with the IND-CCA (Indistinguishability under particular picked figure content attack) secure confirmed encryption plot which yields our IND-CPA (Indistinguishability under specific picked plaintext assault) secure mixture VD-CPABE conspire. In future, we additionally actualize multi-direct decisional Diffie–Hellman suppositions.

References

1. Lai J, Deng RH, Guan C, Weng J (2013) Attribute-based encryption with verifiable outsourced decryption. IEEE Trans Inf Forensics Secur 8(8):1343–1354
2. Waters B (2011) Ciphertext-policy attribute-based encryption: an expressive, efficient, and provably secure realization. In: Proceeding of 14th international conference practice theory

public key cryptography, pp 53–70

3. Parno B, Raykova M, Vaikuntanathan V (2012) How to delegate and verify in public: verifiable computation from attribute-based encryption. In: Proceeding of 9th international conference theory cryptographic, pp 422–439

4. Yamada S, Attrapadung N, Santoso B (2012) Verifiable predicate encryptin and applications to CCA security and anonymous predicate authentication. In: Proceeding of international conference practice theory public key cryptography conference, pp 243–261

5. Han J, Susilo W, Mu Y, Yan J (2012) Privacy-preserving decentralized key-policy attribute-based encryption. IEEE Trans Parallel Distrib Syst 23(11):2150–2162

6. Garg S, Gentry C, Halevi S, Sahai A, Waters B (2013) Attribute based encryption for circuits from multilinear maps. In: Proceeding of 33rd international cryptology conference, pp 479–499

7. Gorbunov S, Vaikuntanathan V, Wee H (2013) Attribute-based encryption for circuits. In: Proceeding of 45th annual ACM symposium theory computing, pp 545–554

8. Sahai A, Waters B (2005) Fuzzy identity based encryption. In: Proceeding of 30th annual international conference theory applications cryptographic techniques, pp 457–473

9. Goyal V, Pandey O, Sahai A, Waters B (2006) Attribute-based encryption for fine-grained access control of encrypted data. In: Proceeding of 13th ACM conference computer communication security, pp 89–98

10. Cramer R, Shoup V (1998) A practical public key cryptosystem provably secure against adaptive chosen ciphertext attack. In: Proceeding of 18th international cryptology conference, pp 13–25

11. Li J, Huang X, Li J, Chen X, Xiang Y (2013) Securely outsourcing attribute-based encryption with checkability. IEEE Trans Parallel Distrib Syst 25(8):2201–2210

12. Hur J, Noh DK (2011) Attribute-based access control with efficient revocation in data outsourcing systems. IEEE Trans Parallel Distrib Syst 22(7):1214–1221

Eradication of Rician Noise in Orthopedic Knee MR Images Using Local Mean-Based Hybrid Median Filter

C. Rini, B. Perumal and M. Pallikonda Rajasekaran

Abstract Medical imaging is a historical achievement to expose the patient's internal functions or internal problems are seen in the external X-ray, MRI, and CT images; it gives more efficient and accurate results for the physicians to start further treatments to the patients. The medical field requires for new techniques like reduce the complications, improve the image quality, accuracy, real-time output, allow to early detection of the disease and reduce the human era. Different kind of medical images are available that are affected by various types of noise, and thus the various types of noise removal techniques are available to remove the noise. Noises are affecting the quality of real-time output. In the proposed method, these problems were addressed efficiently which in turn will enable real time procedures. The hybrid median filter (HMF) is used to remove the Rician noise. Normally, HMF was used to remove the impulse noise. This paper recommend a novel approach for denoising the magnetic resonance images that propose the both bareness and self-resemblance properties of MR images. Noise reduction is important issue for further visual examination for MR images. The hybrid median filter improves the image quality, preserves the edges, and also reduces the effect due to Rician noise. The visual analysis and diagnostic quality of the magnetic resonance images are preserved. The quantitative measurements based on the standard metrics like AD, MSE, PSNR, LMSE, SSIM shows that the recommended method is better than the other denoising methods for magnetic resonance images.

Keywords Rician noise · Median filter · MR images

C. Rini (✉) · B. Perumal · M. P. Rajasekaran
Department of Electronics and Communication Engineering, Kalasalingam University,
Krishnankoil 626126, Tamil Nadu, India
e-mail: rinisuresh2006@gmail.com

B. Perumal
e-mail: palanimet@gmail.com

M. P. Rajasekaran
e-mail: m.p.raja@klu.ac.in

© Springer Nature Singapore Pte Ltd. 2019
A. J. Kulkarni et al. (eds.), *Proceedings of the 2nd International Conference on Data Engineering and Communication Technology*, Advances in Intelligent Systems and Computing 828, https://doi.org/10.1007/978-981-13-1610-4_70

1 Introduction

Magnetic resonance imaging (MRI) is a very efficient and powerful diagnostic tool. Prior to segmentation process, the following steps could be adopted; they are 1. image acquisition, 2. image denoising or preprocessing, 3. image compression, 4. image transmission, 5. image reproduction phases of processing. MR images are usually exaggerated with quite a few intricacies such as inferior peak signal-to-noise ratio (PSNR) and different oblique repose time values in overlying resonances [1]. At the course of the image attainment process for the reason that of the low PSNR, one of the appalling situations is Rician noises that greatly reduce image quality and reduce quantitative measurements for further clinical assessment [2, 3]. Averaging the MR signals for the purpose of increase the PSNR and reduces the Rician noise, but this is not a regular process in real-time clinical circumstances, since these progressions raise the acquirement instance of MRI which confines their exploitation in many circumstances where long imitations are not practicable, such as for unbalanced biological components [4]. There is an inherent negotiation sandwiched between high PSNR and illustration quality (resolution) of MR images, acquiring an elevated resolution image with a preferred PSNR augments the acquisition time of MRI. Therefore, subsequent to acquirement, pre-/post-processing, denoising, and enhancement system are appropriate substitute to take away the noise and enhance the accurateness of real-time clinical system. Usually in image data, noise can be Gaussian or Rician dispersed depending upon the nature of an image (complex or magnitude data). In this paper, the magnitude MR data has been investigated. Noise assessment is usually made over magnitude MR images since this is accustomed output of the scanning [5].

Magnetic Resonance Imaging (MRI) technique is a non-obtrusive medical test to investigate and medicate the problems in the knee bone cartilage. It uses a persuasive magnetic field with the pulses of radio frequency waves to procreate the elaborate view of the knee bone and even soft tissues of the trabecular bone. This is done in order to make the evaluation of the interior parts of the cartilage bone easier for the medics. It is the most hypersensitive method of testing the cartilage bone, especially the knee bone. On contrary to the other traditional X-ray methods and scanning techniques such as computed tomography (CT), MRI is a safer method that does not employ ionizing radiation. Hence, it does not make any chemical changes in the tissues of the human body. The images obtained from these scans can be examined from various viewpoints. Also, the femur, tibia, articular cartilage, ligament, and fibula of the knee bone can be easily differentiated with the help of MRI images rather than the CT images.

2 Previous Works

A fully automated system can be made in order to find the osteoarthritis in the knee bone [6]. For this detection, every detail of the knee bone image should be made clear. This can be done mainly by expatriating the noises in the images. He et al. [7] proposed a noise eradicate process for non-local maximum likelihood (NLML) based on NL-means algorithms. An initial assumption such that the pixels in the image has similar distribution with its neighborhood was made. A simulator MR magnitude data was used in this research for comparative analysis.

A MRI denoising strategy based on generalized total variation method adopted with spatially adaptive parameters was designed by Liu et al. [8]. A spatially varying noise map was used in the proposed method. This work can also further be elaborated by introducing the NC-distribution and discrete nature of noise. As presence of noise in MRI image is executed with spatial domain, the MRI denoising approach based on non-local mean values are derived from similarity of fuzzy information which was presented by Sharif et al. [9]. Homogeneous pixels were always considered for measuring the non-local mean in relevant fuzzy information.

2.1 Characteristics of Rician Noise

Consider the values A_r and A_i, the real and imaginary data that has been corrupted with Gaussian noise with mean having the value of zero, stationary noise with σ as the value of standard deviation. Then the PDF of the magnitude image data resolved is Rician distributed in nature [1].

$$P_{\text{mag}}(M) = \frac{M}{\sigma^2} e^{-\frac{M^2+A^2}{2\sigma^2}} I_0\left(\frac{AM}{\sigma^2}\right) \tag{1}$$

Such that

I_0 is the zero order modified Bessel function and

M the pixel variable of magnitude image given by $M = \sqrt{A_r^2 + A_i^2}$ and

A is given by $A = \sqrt{\mu_{A_r}^2 + \mu_{A_i}^2}$.

The corresponding second moment of Rice PDF is given by

$$E[M^2] = A^2 + 2\sigma^2 \tag{2}$$

When the signal-to-noise ratio approaches zero, the Rician distribution seems to be Rayleigh distributed and when the SNR is high, it seems to be Gaussian in nature. The shape of the Rician distribution can be inferred from the value of A/σ.

Rician noise damages the quality and assessable sense of the images and also affects analysis and feature exposure [10]. To reduce the Rician noise, use filters and compare the low SNR for MR images. Estimate the Rician noise from the image, and standard metrics are used to compare the noise ratio. Rician noise affects the MR images in the shapes and locations of defected area [11]. The clinician was not able to visually diagnose the problem clearly.

3 Existing Noise Expatriation Approaches

Image denoising is a very important task of image processing. It can be used as a main process or as one of the constituent of other processes. Even though there are exists so many methods of elimination of noise, it is still a troublesome job to find the approach that is compatible and applicable to the specific application [12, 13]. In this paper, approaches such as Gaussian filter, Wiener filter, Lee filter, Frost filter, median filter, and hybrid median filter are discussed with the images obtained from the MRI.

3.1 Gaussian Filter

Gaussian filter is also erosive in nature, and so it does not cause high-frequency antiquities. It is contemplated as the classic filter of time domain. Also, its support in the realm of time is equivalent to that of their support in the frequency. It does not have a sharp cutoff at some passband frequencies since the Gaussian filter has its Fourier transform as a Gaussian.

3.2 Wiener Filter

Wiener filtering is the method in which the optimal trade-off between the inverse filters and noise smoothening, especially in terms of the mean square error (MSE). It is a superlative frame of reference. The Wiener filtering can be done in such a way to produce the output by the linear time-invariant filtering of the images. It is a restoration filtering technique.

3.3 Lee Filter

Lee filter was derived from the minimum square error (MSE) criteria by introducing the local statistics method in it. It was believed that it can reduce Rician noise while preserving edges. But the Lee filter does not reduce the Rician noise compared to the values of standard metrics with other filters.

3.4 Frost Filters

Frost filters derived from minimum square of error (MSE). The filter kernel value varies with the local statistical value of the images. The filtered images are not clear or not reduced the noise compared to other filters.

3.5 Median Filter

Mcdian filters arc a most popular nonlinear filter and are estimated from maximum likelihood (ML). The median filters are mainly used for preserve the edges and reduce the noises with robustness against impulsive noise.

The outputs of a median filter, an odd number of sample values, are sorted, and the median value is used as the filter output. It is used to filter both past and future values for analyze the current point. However, the median filter is not a correct filter option to reduce the Rician noise. It can cause edges but it will remove the important details in the image signals. This is mainly used for ranking the input data information and deletes the original temporal order information. Thus for an unchanging signal affected by noise, median filters can reduce the noise poorly, but it will preserve the edges. In order to save the information about both rank and temporal input data, so that median hybrid filter (HMF) or hybrid median filter has been developed.

4 Proposed Hybrid Median Filter

Hybrid median filter (HMF) is a most effective median filter competent of removing Rician noise which is preserve edges. HMF has three steps of operation for convolution with 3×3 local window of the image. The primary process of HMF collects class and revives the median value of the center pixels of horizontal and perpendicular. After this process, the median (MD) of the diagonal pixels was designed. Finally, depending on the values calculated above central pixel is arranged systematically and its median replaces the central pixel. This process is recurred for one pixel on the next 3×3 windows moved over.

In the previous section mentioned, the different type's filters and their functions. Hybrid median filter preserves the edges for better and used for visual analysis, but processing times are high. Real-time procedures are preventing by high completion time. The other Rician noise filters revealed provide faster denoises processing but the MR image edges are blurred and hide the valuable image information a clinician cannot see the defects clearly (Figs. 1 and 2).

Fig. 1 Existing hybrid median filter scheme

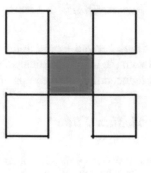

$$\begin{pmatrix} A & B & C \\ D & E & F \\ G & H & I \end{pmatrix}$$

$$h_1 = median(A, E, I)$$
$$h_2 = median(C, E, G)$$
$$h_3 = E$$

Fig. 2 Proposed modified hybrid median filter scheme

$$\begin{pmatrix} A & B & C \\ D & E & F \\ G & H & I \end{pmatrix}$$

$$h_1 = median(A, E, I)$$
$$h_2 = median(C, E, G)$$
$$h_3 = median(A, B, C, D, E, F, G, H, I)$$

5 Results and Discussion

In order to weigh against the MR image output of the projected method, denoised image and real-time images or data sets of normal knee bone real-time MR images have been used which are obtained from government hospital. Denoising tasks are carried out using MATLAB 7.9.0. MR images are affected by Rician noise, and functional ability of refurbishment results is analyzed. Out of the real data sets collected, three input images are shown in Figs. 3 and 4. Input image 1 is the coronal view of right knee of a patient. Input image 2 is the sagittal view of right knee of a patient of age 31. Input image 3 is the coronal view of knee of a patient. The anticipated scheme

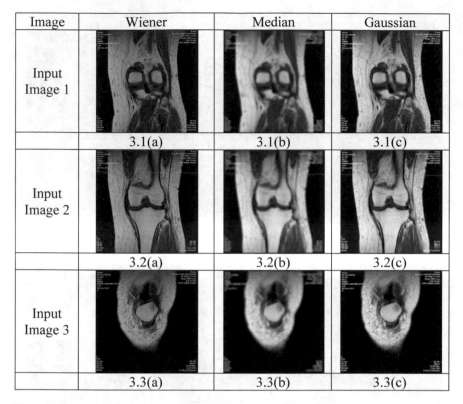

Image	Wiener	Median	Gaussian
Input Image 1	3.1(a)	3.1(b)	3.1(c)
Input Image 2	3.2(a)	3.2(b)	3.2(c)
Input Image 3	3.3(a)	3.3(b)	3.3(c)

Fig. 3 Filtered output images using Wiener, median, and Gaussian filter

has been judged against with existing techniques of denoising filter such as median filter, Wiener filter (WF), Lee filter, Frost filter, and hybrid median filter. The filtered output image using Wiener filter is shown in 3.1(a), 3.2(a), and 3.3(a), median filter output is shown in 3.1(b), 3.2(b), and 3.3(b), and output of Gaussian filter is shown in 3.1(c), 3.2(c), and 3.3(c) of Fig. 3, respectively. The filtered output image using Lee filter is shown in 4.1(a), 4.2(a), and 4.3(a), Frost filter output is shown in 4.1(b), 4.2(b), and 4.3(b) and output of proposed modified hybrid median filter is shown in 4.1(c), 4.2(c), and 4.3(c) of Fig. 4, respectively.

During the course of experimentation, the parameters using the proposed scheme are exposed in Table 1 and these parameters are accustomed empirically for denoising MR images. Mean square error (MSE), root mean squared error (RMSE), peak signal-to-noise ratio (PSNR) [14], and structural comparison index measure (SSIM) were used as qualitative evaluation metrics, and after the parameter comparison, the hybrid median filter is a better filter for removing the Rician noise.

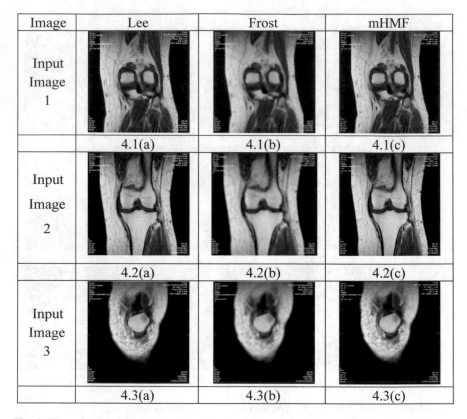

Image	Lee	Frost	mHMF
Input Image 1	4.1(a)	4.1(b)	4.1(c)
Input Image 2	4.2(a)	4.2(b)	4.2(c)
Input Image 3	4.3(a)	4.3(b)	4.3(c)

Fig. 4 Filtered output images using Lee, Frost, and proposed modified hybrid median filter

Table 1 Evaluation of filters with quality metrics

S. no.	Filter\Metrics	AD	MSE	PSNR	LMSE	SSIM
Input image 01	Wiener filter	0.111	0.024	16.238	2.824	0.812
	Mean filter	0.053	0.009	20.698	3.035	0.573
	Gaussian filter	0.025	0.002	26.307	2.809	0.88
	Lee filter	0.026	0.002	26.83	2.715	0.854
	Frost filter	0.042	0.005	23.144	2.946	0.67
	mHMF filter	0.016	0.002	27.322	1.866	0.891
Input image 02	Wiener filter	0.128	0.03	15.3	2.844	0.808
	Mean filter	0.055	0.009	20.499	3.073	0.563
	Gaussian filter	0.026	0.002	26.35	2.828	0.875
	Lee filter	0.027	0.002	26.914	2.726	0.846
	Frost filter	0.042	0.005	23.075	2.979	0.664
	mHMF filter	0.016	0.002	27.392	1.886	0.885

6 Conclusion

We designed a new modified hybrid median filter scheme for removing Rician noise. Successful performance indicators such as PSNR, MSE, AD, LMSE, and SSIM values were used in the study. We experienced the hybrid median filter-based denoising scheme for noisy MRI clinical images of both right and left knees in coronal and sagittal images. The benefit of this scheme is that it is easier to put into practice and its obtaining improved results than the filters in the literature. In our future study, we focus to analyze the effect of denoising in the segmentation of knee and cartilage segmentation and find its usefulness in increasing image quality and also to analyze the use of different soft computing techniques in removing Rician noise.

Acknowledgements The authors thank Dr. Suresh Ramasamy for supporting the research by providing orthopedic knee MR images and essential patient information. Also we show gratitude to the Department of Electronics and Communication Engineering of Kalasalingam University, Tamil Nadu, India, for authorizing to use the computational amenities accessible in Center for Research in Signal Processing and VLSI Design that was set up with the support of the Department of Science and Technology (DST), New Delhi, under FIST Program in 2013 (Reference No: SR/FST/ETI-336/2013 dated November 2013).

References

1. Basu S, Fletcher T, Whitaker R (2006) Rician noise removal in diffusion tensor MRI. In: Larsen R, Nielsen M, Sporring J (eds) Medical image computing and computer-assisted intervention—MICCAI 2006. MICCAI 2006. Lecture Notes in Computer Science, vol 4190. Springer, Berlin, Heidelberg
2. Muneeswaran V, Pallikonda Rajasekaran M (2017) Analysis of particle swarm optimization based 2D FIR filter for reduction of additive and multiplicative noise in images. In: Arumugam S, Bagga J, Beineke L, Panda B (eds) Theoretical computer science and discrete mathematics. ICTCSDM 2016. Lecture Notes in Computer Science, vol 10398. Springer, Cham
3. Muneeswaran V, Pallikonda Rajasekaran M (2018) Beltrami-regularized denoising filter based on tree seed optimization algorithm: an ultrasound image application. In: Satapathy S, Joshi A (eds) Information and communication technology for intelligent systems (ICTIS 2017)—Volume 1. ICTIS 2017. Smart innovation, systems and technologies, vol 83. Springer, Cham
4. Manjon JV, Coupe P, Buades A (2015) MRI noise estimation and denoising using non local PCA. Med Image Anal. https://doi.org/10.1016/j.media.2015.01.004
5. Dam EB, Lillholm M, Marques J, Nielsen M (2015) Automatic segmentation of high- and low-field knee MRIs using knee image quantification with data from the osteoarthritis initiative. J Med Imaging (Bellingham) 2(2). https://doi.org/10.1117/1.jmi.2.2.024001. Epub 20 Apr 2015
6. Dodin P, Martel-Pelletier J, Pelletier JP et al (2011) A fully automated human knee 3D MRI bone segmentation using the ray casting technique. Med Biol Eng Comput 49:1413–1424. https://doi.org/10.1007/s11517-011-0838-8
7. He L, Greenshields IR (2009) A nonlocal maximum likelihood estimation method for Rician noise reduction in MR images. IEEE Trans Med Imaging 28(2):165–172. https://doi.org/10.1109/TMI.2008.927338
8. Liu H, Yang C, Pan N, Song E, Green R (2010) Denoising 3D MR images by the enhanced non-local means filter for Rician noise. Magn Reson Imaging 28:1485–1496
9. Sharif M, Hussain A, Jaffar MA et al (2016) Fuzzy-based hybrid filter for Rician noise removal. Signal Image Video Proc SIViP 10(2):215–224. https://doi.org/10.1007/s11760-014-0729-1

10. Liu RW, Shi L, Huang W, Xu J, Yu SCH, Wang D (2014) Generalized total variation-based MRI Rician denoising model with spatially adaptive regularization parameters. Magn Reson Imaging 32:702–720
11. Chen L, Duan J, Chen CLP (2016) Rician noise removal in MR imaging using multi-image guided filter. In: 2016 3rd International conference on informative and cybernetics for computational social systems (ICCSS), Jinzhou, pp 180–184. https://doi.org/10.1109/iccss.2016.7586446
12. Riji R, Rajan J, Sijbers J et al (2015) Iterative bilateral filter for Rician noise reduction in MR images. SIViP 9(7):1543–1548. https://doi.org/10.1007/s11760-013-0611-6
13. Soyel H, Yurtkan K, Demirel H, McOwan PW (2016) Brain MR image denoising for Rician noise using intrinsic geometrical information. In: Abdelrahman O, Gelenbe E, Gorbil G, Lent R (eds) Information sciences and systems. Lecture Notes in Electrical Engineering, vol 363. Springer, Cham
14. Vishnuvarthanan G, Rajasekaran MP, Subbaraj P, Vishnuvarthanan A (2016) An unsupervised learning method with a clustering approach for tumor identification and tissue segmentation in magnetic resonance brain images. Appl Soft Comput 38:190–212

Segmentation of Tumor Region in Multimodal Images Using a Novel Self-organizing Map-Based Modified Fuzzy C-Means Clustering Algorithm

S. Vigneshwaran, G. Vishnuvarthanan, M. Pallikonda Rajasekaran and T. Arun Prasath

Abstract The significant issue for image segmentation that performs in the realization and analysis of the tumor and lesion region in multimodal such as CT and MRI images lies in the computational time and accuracy values. Recently, tumor region extraction from medical image that executes with quick response of evolution process will result in the aid of clinical surgical application. The automatic process can be ensured from unsupervised image segmentation algorithm, as it provides the clear identification of the tissue and lesion region in CT and MRI image. In specific, the unsupervised image segmentation process comprises of self-organization map (SOM)-based Modified Fuzzy C-Means Clustering (MFCM) algorithm that results in exact tumor identification and clear segmentation of tumor involved in organ such as liver, lung, brain, and thorax region. The proposed SOM-based Modified Fuzzy C-Means Clustering algorithm is an approach that refers to enhance the image quality measures such as mean squared error (MSE), peak signal-to-noise ratio (PSNR), Jaccard index, and dice overlap index (DOI). Modified Fuzzy K-Means (MFKM) algorithm and the self-organization map (SOM) based fuzzy K-Means (FKM) algorithm are evaluated, and it was finalized that better results are obtained from SOM-based Modified Fuzzy C-Means Clustering algorithm.

Keywords Tissue segmentation · Tumor identification · Lesion segmentation SOM-based MFCM algorithm

S. Vigneshwaran (✉) · G. Vishnuvarthanan · M. Pallikonda Rajasekaran · T. Arun Prasath
Kalasalingam Academy of Research and Education, Krishnankoil, Virudhunagar 626126,
Tamil Nadu, India
e-mail: s.vigneshwaran@klu.ac.in

G. Vishnuvarthanan
e-mail: g.vishnuvarthanan@klu.ac.in

M. Pallikonda Rajasekaran
e-mail: m.p.raja@klu.ac.in

T. Arun Prasath
e-mail: t.arunprasath@klu.ac.in

© Springer Nature Singapore Pte Ltd. 2019
A. J. Kulkarni et al. (eds.), *Proceedings of the 2nd International Conference on Data Engineering and Communication Technology*, Advances in Intelligent Systems and Computing 828, https://doi.org/10.1007/978-981-13-1610-4_71

1 Introduction

In early year, tumor disease is diagnosed through the CT and MRI image, which are mostly helpful for radiotherapy treatment and medical research. The most segmentation method was performed upon graylevel information process, in which similar intensities of several organs make the tumor region complicated to identify. Kohonen [9] introduced an unsupervised approach of SOM neural network with automatic topological mapping process, which provides one- or two-dimensional reduction using competitive learning approach. So far, the image segmentation process offered by SOM neural network tends to be better attractive, significantly due to the topological mapping process. Jiang [7] suggested an automatic classification process of SOM neural network by establishing feature vector (color, shape, texture, etc.) that offers assistance of color image segmentation process. Ong [10] proposed fixed order of two-stage classification techniques that uses SOM neural network in image segmentation process, and it is found that noise removal process for unsupervised approach cannot be offered. Khan [8] employed SOM neural network that instantaneously performs with fuzzy clustering approach resulting in an automatic optimal cluster center. This system also gets assistance from trained network to get offered with dimensionality reduction processes during unsupervised segmentation. The segmentation of SOM with fuzzy clustering approach is compared with efficient graph (EG)-based segmentation approach, in terms of center initialization process. The research group headed by Halder [6] recommended an automatic classification process prepared with SOM neural network, modal analysis, and mutational agglomeration. This segmentation approach relies on mapping process and employed with dimensional reduction, and the quality of classification was quite low. Torbati et al. [14] suggested the 2D discrete wavelet transform that performs with moving average SOM (MASOM) algorithm to achieve efficient segmentation, as it ascertains the least participation of human interaction in the modification of the variables that instigate segmentation process. The multilayer clustering process with a novel hierarchical self-organizing map (HSOM) and vector quantization process achieves efficient classification upon the input image [2]. To solve the time efficiency problem, Gular [5] introduced a novel automatic SOM approach that uses the pixel intensity characteristics of input image for effective segmentation. Hybrid approach utilizing automatic process of image segmentation in tissue and segmentation of tumor region has been proposed by Ortiz. Subsequently, this fully automatic process of image segmentation utilizes hybrid SOM and genetic algorithm that rely on feature vector such as entropy and gradient for clustering MR brain images [11]. Vishnuvarthanan et al. [15] recommended SOM-based FKM algorithm that provides brain tumor identification and segmentation of MR input image. This methodology serves to be an automatic process, and it fuses clustering with self-organizing map to attain effective segmentation of MRI images. The major hindrance of this approach during the segmentation process is the more consumption of computational time.

2 Proposed Method

Combination of SOM and MFCM algorithms can be of potentially important for the automatic unsupervised process in image segmentation. Figure 1 exposes the execution process of SOM-based MFCM algorithm. The initial process for SOM-based MFCM algorithm includes clustering vector followed by the mapping and classification process. This process was extensively executed for the feature extraction in terms of mean and standard deviation of the input image [15]. The significant process of feature extraction is adopted to overcome the over fitting problem. Thus, the mapping process uses 8×8 local window to support the membership function acquisition from the nearest neighbor. The SOM prototype w_x will be established from the minimum Euclidean distance with nearest neighbor pixel. Let $X \epsilon J^r$, where X represents the input vector and the input data is indicated as J^r.

$$D_{\min}(z) = \min\left\{ \sum_j \left(x_j(z) - w_j(z) \right)^2 \right\}$$ (1)

From Eq. (1), the nearest minimum neighbor pixel was obtained such that $x \epsilon X$, where $x(z)$ refers to the input vector at instants of time 'n' and $w(z)$ represents the sample vector function, which are nearer to the input vector and associated with the membership function value. In addition, from the nearest neighbor values, input vector and weight function, best matching unit (BMU) of nearest neuron was found. But, the trained network is still initiated by the input vector and it is updated with BMU from the nearest neighbor within time, which relies on exponential decay function. Thus, updating SOM prototype is expressed as:

$$w_x(z+1) = w_x(z) + \alpha(z) * (i - w_x(z))$$ (2)

Fig. 1 Flowchart of SOM-based MFCM algorithm

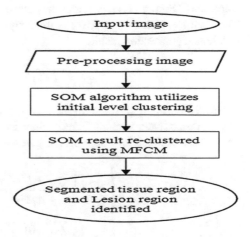

Here, $\propto (z)$ refers to the learning factor of exponential decay at instants of time 'z'. In order to provide the efficient classification process of input information, the updated SOM prototype is used. The updated SOM prototype provides two-dimensional information for classification process. MFCM algorithm receives inputs from the updated SOM prototype for the entire clustering process. The standard MFCM objective function introduced by Ahmed et al. [1] for clustering process is describes as

$$J(U, M) = \sum_{i=1}^{c} \sum_{c=1}^{N} (u_{ic})^m \|x_c - v_i\|^2$$

$$+ \frac{\propto}{N_R} \sum_{i=1}^{c} \sum_{c=1}^{N} (u_{ic})^m \left(\sum_{x_R \in N_k} \|x_R - v_i\|^2 \right) \qquad (3)$$

Let N_k refers to the nearest neighbor function that corresponds to each other of x_c and N_R. The parameter refers to the support for assessment of the nearest neighbor that provides the effective segmentation result. In that process, the parameter $\propto$ value is realized from 0 to 100. The value of parameter $\propto$ in clustering process must be chosen as high as possible due to its inversely proportional characteristic with the SNR value. Chen and Zhang [4] proposed the advance clustering process for finding the nearest neighbors and also provided the lowest processing time. In order to achieve the objective function, the term $\|\bar{x}_c - v_i\|^2$ was substituted in Eq. (3) instead of $\frac{1}{N_R} \sum_{x_R \in N_k} \|x_R - v_i\|^2$. Then, the objective function of MFCM algorithm becomes

$$J(U, M) = \sum_{i=1}^{c} \sum_{c=1}^{N} (u_{ic})^m \|x_c - v_i\|^2 + \propto \sum_{i=1}^{c} \sum_{c=1}^{N} (u_{ic})^m \|\bar{x}_c - v_i\|^2 \qquad (4)$$

Thus, the SOM prototype provides the automatic cluster center for clustering process. The final segmentation process is done by MFCM algorithm, when provided the efficient segmentation result.

$$\min_{(U,V)} J(U, M) \qquad (5)$$

where '$U = (u_{i,c})_{k,N}$' refers to the membership matrix, '$V = (v_1, v_2, \ldots v_c)$' refers to the cluster center vectors. Finally, they update the values of membership function and cluster center, which supports of given equation is

$$u_{ic} = \left\{ \sum_{l=1}^{N} \left[\frac{\|x_c - v_i\|^2 + \propto \|\bar{x}_c - v_i\|^2}{\|x_c - v_l\|^2 + \propto \|\bar{x}_c - v_l\|^2} \right]^{\frac{1}{m-1}} \right\}^{-1} \qquad (6)$$

$$v_i = \frac{\sum_{c=1}^{N} (u_{ic})^m (x_c + \propto \bar{x}_c)}{(1 + \propto) \sum_{c=1}^{N} (u_{ic})^m} \qquad (7)$$

3   Result and Discussion

Table 1 explicates the average values of image quality parameter offered by the proposed SOM-based MFCM algorithm when compared with other traditional image segmentation algorithm, which helps to verify the effectiveness of the proposed SOM-based MFCM algorithm.

Figure 1 represents the clear understanding of the proposed SOM-based MFCM algorithm. Generally, SOM neural network is trained by the spatial characteristics in the input medical image. An effective classification process of pixels can be derived a 8×8 map formation, and this map formed by the SOM is offered as the input to MFCM algorithm. Thus, classification derived from map structure becomes a strong motivation for the re-clustered process to obtain the automatic cluster center. Finally, the tumor and lesion regions were explicitly defined in the input and segmented images.

Figure 2 briefly discusses the effective segmentation provided by SOM-based MFCM algorithm. Figure 2a illustrates the larger cyst tumor affecting the right liver lobe region, which is efficiently identified with aid of the proposed SOM-based MFCM algorithm. Figure 2b demonstrates the lesion affecting the right lung lobe which is segmented by the proposed SOM-based MFCM algorithm, which greatly proves the successfulness of lesion region identification.

Figure 3 exhibits the brain tumor of the patients who have been affected by glioma and metastatic bronchogenic carcinoma. Figure 3a, b represents the input image of the proposed SOM-based MFCM algorithm. A concise demarcation between the exact tumor region and the gray matter (GM) and white matter (WM) was obtained using the proposed method.

Table 2 exhibits the average values of evaluation parameter determined for the competitive soft computing algorithms. The proposed SOM-based MFCM algorithm provides the average performance values when compared the other traditional algorithms. Vishnuvarthanan et al. [15] have offered the average estimation values of SOM-based FKM algorithm, which are evaluated for pixel size of input brain image.

Table 1 Performance evaluation for the proposed SOM-based MFCM algorithm

Clinical images	MSE	PSNR in dB	Jaccard (in %)	DOI (in %)	Computational time in seconds	Memory requirement in bytes
1	0.0392	62.1990	57.75	73.22	1.7318	1.94E+13
2	0.0298	63.3883	55.83	71.66	1.9124	2.54E+13
3	0.0038	72.3192	54.09	70.20	1.8050	2.10E+13
4	0.0411	61.9930	61.75	76.35	1.6520	1.86E+13

Input Image	MFKM	SOM-MFCM Resulting Image

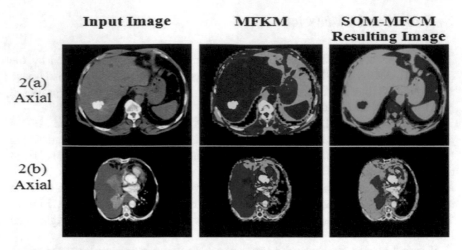

2(a)
Axial

2(b)
Axial

Fig. 2 Segmented results of patient suffering from liver and lung tumor

Input Image	Skull Stripped Image	MFKM	SOM-MFCM Resulting Image

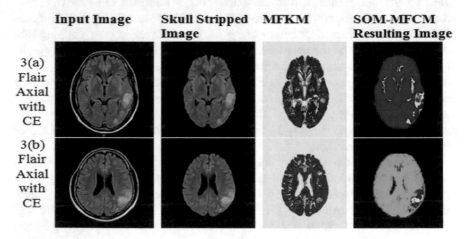

3(a)
Flair
Axial
with
CE

3(b)
Flair
Axial
with
CE

Fig. 3 Segmented results of patient suffering from brain tumor

3.1 Mean Square Error (MSE)

MSE value represents an estimate of error in image quality that extends the deliberation of input and segmented image. It is a theoretical paradigm for an algorithm to provide better segmentation result. MSE value provided by an algorithm must approach nearby zero [3, 15].

$$\text{MSE} = \frac{1}{ij} \sum_{m=1}^{i-1} \sum_{n=1}^{j-1} [x(m,n) - y(m,n)]^2 \qquad (8)$$

Table 2 Performance evaluation for soft computing algorithms

Algorithm	MSE	PSNR (dB)	Jaccard (%)	DOI (%)	Computational time (s)	Memory requirement (bytes)
MFKM	0.3356	55.5395	86.68	92.84	3.4972	1.50 E+14
SOM-based FKM [15]	2.1518	41.85	31.54	47	2.716	3.24 E+13
SOM-based MFCM	0.0285	64.9748	57.35	72.85	1.7753	2.11E+13

Fig. 4 Comparison of MSE values of soft computing algorithms

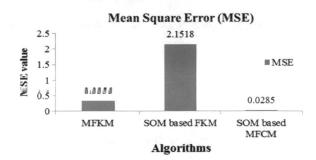

Figure 4 briefly explains the average values of MSE values for soft computing algorithm. Thus, the average MSE value presented by the proposed SOM-based MFCM algorithm is lesser than the MFKM and SOM-based FKM algorithms.

3.2 Peak Signal-to-Noise Ratio (PSNR)

PSNR explains the image quality extent of evaluating the ratio of utmost possible pixels present in the input image and the relevant MSE values [15]. PSNR value is described as

$$PSNR = 10 \log_{10} \frac{MAX_J^2}{MSE} = 20 \log_{10} \frac{MAX_J}{\sqrt{MSE}}$$
$$= 20 \log_{10}(MAX_J) - 20 \log_{10}\left(\sqrt{MSE}\right) \qquad (9)$$

3.3 Jaccard (Tanimoto) Index

Jaccard index is significantly used in for the evaluation of identical extent of pixels presence in the input and segmented image [13, 15]. It has been explicated as

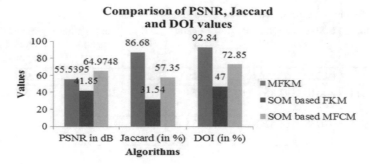

Fig. 5 Comparison of PSNR, TC, and DOI values of soft computing algorithms

$$\text{Jaccard Index} = \frac{S(I(i, j) \cap O(i, j))}{S(I(i, j) \cup O(i, j))} \tag{10}$$

3.4 Dice Overlap Index (DOI)

DOI represents the overlap function and is interrelated with the Jaccard index while resolving the segmentation accuracy [12, 15]. The expansion of DOI value is indicated by,

$$\text{DOI} = 2 \times \frac{\text{Jaccard Index}}{1 + \text{Jaccard Index}} \tag{11}$$

Figure 5 exposes the estimation of average values of PSNR, Jaccard index, and dice overlap index for the MR brain image segmentation algorithm. The clear-cut proposed segmentation process provides higher PSNR value, and it is par above the PSNR values rendered by other segmentation algorithms (40 dB). The proposed SOM-based MFCM algorithm achieves 64.97 dB, and it is higher than the PSNR results of MFKM and SOM-based FKM algorithms. The average Jaccard index of the proposed SOM-based MFCM algorithm is 57.35%, which is higher than the SOM-based FKM algorithm. Efficient segmentation accuracy was achieved by proposed SOM-based MFCM algorithm, and it offers a dice overlap index of 72.85%, which is greater than the SOM-based FKM algorithm, and can be verified in Fig. 5.

3.5 Computational Time

The significant process that achieves the segmentation results must meet the time requirement, which is usually indicated by seconds. Thus, a segmentation algorithm is considered useful, when it provides effective segmentation result satisfying the

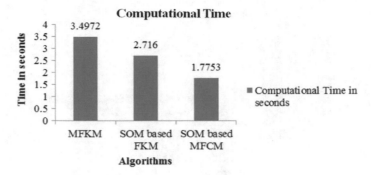

Fig. 6 Comparison of time requirement of soft computing algorithms

condition that execution time should be minimal. The proposed algorithm requires the lowest execution time and also provides effective segmentation results.

Figure 6 reveals the average computational time for soft computing algorithms. The computational time used to process a four number of patient's medical image was used as factor of comparison, in which the average computational time of the proposed SOM-based MFCM algorithm is affordable than the SOM-based FKM and MFKM algorithms.

3.6 Memory Requirement

As the demand for storage is potentially important in future, the segmentation process must be carefully designed such that the resultant image must consume lesser bytes. The proposed segmentation algorithm is in need of minimum memory space for providing the segmentation results.

Figure 7 exhibits the average memory requirement for soft computing algorithms. The proposed SOM-based MFCM algorithm consumes less memory space for executing the input images. Figure 7 clearly exposes that the proposed SOM-based MFCM algorithm requires 2.11E+13 bytes for performing the segmentation upon the four patient's medical image, which is the smallest among the soft computing algorithms.

4 Conclusion

In this paper, image segmentation using the novel SOM-based MFCM algorithm was discussed. Combination of SOM and MFCM algorithm provides efficacious segmentation result when compared with the MFKM and SOM-based FKM algorithms. The proposed algorithm, for the segmentation of medical image processing,

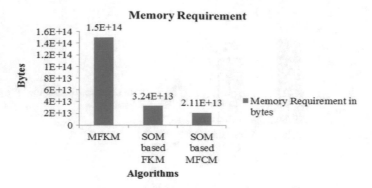

Fig. 7 Comparison of memory requirement of soft computing algorithms

provides better PSNR values among SOM-based FKM and MFKM algorithms. The SOM-based MFCM algorithm enhances the poorly available medical image through segmentation, thus augmenting the image visualization and quality. The proposed SOM-based MFCM algorithm provides segmentation results in lesser processing time when compared with other traditional algorithms. So, the proposed SOM-based MFCM algorithm is recommended as a more accurate process for the identification of tumor regions. Finally, the obtained segmentation results describe that the functioning of the SOM-based MFCM algorithm is better than the SOM-based FKM and MFKM algorithms.

Acknowledgements The authors thank Dr. K.G. Srinivasan, MD, RD, Consultant Radiologist and Dr. K.P. Usha Nandhini, DNB, KGS Advanced MR and CT Scan—Madurai, Tamil Nadu, India, for supporting the research with the patient information. Also, the authors thank the Department of Electronics and Communication Engineering of Kalasalingam University (Kalasalingam Academy of Research and Education), Tamil Nadu, India, for permitting to use the computational facilities available in Center for Research in Signal Processing and VLSI Design which was set up with the support of the Department of Science and Technology (DST), New Delhi, under FIST Program in 2013 (Reference No: SR/FST/ETI-336/2013 dated November 2013).

References

1. Ahmed MN, Yamany SM, Mohamed N, Farag AA, Moriarty T (2002) A modified fuzzy C-means algorithm for bias field estimation and segmentation of MRI data. IEEE Trans Med Imaging 21:193–199
2. Bhandarkar SM, Koh J, Suk M (1997) Multiscale image segmentation using a hierarchical self-organizing map. Neurocomputing 14:241–272
3. Chang CY, Lei YF, Tseng CH, Shih SR (2010) Thyroid segmentation and volume estimation in ultrasound images. IEEE Trans Biomed Eng 57:1348–1357. https://doi.org/10.1109/TBM E.2010.2041003
4. Chen S, Zhang D (2004) Robust image segmentation using FCM with spatial constraints based on new kernel-induced distance measure. IEEE Trans Syst Man Cybern 34:1907–1916. https://doi.org/10.1109/TSMCB.2004.831165

5. Guler I, Demirhan A, Karakis R (2009) Interpretation of MR images using self-organizing maps and knowledge-based expert systems. Digit Signal Process 19:668–677. https://doi.org/10.1016/j.dsp.2008.08.002

6. Halder A, Dalmiya S, Sadhu T (2014) Color image segmentation using semi-supervised self-organization feature map. In: Advances in intelligent systems and computing, vol 264. Springer, pp 591–598

7. Jiang Y, Chen KJ, Zhou ZH (2003) SOM based image segmentation. In: Wang G, Liu Q, Yao Y, Skowron A (eds) Rough sets, fuzzy sets, data mining, and granular computing. RSFDGrC 2003. Lecture Notes in Computer Science, vol 2639, Springer, Berlin, Heidelberg, pp 640–643. https://doi.org/10.1007/3-540-39205-x_107

8. Khan A, Jaffar MA, Choi TS (2013) SOM and fuzzy based color image segmentation. Multi Med Tools Appl 64:331–344. https://doi.org/10.1007/s11042-012-1003-6

9. Kohonen T (1982) Self-organized formation of topologically correct feature maps. Biol Cybern 43:59–69

10. Ong SH, Yeo NC, Lee KH, Venkatesh YV, Cao DM (2002) Segmentation of color images using a two-stage self-organizing network. Image Vis Comput 20:279–289

11. Ortiz A, Gorriz JM, Ramirez J, Salas-Gonzalez D (2014) Improving MR brain image segmentation using self-organising maps and entropy-gradient clustering. Inf Sci 262:117–136. https://doi.org/10.1016/j.ins.2013.10.002

12. Song J, Yang C, Fan L, Wang K, Yang F, Liu S, Tian J (2016) Lung lesion extraction using a Toboggan based growing automatic segmentation approach. IEEE Trans Med Imaging 35:337–353. https://doi.org/10.1109/TMI.2015.2474119

13. Sun S, Guo Y, Guan Y, Ren H, Fan L, Kang Y (2014) Juxta—Vascular nodule segmentation based on flow entropy and geodesic distance. IEEE J Biomed Health Inf 18:1355–1362

14. Torbati N, Ayatollahi A, Kermani A (2014) An efficient neural network based method for medical images segmentation. Comput Biol Med 44:76–87

15. Vishnuvarthanan G, Rajasekaran MP, Subbaraj P, Vishnuvarthanan A (2015) An unsupervised learning method with a clustering approach for tumor identification and tissue segmentation in magnetic resonance brain images. Appl Soft Comput 38:190–212. https://doi.org/10.1016/j.asoc.2015.09.016

Author Index

© Springer Nature Singapore Pte Ltd. 2019
A. J. Kulkarni et al. (eds.), *Proceedings of the 2nd International Conference on Data Engineering and Communication Technology*, Advances in Intelligent Systems and Computing 828, https://doi.org/10.1007/978-981-13-1610-4

Printed in the United States
By Bookmasters